2009
Information and Communications for Development

2009
Information and Communications for Development

Extending Reach and Increasing Impact

THE WORLD BANK
Washington, DC

©2009 The International Bank for Reconstruction and Development / The World Bank
1818 H Street, NW
Washington, DC 20433
Telephone: 202-473-1000
Internet: www.worldbank.org
E-mail: feedback@worldbank.org

All rights reserved

1 2 3 4 12 11 10 09

This volume is a product of the staff of the International Bank for Reconstruction and Development / The World Bank. The findings, interpretations, and conclusions expressed in this volume do not necessarily reflect the views of the Executive Directors of The World Bank or the governments they represent.

The World Bank does not guarantee the accuracy of the data included in this work. The boundaries, colors, denominations, and other information shown on any map in this work do not imply any judgement on the part of The World Bank concerning the legal status of any territory or the endorsement or acceptance of such boundaries.

Rights and Permissions

The material in this publication is copyrighted. Copying and/or transmitting portions or all of this work without permission may be a violation of applicable law. The International Bank for Reconstruction and Development / The World Bank encourages dissemination of its work and will normally grant permission to reproduce portions of the work promptly.

For permission to photocopy or reprint any part of this work, please send a request with complete information to the Copyright Clearance Center Inc., 222 Rosewood Drive, Danvers, MA 01923, USA; telephone: 978-750-8400; fax: 978-750-4470; Internet: www.copyright.com.

All other queries on rights and licenses, including subsidiary rights, should be addressed to the Office of the Publisher, The World Bank, 1818 H Street, NW, Washington, DC 20433, USA; fax: 202-522-2422; e-mail: pubrights@worldbank.org.

Cover photo credits (clockwise from top left): ©iStockimage/Cliff Parnell; Eric Lafforgue; Tenzin Norbhu and Peter Silarszky/World Bank; ©iStockimage/Morgan Lane.

ISBN: 978-0-8213-7605-8
eISBN: 978-0-8213-7606-5
DOI: 10.1596/978-0-8213-7605-8

Table of Contents

Foreword — *xi*
Preface — *xiii*
Acknowledgments — *xv*
Abbreviations — *xvii*

PART I

Chapter 1 Overview — 3
Mohsen Khalil, Philippe Dongier, and Christine Zhen-Wei Qiang

Chapter 2 Nothing Endures but Change: Thinking Strategically about ICT Convergence — 19
Rajendra Singh and Siddhartha Raja

Chapter 3 Economic Impacts of Broadband — 35
Christine Zhen-Wei Qiang and Carlo M. Rossotto with Kaoru Kimura

Chapter 4 Advancing the Development of Backbone Networks in Sub-Saharan Africa — 51
Mark D. J. Williams

Chapter 5 How Do Manual and E-Government Services Compare? Experiences from India — 67
Deepak Bhatia, Subhash C. Bhatnagar, and Jiro Tominaga

Chapter 6	**National E-Government Institutions: Functions, Models, and Trends**	**83**
	Nagy K. Hanna and Christine Zhen-Wei Qiang	
	with Kaoru Kimura and Siou Chew Kuek	
Chapter 7	**Realizing the Opportunities Presented by the Global Trade in IT-Based Services**	**103**
	Philippe Dongier and Randeep Sudan	

PART II

Key Trends in ICT Development	**125**
David A. Cieslikowski, Naomi J. Halewood, Kaoru Kimura, and Christine Zhen-Wei Qiang	
ICT Performance Measures: Methodology and Findings	**133**
User's Guide to ICT At-a-Glance Country Tables	**155**
At-a-Glance Country Tables	**161**
Key ICT Indicators for Other Economies, 2007	**311**
Contributors	*313*

BOXES

1.1	Broadband Raises Rural Incomes in Developing Countries	6
2.1	Examples of Convergence in Developing Countries	22
2.2	The Impact of Voice-over-Internet Protocol on International Calling Prices	23
2.3	The Impact of an Enabling Environment for Convergence: Wireline Telephony and Job Creation	29
3.1	Broadband's Effects on Firms' Behavior to Increase Competitiveness	38
3.2	Broadband's Role in Raising Rural Incomes in Developing Countries	40
3.3	The Republic of Korea's Experiences with Broadband	41
3.4	Broadband-Enabled Telemedicine	42
3.5	Broadband-Enhanced Trade Facilities in Ghana and Singapore	44
4.1	Alternative Infrastructure Providers in Morocco	58
4.2	Spain's Provision of Passive Infrastructure for Fiber-Optic Networks	59
4.3	Sharing Network Infrastructure in Uganda	59
4.4	Developing Infrastructure by Aggregating Demand in the Republic of Korea	60
4.5	A Shared Model for Backbone Infrastructure Development in East Africa	61
4.6	A Public-Private Partnership for Backbone Infrastructure in France	62
4.7	Sweden's Incentive-Based Mechanisms for Developing Backbone Networks	63
5.1	Explaining the Bhoomi Project's Success in Reducing Corruption	73
6.1	The Functions of E-Government Institutions	86
6.2	Kenya's and Mexico's Experiences with Formulating E-Government Strategy Using Interministerial Steering Committees	87
6.3	The Need to Build Strong Demand for E-Government Services and Institutions	90

6.4	Singapore: Pioneering a Centrally Driven Public ICT Agency	93
6.5	Sri Lanka: Pursuing Institutional Innovation in a Turbulent Political Environment	94
6.6	Chief Information Officer Councils in Various Countries	95
7.1	Government and University Initiatives in Skills Development for IT Services and IT-Enabled Services	117

FIGURES

Part I

1.1	Growth Effects of ICT	6
1.2	Global Distribution of Offshore IT Services and IT-Enabled Service Markets	7
1.3	ICT Expenditure in Europe by Sector, 2006	8
1.4	Impact of ICT on Corruption and Service Denial in E-Government Projects in India	9
1.5	Relation between the Country ICT Performance Measures (for Access and Applications) and Income per Capita, Developing Countries	15
2.1	Household Penetration of Wireline Telephone, Cable Television, and Electricity Networks in Selected Countries	24
3.1	Growth Effects of ICT	45
4.1	The Supply Chain for Communications	52
4.2	Economic Impact of Backbone Networks	53
4.3	Population Covered by Incumbent and Competing Networks in Four African Countries, 2007	55
5.1	User Report of Number of Trips Saved, by Project	71
5.2	User Reports of Changes in Travel Costs, by Project	71
5.3	User Reports of Reductions in Waiting Times, by Project	72
5.4	User Reports of Reductions in Bribe Payments, by Project	72
5.5	User Perceptions of Increased Service Quality, by Project	74
5.6	Changes in User Perceptions of Overall Composite Score, by Project	74
5.7	System Breakdowns, according to Service Delivery Site Operators, by Project	76
5.8	Computerization-Driven Changes in Operating Costs for Implementing Agencies	77
5.9	Impact of Computerization on Business Process Change	77
5.10	Impact of Computerization on Corruption and Service Denial	77
7.1	Global Opportunities for IT Services and IT-Enabled Services	105
7.2	India's Addressable Market for Vertical and Horizontal IT-Enabled Service Functions	105
7.3	Global Distribution of Offshore IT Services and IT-Enabled Service Markets	106

Part II

1	Mobile Phone Subscriptions in Developing and Developed Countries, 2000–07	126
2	Status of Competition in Fixed and Mobile Telephony in Developing and Developed Countries, 2007	126

3	Mobile Telephony Penetration before and after the Introduction of Competition	126
4	Number of Internet Users by Region, 2000 and 2007	127
5	International Bandwidth in Developing Regions, 2000–07	128
6	Broadband Penetration and Gross National Income in Various Economies, 2007	128
7	Average Annual Change in Price of Mobile Phone Services in Various Countries, 2004–06	129
8	Monthly Price of Internet Services in Various Sub-Saharan African Countries, 2005–07	129
9	ICT Service Exports as a Percentage of Total Service Exports for the Top-Five Countries, 2000–06/07	130
10	Relation between the Country ICT Performance Measures (for Access and Applications) and Income per Capita, Developing Countries	136
11	Average Country ICT Performance Measures, by Region	137
A1	Internet Use by Businesses and Employees, Selected Countries, 2005 and 2006	144

TABLES

Part I

1.1	Models for E-Government Institutions in Various Countries	12
2.1	Different Forms of Convergence	20
2.2	Policy Responses to Convergence around the World	27
3.1	Impacts of Broadband on Economic Activities in U.S. Communities	39
3.2	Growth Regression Separating Effects of Broadband Penetration	44
3A.1	Definition of Variables	46
3B.1	Regression for per Capita Growth	47
4.1	Policy Options for Expanding Backbone Networks	57
5.1	Performance Measures and Indicators for ICT Projects for Users and Service Providers	70
5.2	User Reports of Bribes Paid to Functionaries and Intermediaries/Agents, by Project	73
5.3	Users' Top-Four Desired Features of Services, by Project	75
5.4	Variations in Service Quality across Seven Bhoomi Project Kiosks	75
5.5	Impact of Computerization on Agency, by Project	76
5.6	Summary Results of User Survey, by Project	78
5.7	An Assessment Framework for E-Government	79
5A.1	Profile of Respondents to User Surveys on E-Government Projects	80
5A.2	Summary of Findings of User Surveys on E-Government Projects	80
6.1	Models for E-Government Institutions in Various Countries	91
6A.1	Characteristics of E-Government Institutions in Selected Countries	101
7.1	Types of IT Services and IT-Enabled Services	104
7.2	Frameworks for Assessment of Locations for IT Services and IT-Enabled Services	108

7.3	Relative Percentage of Components in the Total Cost of Offshoring	109
7.4	NASSCOM IT-Enabled Services Skill Competence Testing Themes	115

Part II

1	Measures of E-Government and E-Commerce in Developing and Developed Countries	131
2	Indicators for the Country ICT Performance Measures	134
3	Example of How a Country ICT Performance Measure Is Calculated (Mauritius)	134
4	Average Country ICT Performance Measures, by Income Level	135
5	Country ICT Performance Measures, by Income Level and Economy, 2007	140
A1	Use of the Internet by Individuals and Businesses, 2005–07	145
A2	Location of Internet Use by Individuals	147
A3	Frequency of Internet Use by Individuals, 2005–07	149
A4	Proportion of Businesses Using the Internet, by Type of Activity, 2005–06	150
A5	Supplemental ICT Data, 2007	152
6	Classification of Economies by Income and Region, FY2009	157

Foreword

Information and communication technology (ICT) is transforming interactions between people, governments, and firms worldwide. In developing countries, farmers receive updated crop prices and public health officials monitor medical inventories by text messages. Women are empowered to make decisions and access new opportunities through online information. Entrepreneurs obtain business licenses in a fraction of the standard time by applying for them through municipal government Web sites. And in an increasingly integrated global economy, ICT enables people to access and share knowledge and services around the world.

Information and Communications for Development 2009: Extending Reach and Increasing Impact is the second report in the World Bank's flagship series on ICT. This report analyzes the benefits of extending access to ICT in the developing world; of mainstreaming ICT applications, particularly in the area of e-government; and of trading services based on information technology. It also features at-a-glance tables for 150 economies of the latest available data on ICT sector indicators. Country ICT performance measures for access, affordability, and applications in government and business are also introduced, so that policy makers can assess their countries' ICT capacities relative to those of other countries.

It is our hope that this report will provide some emerging good-practice principles for policy, regulatory, and investment frameworks in this complex and constantly evolving sector. Ultimately, applying these principles will allow developing countries to leverage ICT to create and sustain opportunities across the various disciplines of economic and social development.

Katherine Sierra
Vice President, Sustainable Development
The World Bank

Preface

The World Bank Group's engagement in the information and communication technology (ICT) sector is driven by three strategic priorities: access to ICT infrastructure, including affordable broadband and rural services; mainstreaming of ICT to improve the delivery of public and private services across various economic and social sectors; and innovation to foster grassroots entrepreneurship and support the development of local information technology (IT) industries.

Since the late 1990s, there has been an unprecedented increase in access to ICT. With almost four billion subscribers, mobile networks now constitute the world's largest distribution platform, promising to deliver services and information by innovative methods. Already, the mobile platform enables access to financial services, market information, and health services for people living in remote or rural areas. Broadband networks provide the framework for the delivery of services ranging from telephony to Internet access to media. In the process, economic and social activities are being transformed, and new opportunities are arising for a growing portion of the world's population. Consequently, ICT deserves a central role in any national development strategy.

This second *Information and Communications for Development (IC4D)* report illustrates the new opportunities offered by mobile, broadband, IT-based services, and various ICT applications. It also discusses how these applications contribute to economic development and sustainable growth. The fact that innovative ICT solutions often emerge in developing countries is an encouraging sign that in the new information era, these countries can accelerate development processes from within. The report also tries to answer the question of how to ensure that developing countries realize the development benefits offered by ICT while providing policy makers and regulators with options for dealing with the challenges of convergence at the level of services and networks.

The cross-cutting and constantly evolving nature of the ICT sector requires timely policy and regulatory responses; leadership and institutional arrangements that cut across all sectors and levels of government; and private-public partnerships that can harness the capabilities of the private sector to meet public policy objectives. From an industry perspective, there will also be the need to adjust business models to market realities. The World Bank Group is committed to continuing its analytical and lending operations to support progress and the sharing of best practices and knowledge across these areas, as well as to expand its investments in private ICT companies to further sector growth, competition, and the availability of better-quality, affordable ICT services to the widest population in developing countries.

Mohsen A. Khalil
Director, Global Information and Communication
  Technologies Department
The World Bank Group

Acknowledgments

This report was prepared by the Global Information and Communication Technologies (GICT) Department and the Development Economics Data Group (DECDG) of the World Bank Group.

Preparation of this report was led by an editorial committee consisting of Mohsen A. Khalil (joint director, World Bank and International Finance Corporation (IFC), GICT Department), Philippe Dongier, Valerie D'Costa, Christine Zhen-Wei Qiang (team leader), Peter L. Smith, Randeep Sudan, Eric Swanson, and Björn Wellenius. The principal authors of the chapters in Part I of the report were Deepak Bhatia, Subhash C. Bhatnagar, Philippe Dongier, Nagy K. Hanna, Christine Zhen-Wei Qiang, Siddhartha Raja, Carlo M. Rossotto, Rajendra Singh, Randeep Sudan, Jiro Tominaga, and Mark D. J. Williams, with substantial inputs from Kaoru Kimura and Siou Chew Kuek. Part II was prepared by David A. Cieslikowski, Naomi J. Halewood, Buyant Erdene Khaltarkhuu, Kaoru Kimura, William Prince, and Christine Zhen-Wei Qiang.

The Information for Development Program (*info*Dev), a multidonor grant facility housed at GICT, is devoted to creating and leveraging knowledge and best practices in ICT for Development for the benefit of developing countries. *Info*Dev cofunded the background research of chapters 4 and 7 of this report (http://www.infodev.org). The Information Solutions Group and the World Bank Institute cofinanced the background studies of chapters 5 and 6.

Inputs, comments, guidance, and support at various stages of the report's preparation were received from the following colleagues: Mavis Ampah, Seth Ayers, Kareem Abdel Aziz, Shaida Badiee, Laurent Besancon, Jerome Bezzina, Cecilia Briceno-Garmendia, Yann Burtin, George R. Clarke, Eric Crabtree, Jeffrey Delmon, Roberto Foa, Vivien Foster, Boutheina Guermazi, Manju Haththotuwa, Tim Kelly, Charles Kenny, Ioannis Kessides, Mohan Kharbanda, Yongsoo Kim, Aart Kraay, Siou Chew Kuek, Gareth Locksley, Samia Melhem, Arturo Muente-Kunigami, Juan Navas-Sabater, Vincent Palmade, Oleg Petrov, Marta Lucila Priftis, Ismail Radwan, Sandra Sargent, David Satola, Arleen Seed, Paul Noumba Um, Eloy Vidal, Lixin Colin Xu, and Shahid Yusuf.

Additional comments were received from Jean-Pierre Auffret (vice president, International Academy of the CIO; professor, George Mason University), Vineeta Dixit (chief executive officer, SW Applications India), Soumitra Dutta (dean of external relations, Roland Berger Professor of Business and Technology, INSEAD), M'Backe Fall (regulator, Mauritania), Sherille Ismail (senior counsel, Office of Strategic Planning and Policy Analysis, U.S. Federal Communications Commission), Mike Jensen (independent expert), Peter Knight (coordinator, e-Brasil Project; president, Telemática e Desenvolvimento), Bruno Lanvin (executive director, of eLab at INSEAD), Jonathan Levy (deputy chief economist, U.S. Federal Communications Commission), Shirin Madon (professor, London School of Economics), Patrick Masambu (executive director, Uganda Communications Commission), Larry Meek (former chief information officer, Vancouver, Canada), Nripendra Misra (chairman, Telecom Regulatory Authority of India), Lorenzo Pupillo (executive director, Public Affairs, Telecom Italia), Enrique Rueda-Sabater (director, business development, emerging markets, Cisco Systems Inc.), Russell Southwood (independent expert), Nidhi Tandon (director, Networked Intelligence for Development), Trond Arne Undheim (director, standards strategy and policy, Europe, Middle East, and Africa, Oracle Corporation), Jeongwon Yoon (director, National Information Society Agency, Republic of Korea), and McKinsey & Company.

The database development team included Reza Farivari, Raymond Muhula, Shahin Outadi, and William Prince of DECDG.

Paul Holtz and Sara Roche edited the report. Special thanks are due to Martin Dimitrov, Alexandra Klopfer, Rani Kumar, Marta Priftis, and Rajesh B. Pradhan from GICT, and to Denise Bergeron, Jose De Buerba, Stephen McGroarty, Santiago Pombo-Bejarano, and Janet Sasser from the World Bank Office of the Publisher for oversight of the editorial production, design, printing, and dissemination of the book.

Abbreviations

$	All dollar amounts are U.S. dollars unless otherwise indicated.
AAG	at-a-glance
AGIMO	Australian Government Information Management Office
AMITI	Asociación Mexicana de la Industria de Tecnologías de Información
AMC	Ahmedabad Municipal Corporation
APFIRST	Agency to Promote and Facilitate Investments in Remote Services and Technology (Andhra Pradesh, India)
BASSCOM	Bulgarian Association of Software and Services Companies
BBC	British Broadcasting Corporation
BPAP	Business Processing Association of the Philippines
BPO	business process outsourcing
bps	bits per second
BRASSCOM	Brazilian Association of Information Technology and Communication Companies
CARD	Computer-aided Administration of Registration Department (India)
CECI	Canadian Center for International Cooperation
CIMA	Chartered Institute of Management Accountants
CIO	chief information officer
CMMI	Capability Maturity Model integration
CMU	Carnegie Mellon University
COO	chief operating officer
COPC	Customer Operations Performance Center Inc.
CORFO	Corporación de Fomento de la Producción de Chile
CRISIL	Credit Rating Information Services of India Limited
DECDG	Development Economics Data Group (World Bank)
DORSAL	Développement de l'Offre Régionale de Services et l'Aménagement de Télécommunications en Limousin (France)
DSL	digital subscriber line
DVB	digital video broadcasting
EASSy	Eastern Africa Submarine Cable System
EDI	electronic data interchange
eGEP	eGovernment Economics Project (India)
e-government	electronic government
EU	European Union

FDI	foreign direct investment	LIC	low-income country
GAAP	generally accepted accounting principles	LMIC	lower-middle-income country
GASSCOM	Ghana Association of Software and Services Companies	MAREVA	méthode d'analyse et de remontée de la valeur (method of analysis and value enhancement)
GDP	gross domestic product		
GICT	Global Information and Communication Technologies Department (World Bank Group)	MATRADE	Malaysia External Trade Development Corporation
		MBOI	Mauritius Board of Investments
GNI	gross national income	MDGs	Millennium Development Goals
GPS	Global Positioning System	MSC	Multimedia Super Corridor (Malaysia)
GRIPS	National Graduate Institute for Policy Studies (Japan)	NAC	NASSCOM assessment of competency
GSMA	GSM Association	NASSCOM	National Association of Software and Services Companies (India)
IC4D	*Information and Communications for Development*	NGN	next-generation network
ICA-IT	International Council for Information Technology in Government Administration	NGO	nongovernmental organization
		NTT	Nippon Telegraph and Telephone (Japan)
ICT	information and communication technology	NTIA	National Telecommunications and Information Administration (United States)
ICTA	Information and Communication Technology Agency (Sri Lanka)		
IDA	Industrial Development Agency (Ireland)	OANA	Organisation of Asia-Pacific News Agencies
IDABC	Interoperable Delivery of European eGovernment Services to Public Administrations, Business and Citizens (Malta)	OECD	Organisation for Economic Co-operation and Development
		Ofcom	Office of Communications (United Kingdom)
IFC	International Finance Corporation		
IFSC	International Financial Services Center (Ireland)	PPP	public-private partnership
		PSDC	Penang Skills Development Centre
IIIT	International Institute of Information Technology (India)	R&D	research and development
		RTC	rights, tenancy, and crop inspection register
IIMA	Indian Institute of Management, Ahmedabad		
*info*Dev	Information for Development Program	SEDB	Singapore Economic Development Board
IP	Internet protocol	SG&A	selling, general, and administrative expenses
IPTV	Internet protocol television		
ISB	Indian School of Business	SMS	short message service
ISG	Information Solutions Group (World Bank)	STPI	Software Technology Parks of India
ISP	Internet service provider		
ISTAG	Information Society Technologies Advisory Group	TESDA	Technical Education and Skills Development Agency
IT	information technology	TRAI	Telecom Regulatory Authority of India
ITB	Industrial Training Board (Singapore)	TRIPS	Trade-Related Aspects of Intellectual Property Rights
ITES	IT-enabled services		
ITU	International Telecommunication Union		
KAVERI	Karnataka Valuation and E-Registration (India)	UIS	UNESCO Institute for Statistics
		UMC	upper-middle-income country

UNCTAD	United Nations Conference on Trade and Development	UNPAN	United Nations Online Network in Public Administration and Finance
UNDESA	United Nations Department of Economic and Social Affairs	VAT	value added tax
UNDP	United Nations Development Programme	VLBI	very long baseline interferometry
UNECA	United Nations Economic Commission for Africa	VLSI	very large scale integration
		VoIP	voice-over-Internet protocol
UNECLAC	United Nations Economic Commission for Latin America and the Caribbean	WiFi	wireless fidelity
		WiMAX	worldwide interoperability for microwave access
UNESCAP	United Nations Economic Commission for Asia and the Pacific	WIPO	World Intellectual Property Organization
UNESCO	United Nations Educational, Scientific, and Cultural Organization	WSIS	World Summits on the Information Society
UNESCWA	United Nations Economic Commission for Western Asia	XLRI	Accenture-Xavier Labor Research Institute

Part I

Chapter 1

Overview

Mohsen Khalil, Philippe Dongier, and Christine Zhen-Wei Qiang

nformation and Communications for Development (IC4D) is a regular publication of the World Bank Group on the diffusion and impact of information and communication technology (ICT), available at http://www.worldbank.org/ic4d. The first report, *IC4D 2006: Global Trends and Policies*, analyzed lessons on developing access to ICT, examined the roles of the public and private sectors in this process, and identified the benefits and challenges of adopting and expanding ICT use in businesses.

This second report, *IC4D 2009: Extending Reach and Increasing Impact*, takes a close look at mobile and broadband connectivity. It analyzes the development impact of high-speed Internet access in developing countries and provides policy options for rolling out broadband networks and addressing the opportunities and challenges of convergence between telecommunications, media, and computing. The report also presents a framework of e-government applications and discusses various country experiences with the institutional and policy arrangements for e-government and for the development of the local information technology (IT) and IT-enabled services (ITES) industries. The common thread running through these topics is the development impact of ICT. Finally, the report presents summary tables on ICT sector indicators in 150 economies and introduces new performance measures in terms of access, affordability, and ICT adoption in government and business.

Impact of ICT in an Increasingly Knowledge-Based World

Knowledge-based activities have become increasingly important and pervasive worldwide. ICT is the foundation of this knowledge-based world. It allows economies to acquire and share ideas, expertise, services, and technologies locally, regionally, and across the world. It also contributes to making the global economy more integrated than ever before.

ICT can help create and sustain new opportunities for economic development. Accelerated knowledge transfer and technological diffusion amplify the competitive advantages of fast-learning economies. As the information requirements for innovation in economic and social activities increase, the importance of ICT for the development agenda will continue to expand.

Mobile Networks Now Constitute the World's Largest Distribution Platform and Create a Major Development Opportunity

The past 15 years have brought an unprecedented increase in access to telephone services. This growth has been driven primarily by wireless technologies and liberalization of telecommunications markets, which enabled faster and less costly network rollout. The total number of mobile phones in the world surpassed the number of fixed telephones in

2002; by the end of 2008, there were an estimated 4 billion mobile phones globally (Wireless Intelligence 2008).[1] No technology has ever spread faster around the world (*The Economist* 2008a). Mobile phones now represent the world's largest distribution platform.

The mobile phone market is especially important for developing countries, where it is growing most rapidly and where it is seen as a "leapfrogging" tool. New telephone connections in low- and lower-middle-income countries have outnumbered those in upper-middle- and high-income countries since 1998 (World Bank 2008c). Virtually all new mobile customers in the coming years will be in developing countries (GSMA 2008).

Mobile communications have a particularly important impact in rural areas, which are home to nearly one-half of the world's population and 75 percent of the world's poor (World Bank 2007). The mobility, ease of use, flexible deployment, and relatively low and declining rollout costs of wireless technologies enable them to reach rural populations with low levels of income and literacy. The next billion mobile subscribers will consist mainly of the rural poor. Mobile operators are thus taking innovative approaches to reach rural customers, such as offering village phone programs in Bangladesh, low-denomination recharges for prepaid phones in East Africa, and combined voice and agricultural information services in China (*The Economist* 2008a).

An important use of mobile phones in rural areas is to access market information. TradeNet, a Ghana-based trading platform, allows users to sign up for short message service (SMS) alerts for commodities and markets of their choice and receive instant alerts for offers to buy or sell when anyone else on the network has submitted an offer by mobile phone. Users can also request and receive real-time prices for more than 80 commodities from 400 markets across West Africa. The Ghana Agricultural Producers and Traders Organization[2] is a major beneficiary: in 2006 it concluded trade deals worth $60,000 with other producer and trader organizations in Burkina Faso, Mali, and Nigeria. These deals involved purchasing tomatoes, onions, and potatoes without middlemen, thereby substantially reducing transaction costs between buyers and sellers (World Bank 2007). In India, access to market information through mobile phones has allowed fishermen to respond faster to market demand and has increased their profits (Jensen 2007); in Niger, it has reduced price disparities in grain markets (Aker 2008).

Once legal frameworks are in place, banking and payment services provided through mobile phones can bring many more people into the formal financial system (World Bank 2007). Mobile banking services offered by Wizzit in South Africa, Safaricom (M-PESA) in Kenya, and Globe Telecom and Smart in the Philippines are such examples. These services allow mobile phone users to pay for purchases in stores and transfer funds, significantly reducing transaction costs. In Sierra Leone, workers in the cities have cut out intermediaries and now transfer money almost instantly to relatives in remote villages (World Bank 2008b).

Mobile phones can improve the effectiveness and reach of health programs. In many countries, health care is one of the largest public budgetary expenses. Improved information systems that track service delivery, establish accountability, and manage patients for better health outcomes can result in major efficiency gains. For example, using the mobile phone as an interface, Voxiva's HealthNet system allows drug inventory management, while its HealthWatch supports monitoring programs. These systems have been deployed in India, Peru, and Rwanda.[3] SIMpill, deployed in South Africa, uses mobile phone technology to ensure patients take medications on time, notifying health professionals if a patient does not appear to be taking his or her medication as prescribed.[4]

The development potential of the wireless platform is enormous. Mobile communications are evolving from simply voice services and text messaging to more broad-bandwidth intelligent systems that enable a diverse range of applications in locations where conventional services are not available in developing countries. "Smart" wireless phones now allow users to also browse the Internet, download music, and access information services. This opportunity is especially exciting given that the developing world missed out on much of the initial Web revolution because it did not have adequate Internet infrastructure (*The Economist* 2008b).

Broadband Increases Productivity and Contributes to Economic Growth, for Which It Deserves a Central Role in Development Strategies

Broadband networks, both fixed and mobile, are necessary to deliver modern communication and information services that require high rates of data transmission. Enterprise file transfer, television, and high-speed Internet are examples of such services. High-speed Internet connections provide ready access to a wide range of services, such as voice, video, music, film, radio, games, and publishing. Broadband networks enhance the efficiency and reach of existing services and provide spare capacity for unknown future applications. Indeed, broadband networks are key to the ongoing transformation of the ICT sector through the convergence of telecommunications, media, and computing. The convergence process comprises service convergence, which enables providers to use a single network to provide multiple services; network convergence, which allows a service to travel over any combination of networks; and corporate convergence, by means of which firms merge or collaborate across sectors. Driven by technology and demand, convergence is resulting in major changes in market structures and business models.

Broadband has considerable economic impact at all levels of individuals, firms, and communities. Individuals increasingly use broadband to acquire knowledge and skills to increase their employment opportunities. Where broadband has been introduced in rural areas of developing countries, villagers and farmers have gained better access to crop market prices, training, and job opportunities (Qiang and Rossotto 2009, chapter 3 in this volume). In developed countries and urban areas in developing countries, an increasing number of individuals build up social networks through broadband-enabled, peer-to-peer Web-based groups that facilitate economic integration and drive development. Blogs (Web logs, or online diaries), wikis (Web sites where users can contribute and edit content), video sharing sites, and the like allow new, decentralized, and dynamic approaches to capturing and disseminating information that allows individuals to become better prepared for the knowledge economy (Johnson, Manyika, and Yee 2005).

Access to broadband supports the growth of firms by lowering costs and raising productivity. Realizing these performance improvements, however, depends on firms' ability to integrate their technological, business, and organizational strategies. When fully absorbed, broadband drives intense, productive uses of online applications and services, making it possible to improve processes, introduce new business models, drive innovation, and extend business links. A study involving business and technology decision makers in 1,200 companies in six Latin American countries—Argentina, Brazil, Chile, Colombia, Costa Rica, and Mexico—showed that broadband deployment was associated with considerable improvements in business organization, including speed and timing of business and process reengineering, process automation, data processing, and diffusion of information within organizations (Momentum Research Group 2005).

Firms in the media, export, and other information-intensive sectors have benefited most from integrating broadband into their business processes. Clarke and Wallsten (2006), in a study of 27 developed and 66 developing countries, found that a 1-percentage-point increase in the number of Internet users is correlated with a boost in exports of 4.3 percentage points. Increases of 25 percent or more in the efficiency of claims processed per day have been documented by U.S. insurance companies that have adopted wireless broadband (Sprint 2006). Other industries that have benefited significantly include consulting, accounting, marketing, real estate, tourism, and advertising.

Local communities around the world have realized considerable economic gains and new opportunities from broadband services. Studies from Canada, the United Kingdom, and the United States find that broadband connectivity has a positive economic impact on job creation, community retention, retail sales, and tax revenues (Ford and Koutsky 2005; Kelly 2004; Strategic Networks Group 2003; Zilber, Schneier, and Djwa 2005). In rural areas of developing countries, communities have recently begun to launch broadband services and applications that give local populations access to new markets and services. Facilitating information exchange and value creation between buyers and sellers of agricultural products, which has improved income and livelihoods in rural areas (box 1.1), is a prime example of this. Previously, such opportunities were available only in the largest or wealthiest localities.

According to a recent World Bank econometrics analysis of 120 countries, for every 10-percentage-point increase in the penetration of broadband services, there is an increase in economic growth of 1.3 percentage points (Qiang 2009). This growth effect of broadband is significant and stronger in developing countries than in developed economies, and it is higher than that of telephony and Internet (figure 1.1). The impact can be even more robust once the penetration reaches a critical mass.

Because broadband networks have the potential to contribute so much to economic development, they should be widely available at affordable prices and should become an integral part of national development strategies.

> **Box 1.1** Broadband Raises Rural Incomes in Developing Countries
>
> Experience shows that access to broadband networks has had a positive impact on rural incomes in developing countries. In India, the E-Choupal program was started by ITC, one of India's largest agricultural exporters, in 2000. The program operates in traditional community gathering venues (*choupals*) in farming villages, using a common portal that links multimedia personal computers by satellite. Training is provided to the hosts, who are typically literate farmers with a respected role in their communities. The computers give farmers better access to information such as local weather forecasts, crop price lists in nearby markets, and the latest sowing techniques. Collectively, these improvements have resulted in productivity gains for the farmers. E-Choupal also enables close interaction between ITC and its rural suppliers, which increases the efficiency of the company's agricultural supply chains, eliminates intermediaries, and improves terms of business. The fact that ITC pays a higher price than its competitors for exportable products has encouraged farmers to sell their increased output to the company. By 2008, E-Choupal had reached millions of small farmers in more than 40,000 villages, bringing economic and other benefits. It aims to reach 100,000 villages by 2010.
>
> Another program, launched by the Songtaaba Association, has allowed female agricultural producers in Burkina Faso to become economically empowered through broadband. Songtaaba, an organization manufacturing skin care products, provides jobs to more than 3,100 women in 11 villages. In order to provide its members with regular access to useful information and improve the marketing and sales of their products, the association set up telecenters in two villages equipped with cell phones, Global Positioning System (GPS), and computers with high-speed Internet connections. The telecenters, managed by trained rural women, help the association run its businesses more efficiently. The organization also maintains a Web site that offers its members timely information about events where they can promote or sell their products. In the two years following the establishment of the telecenters and the launch of the Web site in 2005, orders have increased by about 70 percent, and members have more than doubled their profits.
>
> **Source:** Qiang and Rossotto 2009 (chapter 3 in this volume).

Figure 1.1 Growth Effects of ICT

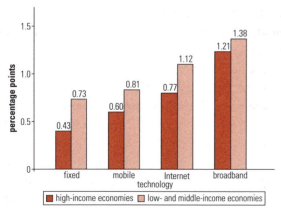

Source: Qiang 2009.

Note: The y axis represents the percentage-point increase in economic growth per 10-percentage-point increase in telecommunications penetration. All results are statistically significant at the 1 percent level except for those for broadband in developing countries, which are significant at the 10 percent level.

Currently though, few people in developing economies have access to broadband networks. In 2007, an average of less than 5 percent of the population of low-income economies was connected to broadband networks, and that was mostly in urban centers. In this light, developing countries are missing a great development opportunity.

The Global Market for IT and ITES Is Expanding, and Developing Countries Are Seizing the Opportunity to Build Local Industries

The services sector is growing globally—it already accounts for 70 percent of employment and 73 percent of gross domestic product (GDP) in developed countries and for 35 percent of employment and 51 percent of GDP in developing countries (UNCTAD 2008). IT services, a component of the services sector, represents a $325 billion annual potential market, according to McKinsey & Company estimates. IT services include hardware and software maintenance, network administration and system integration, help desk services, application development, and consulting, as well as activities in engineering, such as mechanical design, production, and software engineering.

Another component of the services sector is ITES. ITES are services that can be delivered remotely using telecommunications networks. Estimates of the size of the ITES market vary.[5] While analysis by McKinsey & Company suggests the annual potential market for ITES was $150 billion in 2007, Gartner Research (2008) expects the global market to grow from $171 billion in 2008 to $239 billion in 2012. Even more optimistic is an estimate from NASSCOM-Everest (2008), which suggests an ITES market of $700 billion–$800 billion by 2012. In the ITES market, services for industries such as banking, insurance, and telecommunications account for close to two-thirds of the potential market, while services for functions that exist across industries, such as human resources management, finance, administration, and marketing, account for about one-third.

Developing countries have been very successful in IT services and ITES. Undoubtedly, India is the global leader in both industries. However, China, Mexico, and the Philippines are also emerging as potential players in this space. In addition, transition economies in Central and Eastern Europe (the Czech Republic, Hungary, Moldova, Poland, Romania, and the Russian Federation) have developed their capacity in IT services and ITES, though on a smaller scale (figure 1.2).

For these countries, the expansion of IT services and ITES creates significant economic and social benefits. India, for instance, exported more than $40 billion worth of IT services and ITES in 2007, a figure that represents one-quarter of the country's total exports and nearly half of its service exports. In the Philippines, IT services and ITES employed 345,000 people as of mid-2008 and are projected to directly employ close to 1 million people by the end of 2010. Employment of this scale means that the sector would account for 27 percent of all new jobs created in the Philippines by 2010 (BPAP 2007).

Another important positive impact of the growth of IT services and ITES is on the status of women. Women account for about 65 percent of the total professional and technical workers in IT services and ITES in the Philippines. In India, women make up 30 percent of the IT services and ITES workforce—a much higher rate of female participation than in the services sector in general—and this share is expected to grow to 45 percent by 2010. More than half of call center employees are women. In both countries, women fill a greater number of high-paying jobs in IT services and ITES than in most other sectors of the economy.

Given the large potential market for IT services and ITES, there is an important opportunity for more countries to participate and benefit.[6] The opportunity is especially attractive because only about 15 percent of the potential market, or about $65 billion in 2007, has been exploited so far. There also remains significant room for growth from new entrants: estimates by McKinsey suggest that only about 27 percent of

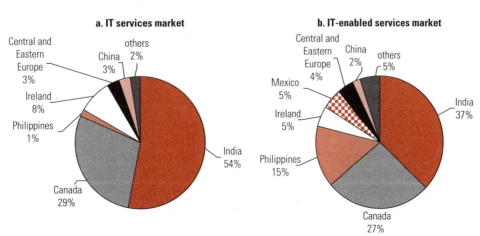

Figure 1.2 Global Distribution of Offshore IT Services and IT-Enabled Service Markets

Sources: McKinsey & Company 2008; NASSCOM-Everest 2008; Tholons 2006.

the market potential will be realized by 2010. Countries that meet the requirements of the untapped IT and ITES market are likely to experience rapid growth in these industries.

An increasing number of countries are beginning to develop IT services and ITES into major potential sources of economic growth. South Africa, for example, is emerging as an attractive ITES location by leveraging its English-speaking workforce. Similarly, the Arab Republic of Egypt, Morocco, and Tunisia are developing a range of ITES operations, including call centers. Israel, Malta, and Mauritius are beginning to fill niche segments such as packaged application development (Israel), remote gaming (Malta), and higher-value-added activities such as advisory, design, and legal services (Mauritius).

E-Government Can Lead the Way to Mainstream ICT Applications

E-government is the most cited and high-profile of all ICT applications, given its importance in underpinning development efforts. In many countries, developed and developing alike, there has been significant government expenditure on IT. In Europe, IT spending in government is growing faster than in most other sectors (figure 1.3). Since 1999, China has embarked on major initiatives in this area. Total e-government spending is expected to increase to more than $10 billion in 2008, from $7 billion in 2006. China's investment in the State Economic Management Information Systems program alone amounts to about $2.5 billion (Zhou 2007). India is also planning large investments; its National e-Government Program will receive $5.5 billion in funding between 2007 and 2012.

Such significant funding for e-government reflects growing recognition of its benefits for the delivery of public services. Users rank improved transactional efficiency (as reflected in a reduced number of visits and less waiting time), reduced corruption, and better quality of service (such as reduced error rates and increased convenience) as most important in their dealings with public services. Nondiscriminatory treatment and an effective complaint handling system are also desired features (Bhatia, Bhatnagar, and Tominaga 2009, chapter 5 in this volume).

Successful e-government projects have reduced transaction costs and processing time and increased government revenues. For instance, the e-Customs System in Ghana (GCNet) increased customs revenues by 49 percent in its first 18 months of operation and reduced clearance times from three weeks to two days (De Wulf and Sokol 2004). An e-procurement system in Brazil cost only $1.6 million, yet it enabled savings of $107 million for the state in 2004 alone as a result of improved process efficiency and lower prices for goods and services procured. The fully automated tendering process launched as part of the same system in Brazil saved suppliers an estimated $35 million (Crescia 2006).

Some e-government projects have also improved governance by reducing corruption and abuse of discretion, thereby making vital contributions to development. In India, a survey found that fewer users were required to pay bribes to accelerate service delivery under e-government

Figure 1.3 ICT Expenditure in Europe by Sector, 2006

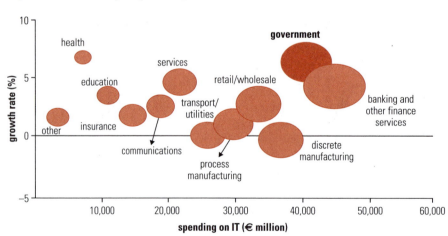

Source: Information Society Technologies Advisory Group 2006.

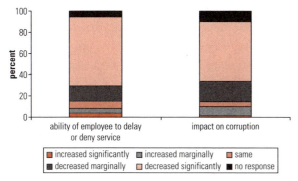

Figure 1.4 Impact of ICT on Corruption and Service Denial in E-Government Projects in India

Source: Bhatia, Bhatnagar, and Tominaga 2009 (chapter 5 in this volume).

projects than under manual systems, and that the frequency of paying bribes to service officials has fallen. For example, the land registration system in the state of Karnataka in India is estimated to have cut bribes by about $18 million annually. Furthermore, an overwhelming proportion of supervisors sense that abuse of discretionary power through means such as denying services to citizens has narrowed (figure 1.4). They are also more aware of the need to comply with service standards specified in citizen charters.

The potential to access public services at home or at a local center also empowers women and minorities. For example, among the users of e-government services, women are usually in charge of dealing with public administrations at the household level. The delivery of e-government services translates to easier access and less time than traveling to or queuing up at government departments. For minorities, ICT facilitates access to relevant public information on rights and benefits, inheritance and family laws, health care, and housing, allowing the public to make informed decisions on issues of importance.

Some Policy Directions

The agenda of ICT for development is rather new and still in flux. A good case can be made for ICT as a factor of economic development. But how to use ICT cost-effectively to meet private and public objectives is less clear. There are examples of failures as well as successes. The technology and its products continue to change at a fast pace. The market responses are hard to predict. The jury is still out on the extent to which the evolving ICT sector should be regulated. Both developing and developed economies are struggling to understand the difficulties, constraints, and uncertainties and how best to handle them. The following are some of the most challenging issues faced by developing countries:

- Convergence does not fit easily into established sector frameworks. Translating a broad vision into specific policies and regulations is likely to be difficult. Although a proactive response could yield the greatest development benefits, some countries adopt a "wait and watch" strategy while policy options become clearer or until the issues gain urgency.

- Broadband networks are developing mainly in the potentially profitable cities and intercity corridors. People living in commercially less-attractive provincial and rural areas, as well as low-income groups in urban areas, tend to be left behind.

- Although there are a number of success stories, a high rate of failure has been reported in the adoption of e-government by developing countries. Ensuring robust performance from new large-scale information systems has proven to be a challenge even for countries with sophisticated technical skills.

- Many countries have major gaps in their ability to compete in the IT services and ITES markets, most notably in relation to scarcity of skilled labor.

To address these challenges, elements of good practice are emerging for policy, regulatory, and investment frameworks to extend the reach and increase the impact of ICT on development.

Policy Responses to Convergence Facilitate ICT Development

The technological drivers for convergence are in place or are quickly diffusing. Fundamental to convergence are the digitization of communication and the rapidly falling cost of computing. More recently, expanding use of Internet protocol (IP)–based networking has made interconnection among diverse networks, devices, and applications possible. These developments have led to the deployment of broadband next-generation networks that deliver a wide range of services interactively over any combination of communication networks anytime and anywhere.

As these technical factors fall into place, convergence finds market traction with service providers seeking to diversify their businesses, increase revenues, and cut costs.

Globally, service providers are embracing convergence by investing in broadband networks. They are entering new markets and improving business prospects by consolidating content and services and by adopting new business models. An evolving set of providers offers innovative services with high quality and maximum choice at low prices.

As users adopt new ICT services and applications, demand for greater access and content is growing with previously unmatched breadth and speed in both developing and mature economies. Since its first release in 2003, the Internet telephony service provider Skype has attracted more than 300 million subscribers in 225 countries and territories (Skype 2008). In 2007 alone, Skype carried an estimated 27 billion minutes of computer-to-computer calls (TeleGeography 2007). By the end of 2008, there will be an estimated 40 million "triple play" subscribers worldwide—all of whom receive most of their telephony, video, and Internet services over broadband networks (Pyramid Research 2007). Further, consumers are now also participating in content creation. Growing access to a broader variety of services and applications through a range of new and constantly improving devices, including mobile phones that go well beyond voice services, has stimulated enormous social interaction and exchange. This has led to higher demand for advanced ICT networks and services that have the ability to support new applications.

As supply and demand align, the technical and market factors driving convergence are visible in markets at all stages of economic development, from Brazil to France and from India to Nigeria. While high-income countries have had these conditions in place for some time, there are clear indications that even low-income countries and population groups are now recognizing these forces and the resulting convergence as a reality (Singh and Raja 2009, chapter 2 in this volume).

The greatest benefits of convergence are realized in markets that enable it promptly. Convergence, though, typically does not fit easily into traditional policy frameworks. Attempting to stick to existing policies creates regulatory uncertainty and inconsistency—the economic costs of which will increase over time in a way that will hinder technological progress and market evolution. Consequently, countries that resist convergence or adopt a "wait and watch" approach will ultimately miss the benefits of improved ICT networks and services.

As understanding of convergence evolves, emerging trends point to three global good practice principles for regulatory frameworks to enable convergence. First, regulatory frameworks must promote competition. Service providers can deploy converged networks and services only if regulatory frameworks lower entry barriers in order to increase competition, reduce prices, and drive growth. However, it is equally important that regulators intervene in cases of market failures and do not allow abuse of market dominance. Hence, regulatory frameworks that establish and effectively enforce competitive, level playing fields will result in the greatest benefits for users.

Second, policy makers should rely more on market forces and less on regulation. Maintaining legacy regulatory frameworks will likely stifle the growth of convergence. Instead, regulation can move toward allowing innovation and the entry of value-added service providers to promote the development of content and ICT-enabled businesses and social services.

Finally, policy and regulatory frameworks should allow new technologies to contribute everything they have to offer. Regulatory frameworks that are technology neutral and allow flexibility in service provision will encourage investments and innovation. When service providers are able to use their networks to the fullest extent, they can reduce costs, increase business viability, and ultimately encourage markets that are more efficient. Users of the networks will benefit from lower prices, more choices, and better quality.

Policy makers seeking to respond to and enable convergence will find that doing so enhances the effects of earlier sector liberalization efforts and supports innovation in services that benefit the ICT sector. Countries that begin these second-generation reforms in the ICT sector will find themselves better off for it.

Public-Private Partnerships Can Leverage the Private Sector to Meet Public Policy Objectives

In market economies, responsibility for providing ICT infrastructure and services rests primarily with the private sector. This market-based approach has proven very successful in extending the reach of voice services. Competition is particularly effective among wireless mobile operators, resulting in the rapid rollout of networks across urban areas and, more recently, into rural areas at constantly declining cost.

The role of the public sector in achieving this outcome has been twofold: first, it has liberalized the market and regulated competition; and second, it has established mechanisms such as universal service funds and output-based aid that offer incentives for operators to provide services in areas of the country that otherwise would not be commercially viable.

The role of the public sector in providing access to broadband networks in developing countries is also likely to be a combination of market reforms and targeted incentives. Establishing effective competition over time among broadband networks and service providers is often a central part of a successful policy approach. This will require removal of legal and regulatory barriers to investment, entry, and competition. It will also require that competing operators are able to interconnect with the incumbent operators' infrastructure, thereby avoiding economic and technical bottlenecks as well as inefficient duplicate investments. Such a policy approach has been successfully applied in countries such as Brazil and Nigeria, where multiple broadband network companies have developed in competition with each other (World Bank 2008a). Competition is also emerging among companies that install submarine fiber-optic cables providing international broadband connectivity to developing countries, such as off the east coast of Africa, where three submarine fiber-optic cables are currently under development (Technology Review 2007).

Investment in broadband networks by private operators has primarily focused on urban areas. If these networks are to be rolled out to smaller towns and into rural areas, some form of more direct public sector support is likely to be required. In many high-income countries, such support was initially provided through state ownership of the incumbent operator. However, state-owned operators have often proven to be ineffective in many developing countries (Williams 2009, chapter 4 in this volume). Alternative methods of channeling public support will therefore be required. The most effective means of doing this will be through public-private partnerships (PPPs), which are able to harness the investment resources and technical expertise of the private sector to meet policy objectives, such as ensuring that networks are developed in otherwise commercially unviable areas. Although public support or incentives may initially be required, the objective should continue to be the development of a sector that is in the long run, commercially viable on a stand-alone basis and preferably that is in a competitive environment as demand picks up.

There are many different models for PPPs. Competitive subsidy or cost-sharing mechanisms are one type of model that has traditionally been used to encourage the rollout of voice networks into underserved areas (*info*Dev and ITU 2008), and which has been recently applied in the rollout of broadband infrastructure. In France, for example, the government has launched a scheme to provide broadband infrastructure in Limousin, a rural region in central France with limited broadband services. The project is structured as a 20-year concession to build and operate a backbone network and to construct a broadband wireless network with the costs being shared between the public and private sectors (ICEA 2008). The Eastern African Submarine Cable System (EASSy), a project to build a submarine fiber-optic cable that will stretch from South Africa to Sudan with connections to all the countries along its route, is an example of a different type of PPP. EASSy is owned by a consortium of private operators but financed by development finance institutions with no subsidies or support from governments. The partnership has ensured that the cable will be operated on an open-access basis, allowing all operators and service providers in the region to obtain access to affordable capacity by having access to competing cables and providers of capacity.[7]

These PPPs in broadband networks are new, and governments are experimenting with different models. The key to the success of these projects will be ensuring that they are structured so that the private sector has sufficient incentives to invest and operate networks efficiently while also achieving the governments' policy objectives of broadband network rollout.

Cross-Sector Leadership and Institutions Are Essential to Realize the Benefits from Investing in E-Government

E-government often entails institutional and political reform facilitated by technology. Competent leaders and institutions are essential to overcome resistance and inertia, to make timely policy choices, and to implement policy effectively. The cross-cutting nature of e-government makes it impossible to use traditional institutional arrangements that assign the entire agenda to a single ministry. Rather, e-government requires coordination among various government agencies. Public leadership needs to shift away from focusing on individual agencies and turf protection and toward management through collaboration across agencies. Moreover, e-government is a continuous process of policy development, investment planning, innovation, learning, and change management. This process must fit with and respond to a dynamic development strategy that supports evolving national goals and creates sustained institutional reforms and public service improvements while providing frameworks and structures that ensure continuity and proper institutional coordination.

In order to realize development gains from e-government investments, a sufficiently influential institutional structure that attracts strong commitment and support is essential. Many countries have moved toward direct, institutionalized engagement of top public leaders to position the coordination of government transformation under the highest authority. Often the office of the prime minister or the head of state hosts a coordinating unit and chairs the interministerial e-government steering committee. Such an approach has three benefits. First, the head of this coordinating unit becomes a visible leader. Second, this leader can use e-government as a core component of a public management reform agenda and, more broadly, as a key to shifting to a knowledge-based, innovation-driven economy. Third, the coordination unit encourages work across ministries and levels of government to implement e-government programs.

A survey of 30 developing and developed countries found four basic models of a national institutional framework to lead their e-government agenda and fulfill the key functions of governance and coordination (table 1.1). In practice, though, these frameworks are more diverse and complex than suggested by these basic models and may evolve over time, shifting from one model to another or becoming hybrids.

These four institutional models focus on the leading or central institution for e-government strategy and policy making, and on governance and coordination. But in terms of facilitating implementation, governments have increasingly experimented with new arrangements outside the ministerial structure to overcome sectoral fragmentation and civil service constraints and to expand e-government institutional capability. Countries such as Bulgaria, Ireland, and Singapore now have dedicated executive ICT agencies in their civil services. These agencies have special autonomy and salary structures to attract and motivate the best technical talent. Others, including India and Sri Lanka, are experimenting with ICT agencies that have a government-appointed board of directors and representatives of key stakeholders from the private sector and civil society.

Institutional innovation in these countries has a number of advantages. Apart from being shielded from the larger bureaucracy and having the flexibility to react swiftly to changing demands, agencies dedicated to e-government can hire personnel at competitive wages, provide shared services (such as network infrastructure) to the government, and outsource tasks to the private sector. Active private sector participation helps the agencies operate in an agile, businesslike way and accelerate e-government financing and

Table 1.1 Models for E-Government Institutions in Various Countries

Model	Countries	Benefits	Drawbacks
Policy and investment coordination (cross-cutting ministry such as finance, treasury, economy, budget, or planning)	Australia, Brazil, Canada, Chile, China, Finland, France, Ireland, Israel, Japan, Rwanda, Sri Lanka, United Kingdom, United States	Has direct control over funds required by other ministries to implement e-government. Helps integrate e-government with overall economic management.	May lack the focus and technical expertise needed to coordinate e-government and facilitate implementation.
Administrative coordination (ministry of public administration, services, affairs, interior, state, or administrative reform)	Bulgaria, Arab Republic of Egypt, Germany, Republic of Korea, Mexico, Slovenia, South Africa	Facilitates integration of administrative simplification and reforms into e-government.	May lack the technical expertise required to coordinate e-government or the financial and economic knowledge to set priorities.
Technical coordination (ministry of ICT, science and technology, or industry)	Ghana, India, Jordan, Kenya, Pakistan, Romania, Singapore, Thailand, Vietnam	Ensures that technical staff is available; eases access to nongovernmental stakeholders (firms, NGOs, and academia).	May be too focused on technology or industry and disconnected from administrative reform.
Shared or no coordination	Russian Federation, Sweden, Tunisia	Least demanding and with little political sensitivity (does not challenge the existing institutional framework and responsibilities of ministries).	May lead to rivalries among ministries. No cross-cutting perspective. Fails to exploit shared services and infrastructure and economies of scale.

Source: Hanna and Qiang 2009 (chapter 6 in this volume).

implementation, making the best use of scarce public resources and relevant expertise.

One disadvantage, however, is the potential struggle to obtain political weight and financial resources if the new entity lacks institutional links to powerful ministries. On the other hand, if such links are too strong, the government bureaucracy might assert control over the agency and undermine the effectiveness or businesslike culture of agency staff. Hence, the viability of these agencies depends on political leaders giving the agency the autonomy needed to act in an agile manner and avoid interference in staffing and day-to-day management.

Public Sector Interventions to Promote IT Services and ITES Can Be Good Investments Irrespective of Success of the IT Initiatives

In countries that have succeeded in IT services and ITES, governments have generally adopted a proactive role in promoting the sector. Such support can often be provided with low levels of public funding by leveraging private sector investments. Most of the public interventions to promote these industries—such as improving education, providing adequate infrastructure, or catalyzing regulatory reforms—contribute to the broader business environment and benefit many other sectors of the economy whether related to the IT industries or simply able to realize efficiencies as a result of IT applications. In this sense, government support for IT services and ITES is consistent with the argument that public interventions should create positive externalities.

Locations that have successfully developed IT services and ITES typically have empowered industry development institutions to identify approaches that adapt to the rapidly evolving needs of the local and global economies. They achieve this through ongoing engagement with IT services and ITES companies, as opposed to adopting a policy approach with a predetermined strategic blueprint. The private sector can provide governments with invaluable information and insights on available opportunities, market trends, and future skill requirements (Dongier and Sudan 2009, chapter 7 in this volume).

Ireland's government-sponsored Industrial Development Agency (IDA), for example, has achieved significant success in attracting IT services and ITES investments. The agency deals with a multitude of aspects of inward investment: marketing, managing investment proposals, providing financial incentives and property solutions, helping investors get started, and working with investors to maximize their contribution to Ireland's economy (IDA 2006). Nine of IDA's 13 board members are from the private sector. IDA's investment program has been a major driving force behind the growth of IT services and ITES.

Given the importance of skills as a driver of growth of IT services and ITES, a focus on expanding the talent pool in close alignment with local and global industry needs is essential. Partnerships composed of leading companies, industry associations, and universities have successfully aligned education and skills competencies with the needs of the industries in several countries. Singapore has been one of the most proactive examples in this regard. Its Industrial Training Board (ITB) established an extensive system of training advisory committees with industry participation and introduced industry-based training schemes in partnership with companies. It also established arrangements for keeping training staff abreast of the latest technological developments (Lee and others 2008). In addition, the Info-Comm Development Agency of Singapore has been active in forging global partnerships to improve ICT sector skills. It collaborated, for instance, with Carnegie Mellon University's Entertainment Technology Center and the National University of Singapore's School of Computing in 2006 to develop a degree program in interactive digital media (CMU 2006).

Developing globally benchmarked skills in partnership with leading standards organizations helps not only maintain a certain level of quality, but also align skills with industry requirements. Universities in the Philippines, for example, offer courses in finance and accounting modeled after the U.S. Generally Accepted Accounting Principles (GAAP). This has made the Philippines a natural choice for U.S. banks and financial institutions seeking to offshore portions of their operations. Similarly, the Chartered Institute of Management Accountants (CIMA), one of the world's largest professional accounting bodies, has its second-largest number of management accountants in Sri Lanka, after the United Kingdom, making Sri Lanka an attractive offshoring destination.

Structure of the Report

IC4D 2009 has two parts. Part I analyzes key aspects of extending access to ICT infrastructure and services (chapters 2, 3, and 4), mainstreaming ICT applications featuring e-government (chapters 5 and 6), and developing IT services and ITES (chapter 7).

Chapter 2 sets the sector context for the report. It explains convergence and shows that convergence is already a widespread, market-driven reality. It also discusses some of the main opportunities and challenges convergence poses to businesses, users, and governments, while advocating for proactive government responses.

Chapters 3 and 4 cover two important aspects of broadband. Chapter 3 summarizes the findings from the literature on the economic impacts of broadband on individuals, firms, communities, and the overall economy. It also introduces a cross-country empirical model for analyzing the impact of broadband on economic growth. It concludes that broadband has a significant impact on growth and deserves a central role in country development and competitiveness strategies. Chapter 4 outlines a market-based approach to policy for developing backbone network infrastructure, an essential part of providing broadband services. Building on the model of infrastructure competition, this approach seeks to harness both the investment resources and the operational expertise of the private sector to help meet the public policy objective of extending affordable access, thereby minimizing the financial and operational burden on the public sector.

Chapters 5 and 6 focus on e-government. Chapter 5 proposes a framework for assessing ICT investment in public administrations and applies it to five e-government projects in India that shifted from manual to electronic service delivery. It analyzes positive and negative changes perceived by the users and implementing agencies in terms of cost and quality of accessing public services and improvements in governance. Chapter 6 highlights the importance of institutional development for e-government programs. It outlines key functions of effective e-government institutions and identifies basic models that countries have used to fulfill these functions.

Chapter 7 covers development impact and policy options related to the expanding IT services and ITES. It aims to help policy makers take advantage of the opportunities presented by increased cross-border IT services and ITES, demonstrating the benefits for countries that have seized the opportunities and examining the potential competitiveness of small economies and least developed countries. Factors crucial to the competitiveness of a country or location—skills, cost advantages, infrastructure, and a hospitable business environment—are also analyzed.

Part II of this report presents a trend analysis of ICT sector performance and the World Bank ICT at-a-glance (AAG) tables for 150 economies. The analysis uses data on nearly 30 ICT indicators from the AAG tables both to demonstrate the progress that many developing countries have made in recent years in improving ICT access, use, quality, affordability, trade, and applications and to show how that progress relates to enabling policies and regulations.

Country ICT performance measures are introduced in this report to provide a quick and effective way for policy makers to assess their countries' ICT capacities in comparison with other countries, as well as to benchmark their countries' progress along three key dimensions of ICT development over time. The economies have been evaluated and given a score on a scale from 1 to 10, corresponding to the performance deciles, for each of the dimensions of ICT sector performance: (1) access to ICT services, (2) affordability of ICT services, and (3) adoption of ICT applications in government and business.

Overall, as would be expected, there is a close relationship between the country ICT performance measures and income levels. The leading economies in ICT performance are mainly developed economies. Among developing countries, some stand out as better ICT performers than their incomes would suggest, such as Serbia, Croatia, Ukraine, Macedonia, Syria, Jordan, Vietnam, and Moldova in terms of access and Malaysia, Jordan, Peru, Guatemala, India, and Mongolia in terms of adoption of ICT applications (figure 1.5).

Recognizing that comparable ICT data provide a good basis for sound policy, a global Partnership on Measuring ICT for Development was launched in 2004. The members include the United Nations Conference on Trade and Development (UNCTAD), the International Telecommunication Union (ITU), the Organisation for Economic Co-operation and Development (OECD), the United Nations Educational, Scientific, and Cultural Organization [UNESCO] Institute for Statistics (UIS), the United Nations Regional Economic Commissions, Eurostat, and the World Bank. The partnership has been assisting statistical agencies in developing countries with their ICT data collection and dissemination efforts, and conducting workshops at the regional level to exchange national experiences and discuss definitions, methodologies, survey vehicles, and results analysis.

Figure 1.5 Relation between the Country ICT Performance Measures (for Access and Applications) and Income per Capita, Developing Countries

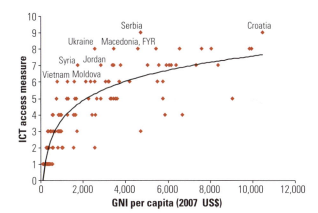

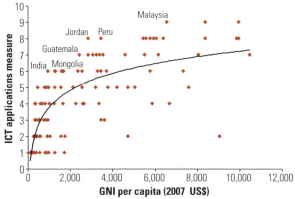

Source: World Bank staff.

One area that is receiving increasing attention in all countries, but that remains weak in most, is impact evaluation. Measuring the impact of ICT on development and evaluating the outputs of ICT interventions not only reveal the magnitude of the impact of ICT on development outcomes but are perhaps the most rigorous way to deal with the issue of attribution, or pointing to the effects of specific development interventions. Impact evaluation is essential to ensure that ICT strategies are relevant and to be able to hold governments accountable for their implementation.

More research is needed on the socioeconomic impact of ICT on development, the cost-effectiveness of ICT strategies and programs, and the economic justification of public sector intervention. Are these investments cost-effective in meeting social and development goals? How do they compare with alternative uses of public resources; for example, education and health? Rather than relying primarily on anecdotal evidence, there has been a clear call for rigorous and robust impact evaluation as a basis for guiding policy development and public investment decisions.

Notes

1. Nevertheless, it is important to note that the sale of 4 billion mobile phones worldwide implies far fewer than 4 billion individual users. The phenomenon of multiple mobile phone ownership is growing in many economies, including some in Africa. As evidence of this trend, penetration rates for mobile phones have risen to more than 100 percent in some markets. On the other hand, shared usage of mobile phones in rural areas of developing countries is also growing, suggesting that the footprint of mobile phone coverage is wider than for other ICT services.

2. See http://www.tradenet.biz/gapto for more information.

3. See http://www.voxiva.com/solutionslist.php?catname=Health for more information.

4. See http://www.simpill.com/index.html for more information.

5. Estimating the size of the IT services and ITES is difficult because of definitional issues and the relative novelty of the industries. Official statistics are often not available or not reliable, and calculations based on balance of payments and trade in services do not accurately isolate IT services and ITES. As a result, much of the data on the size of the current market comes from private surveys, consulting firms, and anecdotal evidence.

6. According to NASSCOM (National Association of Software and Services Companies, India), the global financial crisis is expected to result in reduced technology-related spending for the first two to three quarters of 2009, but it is expected to pick up in 2010; and "greater focus on cost and operational efficiencies in the recessionary environment is expected to enhance global sourcing" (NASSCOM 2009).

7. See http://go.worldbank.org/GKHOFFDJB0 for more information.

References

Aker, Jenny C. 2008. "Does Digital Divide or Provide? The Impact of Cell Phones on Grain Markets in Niger." Department of Agricultural and Resource Economics, University of California, Berkeley. http://papers.ssrn.com/sol3/papers.cfm?abstract_id=1093374.

Bhatia, Deepak, Subhash C. Bhatnagar, and Jiro Tominaga. 2009. "How Do Manual and E-Government Services Compare? Experiences from India." In *Information and Communications for Development 2009: Extending Reach and Increasing Impact*, 67–82. Washington, DC: World Bank.

BPAP (Business Processing Association of the Philippines). 2007. *Offshoring and Outsourcing, Philippines Roadmap 2010*. Makati City. http://www.bpap.org/bpap/index.asp?roadmap.

Clarke, George, and Scott Wallsten. 2006. "Has the Internet Increased Trade? Evidence from Industrial and Developing Countries." *Economic Inquiry* 44 (3): 465–84.

CMU (Carnegie Mellon University). 2006. "Carnegie Mellon Collaborates with National University of Singapore to Create Concurrent Digital Media Degree." Press release, CMU, Pittsburgh, PA. http://www.cmu.edu/news/archive/2006/november/nov.-9---etc,-singapore-join-forces.shtml.

Crescia, Elena. 2006. "Measuring e-gov Impact: The Experience of São Paulo, Brazil." Presentation made to the E-Development Thematic Group at the World Bank, Washington, DC, February 9. http://siteresources.worldbank.org/INTEDEVELOPMENT/Resources/Measuring-Impact.pdf.

De Wulf, Luc, and José B. Sokol. 2004. *Customs Modernization Initiatives: Case Studies*. Washington, DC: World Bank.

Dongier, Philippe, and Randeep Sudan. 2009. "Realizing the Opportunities Presented by the Global Trade in IT-Based Services." In *Information and Communications for Development 2009: Extending Reach and Increasing Impact*, 103–22. Washington, DC: World Bank.

The Economist. 2008a. "Halfway There: How to Promote the Spread of Mobile Technologies among the World's Poorest." May 29. http://www.economist.com/business/displaystory.cfm?story_id=11465558.

———. 2008b. "The Meek Shall Inherit the Web." September 4. http://www.economist.com/science/tq/displaystory.cfm?story_id=11999307.

Ford, George S., and Thomas M. Koutsky. 2005. "Broadband and Economic Development: A Municipal Case Study from Florida." *Applied Economic Studies* (April): 1–17. http://www.freepress.net/docs/broadband_and_economic_development_aes.pdf.

Gartner Research. 2008. "Gartner on Outsourcing, 2008–2009." Stamford, CT. http://www.gartner.com/resources/164200/164206/gartner_on_outsourcing_20082_164206.pdf.

GSMA (GSM Association). 2008. "The GSMA Development Fund Top 20: Research on the Economic and Social Impact of Mobile Communications in Developing Countries." GSMA Development Fund.

Hanna, Nagy R., and Christine Zhen-Wei Qiang. 2009. "National E-Government Institutions: Functions, Models, and Trends." In *Information and Communications for Development 2009: Extending Reach and Increasing Impact*, 83–102. Washington, DC: World Bank.

ICEA (Ingénieurs Conseil et Économistes Associés). 2008. "Strategies for the Promotion of Backbone Communications Networks in Sub-Saharan Africa." Study commissioned by the World Bank, Washington, DC.

IDA (Industrial Development Agency), Ireland. 2006. "Guide to IDA Ireland's Legislation, Structure, Functions, Rules, Practices, Procedures and Records." IDA Ireland, Dublin. http://www.idaireland.com/uploads/documents/IDA_Publications/FOI_Manual_November_06_2.pdf.

*info*Dev and ITU. 2008. "ICT Regulation Toolkit." Washington, DC. http://www.ictregulationtoolkit.org/en/Section.618.html.

Information Society Technologies Advisory Group (ISTAG). 2006. "Shaping Europe's Future through ICT." European Union, Brussels. http://ec.europa.eu/information_society/tl/research/key_docs/documents/istag.pdf.

Jensen, Robert. 2007 "The Digital Provide: Information (Technology), Market Performance and Welfare in the South Indian Fisheries Sector." *The Quarterly Journal of Economics* 122 (3): 879–924.

Johnson, B., J. M. Manyika, and L. A. Yee. 2005. "The Next Revolution in Interactions." *McKinsey Quarterly* 4: 20–33.

Kelly, D. J. 2004. "A Study of Economic and Community Benefits of Cedar Falls, Iowa's Municipal Telecommunications Network." Iowa Association of Municipal Utilities, Ankeny, Iowa. http://www.baller.com/pdfs/cedarfalls_white_paper.pdf.

Lee, Sing Kong, Goh Chor Boon, Birger Fredriksen, and Tan Jee Peng. 2008. *Toward a Better Future: Education and Training for Economic Development in Singapore Since 1965*. Washington, DC, and Singapore: World Bank and National Institute of Education.

McKinsey. 2008. "Development of IT and ITES Industries—Impacts, Trends, Opportunities, and Lessons Learned for Developing Countries: Exhibits to Economic Impact Discussion." Presentation by McKinsey & Co. at the World Bank, Washington, DC, June 2008.

Momentum Research Group. 2005. "Net Impact Latin America: From Connectivity to Productivity." Momentum Research Group, Austin, TX. http://www.netimpactstudy.com/nila/pdf/netimpact_la_full_report_t.pdf.

NASSCOM (National Association of Software and Services Companies). 2009. "Indian IT-BPO Industry Factsheet." http://www.nasscom.org/Nasscom/templates/NormalPage.aspx?id=53615.

NASSCOM-Everest. 2008. "Roadmap 2012—Capitalizing on the Expanding BPO Landscape." NASSCOM-Everest. http://www.nasscom.in/Nasscom/templates/NormalPage.aspx?id=53361.

Pyramid Research. 2007. "From Triple-play to Quad-play." Pyramid Research, Cambridge, MA.

Qiang, Christine Zhen-Wei. 2009. "Telecommunications and Economic Growth." Unpublished paper, World Bank, Washington, DC.

Qiang, Christine Zhen-Wei, and Carlo M. Rossotto. 2009. "Economic Impacts of Broadband." In *Information and

Communications for Development 2009: Extending Reach and Increasing Impact, 35–50. Washington, DC: World Bank.

Singh, Rajendra, and Siddhartha Raja. 2009. "Nothing Endures but Change: Thinking Strategically about ICT Convergence." In *Information and Communications for Development 2009: Extending Reach and Increasing Impact*, 19–34. Washington, DC: World Bank.

Skype. 2008. "Skype Appoints New Chief Operating Officer." Press release, July 1.

Sprint. 2006. "Sprint Mobile Broadband: Enhancing Productivity in the Insurance Industry and Beyond." Sprint. http://www.sprint.com/business/resources/065455-insurancecs-1g.pdf.

Strategic Networks Group. 2003. "Economic Impact Study of the South Dundas Township Fiber Network." Prepared for the U.K. Department of Trade and Industry, Ontario. http://www.berr.gov.uk/files/file13262.pdf.

Technology Review. 2007. "Race is On to Lay Undersea Fiber Optic Cable on Eastern Africa Coast." Massachusetts. http://www.technologyreview.com/Wire/18814/?a=f.

TeleGeography. 2007. *Voice Report*. Washington, DC: TeleGeography.

Tholons. 2006. "Emergence of Centers of Excellence." Unpublished report.

UNCTAD (United Nations Conference on Trade and Development). 2008. *Globalization for Development: The International Trade Perspective*. New York: United Nations. http://www.unctad.org/en/docs/ditc20071_en.pdf.

Williams, Mark D. J. 2009. "Advancing the Development of Backbone Networks in Sub-Saharan Africa." In *Information and Communications for Development 2009: Extending Reach and Increasing Impact*, 51–66. Washington, DC: World Bank.

Wireless Intelligence. 2008. Wireless Intelligence database. London: Wireless Intelligence. http://www.wirelessintelligence.com.

World Bank. 2007. *World Development Report 2008: Agriculture for Development*. Washington, DC: World Bank. http://go.worldbank.org/ZJIAOSUFU0.

———. 2008a. "Broadband for Africa. Policy for Promoting the Development of Backbone Networks." Unpublished report. Global Information and Communications Technologies Department, World Bank, Washington DC.

———. 2008b. "Sending Money Home: How It Works in Sierra Leone." World Bank, Washington, DC. http://go.worldbank.org/X31JDSTUM0.

———. 2008c. World Development Indicators (WDI) database. Washington, DC: World Bank.

Zhou, Hongreng. 2007. "E-government Funding in China." Paper presented at the 7th Global Forum on Reinventing Government, Vienna, Austria, June 26–29. http://unpan1.un.org/intradoc/groups/public/documents/unpan/unpan025948.pdf.

Zilber, Julie, David Schneier, and Philip Djwa. 2005. "You Snooze, You Lose: The Economic Impact of Broadband in the Peace River and South Similkameen Regions." Prepared for Industry Canada, Ottawa.

Chapter 2

Nothing Endures but Change: Thinking Strategically about ICT Convergence

Rajendra Singh and Siddhartha Raja

Countries that adopt policies enabling convergence among telecommunications, media, and computing services will enhance the impact of information and communication technology (ICT) on economic development. Technological innovation and market demand are driving the ICT sector toward convergence. This trend matters because convergence can lower entry barriers, allow service providers to try new business models, promote competition, reduce costs for service providers and users, and broaden the range of services and technologies available to users. At the same time, convergence can also lead to market consolidation, reduced competition, and new barriers to entry. This trend and its implications apply to countries at all development stages, from mature to low-income economies.

This chapter explains convergence and its main forms; shows that convergence is already a widespread, market-driven reality; discusses some of the main opportunities and challenges it poses to businesses, users, and governments; and describes potential government policy responses—along with the likely outcomes and potential benefits and risks.[1]

Understanding Convergence

Convergence is shorthand for several changes occurring in the ICT sector. Broadly speaking, convergence refers to the erosion of boundaries among previously separate services, networks, and business models in the sector.

There are three main forms of convergence. The first, service convergence, or "multiple play," allows a firm to use a single network to provide several communication services that traditionally required separate networks. The second form is network convergence, where a common standard allows several types of networks to connect with each other. Consequently, a communication service can travel over any combination of networks. While these two forms of convergence are technological, the third form, corporate convergence, results from mergers, acquisitions, or collaborations among firms. Under the third form, newly organized business entities offer multiple services and address different markets. Table 2.1 summarizes the three forms of convergence and associated benefits, risks, and policy implications.[2]

Convergence Is Reality

Convergence is primarily a process that results from service providers adopting new technologies and business models allowed by technology and driven by demand. The factors pushing service providers toward converged business models are increasingly common worldwide, including in developing countries.

The fundamental technology drivers for convergence have been the digitization of communication and the falling costs of computing power and memory. Both factors have increased a network's capacity to carry information while bandwidth remains fixed. Consequently, the capacities of

Table 2.1 Different Forms of Convergence

Dimension	Service convergence	Network convergence	Corporate convergence
Definition	Service providers use a single network to provide multiple services.	A service can travel over any combination of networks.	Firms in one sector acquire, merge, or collaborate with firms in other sectors.
Examples	Service providers offer telephony, television, and Internet services using telephone, cable television, or fixed wireless networks. Examples include providers in Chile, Arab Republic of Egypt, India, Poland, and Ukraine.	Internet telephony services such as Skype and Jajah carry voice telephony using the Internet and traditional networks. In the United Kingdom, BT's Fusion service carries calls over WiFi (wireless fidelity) and cellular networks.	Internet, media, and telecommunications firms partner, merge, or expand their range of services. Such developments have occurred in Brazil, Nigeria, and Sri Lanka.
Benefits	Service providers can enter new sectors, use their networks more efficiently, offer discounts, and increase access to new ICT services.	Reduced costs can lower tariffs. Network integration permits mobility for consumers and expands coverage.	Mergers create opportunities for new services and markets, lower costs and tariffs, and increase the coverage of individual firms.
Risks	Subscribers could be locked into one service provider. Smaller firms, especially those without their own broadband networks, might get pushed out of the market.	Service providers could reduce investments in network infrastructure, slowing build-out.	Mergers could lead to market dominance, less competition, and less diversity of media content.
Policy implications	Multiple play changes the scope and boundaries of markets and alters entry barriers.	Connecting different networks allows location- and network-independent service provision.	Mergers create new business models and alter the market structure, changing the dynamics of the sector.

Source: Authors' analysis.

telephone, cable TV, and wireless networks have grown steadily. More recently, the growing use of Internet protocol (IP)-based packet-switched data transmission has made it possible for different devices and applications to use any one of several networks and for previously separate networks to interconnect. Together, these factors have facilitated the growth of multimedia communication. This has reduced costs and eased the design and deployment of multimedia access devices, and has thus led to a proliferation of increasingly inexpensive digital devices. For example, the personal computer or mobile telephone can now receive and transmit different types of media and services because of enhanced processing power and memory capacity.

With these technical factors evolving, convergence has found significant market traction with service providers seeking to increase revenues and cut costs of service provision. Service providers around the world are embracing convergence through investment in all-IP networks—estimated to reach a cumulative total of $200 billion in 2015—and in converged business models. In an indication of this expanding underlying technological base, one analyst estimates that the global IP switch and router market grew about 10 percent in 2007, to $11 billion (*Marketwire* 2008). One major IP network equipment manufacturer has seen sales in emerging markets double since 2005, well above its worldwide sales growth of 40 percent (Cisco Systems 2007).

The deployment of broadband networks is another market factor supporting convergence. Broadband connectivity facilitates convergence because it allows the provision of multimedia content—such as compact-disc-quality audio and streaming video—at reasonable prices. As of 2007, broadband was commercially available in 166 countries (ITU 2006), and nearly a quarter of the 300 million subscribers were in middle-income countries (Internet World Stats 2007).

Indeed, demand for converged services is quite evident. By late 2007, there were more than 30 million "triple-play" subscribers worldwide—typically receiving telephony, video, and Internet services over broadband networks (Pyramid Research 2007). Skype, an Internet telephony service, has more than 300 million subscribers in 225 countries and territories (Skype 2008) and carried 13.75 billion minutes of international computer-to-computer calls in 2007 (TeleGeography 2007).

There has also been consolidation in the development and provision of content and services. Investments, mergers, and cross-holdings in the media and telecommunications industries have increased the number of content creators and network operators with access to content and delivery mechanisms. The development of online advertising has allowed many content providers to offer their services free or well below cost. Such arrangements allow consumers to sample content and find uses for it, even if only in a limited manner. As a result, consumers create demand for content, resulting in higher demand for services that can support it.

These technical and market factors driving convergence are visible in countries across the full range of economic development stages. While high-income countries have had these conditions in place for some time, there are clear indications that even low-income countries and population groups are now recognizing these forces and the resulting convergence as a reality. Box 2.1 provides examples of convergence in emerging markets and low-income countries.

Implications of Convergence for the ICT Sector

Convergence has far-ranging implications for ICT service providers and users. It changes business models, expands markets, increases the range of services and applications available to users, and alters market structure and dynamics. Furthermore, given that ICT is a critical input to economic and social activities,[3] convergence has an indirect effect on social and economic development.

There are also risks and challenges. Most prominently, convergence may lead to monopolization, allowing larger firms to extend their reach into new markets or raising entry barriers for new entrants. Hence, policy makers must think strategically about convergence and its role in their economy in order to enhance its benefits and contain risks.

Changing Business Models for Service Providers

Service providers in both the telecommunications and media sectors have seen convergence as a powerful way of increasing revenues and reducing costs.

Increased revenues. By offering a wider range of services, service providers can capture more revenues from their subscribers. A major U.S. cable TV operator saw its average monthly revenues per subscriber jump from $42 in 1998 to $102 in 2007, with non-TV services such as telephony or broadband Internet now contributing one-third of its total revenues (Comcast 2007). In Chile, about 60 percent of VTR's 853,000 cable TV subscribers also use telephone or Internet services, increasing the company's revenues by 44 percent between 2005 and 2007 (Liberty Global 2007). More recently, telecommunications firms are seeing payoffs from diversification. One U.S. firm now derives 25 percent of its retail revenues from broadband and video services, and two of the country's largest telecommunications firms saw their revenues from video service quadruple between 2007 and 2008 (AT&T 2008 and Verizon Communications 2008).

Reduced costs. Service providers also see convergence as a way to cut costs and operating expenses. BT (2006) expects that its all-IP network will help reduce operating expenses by £1 billion a year. Savings are expected because this network replaces the company's 17 separate networks, including its traditional telephony network, with one—integrating a number of operational and network management systems (*Dow Jones International News* 2008). Similarly, Verizon expects that migrating its customers to an all-fiber-optic IP network will save more than $1 billion a year on network maintenance (*Providence Journal* 2007).

Use of standardized IP networks is also driving cost savings. Telecom Italia cut costs by 60 percent by introducing IP technology for calls between Milan and Rome (*The Economist* 2006). Service convergence also cuts costs by increasing network use. Traditionally, telephone and cable TV networks provided only one service. Today these networks can carry multiple services, lowering the cost of each.

However, such cost reductions come at a high upfront price. BT's savings will follow a £10 billion ($16 billion) investment between 2004 and 2011 (*Business Monitor International* 2008a). Similarly, Verizon expects to spend about $23 billion building its U.S. network (*Providence Journal* 2007). The high capital spending required to offer converged services creates a new entry barrier that small or new service providers might not be able to overcome.

Expanded Access through Larger Markets

Convergence expands consumers' access to services because it lowers prices, which in turn increases the addressable market and widens coverage by using multiple infrastructures.

Lower prices for consumers. The reduced costs of operating converged networks and providing multiple services translate into lower prices for consumers. The starkest examples of this phenomenon come from the voice telephony market, where voice-over-Internet protocol (VoIP)

Box 2.1 Examples of Convergence in Developing Countries

Service Convergence

- Argentine cable TV operators Multicanal and Cablevision are investing about $310 million in fiber-optic networks in 2008, with plans to offer triple-play services. In July 2008, however, fixed telephony service providers Telecom Argentina and Telefónica Argentina failed to get court approval to launch triple-play services.

- In 2006, Telefónica Chile began offering Internet protocol television (IPTV) and satellite television services to counter a decline in fixed-line revenues and subscriptions. Cable TV operator VTR saw its triple-play subscriber base double in 2006, and is considering acquiring a third generation (3G) license to add mobile voice services to its portfolio.

- India's incumbent public telecommunications provider, MTNL, started offering IPTV services in Mumbai in 2006. The service now offers about 150 channels, costs about $5 a month, and has a reported 6,000 subscribers. A number of private operators have since begun offering IPTV services.

- In March 2008, Ukraine's Comstar began offering IPTV services over its fiber-based, next-generation network—making Comstar the country's first triple-play voice, video, and Internet provider. The company will soon face competition from Golden Telecom Ukraine and fixed-line operator Ukrtelecom. The IPTV offerings follow broadcaster Viasat's plans to introduce digital satellite TV services in 2008.

- Africa's service providers are beginning to invest in multiple play. In March 2008, Ghana Telecom announced plans to introduce IPTV services. The legalization of voice-over-Internet (VoIP) in Cape Verde in May 2008 has led service provider Cabo Com to announce investments in triple-play services, while looking to compete in broadband pricing.

Network Convergence

- Jajah is a service that carries calls between traditional telephones over the Internet. This combines wide access to telephones with the lower costs of carrying calls over the Internet. Launched in 2006, Jajah now connects customers across 200 countries.

Corporate Convergence

- Since 2006, Brazil has seen convergence among telephone and cable companies. Telemar acquired Way TV, while Telefónica bought a stake in TVA. Convergence is emerging in response to the introduction of triple-play services by NET Serviços, which has an estimated 400,000 subscribers.

- In 2007, MTN Nigeria acquired VGC Communications, a fixed and wireless phone provider. This move came after VGC secured a unified license to offer fixed and wireless telephony, Internet, and value added services in 2006. The chief executive officer of MTN noted that the company made the acquisition with the intention of accessing VGC's infrastructure and staff to achieve convergence.

- Sri Lanka's Dialog Telekom now offers telecommunications and media services. It has become a quadruple-play operator, offering fixed and mobile voice, TV, and Internet service, albeit on separate networks. Its satellite TV service reaches more than 60,000 households, while its mobile phone service has 4.3 million subscribers and will soon include 3G services.

Source: Authors.

technology has significantly changed price structures. VoIP has affected the pricing of international call traffic because it allows carriers to bypass and compete with traditional call pricing regimes. In 2007, one-fifth of international voice telephony traffic (in terms of minutes) used VoIP. In fact, VoIP traffic grew five times faster than did traditional voice traffic (TeleGeography 2007). Services that use the Internet to carry telephone calls offer significant discounts to consumers (box 2.2).

Many service providers also give discounts on bundles of services, charging less than if subscribers paid for each separately. Such discounts can be as high as 40 percent (Pyramid Research 2007, p. 14). Lower prices increase the addressable market and make some services more attractive to users who are price conscious or unsure of the usefulness of new services. In Sweden, for example, a cable TV company offering triple-play services gives subscribers the least expensive service free (OECD 2006).

Wider coverage. Convergence also allows service providers to reach new subscribers. Multiple play allows new services to travel over existing networks, expanding the reach of communication services. One recent report found that telecommunications firms offering IP television (IPTV) have succeeded in countries with relatively low pay TV penetration but high broadband penetration (Telecommunications Management Group 2008).

The evolution of digital video broadcasting (DVB) and mobile TV will enable the use of triple play over wireless networks, further extending the reach of services. The provision of DVB over cellular networks has recognized potential to increase the number of TV viewers in countries such as Kenya and the Philippines.[4] New broadband wireless technologies are also raising expectations. For example, in 2005, Kenya Data Networks began deploying a WiMax system designed to offer converged services, such as voice and data (*All Africa* 2005). Wireless triple play will be especially

Box 2.2 The Impact of Voice-over-Internet Protocol on International Calling Prices

Internet-based VoIP services such as Skype make it possible to have long-distance telephone conversations that are much cheaper than with traditional long-distance services. Lower costs are also possible for telephone-based services conducted over Internet networks—such as Jajah, which uses the Internet to carry phone conversations. If all the international calls made to just the top-10 destinations from the United States used Jajah, the annual savings would top $2.5 billion.

If a country's licensing regime prevents the entry of VoIP-based providers or restricts the type of technology they can use, it reduces the benefits of convergence for consumers. Moreover, countries that have banned these technologies have undermined their technological competitiveness. Failure to legalize VoIP reduces the opportunities for entrepreneurs to develop businesses into a core of fast-growing information technology (IT) startups, which tends to happen in countries where VoIP is legal.

VoIP and Traditional Carrier Charges for Calls from the United States to India

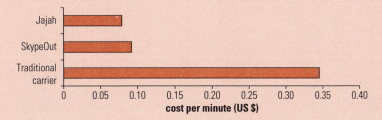

Sources: Authors' estimates based on tariff data from service providers' Web sites, TeleGeography traffic estimates, and Economist Intelligence Unit 2007.

useful in developing countries, where mobile phone subscription is far more common than ownership of personal computers or TVs.

Further, network convergence allows any combination of communication networks to carry services. A lot of nontraditional infrastructure can now carry telephone services, including cable TV and electricity distribution networks. This development can significantly improve coverage, even in low-income countries. In one set of low-income countries, wireline telephones reach an average of just 7 percent of households, while 33 percent of households are electrified and 18 percent subscribe to cable TV (Figure 2.1). Thus, having a combination of networks to carry communication services can move countries closer to universal service.

Broader Range of Services and Applications

Convergence enables ICT users to access a range of services through a wide variety of devices, including mobile phones. Some 3.9 billion mobile phones are in use worldwide, giving these devices enormous potential for providing multimedia services. Already, 27 percent of U.S. mobile phone users between the ages of 25 and 34 watch video on their cell phones (Economist Intelligence Unit 2008). South African media conglomerate Naspers has plans to expand its mobile television services into four new African markets, after introducing it in Namibia, Kenya, and Nigeria (*Reuters* 2008). Similarly, an estimated 66 million mobile phone subscribers in India can access Internet services (TRAI 2008). A 2005 survey of 4,000 mobile phone users found that nearly a third were using their phones for e-mail or Internet browsing (OECD 2007, figure 5.8).

Thus, if service providers build service-converged networks, then financial services, public services, and entertainment applications will be able to reach a far larger portion of the world's population. Similar possibilities arise from the mixed use of cable TV, wireless broadband, and other ICT networks. Access to high-quality, reliable, affordable ICT networks can significantly strengthen governance through e-government applications and provide opportunities for the remote delivery of health information or education services.

Increased demand for content and applications over converged networks drives significant economic development. For instance, media and entertainment expansion into mobile telephony is growing rapidly: mobile gaming is a $4 billion global market, and in 2005 alone, more than 420 million songs were downloaded onto mobile phones around the world (SSKI Research 2007). Creation of these new markets drives employment and investment and catalyzes network growth.

Moreover, online services such as blogs, video repositories, and social networking tools create opportunities for social development. The information and knowledge channels created allow the exchange of ideas and provide a platform

Figure 2.1 Household Penetration of Wireline Telephone, Cable Television, and Electricity Networks in Selected Countries

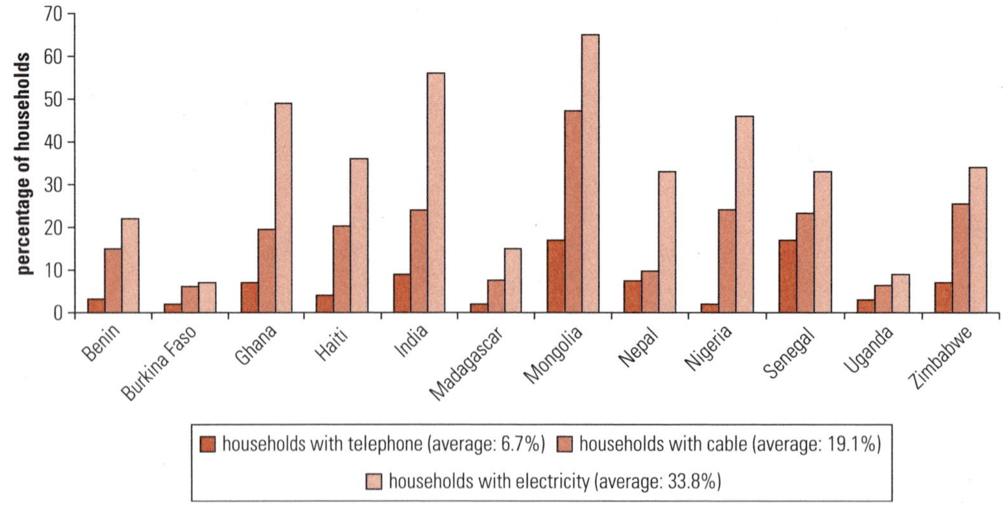

households with telephone (average: 6.7%) ■ households with cable (average: 19.1%)
households with electricity (average: 33.8%)

Sources: UNDP 2007; World Bank 2007.

for creativity. Convergence also significantly alters the structure of the media sector, where content creation and distribution were traditionally in the hands of a few firms or the state. In Myanmar, for example, protesters' unprecedented access to digital video communication over the Internet enabled significant worldwide media coverage of recent pro-democracy demonstrations (*Oxford Analytica* 2007).

Altered Market Structure and Dynamics

Service providers that offer new services and reduce costs alter the structure and dynamics of the markets in which they operate. Thus, convergence can increase competition in a market. However, it also raises the threat of reduced competition. Consequently, convergence has significant implications for competition in the ICT sector.

Potential to increase competition. The different forms of convergence enable greater competition across ICT markets by reducing barriers to market entry and providing market access to new service providers. With multiple play, telephone or cable TV companies can leverage existing networks to offer nontraditional services. Network convergence expands market access by connecting new service and application providers to consumers.

Cable TV companies began to provide Internet and telephone services in the mid- to late 1990s using quickly maturing VoIP technology. For example, as of June 2008, the U.S. cable TV provider Comcast has 5.6 million telephone subscribers (Comcast 2008), almost triple the number it had in 2006, while U.S. telecommunications firm Verizon lost 15 percent of its fixed telephone subscribers due to increased competition from mobile telephones and broadband, especially cable TV broadband over that time (Verizon Communications 2007).

Following the stabilization of IPTV technologies in the mid-2000s, telephone companies are getting into the media business. Hence, while traditional telecommunications companies faced greater competition in their original lines of business in the 1990s, they are now entering and competing in the media space. Telephone companies are deploying new networks to provide triple-play services or have been investing billions of dollars to upgrade their networks. These investments are paying off. In July 2008, Verizon reported that it had 1.4 million video service subscribers on its new fiber-optic network. Indeed, broadband and video services are driving its growth. Revenues in these segments were up by 53 percent even as overall revenues grew by just 1 percent over the previous four quarters (Verizon Communications 2008).

Cable TV companies are now starting to look to network convergence to counter the entry of telecommunications companies into the media business. Comcast has joined a consortium that plans to deploy wireless broadband services—leading to a quadruple-play business model that includes fixed and mobile telephony, cable TV, and Internet services (*FinancialWire* 2008). Similarly, France Telecom has added to its fixed and mobile telephony and Internet service offerings by offering video services, and the United Kingdom's NTL, a cable TV operator, acquired Virgin Mobile—a mobile telephony service provider—to extend its capabilities to wireless (Incode 2006). This move from triple to quadruple play suggests growing competition between telecommunications and media companies.

In competitive markets, service providers will pass lower costs on to users in the form of lower prices. The French Internet service provider Iliad led significant price reductions in the triple-play market by reducing its bundled tariff; the rest of the market soon followed (*The Wall Street Journal* 2006). This would not have been possible without Iliad's converged use of its network to deliver voice and video services.

Competition between networks that offer similar services also drives investments. This is especially apparent from the high levels of capital spending recently seen in the telecommunications sector. Of the 10 countries with the highest broadband penetration, 9 also have strong cable TV infrastructure (Noam 2007). The ability of cable TV infrastructure to carry converged services has driven investments in fiber-optic networks by telecommunications operators.

Singapore is a useful example of the potential for increased competition. In late 2007, StarHub was the territory's monopoly cable TV provider. Now SingTel, the incumbent telecommunications company, has begun to invest in and roll out a new IPTV operation, ushering service convergence into the market. The broadcasting regulator noted that the new service will "inject vibrancy into the Singapore media scene and offer consumers more choices" (*Business Times Singapore* 2006).

Risk of reduced competition. Although convergence has the potential to increase competition and reduce tariffs, it can also reduce or undermine competition (Katz and Woroch 1998). When the Brazilian telephone company Telemar acquired cable TV operator Way TV in 2006, regulator ANATEL responded to concerns about competitive implications by conducting a review. (Ultimately, it found no reason to stop the acquisition.) The country's association of cable TV operators opposed the deal, saying that the entry of the larger operators could impede competition. However, the association's view was also seen as a defensive response to the entry of a new player in the market (*Global Insight Daily Analysis* 2007).

The potential reemergence of natural monopolies is becoming an important issue in the converged broadband era. Natural monopolies may arise when economies of scale or scope are pronounced. IP and broadband networks often require substantial upfront fixed costs and have falling per-unit costs. Incumbent cable TV and telephone companies with the resources to build or upgrade their networks to offer converged services will have a better chance at attaining market power. This leads to an even more challenging environment for new entrants that might not have the networks or resources to challenge a natural monopoly. Such a situation is especially likely in many developing countries, where there are few infrastructure owners to begin with.

Further, if only one service provider owns a backbone or access network, it might not provide competitors with access to it, or charge high interconnection rates. Alternatively, a firm might not have incentives to make significant upfront investments in networks if it believes that competitors might too easily access its facilities. In 2006, German incumbent Deutsche Telekom suggested that it would cut investments in its hybrid fiber very-high-speed digital subscriber line (VDSL) network if forced to open the network to competitors (*Telecom Policy Report* 2006). While the German government gave Deutsche Telekom permission to keep its network closed to protect its return on investment, the European Commission saw this as anticompetitive—benefiting Deutsche Telekom but cutting off potential benefits to the market from increased competition from other service providers (*Tarifica Alert* 2007). Such developments are reflected in growing debates on network neutrality and open network access (Frieden 2006).

Convergence can also reduce competition in other ways. Alternative providers might not be able to replicate pricing options or bundles of services offered by the dominant or larger service provider, increasing costs for subscribers changing services and reducing competition. In addition, concerns about changing telephone numbers, TV channel numbers or programs, or even e-mail addresses might dissuade consumers from seeking out less expensive or higher-quality services.

The merging of telecommunications or media firms might also reduce the diversity of content available to users. In 2003, when the U.S. Federal Communications Commission relaxed restrictions on cross-ownership of media outlets, one of the main reasons for opposition was that it would allow mergers and acquisitions that could reduce the diversity of new and local content (Goldfarb 2003). In a sign of the social implications of advanced ICT, most of the 3 million responses received were by e-mail.

Responding Strategically to Convergence

The discussion above suggests that convergence is likely to gain further momentum around the world. As demand and supply align, advanced ICT networks could emerge as quickly in developing as in developed countries—even in poor countries with a late start. This shift would enable the realization of significant benefits and enhance the development impact of ICT.

For that to happen, it is essential that policy and regulatory frameworks allow markets to function. The well-known success of mobile telephony worldwide has had as much to do with market liberalization as with high demand and low-cost technologies. Research on the diffusion of advanced telecommunications services in developing countries finds that the rate of adoption depends on the existence of an appropriate business environment—which, in turn, is directly dependent on the regulatory and policy environment (Antonelli 1992, p. 11).

A key strategic consideration for governments is the implication of convergence for competition and market structure in the ICT sector. If developing countries seek to maximize the benefits of convergence, they could consider policies that increase access to advanced technologies and innovative, high-quality services by opening markets and removing regulatory barriers to new technologies and business models (Guermazi and Satola 2005, p. 25). Such policy frameworks will create the conditions needed to promote competition. However, consistent oversight might still be required to ensure that convergence does not lead to monopolization.

Governments seeking to maximize the benefits and minimize the risks of convergence will have to think strategically about their policy responses to convergence. If policies restrict convergence from playing out in the market, do not promote competition, or fail to address the risk of monopolization, they will lead to suboptimal outcomes that reduce the development impact of ICT.

Policy Responses

Although convergence is a universal phenomenon, its implications and appropriate policy responses vary by country, depending on prevailing circumstances and legacy factors (Raja and Williams 2007). Still, it is possible to define some useful—if broad—categories of policy responses to convergence. Some countries resist the introduction of convergence. Others "wait and watch," embarking on changes only when they consider them necessary. A third response is to create enabling policies for convergence.

These three categories describe how countries have responded to convergence. Yet the typology is also useful in understanding the implications and outcomes of these different types of responses. Table 2.2 summarizes these responses and the outcomes of each.

Resistance

Governments may believe that convergence will undermine social, political, cultural, or economic goals. In developing countries, VoIP is often perceived as potentially undermining the revenues of incumbent telecommunications operators (and of government, if the incumbent is a state enterprise)—especially when lack of competition has allowed these firms to draw large monopoly rents (ITU 2007, p. 13). Similarly, the political, cultural, and social importance of broadcasting and media often makes governments wary of new providers.

Thus, governments may resist convergence and take steps to prevent new services and providers from entering the market. In 2006, for example, 36 of 54 African countries forbade VoIP (Balancing Act 2006). Moreover, some countries broadly accept the idea of convergence but restrict specific modalities. For example, in the United Arab Emirates, incumbent Etisalat offers a full range of converged telecommunications and video services, but Internet telephony services like Skype were banned in 2006 (*Business Monitor International* 2008b). Concerns involving content regulation have led Bahrain, which has a liberal telecommunications sector, to restrict private participation in media services, preventing fully converged services (Reporters without Borders 2008). And India, in spite of having an open, competitive media sector, does not allow private FM radio stations to broadcast news (TRAI 2008).

Resisting convergence reduces potential benefits, is difficult to enforce, and inevitably leads to pressures for reform. Restrictions cause users to lose potential benefits from innovations and cost reductions. Since Kenya legalized VoIP in 2004, prices for international calls have fallen by up to 80 percent. Legalization of VoIP drove both its own growth and the adoption of broadband and triple play in Kenya, Tanzania, and Uganda (Balancing Act 2007). Where VoIP is

Table 2.2 Policy Responses to Convergence around the World

Indicator	Resist	Wait and watch	Enable
Perceptions	Government believes that convergence may undermine social, political, cultural, or economic goals.	Government believes that existing policies accommodate convergence, or decides not to act.	Government believes that convergence can benefit the ICT sector and economy at large.
Actions	Government takes steps to prevent new services and providers from entering the market.	Government makes no policy changes. Issues are dealt with on a case-by-case basis.	Government updates policies, promotes industry responses, or directly invests.
Outcomes	New services cannot develop legally, but may still defeat restrictions.	Case-by-case decisions allow progress but expose policy inconsistencies.	Market evolves with new services and business models.
	Users lose potential benefits from innovations and cost reductions.	Growing uncertainty discourages investors and operators.	Growth and innovation accelerate.
	Government faces increasing pressure to remove restrictions.	Government faces increasing pressure to revise policies.	Users benefit from increased access and choice, and reduced prices.

Source: Authors' analysis.

permitted, small providers can evolve into IT (information technology) businesses (Economist Intelligence Unit 2007).

Even if new services cannot develop legally, innovators may still defeat restrictions. The global gray market for international voice telephony, accounting for between a quarter and a third of international call revenues, attests to the possibility of service provision regardless of market restrictions.

Resisting convergence may protect the short-term interests of governments and specific players. Nevertheless, the evolution of technology and the potential for provision despite restrictions will ultimately undermine such a policy. Resistance will simply delay convergence and its benefits while undermining policy credibility.

Wait and Watch

Governments may believe that their existing policies accommodate convergence or decide not to act on market developments. Countries seeking to maintain a laissez-faire approach may choose not to regulate for or against convergence. On the other hand, some governments may not have the political capacity to resist or enable convergence, so wait and watch may be their only practical option.

Under the wait-and-watch approach, governments do not change their policies. Instead, they rely on existing policy, legal, and regulatory instruments to deal with issues on a case-by-case basis. In the United States, the Federal Communications Commission and Department of Justice track mergers and acquisitions in the ICT sector. In addition to the Federal Communications Commission's powers, general competition law is used to address issues of monopoly and anticompetitive behavior.

The wait-and-watch approach does not necessarily restrain convergence. Nevertheless, it can lead to confusion and uncertainty. Convergence blurs the boundaries among ICT subsectors, and case-by-case decisions on structural issues might expose inconsistencies due to the different business and regulatory histories of each subsector (Bar and Sandvig 2000). Overlapping or conflicting rules and policies increase regulatory risk and the cost of capital by up to 6 percentage points, depending on the country or region—slowing investment in infrastructure and services (see Estache and Pinglo 2004; Jamison, Holt, and Berg 2005; Kirkpatrick, Parker, and Zhang 2006; and Smith 2000).

The potential for such conflict has grown with service providers adopting new business models. For example, the U.S. "wait and watch" response is leading to conflicts and concerns. A dispute over the introduction of IPTV services in the state of Connecticut led telecommunications operator AT&T to consider cancelling $336 million in investments and suspending 1,300 jobs (*New Haven Register* 2007). The conflict arose because the state required city-level franchising for cable TV operators. Hence, AT&T faced delays and increased costs if its video service was to be treated as a cable TV service because instead of securing one statewide telecommunications license, it would have had to seek licenses in every city. This conflict was resolved in October 2007 after 17 months of uncertainty (*Telecommunications Reports* 2007). The process saw the state cable TV regulator reversing decisions and being challenged in the courts twice. Consequently, not only did the uncertainty cause risk to significant investment and job creation for the state, it also undermined the credibility of regulation.

As conflicts and uncertainty grow, governments face increasing pressure to revise policies. The absence of a response can have a significant negative effect by failing to provide certainty for investors and not providing a way to overcome inconsistencies in legacy frameworks. As such, the United States is now concerned that it is falling behind its European and Asian peers in broadband penetration and low tariffs (EDUCAUSE 2008), leading to calls for government intervention and a national broadband strategy in a market that has typically adopted a laissez-faire approach to the ICT sector (NTIA 2008). Therefore, while a wait-and-watch approach may not prevent convergence, it can lead to suboptimal benefits.

Enable

Some governments believe that convergence can benefit the ICT sector and the economy at large, and choose to create environments that actively promote innovation and competitive service provision. International experiences with the mobile telephone revolution show that when service providers are allowed to offer services, face few government restrictions, and have explicit or implicit government support, such markets can develop very quickly. A similar expansion in investments and access to advanced ICT services can result from the creation of enabling policy environments for convergence.

Enabling policy environments allow markets to evolve with the introduction of new services and business

Box 2.3 The Impact of an Enabling Environment for Convergence: Wireline Telephony and Job Creation

Around the world, the wireline telephony business is stagnating or shrinking because of the shift toward mobile, cable, and broadband telephony. This shift is threatening wireline telephone companies, raising the possibility of job losses.

In April 2008, U.S. telecommunications firm AT&T announced that it would cut 4,600 jobs in its shrinking wireline business. However, the company is simultaneously hiring about as many or more employees to support the rollout and operation of its expanding wireless, video, and broadband services. AT&T's shift into converged and broadband services is allowing it to keep its total headcount approximately the same. In fact, wireline firms worldwide invested more than $36 billion in equipment throughout 2007, up more than 10 percent from 2006, with spending increasing on optical transport and routers as well as on VoIP equipment.

Subscriber Base for Wireline Services around the World

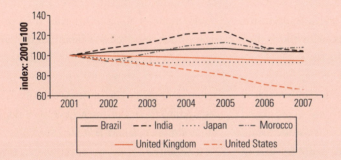

If policy prevents AT&T and similar firms from expanding into new market segments, they cannot build their businesses—leading to negative outcomes such as job losses. Restrictive government policies can prevent new business models and undermine economies. Enabling policies will allow an expansion in economic activity and potential job creation.

Sources: Authors' analysis; DellOro Research 2008; Pyramid Research 2008; *The Wall Street Journal* 2008.

---

models. Box 2.3 discusses the importance of allowing firms to overhaul their business models in response to changing technology and market conditions. Policies that promote convergence accelerate growth and innovation—reducing inconsistencies and artificial barriers, lowering risks and entry costs, and creating better environments for investments. In addition, users benefit from increased access and choices as well as reduced prices.

Creating an enabling environment can involve different levels of government engagement with the ICT sector. First, governments can amend policies to address convergence and remove barriers and restrictions. At minimum, policy responses to convergence will resolve some of the conflicting rules among converging sectors and create level playing fields in the market.

Several countries have reformed their policy and regulatory frameworks to accommodate and enable convergence while simultaneously moving toward a greater focus on market forces. Kenya and Singapore, for example, have moved toward technology-neutral licensing regimes that allow service providers the flexibility to deploy the most efficient networks. Going further, the Republic of Korea,

Malaysia, and the United Kingdom, among others, restructured their legal and regulatory frameworks to align with convergence and allow multiple play without restriction.

Given that the primary implication of convergence is a change in market structure, policy makers have the opportunity to promote competition as they undertake policy reforms. Creating a competitive market on a level playing field has been recognized as the most effective way to drive growth, encourage efficiency (leading to reduced prices and improved quality), and promote investment.

The second level of government involvement may provide incentives for firms to invest in the deployment of advanced ICT services. Japan's government provided interest-free credit, subsidies, preferential tax rates, competition-enhancing rules, and other measures to promote the deployment and use of fiber-optic broadband networks (*Dow Jones International News* 2000). Today, Japan leads the world in fiber-optic home subscriptions, with more than 8 million homes connected (*The New York Times* 2007).

Finally, some governments directly invest in infrastructure and services. Government investment can provide a significant push during the early stages of convergence and make the government's policy stance clear. One study found that fiber-to-home deployments are financially feasible in cities if take-up is at least 25 percent of homes (Sigurdsson 2002). Passive infrastructure accounts for up to 80 percent of these costs (Gauthey 2006). Hence, governments that reduce the costs of rollout by sharing costs or providing rights-of-way can jumpstart development.

As part of their investments, governments can lead development of advanced networks or create open-access infrastructure that can attract private investment, such as France did. By 2006, 40 percent of French households had broadband service, and multiple service providers had benefited from the unbundling of incumbent France Telecom's network. Today, the national and local governments are investing in the rollout of open-access fiber networks, which private service providers will pay to use. Included in this plan is the opening of sewers and conduits to allow competitive service providers to lay their fiber-optic cables (*The Wall Street Journal* 2006). One study estimates that this approach will cut costs by up to 60 percent (Paul Budde Communication Pty Ltd 2008).

Direct investment, however, involves some risks and challenges. One is that public funding of broadband networks can distort the market. To address this issue, the European Commission verifies that interventions are in line with state aid rules.[5] These rules require justification for state intervention and an analysis of the impact of the aid on competition in the market. In areas where competing private operators are present, the European Commission can prohibit state investment if "intervention may crowd out existing and future investments by market players" (Papadias, Riedl, and Westerhof 2006). This implies that governments need to demarcate their roles as investors from possible roles as service providers. Put another way, public investments should not serve as a way for governments to reenter service provision, effectively rolling back the sector reforms of the past two decades.

Governments also risk investing in technologies or services that eventually might not find a mass market, become obsolete, or slow further innovation. France's recent successes in the broadband market came after much criticism of its deployment of the pre-Internet data service Minitel. The government invested $11 billion in the system over 20 years, with service beginning in 1981. At that time, Minitel was an advanced data service and a pioneer in the market, though it remained a policy and business priority in France well beyond its useful life as the Internet took hold in other countries (*International Herald Tribune* 1996).

The three levels of government involvement in ICT convergence can be cumulative. Creating a framework that promotes competition and innovation, however, may need to follow these stages in sequence. Experience suggests that the priority has been to remove policy and regulatory restrictions first, then create new frameworks to address convergence and promote competition and innovation, and finally move toward encouraging or investing in these technologies and services. These might be considered the stages in the creation of a policy framework that enables convergence.

The example of the United Kingdom is one that illustrates these stages. The United Kingdom began creating an enabling policy and regulatory framework in 2003, when it promulgated the Communications Act and created a converged regulator, the Office of Communications. In 2004, the government and regulator began to push incumbent BT to reorganize. The goal was to lead BT toward opening its local networks to competitors; the government believed that this move would promote competition and expand the penetration of broadband services.

Even so, in 2007, the government began discussions about investing in its own national fiber-optic network, at an estimated cost of about £10 billion. The reason for this move was to catch up with other countries investing in fiber-optic infrastructure, "delivering considerably higher bandwidth than is available in the U.K." (BBC 2007). Further, the government formed a "convergence think tank" in 2007 to suggest ways of improving policies given technological and market developments since the last major policy revisions in 2003.[6] Thus, the government continues an evolving engagement with convergence, pointing to a migration through stages from a policy response to working with firms and, most recently, planning direct investment in the ICT sector.

Conclusion

Around the world, service providers embrace ICT convergence to enter new markets, drive growth, and improve their business prospects. Users are also responding, with significant numbers subscribing to innovative services at lower prices. New technologies and the market are driving convergence forward, leading to significant potential benefits.

A policy maker's role is to respond to this changing environment. Different countries have followed different paths in response to convergence. With profound implications for the whole ICT sector, it is essential that policy makers have a firm understanding of the implications of convergence and their decisions.

In the long term, countries that resist are likely to miss the benefits of improved ICT networks and services. Some countries want to wait and watch as they may believe that their existing policies accommodate convergence or that the issue does not merit immediate attention, but risks remain because convergence typically does not fit easily into traditional policy frameworks. With the passage of time, the economic costs of regulatory uncertainty and inconsistency that hinder convergence will increase. The greatest benefits come in markets that enable convergence.

If a country decides to move toward enabling frameworks, it will at minimum need to review policies and regulations, and then implement coordinated measures. Indeed, translating a broad vision for convergence into specific policies and regulations is likely the more difficult task.

Emerging trends suggest some global best practice principles for regulatory frameworks to respond to convergence:

- *Create regulatory frameworks that promote competition.* Service providers can deploy converged services only if regulators lower entry barriers and allow innovation— and, by doing so, increase competition, reduce prices, and drive growth. However, it is equally important that regulators prevent market failures and do not allow monopolization. Hence, regulatory frameworks that establish level competitive playing fields will provide the greatest benefits for users.

- *Rely more on market forces and less on regulation.* Maintaining legacy regulatory frameworks will likely stifle the growth of convergence. Instead, regulation can move toward allowing innovation and competition on a level playing field, then step back from intervening unless there are market failures.

- *Allow new technologies to contribute everything they have to offer.* Regulatory frameworks that are technology neutral and allow flexibility in service provision will encourage investments and innovation. Service providers can fully use their networks and reduce costs, increasing business viability and leading to markets that are more efficient. Users will benefit from lower prices, more choices, and increased competition.

Policy makers seeking to respond to and enable convergence will find that doing so promotes competition and supports innovation in services that benefit the ICT sector. Following these principles will lead to better outcomes for the ICT sector and the economy as a whole. A detailed examination of the specific regulatory issues is provided elsewhere.[7]

As a market phenomenon that can lower prices, expand coverage, and increase investments, convergence will enhance the effects of earlier sector liberalization efforts. Countries that begin on these second-generation reforms in the ICT sector will find themselves, and their economies, better off for it.

Notes

1. For conciseness, this chapter focuses on the supply of ICT services rather than on their demand and use, including content and applications. The chapter presents a selection of the different views on convergence found in current practice, taking into account the interests of ICT policy makers and businesses in the developing world.

2. Another often-discussed aspect of convergence relates to user devices. However, this chapter takes the view that converged multimedia devices are a driver and not a result of convergence.

3. There is a significant amount of literature dedicated to the analysis of the development impact of ICT. See, for example, Wang (1999) and Grace, Kenny, and Qiang (2004).
4. Many countries have also begun to consider digitizing terrestrial broadcasting. Such developments alter the scope of services that can be carried over the broadcast spectrum because they reduce the amount of spectrum needed to carry TV signals. The freed excess spectrum—the "digital dividend"—can be used for broadband and other new wireless services and networks, introducing convergence among wireless technologies. It can also significantly increase coverage, especially since the bands used for broadcasting have wider reach.
5. The Commission's Director General for Competition "monitors state aid to the ICT sector and contributes to the development of State aid policy in this field. State aid is defined as an advantage in any form conferred on a selective basis to undertakings by national public authorities. In view of this definition, a number of measures such as research and development aid or regional aid to ICT companies have to be monitored by the director general in order to avoid market distortions. The director general also clears aid that is beneficial to consumers, by providing new research grants and encouraging the development of new products, such as open source." More information is available at http://ec.europa.eu/comm/competition/sectors/ICT/overview_en.html.
6. More information on this think tank is available at http://www.culture.gov.uk/Convergence/index.html.
7. A more detailed examination of these possibilities, specifically for multiple play, is covered in Singh and Raja (2008).

References

All Africa. 2005. "Firm Imports Sh72m Expansion Kit." January 3.

Antonelli, Christiano. 1992. *The Diffusion of Advanced Telecommunications in Developing Countries*. Paris: Organisation for Economic Co-operation and Development.

AT&T. 2008. "Quarterly Earnings Report for Q1 2008." http://www.att.com/Investor/Financial/Earning_Info/docs/1Q_08_slide_bw.pdf.

Balancing Act. 2006. "Kenya—Legal VoIP Begins to Shake Up the Market and Bring Prices Down." http://www.balancingact-africa.com/news/back/balancing-act_297.html.

———. 2007. "African VoIP Report Reveals Steady Uptake of IP among Carriers and Grey Market Persisting despite Price Falls." http://www.balancingact-africa.com/news/back/balancing-act_342.html.

Bar, François, and Christian Sandvig. 2000. "Rules from Truth: Communication Policy after Convergence." Paper presented at the Telecommunications Policy Research Conference, Alexandria, VA, September 23–25.

BBC (British Broadcasting Corporation). 2007. "Government Mulls Broadband Help." September 19.

BT. 2006. "Delivering the 21st Century Network." http://www.btplc.com/21CN/WhatisBTsaying/Speechesandpresentations/External_Final_iss_6_generic_pres_jan06.ppt.

Business Monitor International. 2008a. "BT's 21CN Behind Schedule." April 4.

———. 2008b. "Market Data Analysis: Fixed Line—Q3 2008." June 28.

Business Times Singapore. 2006. "StarHub Unfazed by Internet TV Challenge." November 9.

Cisco Systems. 2007. "Annual Report." http://www.cisco.com/web/about/ac49/ac20/ac19/ar2007/printable_report/index.html.

Comcast. 2007. "Annual Report." http://www.comcast.com/2007annualreview/index.htm.

———. 2008. "Corporate Overview." http://www.comcast.com/corporate/about/pressroom/corporateoverview/corporateoverview.html.

DellOro Research. 2008. "Service Providers Worldwide Spend Over $36 Billion Fortifying Wireline Networks in 2007." Press release, April 9.

Dow Jones International News. 2000. "Japan Govt May Create National Fiber-Optic Network." November 1.

———. 2008. "BT Shifts Emphasis with 21CN Network, Slows Rollout." May 15.

The Economist. 2006. "The End of the Line." October 12.

Economist Intelligence Unit. 2007. "Africa: Tariffs Tumble, VoIP Rises." March 8.

———. 2008. "USA Consumer Products: TV Viewing Stays Strong." August 7.

EDUCAUSE. 2008. *A Blueprint for Big Broadband*. White paper, EDUCAUSE, Washington, DC.

Estache, Antonio, and Maria Elena Pinglo. 2004. "Are Returns to Private Infrastructure in Developing Countries Consistent with Risks Since the Asian Crisis?" Policy Research Working Paper 3373, World Bank, Washington, DC.

FinancialWire. 2008. "Comcast to Take Stake in Sprint, Clearwire Wireless Broadband Business." May 8.

Frieden, Robert M. 2006. "Network Neutrality or Bias?—Handicapping the Odds for a Tiered and Branded Internet." bepress Legal Series Working Paper 1755, Berkeley, CA. http://law.bepress.com/expresso/eps/1755.

Gauthey, Gabrielle. 2006. "Fibre, a Real Breakthrough." *La Lettre de l'Autorité* 53. l'Autorité de Régulation des Communications Electroniques et des Postes, Paris.

Global Insight Daily Analysis. 2007. "Telefónica Launches Own TV Service in Brazil." August 16.

Goldfarb, Charles B. 2003. *FCC Media Ownership Rules: Issues for Congress*. Washington, DC: U.S. Congressional Research Service.

Grace, Jeremy, Charles Kenny, and Christine Zhen-Wei Qiang. 2004. "Information and Communication Technologies and Broad-Based Development: A Partial Review of the Evidence." World Bank Working Paper 12, World Bank, Washington, DC.

Guermazi, Boutheina, and David Satola. 2005. "Creating the 'Right' Enabling Environment for ICT." In *E-Development: From Excitement to Effectiveness*, ed. Robert Schware, 23–46. Washington, DC: World Bank.

Incode. 2006. "The 'Quad Play'—The First Wave of the Converged Services Evolution." http://www.fixedmobileconvergence.net/whitepapers/fmc-incode.pdf.

International Herald Tribune. 1996. "Beyond Minitel: France on the Internet." January 8.

Internet World Stats. 2007. "Broadband Internet Subscribers." http://www.internetworldstats.com/dsl.htm.

ITU (International Telecommunication Union). 2006. *World Information Society Report*. Geneva: ITU.

———. 2007. *The Status of Voice over Internet Protocol (VoIP) Worldwide*. Geneva: ITU.

Jamison, Mark A., Lynne Holt, and Sanford V. Berg. 2005. "Measuring and Mitigating Regulatory Risk in Private Infrastructure Investment." *The Electricity Journal* 18 (6): 36–45.

Katz, Michael L., and Glenn A. Woroch. 1998. "Introduction: Convergence, Competition, and Regulation." *Industrial and Corporate Change* 6 (4): 701–18.

Kirkpatrick, Colin, David Parker, and Yin-Fang Zhang. 2006. "Regulation and Foreign Direct Investment in Infrastructure: Does Regulation Make a Difference?" *Transnational Corporations* 15 (1): 143–71.

Liberty Global, Inc. 2007. "Annual Report." http://www.lgi.com/pdf/LGI_AR_2007Grayscale.pdf.

Marketwire. 2008. "Infonetics Research: IP Router Sales Up in a Usually Down Quarter." June 4.

New Haven Register. 2007. "Consumers Win on TV Competition." November 9.

The New York Times. 2007. "Unlike U.S., Japan Pushes Fiber over Profit." October 3.

Noam, E. 2007. "State of Telecom 2007." Presented at the Columbia Institute for Tele-Information, "The State of Telecom–2007," New York, October 19.

NTIA (National Telecommunications and Information Administration). 2008. *Networked Nation: Broadband America 2007*. Washington, DC: NTIA.

OECD (Organisation for Economic Co-operation and Development). 2006. *Multiple Play: Pricing and Policy Trends*. Paris: OECD.

———. 2007. *OECD Communications Outlook*. Paris: OECD.

Oxford Analytica. 2007. "Burma: Internet Puts the Junta under Pressure." October 26.

Papadias, Lambros, Alexander Riedl, and Jan Gerrit Westerhof. 2006. "Public Funding for Broadband Network–Recent Developments." *European Competition Policy Newsletter* 3: 13–18.

Paul Budde Communication Pty Ltd. 2008. "France Broadband Market: Fibre and Wireless Services." Paul Budde Communication Pty Ltd., Bucketty, Australia.

Providence Journal. 2007. "Verizon Cutting Copper Ties during Upgrades." July 9.

Pyramid Research. 2007. "From Triple-Play to Quad-Play." Pyramid Research, Cambridge, MA.

———. 2008. "Fixed Communications Forecast." Cambridge, MA.

Raja, Siddhartha, and Mark Williams. 2007. "Converging Media/Diverging Experiences." Paper presented at the Telecommunications Policy Research Conference, Alexandria, VA, September 28–30.

Reporters without Borders. 2008. "Country Report: Bahrain." http://www.rsf.org/article.php3?id_article=26040.

Reuters. 2008. "Naspers May Expand Mobile TV Service in Africa." June 25.

Sigurdsson, Halldor Matthias. 2002. "Techno-Economics of Residential Broadband Deployment." Center for Information and Communication Technologies, Technical University of Denmark, Copenhagen. http://www2.imm.dtu.dk/pubdb/views/edoc_download.php/5443/pdf/imm5443.pdf.

Singh, Rajendra, and Siddhartha Raja. 2008. "Convergence in ICT Services: Regulatory Responses to Multiple Play." Working Paper 44620. World Bank, Washington, DC.

Skype. 2008. "Skype Appoints New Chief Operating Officer." Press release, July 1.

Smith, Warrick. 2000. *Regulatory Infrastructure for the Poor: Perspectives on Regulatory System Design*. Washington, DC: World Bank.

SSKI Research. 2007. *Entertainment and Media*. Bombay.

Tarifica Alert. 2007. "EC Gives Germany a Broadband Ultimatum." May 8.

Telecom Policy Report. 2006. "Deutsche Telekom's VDSL Flap with Regulator in EC Hands." 14 August.

Telecommunications Management Group. 2008. *IPTV: The Killer Broadband Application*. Bromsgrove, U.K.

Telecommunications Reports. 2007. "Court Rules AT&T Can Offer IPTV under New Video Law." November 15.

TeleGeography. 2007. *Voice Report*. Washington, DC: TeleGeography.

TRAI (Telecom Regulatory Authority of India). 2008. "Recommendations on 3rd Phase of Private FM Radio Broadcasting." Telecom Regulatory Authority of India, New Delhi.

———. 2008. "The Indian Telecom Services Performance Indicators: January–March 2008." New Delhi: TRAI. http://www.trai.gov.in/trai/upload/PressReleases/585/reportQE3july08.pdf.

UNDP (United Nations Development Programme). 2007. *Human Development Indicators 2007*. New York: UNDP.

Verizon Communications. 2007. "Annual Report." Verizon Communications Inc., http://investor.verizon.com/financial/quarterly/pdf/07_annual_report.pdf.

———. 2008. "Quarterly Earnings Report for Q1 2008," Verizon Communications Inc., http://investor.verizon.com/news/view.aspx?NewsID=909.

The Wall Street Journal. 2006. "How France Became a Leader in Offering Faster Broadband." March 28.

———. 2008. "Corporate News: AT&T to Cut Jobs in Landline Revamp." April 19.

Wang, Eunice Hsiao-hui. 1999. "ICT and Economic Development in Taiwan: Analysis of the Evidence." *Telecommunications Policy* 23 (3/4): 235–43.

World Bank. 2007. *World Development Indicators 2007*. Washington, DC: World Bank.

Chapter 3

Economic Impacts of Broadband

Christine Zhen-Wei Qiang and Carlo M. Rossotto
with Kaoru Kimura

Broadband has been increasingly recognized as a service of general economic interest in recent years.[1] Broadband's economic significance can be put into context by referring to similar changes in other areas of infrastructure, such as road, rail, and electricity. Each of these infrastructure services transforms economic activities for citizens, firms, and governments; enables new activities; and provides nations with the ability to gain competitive and comparative advantages. Though many of these advantages were unforeseen when original investments were made, they quickly became an essential part of economic lifestyles and activities. A similar assumption about the expected transformative benefits of broadband on economic and social variables has led many governments to set ambitious targets for its deployment.

In making a case for public policy on broadband, many studies have sought to identify and measure broadband's economic benefits. Though some of these studies have found a positive relationship between broadband access and economic development, most of them have been restricted to developed economies and their firms and communities, and to qualitative arguments and case studies. This chapter attempts to fill the gap on the macroeconomic evidence of broadband's impact, in both developed and developing countries.

This chapter has three sections. The first reviews the literature on the economic impacts of broadband—by enhancing the knowledge, skills, and networks of individuals; raising private sector productivity; and increasing community competitiveness. This section also explores broadband's role as an enabling technology in increasing investment payoffs in other sectors, transforming research and development (R&D), facilitating trade in services and globalization, and improving public services to enhance national business environments and competitiveness. The second section introduces a cross-country empirical model for analyzing broadband's impacts on economic growth using data from 120 developing and developed countries. Undertaking this quantitative analysis can provide policy makers with an assessment of the potential benefits of broadband and its impacts on the overall economy. The final section summarizes key results and implications for developing countries. The main conclusion is that broadband has a significant impact on growth and deserves a central role in country development and competitiveness strategies.

This chapter is not intended to debate the technological, regulatory, and content aspects of broadband or to analyze its social and political impacts—an important but complex issue often surrounded by controversies and subject to local cultural contexts. The awareness clearly exists in the academic literature and among industry leaders that broadband is associated with different classes of impacts, ranging from enhanced social networks and interactions to increased political activism and the spread of democracy through the grassroots organization of political movements and the spread and control of information. This broader framework has an indirect impact on economic variables, even though it is not explicitly analyzed in this chapter. Policy and regulatory recommendations on how to facilitate access to and leverage the potential of broadband are also beyond the scope of this chapter; these are in part addressed in chapter 4 of this report.

Existing Literature

Some broadband applications and attributes are available with narrowband (dial-up links), raising the question of whether the benefits of broadband are more than marginal relative to the status quo (Bauer, Muth, and Wildman 2002). Over time, though, the advantages of broadband over narrowband are evident (Saksena and Whisler 2003):

- Progression from "e" (electronic) to "u" (ubiquitous) access: omnipresent, always on, and always aware of the user's location

- Higher connection speeds, which contribute to the spread of Internet protocol (IP) networks

- Enhanced cross-platform security to protect private communications and critical data, as well as embedded security among different platforms, applications, and environments

- Reduced costs for businesses through lower telecommunication costs (relative to the cost of leased lines) and transaction costs (for example, through enhanced customer relationships)

- Enhanced multimedia applications (for example, increased access to online video content)

- Increased development of complementary products (such as outsourcing) due to the global nature of networks and information services, as well as encouragement of global, real-time, and transparent competition.

Understanding the economic benefits of broadband involves several complexities. One is that a critical mass of broadband penetration—a common feature of network infrastructure—has been reached in only a small number of countries, and only quite recently. In this context, many facts are still evolving.

Another issue is that the literature tends to confound benefits with applications or attributes and the activities they enable, such as telecommuting and e-government (electronic government; Firth and Mellor 2005). This creates an impression that the benefits of broadband are unambiguous and does not take into account the investment costs involved—especially if public funding is required.

A third issue is that some of the reports and studies cited in the literature were commissioned by parties with direct commercial or policy interests in the industry's success—for example, telecommunications providers, associations of commercial firms, and government agencies with mandates to promote broadband. Moreover, impacts can take many forms, and causal links are often difficult to disentangle and analyze individually. Hence, findings may need to be interpreted with caution. Still, it is useful to examine how individuals, firms, and communities interact with broadband and how broadband enables transformational changes with each of these three players, as well as how broadband impacts the overall economy.

Individuals and Their Roles in Economic Processes

One crucial impact of broadband is its growing role in improving human capital, a necessary condition for economic growth and competitiveness. The benefits of broadband diffusion for individuals—such as better and more diverse access to information—have been documented in the literature. Broadband users spend up to 64 percent more time on the Internet than do dial-up subscribers (Saksena and Whisler 2003). An analysis of click-stream data on Internet use in 10 U.S. cities, comparing one group with dial-up access with a second group having broadband access, shows that content-intensive and socially interactive sites were used more often by those with broadband access (Rappoport, Kridel, and Taylor 2002).

Individuals can acquire skills (increasing their marketability as workers) and develop social networks through broadband-enabled Web applications, facilitating peer-to-peer communities and their integration with the economy. Blogs (online diaries), wikis (Web sites where users can contribute and edit content), and the like have created new, decentralized, dynamic approaches for capturing and disseminating the knowledge needed for individuals to become better prepared for the knowledge economy (Johnson, Manyika, and Yee 2005). It is also believed that broadband can enhance a city's or a country's appeal to the "creative class" of knowledge workers and attract human capital amid intensifying global competition for talented workers (Dutta and Mia 2008).

Broadband diffusion enables individuals outside the boundaries of traditional institutions and hierarchies to innovate to produce content, goods, and services. The role of network users in the innovation process increases as they generate or contribute to new ideas (user-led innovation, or "the democratization of innovation"; von Hippel 2005) and collectively develop new products (such as open source software). But few of the above claims have been empirically substantiated in the literature.

Firm Efficiency and Productivity

Most studies looking directly at the impacts of broadband tend to be conducted at the firm level. Evidence of broadband promoting firm growth has been fairly well documented in developed countries, particularly its ability to lower costs and raise productivity. Internet business solutions have enabled private companies to cut costs (by $155 billion in the United States and a collective $8.3 billion in France, Germany, and the United Kingdom) and increase revenues (by a collective $79 billion in France, Germany, and the United Kingdom)—suggesting that the companies focused their Internet solutions on growth rather than just on cost savings (Varian and others 2002).

For example, British Telecommunications (BT) had about 8,500 workers who worked from home using broadband in 2004, a setup that provides significant financial benefits to the company. On average, each worker saved the company accommodation costs of about £6,000 a year; had a productivity rate increase of 15–31 percent (averaging 20 percent); and took an average of only 3 days of sick leave a year, compared with an industry average of 12 days. All this added up to annual savings of more than £60 million for the company. BT has also extended flexible working arrangements to its engineers. The latest data from a trial of 3,000 engineers show that service quality has risen by 8 percent. The engineers in the trial worked an average of two hours less per week but earned more, and BT was saving money through the elimination of overtime payments (Broadband Stakeholder Group 2004).

Such improvements in performance depend on firms' ability to conjoin their technological, business, and organizational strategies. When fully absorbed, broadband drives intensive, productive uses of information and communication technology (ICT) and online applications and services, making it possible to improve processes, introduce new models and structures, drive innovation, and extend business links (box 3.1). Forman, Goldfarb, and Greenstein (2005) distinguish between "IT (information technology) using" and "IT enhancing" firms, and find that the changes broadband delivers to firm behavior generally lie on a spectrum—with the highest productivity increases appearing in firms that commit most intensively to integrating broadband, or IT in general, with new business processes.

On one end of this spectrum are firms where broadband diffusion is affecting their entire industries. Broadband connectivity and speed remove the need for proximity to customers and facilitate hyper-differentiation of customers and choice, in line with the popularization of "long tail" marketing strategies of niche products (Allaire and Austin 2005). For example, broadband is revolutionizing the print, movie, music, gaming, and advertising industries by enabling direct involvement by users in creating digital content (through the emergence of Web 2.0, Wikipedia, Facebook, YouTube, Blogger.com, and MySpace, among others) and high-speed downloads, and peer-to-peer distribution and interaction, reducing transaction costs and saving time for customers and producers (Heng 2006). Between 2000 and 2003, DVD sales rose $14.1 billion in the United States, of which 9 percent—$1.3 billion in revenues and $630 million in profits—was attributed to online sales enabled by increased broadband connectivity (Smith and Telang 2006). Developing broadband networks has been a key strategy for the Japanese animation industry to maintain its international competitiveness and remain a global market leader (Government of Japan 2003).

Export-oriented firms also benefit considerably from broadband use. Broadband lowers the costs of international communications and improves the availability of information, enabling companies to access foreign markets more easily and become more competitive. Clarke and Wallsten (2006), in a study of 27 developed and 66 developing countries, found that a 1 percentage point increase in the number of Internet users is correlated with a boost in exports of 4.3 percentage points and an increases in exports from low-income to high-income countries of 3.8 percentage points. Although this study was not broadband-specific, it is safe to infer that broadband would have an even bigger positive impact.

Broadband is also particularly important for firms in information-intensive service sectors, such as financial markets, insurance, and accounting. For example, some

> **Box 3.1** Broadband's Effects on Firms' Behavior to Increase Competitiveness
>
> A number of examples show how broadband can help transform processes, business models, and relationships in the private sector.
>
> - A study involving business and technology decision makers from 1,200 companies in six Latin American countries (Argentina, Brazil, Chile, Colombia, Costa Rica, and Mexico) showed that broadband deployment was associated with considerable improvements in business organization, including speed and timing of business and process reengineering, process automation through network integration, and better data processing and diffusion of information and knowledge within organizations (Momentum Research Group 2005).
>
> - A 2005 study by McKinsey highlighted the growing importance of broadband to companies' competitiveness through new ways of structuring work. Modern companies build distinctive capabilities based on a mix of talent and technology; they specialize in core activities and outsource the rest. Broadband helps allocate activities more efficiently between workers tackling complex, highly dynamic tasks and more traditional, transactional workers. It is also a key component in raising the productivity of employees whose jobs cannot be automated, and in doing so cost-effectively (Johnson, Manyika, and Yee 2005).
>
> - Companies that adopt broadband and ICT to transform their supply chains prompt other companies in their value and distribution chains to adopt new technologies and interoperable IT systems (Atkinson and McKay 2007). For example, the automobile industry's entire technology-intensive supply chain is linked through broadband networks and high-power computing. Broadband networks are essential to engineering design service firms to test and implement design options directly with car and parts manufacturers.
>
> - Studies in the United Kingdom indicate that enterprises using broadband are more likely to have multiple business links. For example, they use e-mail and the Internet to raise the quality and lower the costs of gathering market intelligence and to communicate with suppliers and business partners. Enterprises with more links tend to have higher labor productivity (Clayton and Goodridge 2004).

98 percent of small and medium-size enterprises in Australia's finance and insurance industries use broadband (Australian Communications and Media Authority 2008). Increases of 25 percent or more in the number of claims processed each day have been documented by U.S. insurance companies that have adopted wireless broadband (Sprint 2006). Other examples include consulting firms, marketing, real estate, travel and tourism, advertising, and graphic design.

As wireless broadband applications become increasingly mature, the health care sector and small businesses are reaping big gains, especially from the mobility aspects such as not having to staff a head office, more ready access to corporate information, field service automation, and sales force automation. In 2005, productivity improvements because of the use of mobile broadband solutions across the U.S. health care industry were estimated to be worth $6.9 billion (Entner 2008).

Community Competitiveness

Deploying broadband networks at the community and municipal levels has become an important factor in allowing local businesses to grow and remain competitive. An often-cited 2005 study by the Massachusetts Institute of Technology of a broad range of U.S. communities where broadband had been deployed since December 1999 found that it benefits economic activity in ways consistent with the qualitative stories told by broadband advocates. Between 1998 and 2002, U.S. communities that were among the early adopters of mass-market broadband experienced faster growth in employment, number of

Table 3.1 Impacts of Broadband on Economic Activities in U.S. Communities

Indicator	Results
Employment	Broadband added 1.0–1.4 percentage points to the growth rate in the number of jobs during 1998–2002.
Number of businesses	Broadband added 0.5–1.2 percentage points to the growth rate in the number of firms during 1998–2002.
Housing rental rates (proxy for property values)	Rates were more than 6 percent higher in 2000 in zip codes where broadband was available by 1999.
Industry mix	Broadband added 0.3–0.6 percentage point to new business creations in IT-intensive sectors in 1998–2002.
	Broadband reduced the share of small business (those with fewer than 10 employees) by 1.3–1.6 percentage points in 1998–2002.

Source: Gillett and others 2006.

businesses, and businesses in IT-intensive sectors, as well as higher market rates for rental housing, than communities where broadband was adopted later (Gillett and others 2006; table 3.1).

Other studies of community-level broadband experiences include the following:

- A case study of a municipal fiber network built in 2000–01 in South Dundas Township, Ontario (Strategic Networks Group 2003)

- A study comparing Cedar Falls, Iowa (which launched a municipal broadband network in 1997) with its otherwise similar neighboring community of Waterloo (Kelly 2004)

- A study comparing per capita retail sales growth in Lake County, Florida, with 10 other Florida counties selected as controls based on their similar retail sales level prior to Lake County's broadband rollout (Ford and Koutsky 2005).

All these studies found that broadband connectivity had positive impacts on job creation, company and community retention, retail sales, and tax revenues. A more recent survey conducted for Industry Canada of subscribers to two remote, rural broadband networks in British Columbia found that about 80 percent of business respondents believed that they would be at a major disadvantage if they did not have broadband access (Zilber, Schneier, and Djwa 2005).

Even in rural areas of developing countries, broadband diffusion is making existing markets function better by reducing information asymmetry and creating a range of economic opportunities for communities—contributing to income diversification and rural nonagricultural employment as well as increasing incomes from agricultural jobs. In recent years, communities in developing countries have launched broadband-enabled services and applications to give local populations access to new markets and services and facilitate information exchange and value creation between buyers and sellers of agricultural products (box 3.2). Before that, many of these opportunities had been available only in the largest or wealthiest communities.

The Overall Economy

Broadband is not just an infrastructure. It is a general-purpose technology that can fundamentally restructure an economy. Thus, examining the overall economic impact is a logical way to assess the implications of broadband diffusion because it takes a more comprehensive view than looking only at impacts on individuals, firms, or communities.

The first generation of country-level studies on broadband appeared before the technology had been significantly adopted even in developed countries. In a study commissioned by Australia's National Office for the Information Economy (replaced by the Australian Government Information Management Office in 2004), Allen Consulting Group (2002) estimated that broadband would add 0.6 percentage point to Australia's gross domestic product (GDP) growth rate each year through 2005. According to 2003 estimates by Accenture, next-generation broadband has the potential to contribute $500 billion to GDP in the United States and from $300 billion to $400 billion in Europe, and was likened to water and electricity as the "next great utility" (Saksena and

> **Box 3.2** Broadband's Role in Raising Rural Incomes in Developing Countries
>
> Experience shows that access to broadband networks has had a positive impact on rural incomes in developing countries. In India, the E-Choupal program was started in 2000 by ITC, one of India's largest agricultural exporters. The program operates in traditional community gathering venues (*choupals*) in farming villages, using a common portal that links multimedia personal computers by satellite. Training is provided to the hosts, who are typically literate farmers with a respected role in their communities. The computers give farmers better access to such information as local weather forecasts, crop price lists in nearby markets, and the latest sowing techniques. Collectively, these improvements have resulted in productivity gains for the farmers. E-Choupal also enables close interaction between ITC and its rural suppliers, which increases the efficiency of the company's agricultural supply chains, eliminates intermediaries, and improves terms of business. The fact that ITC pays a higher price than its competitors for exportable products has encouraged farmers to sell their increased output to the company. By 2008, E-Choupal had reached millions of small farmers in more than 40,000 villages, bringing economic and other benefits. It aims to reach 100,000 villages by 2010.
>
> Another program, launched by the Songtaaba Association, has allowed female agricultural producers in Burkina Faso to become economically empowered through broadband. Songtaaba, an organization manufacturing skin care products, provides jobs to more than 3,100 women in 11 villages. In order to provide its members with regular access to useful information and improve the marketing and sales of their products, the association set up telecenters in two villages equipped with cell phones, Global Position System, and computers with high-speed Internet connections. The telecenters, managed by trained rural women, help the association run its businesses more efficiently. The organization also maintains a Web site that offers its members timely information about events where they can promote or sell their products. In the two years following the establishment of the telecenters and the launch of the Web site in 2005, orders have increased by about 70 percent, and members more than doubled their profits.
>
> **Sources:** Agenda 2007; Bhatnagar and others 2002; ITC 2008; M. S. Swaminathan Research Foundation 2008; Shore 2005; UNCTAD 2006a.

Whisler 2003). Criterion Economics reached a similar conclusion, estimating that universal broadband access in the United States could account for $300 billion–$500 billion of GDP by 2006 (Crandall and Jackson 2001).

Some sources—including the Organisation for Economic Co-operation and Development (OECD; 2001)—have supported broadband's importance in terms of its potential and actual impacts on the overall economy, while others have suggested that the macroeconomic impacts of broadband are still to come (Galbi 2001). In the early 2000s broadband was still rare. All of these first-generation studies were hypothetical and forward-looking, and had little evidence on which to base their analyses.

More recent studies have analyzed actual broadband experiences. The Republic of Korea is often cited as a country where economic and employment growth resulted from proactive broadband policies (box 3.3). Broadband seems to have played a significant role in transforming Korea's overall economy and improving its global integration and competitiveness.

Broadband is also believed to play an important role in enabling innovation and R&D, important factors that contribute to sustainable economic growth. Broadband-enabled combinations of ICT and other technologies—such as biotechnology and nanotechnology—are considered essential to generating inventions and innovations in numerous fields (Carlaw, Lipsey, and Webb 2007). For example, according to the Commonwealth Scientific and Industrial Research Organization, broadband networks are transforming astronomy by allowing telescopes to be operated remotely. VLBI (very long baseline interferometry)-over-broadband is a new technique being tested and used

> **Box 3.3** The Republic of Korea's Experiences with Broadband
>
> Korea is a well-known leader in broadband. Its government has both catalyzed rapid rollout of broadband infrastructure and facilitated the uptake of broadband services by citizens, businesses, and the public sector. In 2007, 99 percent of the country's households had access to high-speed Internet. About 90 percent subscribed to broadband, with half enjoying connection speeds of 50–100 megabits per second. Korea was ranked first in the International Telecommunication Union's (ITU's) Digital Opportunity Index that year.
>
> In 1995, as part of its Information Infrastructure Plan, the Korean government stated its vision for nurturing a knowledge-based economy and identified robust broadband as the first step. It implemented its broadband strategy through a combination of deregulation, facilities-based competition, and privatization of the incumbent telecommunications provider Korea Telecom. The government also invested in the construction of backbone networks and provided subsidized loans to telecom operators to facilitate the development of local access networks. In 1995–2005, public investments totaled $900 million, triggering $32.6 billion in private investment.
>
> The government also implemented broadband promotion policies designed to stimulate Internet use and aggregate demand among the population. In 2000, the government set up the Internet Education to Ten Million People Project, aimed at providing IT literacy training for all citizens. In addition to the training and awareness raising, the government vigorously promoted e-business incubation and IT use in public administration (e-government). Such policies and programs led to an explosion in demand.
>
> The rapid deployment of broadband provided important opportunities for Korea's ICT industry. Some 300,000 jobs have been created in ICT, and the sector is growing three times faster than the rest of the economy. Particularly fast-growing areas of the sector include development of search engines and local content. In addition, Korea has developed a competitive advantage in certain niches of the ICT industry—such as online gaming, where Korean companies are the biggest global players.
>
> Broadband has played a significant role in transforming Korea's overall economy and improving its global competitiveness. Korea's media, automotive, and banking industries have benefited from the introduction of broadband, changing their business models and production and supply chain management. According to the National Statistics Office, Korea's e-commerce market more than doubled between 2002 and 2006, from $178 billion to $414 billion. Moreover, a much larger share of the population in Korea accesses news information through broadband than in Europe, Japan, and the United States. This helps develop a well-informed population that is ready for global integration and competition. ITU (2005) attributes the rising share of ICT as a share of GDP in Korea in part to the country's early leadership in the broadband field, both fixed and mobile. In 1992, Korea's ICT industry contributed around 2 percent of its GDP, a percentage close to the world average. However, a decade later, ICT's share of GDP in Korea had risen to around 4.6 percent, almost twice the global average.
>
> **Sources:** Authors' analysis; Ahonen and O'Reilly 2007; ITU 2005; Kelly, Gray, and Minges 2003; Korea National Statistics Office 2007.

in Australia, Europe, and the United States to process data from telescopes, allowing for new experiments and inferences that would not have been possible without high-speed broadband networks.[2] In the area of bioinformatics, the government of India has invested in the Biotechnology Information Systems Network (BISnet), a broadband network linking 57 research centers using a high-speed computer network, to tap synergies between ICT and biotechnology (Government of India 2005).

Broadband, as well as the digitalization of scientific content, also has a transformative effect on innovation processes. It allows around-the-clock R&D and concurrent development on multiple phases and projects in different locations. New forms of ICT-related innovation processes are emerging and fundamentally changing how science and research are conducted. Broadband enables faster diffusion of codified knowledge and ideas, linking science more closely to business. It also lowers barriers to product and process innovation, fosters startups, improves business collaboration, enables small businesses to expand their R&D and collaborate with larger R&D consortiums, and encourages greater networking among the community of researchers (Van Welsum and Vickery 2007).

Many developing countries expect public investments in broadband—for education, health care, and the overall economy—to pay off. In 2006, the government of the Arab Republic of Egypt, in partnership with the private sector, installed a WiMax network to connect two public schools, a mobile health care center, a municipal building, and an e-government services kiosk in rural Oseem.[3] In April 2008, the government of Brazil formalized an arrangement with five fixed-line operators to build a broadband network to connect public schools in 3,440 municipalities by the end of 2010 (*Gazeta Mercantil* 2008). Even though the evidence in terms of measurable outcomes such as improved educational achievements and medical treatment results is not yet conclusive, there are numerous case examples pointing to the potential transformational impact of broadband connectivity on the effectiveness and efficiency of health and education service delivery (box 3.4). The greatest impact may be achieved in remote areas without direct access to critically needed medical specialists and qualified teachers.

Moreover, many governments believe that broadband is becoming increasingly important for globalization. This is beginning to have a fundamental impact on how economies work and on the global allocation of resources—especially for developing countries, due to their greater integration in global value chains than ever before (OECD 2008).

Box 3.4 Broadband-Enabled Telemedicine

The use of broadband-enabled telemedicine is widespread, both in developed and developing countries, yet there are few high-quality studies assessing its diagnostic efficacy and outcome capabilities. Many studies do not separate new opportunities offered by telemedicine from evaluations, which is required to adhere to standards of high-quality evidence. The best evidence for the effectiveness of telemedicine is in medical specialties for which verbal interactions are a key component of the patient assessment and when medical results comparable to in-person encounters can be achieved.

In remote areas without direct access to critically needed medical specialists, broadband networks allow health professionals to care for patients living in different rural locations using videoconference facilities. In such cases, even if the evidence of improved medical outcome is not proven, rapid diagnosis and treatment, reduced costs and travel time for patients, and decreased medical errors are tangible benefits, especially in the context of the ever-falling costs of broadband connectivity solutions.

The Aravind Eye Hospital in the southern Indian state of Tamil Nadu provides such a case in point. Using a wireless broadband network with speeds 100 times faster than a dial-up network, the hospital was able to connect five of its rural clinics in 2004 to provide eye services to thousands of rural residents. With high-speed links to the hospital, the clinics screen about 1,500 patients each month through a Web camera consulting with an Aravind doctor. This videoconferencing system enabled patients to have minor eye problems diagnosed and resolved locally; only those with more serious problems had to travel long distances to a hospital. Hence, patients were able to save on unnecessary travel time and cost, and avoided the corresponding loss in income. The hospital's study also showed that 85 percent of the men and 58 percent of the women who had lost their jobs due to sight impairment were reintegrated into the workforce after treatment. Thereby, the hospital was also able to address the shortage of rural doctors through this broadband system, and the pilot project proved so successful after just 17 months that plans have been made to implement a similar system in 50 clinics serving half a million patients each year.

Sources: Authors' analysis; University of California Berkeley 2006.

Broadband enables economic integration and encourages greater international competition in sectors and jobs that were previously uncontested. Rapid broadband diffusion and increasing speeds and bandwidth, along with the ongoing liberalization of trade and investment in services,[4] have increased the tradability of many service activities—especially business services—and created new kinds of tradable services. The availability, quality, and affordability of broadband services are now important factors for international investors when deciding whether to invest in a specific country. Developing countries with better ICT infrastructure attract more offshoring, outsourcing, and foreign investment (Abramovsky and Griffith 2006) and, as a result, trade more (UNCTAD 2006b).

The boom in broadband-enabled IT services and their clear contribution to GDP, employment, and exports have been well documented (see chapter 7). In India, software exports jumped from less than $1 billion in 1995 to more than $32 billion in 2007; the software industry now accounts for more than three-quarters of the country's services exports and employs 1.6 million people. Other developing countries, such as China, Costa Rica, the Philippines, and some small Caribbean islands, have benefited from the global outsourcing of IT services (Li 2003; World Bank 2008a). Because traded services are often transmitted through high-speed data networks, the advantage of broadband over narrowband is evident in this case. For example, electronic delivery of software, testing, and remote product updating, as well as training and technical assistance for customized software, cannot be performed without broadband.

At the same time, governments have increasingly leveraged broadband to have efficient, reliable, cost-effective public services that contribute to an attractive business environment and national competitiveness. Many good examples of e-government applications are in place today (see chapter 5). Ghana and Singapore, for example, have been successful in using broadband in customs and trade facilities (box 3.5).

Cross-Country Growth Analysis

The literature review above suggests that the presumed economic impacts of broadband are real and, in many cases, measurable. But most results have been restricted to developed economies, their firms and communities, and to case studies. Because broadband's benefits are pervasive, the best way to look at its impact is by focusing on economic growth, despite the complications that such an effort entails.

This section uses an endogenous growth model (Barro 1991) to test the impact of broadband penetration on the average growth rate of per capita GDP between 1980 and 2006. Such a macro-level econometric analysis makes it possible to control for other factors that may have similar impacts on growth, and thus explores the effects of broadband access specifically. In addition, a cross-country analysis sheds light on developing countries where empirical evidence is lacking. Annex 3A provides more details on the definition of variables, methodology, results, and limitations of the analysis.

The average growth rate of per capita GDP between 1980 and 2006 was used as the dependent variable and regressed onto the following variables, selected as representative of conditioning variables in the growth literature:

- Per capita GDP in 1980 (GDP_{80})[5]

- Average ratio of investment to GDP between 1980 and 2006 (I/Y_{8006})

- Primary school enrollment rate in 1980 ($PRIM_{80}$) (a proxy for human capital stock)[6]

- Average penetration of broadband and other telecommunications services between 1980 and 2006 for developed (BBNDH) and developing (BBNDL) countries (a proxy for technological progress and the focus of this analysis)

- Dummy variables for countries in the Sub-Saharan Africa (SSA) and Latin America and Caribbean (LAC) Regions.[7]

Data for all the ICT-specific variables are from the ITU (2007) and World Bank (2008b). The sample consisted of 120 countries, the majority of which are developing countries.

Table 3.2 summarizes the results from the growth regression, which are consistent with the literature on endogenous growth: the average growth rate of per capita GDP between 1980 and 2006 was negatively correlated with initial GDP per capita (GDP_{80}) and positively correlated with the average share of investment in GDP (I/Y_{8006}). Consistent with the convergence implication of the neoclassical growth model, the coefficients on these variables were significant.

The coefficient on average broadband penetration for high-income countries (BBNDH) was positive and significant. This result suggests a robust and noticeable growth dividend from broadband access in developed countries: all

Box 3.5 Broadband-Enhanced Trade Facilities in Ghana and Singapore

Ghana. In 2003, Ghana introduced the GCNet customs system as an ICT-based solution to foster trade development and facilitation and ensure effective mobilization of customs revenues. The electronic data interchange system links all the main players in the clearing process, enabling quick online processing of customs clearance documentation and facilitating clearance of goods through ports. Among other features, the system allows around-the-clock submission of customs documents, provides a one-stop platform for processing and verifying trade documents, and enables systematic monitoring of consignment movements. Within its first 18 months, GCNet increased customs revenues by 49 percent and substantially reduced clearance times.

The system's backbone is a private broadband communication network that consists of a fiber-optic broadband link between the GCNet office and the Customs, Excise, and Preventive Services Department, and is complemented by radio links and leased lines to the department's offices throughout Ghana. By increasing the speed, reliability, and transparency of the clearing process and revenue accrual, broadband contributes directly to the country's competitiveness and economy.

Singapore. SingaporeONE was launched in 1998 to connect citizens, firms, and the government in a single broadband network. A public-private consortium, 1-Net Singapore, was formed to run SingaporeONE's backbone. Infocomm@SeaPort is one of the programs that uses SingaporeONE's broadband capacities. It was launched in 2007 to enhance the capabilities and efficiencies of Singapore's ports and improve the port community's infrastructure. One of its first projects is WISEPORT, a mobile wireless broadband network providing low-cost, high-bandwidth, secure access within 15 kilometers of Singapore's southern coastline. By the end of 2008, all ships in Singapore will have access to mobile wireless broadband, allowing real-time and data-intensive communications between the ships and their customers and business partners. The parties involved will be able to perform multiple tasks remotely, including regulatory filings and real-time access to navigational data. This network is aimed at maintaining Singapore's competitiveness relative to other growing ports in the region.

Sources: de Wulf and Sokol 2004; Ghana Shippers' Council 2008; Keng and others 2008.

Table 3.2 Growth Regression Separating Effects of Broadband Penetration

Variable	Coefficient	t-Statistic
GDP_{80}	−0.100	3.86
I/Y_{8006}	0.164	5.46
$PRIM_{80}$	0.001	−0.18
BBNDH	0.121	2.87
BBNDL	0.138	−1.96
SSA dummy	−1.018	2.19
LAC dummy	−0.655	−1.55
Constant	−1.726	−1.83

Source: Authors' analysis.

Note: BBNDH = average broadband penetration for high-income countries between 1980 and 2006; BBNDL = average broadband penetration for middle- and low-income countries between 1980 and 2006; GDP_{80} = per capita gross domestic product (GDP) in 1980; I/Y_{8006} = average ratio of investment to GDP between 1980 and 2006; LAC = Latin America and the Caribbean Region; $PRIM_{80}$ = primary school enrollment rate in 1980; SSA = Sub-Saharan Africa Region.

else equal, a high-income economy with an average of 10 broadband subscribers per 100 people would have enjoyed a 1.21 percentage point increase in per capita GDP growth. This potential growth increase is substantial given that the average growth rate of developed economies was just 2.1 percent between 1980 and 2006.

The growth benefit that broadband provides for developing countries was of similar magnitude as that for developed economies—about a 1.38 percentage point increase for each 10 percent increase in penetration. But the coefficient on average broadband penetration for middle- and low-income countries (BBNDL) was statistically significant at 10 percent but not at 5 percent, perhaps reflecting that broadband is a recent phenomenon in developing countries and penetration has not yet reached a critical mass to generate aggregate effects as robust as in developed countries. In 2006, 3.4 percent of the population in low-income countries and 3.8 percent in middle-income countries had broadband, compared with 18.6 percent in developed economies.

Despite its shorter history, broadband seems to have a higher growth impact relative to communications technologies such as fixed and mobile telephony and the Internet (figure 3.1). Thus, current differences in broadband penetration among countries may generate significant long-run growth benefits for early adapters. Moreover, the significant and stronger growth effects of other technologies in developing countries than in developed countries suggest that the growth benefit of broadband in developing countries could be on a similar path.

Conclusion

Broadband is a significant technological development, providing users with fast, always-on access to new services, applications, and content. Much of the research on the relationship between broadband adoption and its economic impacts has been in the form of qualitative arguments, anecdotes, and limited case studies. Formal empirical studies have focused on developed countries, and firm- and community-level studies in those countries confirm the high potential economic gains from broadband—including higher productivity, lower costs, new economic opportunities, job creation, innovation, and increased trade and exports.

Filling the gap in assessing the macroeconomic impact of broadband and in empirical evidence for developing economies, this chapter represents a first attempt at macro-econometric analysis and validation of the positive impacts that broadband, as a proxy for the more pervasive role of networks, can have on economic growth. The empirical findings here suggest that broadband's benefits are major and robust for both developed and developing countries, although the significance is higher for the former, which have a longer track record of broadband diffusion. As the number of broadband subscribers increases and the applications supported by broadband reach a critical mass, the benefits could show the same statistical significance in developing economies, as with all other communications technologies.

Whether this great potential to contribute to growth and competitiveness is realized will depend on whether governments understand the opportunity and ensure that supportive conditions are in place through regulatory and policy reforms as well as strategic investments and public-private partnerships. Realizing the benefits of broadband also requires development of new content, services, and applications, as well as increased human capacity to adapt the technology in economic activities. Broadband clearly deserves a central role in national development strategies.

As this chapter indicates, several areas for future research would be fruitful. First, once longer periods of data are available on broadband penetration, a subperiod analysis could be conducted, thus establishing multiple data points for each economy in an endogenous growth regression. This

Figure 3.1 Growth Effects of ICT

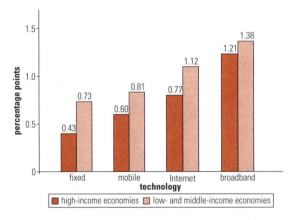

Source: Qiang 2009.

Note: The y axis represents the percentage-point increase in economic growth per 10-percentage-point increase in telecommunications penetration. All results are statistically significant at the 1 percent level except for that of broadband in developing countries, which is at the 10 percent level.

approach would make it possible to study individual dynamics, give information on the ordering of events, and control for individual unobserved heterogeneity, addressing the endogeneity issue better. Second, it may be useful to look at how broadband affects key variables that are good for growth, such as trade, foreign investment, education, and innovation. Third, the microeconomic foundations of broadband's impact should be further explored with specific reference to developing countries. The current largely anecdotal analysis needs to advance to systematic impact evaluation. The effect of broadband on the creation and expansion of social and political networks could also be explored further in both developed and developing countries.

Annex 3A: Statistical Note

Initially, the four-equation simultaneous model (namely, the output equation or economywide production function; demand function for telecommunications; equation determining investment in telecommunications infrastructure; and equation relating investment to increased rollout) used by Röller and Waverman (2001) was considered to analyze the broadband impact on growth. However, this approach uses annual data, so errors or missing data cause significant difficulties.[8]

The Barro cross-sectional endogenous growth model was then used to look at long-term average growth rates. Waverman, Meschl, and Fuss (2005) used a similar model to test the impact of mobile telephony on economic growth in developing countries. This endogenous technical change approach, which uses period averages and initial values, is therefore less prone to data errors. Given the poor data availability in a large number of developing countries, this model may prove more successful in obtaining sensible estimates.

The following equation was used to test the impact of telecommunications penetration, including broadband, on the growth rates:

$$GDP_{8006} = \alpha_0 + \alpha_1 * GDP_{80} + \alpha_2 * (I/Y_{8006}) + \alpha_3 * TELEPEN_{8006} + \alpha_4 * PRIM_{80} + \alpha_6 * SSA + \alpha_7 * LAC + \mu.$$

Applying the same regression specification to various telecommunications services makes the results very comparable.

Table 3A.1 provides the variable description. The data come from the ITU (2007) and World Bank (2008b). Our econometrics analysis covers about 120 countries for the time period 1980 to 2006 to reflect the long-term growth perspectives.

We also divided the sample into developed and developing economies (the latter including both middle-income and low-income countries according to the World Bank country classifications), created dummy variables, and generated the new variables TELEPENH and TELEPENL (the product of the dummy variables and the telecommunications penetration variables). In this way, we intended to differentiate the

Table 3A.1 Definition of Variables

Variable	Description
GDP_{8006}	Average growth rate of real GDP per capita in US$ over 1980–2006
GDP_{80}	Level of real GDP per capita in 1980
I/Y_{8006}	Average share of investment in GDP for 1980–2006

$TELEPEN_{8006}$
  FIXED
  MOBILE
  INTERNET
  BBND
  H
  L
	Average telecommunications penetration per 100 people over 1980–2006
	  Number of main lines
	  Mobile subscribers
	  Internet users
	  Broadband subscribers
	  High-income countries (developed)
	  Low- and middle-income countries (developing)

$PRIM_{80}$	Primary school enrollment rate in 1980
SSA	Dummy variable for countries in the Sub-Saharan Africa Region
LAC	Dummy variable for countries in the Latin America and Caribbean Region

Source: Authors' analysis.

Table 3B.1 Regression for per Capita Growth

Indicator	Dependent variable: GDP_{8006}				
	−1	−2	−3	−4	−5
GDP_{80}	0.098	−0.129	−0.123	−0.128	−0.100
	(2.82)**	(4.62)**	(4.49)**	(4.46)**	(3.86)**
I/Y_{8006}	0.171	0.177	0.155	0.164	0.164
	(6.02)**	(6.59)**	(5.64)**	(5.96)**	(5.46)**
FIXEDH		0.043			
		(4.11)**			
FIXEDL		0.073			
		(3.18)**			
MOBILEH			0.060		
			(4.03)**		
MOBILEL			0.081		
			(3.54)**		
INTERNETH				0.077	
				(3.69)**	
MOBILEL				0.112	
				(2.91)**	
BBNDH					0.121
					(2.87)**
BBNDL					0.138
					−1.96
$PRIM_{80}$	0.003	−0.002	−0.004	−0.001	0.001
	−0.43	−0.36	−0.58	−0.1	−0.18
SSA	−1.382	−0.693	−0.859	−0.903	−1.018
	(3.35)**	−1.66	(2.11)*	(2.20)*	(2.19)*
LAC	−0.973	−0.686	−0.839	−0.861	−0.655
	(2.35)*	−1.68	(2.03)*	(2.08)*	−1.55
Constant	−1.786	−2.285	−1.86	−1.982	−1.726
	(2.09)*	(2.82)**	(2.31)*	(2.43)*	−1.83
Number of countries	120	120	120	120	119
R^2	0.44	0.52	0.52	0.51	0.49

Source: Authors' analysis.

Note: The numbers in parentheses are the absolute value of t-statistics. See table 3A.1 for the definitions of variables used in this table.

** Significant at the 1 percent level; * significant at the 5 percent level.

growth effects in developing countries from those in high-income countries. Table 3B.1 presents a summary of the growth regression results.

Because demand for telecommunications services rises with wealth, the impact of increased telecommunications penetration on economic growth and the impact of rising per capita GDP on the demand for telecommunications constitute the so-called two-way causality issue. Especially in the cases of mobile, Internet, and broadband, where data are only available starting 1990 or later (1998 in the case of broadband), penetration rates are potentially endogenous. The growth model approach does not deal with this problem explicitly—we performed a Hausman test to determine whether any reverse causality is present. We used the initial values of telecommunications penetrations as the instruments. The null hypothesis of ordinary least squares being consistent and efficient could not be rejected.

Notes

1. Broadband can be delivered over fixed and wireless networks. This chapter illustrates the benefits achieved by broadband regardless of the type of network delivery.
2. See http://www.csiro.au/news/BroadbandNetworks.html.
3. See "Egypt Unveils First 'Digital Village'" at http://www.egovnews.org/?m=200612&paged=2.

4. Services now account for around two-thirds of output and foreign direct investment in most developed countries, and for up to a quarter of total international trade.

5. Barro (1991) found that, conditional on the initial human capital stock, average per capita GDP growth was negatively correlated with initial per capita GDP. Thus, all else being equal, there should be convergence in income levels between poor countries and rich countries, although this only takes place over long periods.

6. Barro (1991) also found that the initial level of human capital stock was positively correlated with per capita GDP growth, so countries that were initially rich might grow faster than poor countries if there were large differences between their initial endowments of human capital.

7. A common assumption is that countries in the Sub-Saharan Africa and the Latin America and Caribbean Regions have poorer telecommunications sector performance than countries elsewhere. Our results confirmed this assumption for sub-Saharan African countries. However, the negative coefficient of the Latin America and Caribbean dummy—smaller than that of the Sub-Saharan Africa Region dummy—is statistically insignificant.

8. This technical note draws from Qiang (2009).

References and Other Resources

Abramovsky, Laura, and Rachel Griffith. 2006. "Outsourcing and Offshoring of Business Services: How Important Is ICT?" *Journal of the European Economic Association* 4 (2–3): 594–601.

Agenda (online journal). 2007. "ICTs—Women Take a Byte." Volume 71, Durban, South Africa. http://www.agenda.org.za/content/blogcategory/2/88889071/.

Ahonen, Tomi T., and Jim O'Reilly. 2007. "Digital Korea." Futuretext, London.

Allaire, J., and R. D. Austin. 2005. "Broadband and Collaboration." In R.D. Austin and S. P. Bradley, eds., *The Broadband Explosion.* Cambridge, MA: Harvard Business School Press.

Allen Consulting Group. 2002. *Australia's Information Economy: The Big Picture.* Prepared for Australia's National Office for the Information Economy, Canberra. http://unpan1.un.org/intradoc/groups/public/documents/APCITY/UNPAN004001.pdf.

Arellano, M., and S. Bond. 1991. "Some Tests of Specification for Panel Data: Monte Carlo Evidence and an Application to Employment Equations." *Review of Economic Studies* 58: 277–97.

Atkinson, R. D., and A. S. McKay. 2007. "Digital Prosperity: Understanding the Economic Benefits of the Information Technology Revolution." The Information Technology and Innovation Foundation. http://www.itif.org/index.php?id=34.

Australian Communications and Media Authority. 2008. "Telecommunications Today Report 2: Take-up and Use by Small and Medium Enterprises." Canberra.

Barro, Roberto J. 1991. "Economic Growth in a Cross Section of Countries." *Quarterly Journal of Economics* 106 (2): 407–43.

Bauer, J., P. Gai, J. Kim, T. Muth, and S. Wildman. 2002. "Broadband: Benefits and Policy Challenges." Prepared for Merit Inc. James H. and Mary B. Quello Center for Telecommunication Management and Law, Michigan State University. http://quello.msu.edu/reports/broadband-report-final.pdf.

Bhatnagar, Subhash, Ankita Dewan, Magüi Moreno Torres, and Parameeta Kanungo. 2002. "M. S. Swaminathan Research Foundation's Information Village Research Project (IVRP), Union Territory of Pondicherry." Indian Institute of Management, Ahmedabad, and World Bank, Washington, DC. http://siteresources.worldbank.org/INTEMPOWERMENT/Resources/14654_MSSRF-web.pdf.

Broadband Stakeholder Group. 2004. "Impact of Broadband-Enabled ICT, Content, Applications and Services on the UK Economy and Society to 2010." Briefing paper. http://www.broadbanduk.org/component/option,com_docman/task,doc_view/gid,111.

Carlaw, K. I., R. G. Lipsey, and R. Webb. 2007. "The Past, Present and Future of the GPT-Driven Modern ICT Revolution." Final (Blue) Report. Industry Canada, Ottawa.

Clarke, George, and Scott Wallsten. 2006. "Has the Internet Increased Trade? Evidence from Industrial and Developing Countries." *Economic Inquiry* 44 (3): 465–84.

Clayton T., and P. Goodridge. 2004. "E-business and Labor Productivity in Manufacturing and Services." *Economic Trends* 609: 47–53.

Crandall, R., and C. Jackson. 2001. "The $500 Billion Opportunity: The Potential Economic Benefit of Widespread Diffusion of Broadband Internet Access." Criterion Economics, Washington, DC.

de Wulf, Luc, and José B. Sokol. 2004. "Customs Modernization Initiatives: Case Studies." World Bank, Washington, DC.

Dutta, Soumitra, and Irene Mia. 2008. *The Global Information Technology Report 2006–2007: Connecting to the Networked Economy.* Basingstoke, U.K.: Palgrave Macmillan.

Entner, Roger. 2008. "The Increasingly Important Impact of Wireless Broadband Technology and Services on the U.S. Economy." An Ovum Study for CTIA-The Wireless Association. http://files.ctia.org/pdf/Final_OvumEconomicImpact_Report_5_21_08.pdf.

Firth, Lucy, and David Mellor. 2005. "Broadband: Benefits and Problems." *Telecommunications Policy* 29 (2–3): 223–36.

Ford, George S., and Thomas M. Koutsky. 2005. "Broadband and Economic Development: A Municipal Case Study from

Florida." *Applied Economic Studies* (April): 1–17. http://www.freepress.net/docs/broadband_and_economic_development_aes.pdf.

Forman, C., A. Goldfarb, and S. Greenstein. 2005. "Geographic Location and the Diffusion of Internet Technology." *Electronic Commerce Research and Applications* (4): 1–113.

Galbi, D. 2001. "Growth in the 'New Economy.'" *Telecommunications Policy* 25 (1–2): 139–54.

Gazeta Mercantil. 2008. "Telecommunications Operators Will Take Broadband to Schools." May 26.

Ghana Shippers' Council 2008. "Boankra Inland Port." http://www.ghanashipperscouncil.org/project.htm.

Gillett, Sharon E., William H. Lehr, Carlos A. Osorio, and Marin A Sirbu. 2006. "Measuring the Impact of Broadband Deployment." Prepared for the U.S. Department of Commerce, Economic Development Administration, Washington, DC. http://www.eda.gov/ImageCache/EDAPublic/documents/pdfdocs/mitcmubbimpactreport_2epdf/v1/mitcmubbimpactreport.pdf.

Government of India. 2005. "National Biotechnology Development Strategy." Prepared by the Department of Biotechnology, Ministry of Science and Technology. http://www.mindfully.org/GE/2005/India-Biotech-Strategy10apr05.htm.

Government of Japan. 2003. "Report on the Promotion of the Animation Industry." Prepared by the Bureau of Industrial and Labor Affairs. http://www.metro.tokyo.jp/ENGLISH/TOPICS/2003/03041002.htm.

Hausman, Jerry, and Zvi Griliches. 1986. "Errors in Variables in Panel Data." *Journal of Econometrics* 31 (1): 93–118.

Heng, Stephan. 2006. "Media Industry Facing Biggest Upheaval Since Gutenberg, Media Consumers Morphing into Media Makers." Deutsche Bank, Frankfurt.

ITC. 2008. "About e-Choupal." http://www.echoupal.com.

ITU (International Telecommunication Union). 2007. World Telecommunication Indicators Database. Geneva.

———. 2005. "Ubiquitous Network Societies: The Case of the Republic of Korea." http://www.itu.int/osg/spu/ni/ubiquitous/Papers/UNSKoreacasestudy.pdf.

Johnson, B., J. M. Manyika, and L. A. Yee. 2005. "The Next Revolution in Interactions." *McKinsey Quarterly* 4: 20–33.

Kelly, Doris J. 2004. "A Study of Economic and Community Benefits of Cedar Falls, Iowa's Municipal Telecommunications Network." Iowa Association of Municipal Utilities, Ankeny, Iowa. http://www.baller.com/pdfs/cedarfalls_white_paper.pdf.

Kelly, T., V. Gray, and M. Minges. 2003. "Broadband Korea: Internet Case Study." ITU, Geneva. http://www.itu.int/ITU-D/ict/cs/korea/material/CS_KOR.pdf.

Keng, Ng Cher, Ong Ling Lee, Tanya Tang, and Soumitra Dutta. 2008. "Singapore: Building an Intelligent Nation with ICT." In Soumitra Dutta and Irene Mia, eds., *Global Information Technology Report: Fostering Innovation through Networked Readiness.* 121–32. Hampshire, U.K.: Palgrave Macmillan.

Korea National Statistics Office. 2007. http://www.nso.go.kr/eng2006/emain/index.html.

Li, Zhongzhou. 2003. "The Impact of ICT and E-business on Developing Country Trade and Growth." Presented at the OECD-APEC Global Forum: Policy Frameworks for the Digital Economy, January 15, Honolulu, HI. http://www.oecd.org/dataoecd/20/9/2492709.pdf.

Marine, S., and T. Albrand. 2006. "Universal Broadband Access in Emerging Economies: A Key Factor to Bridge the Digital Divide." http://www1.alcatel-lucent.com/com/en/apphtml/atrarticle/2006q3universalbroadbandaccessinemergingeconomiesakeyfactortcm1721109951635.jhtml?_DARGS=/com.

Momentum Research Group. 2005. "Net Impact Latin America: From Connectivity to Productivity." Austin, TX. http://www.netimpactstudy.com/nila/pdf/netimpact_la_full_report_t.pdf.

M. S. Swaminathan Research Foundation. 2008. "Basic Overview." http://www.mssrf.org/index.htm.

OECD (Organisation for Economic Co-operation and Development). 2001. *Science, Technology and Industry Outlook: Drivers of Growth: Information Technology, Innovation, and Entrepreneurship.* Paris: OECD.

———. 2008. Staying *Competitive in the Global Economy. Moving Up the Value Chain.* Paris: OECD.

Qiang, Christine Z. 2009. "Telecommunications and Economic Growth." Unpublished paper. World Bank, Washington, DC.

Rappoport, Paul N., Donald. J. Kridel, and Lester D. Taylor. 2002. "The Demand for Broadband: Access, Content and the Value of Time." In *Broadband: Should We Regulate High-Speed Internet Access?* 57–82. Washington, DC: Brookings Institution Press. http://aei-brookings.org/admin/authorpdfs/redirect-safely.php?fname=../pdffiles/phpvw.pdf.

Röller, Lars-Hendrik, and Leonard Waverman. 2001. "Telecommunications Infrastructure and Economic Development: A Simultaneous Approach." *American Economic Review* 91 (4): 909–23.

Saksena, Asheesh, and Arnim E. Whisler. 2003. "Igniting the Next Broadband Revolution." *Accenture Outlook Journal* (January). http://www.accenture.com/Global/Research_and_Insights/Outlook/By_Alphabet/IgnitingRevolution.htm.

Shore, Keane J. 2005. "Work in Progress—Rural Pondicherry's Wireless Internet." International Development Research Center, Ottawa. http://www.idrc.ca/en/ev-47023-201-1-DO_TOPIC.html.

Smith, Michael D., and Rahul Telang. 2006. "Piracy or Promotion? The Impact of Broadband Internet Penetration on DVD Sales."

Carnegie Mellon University, Pittsburgh, PA. http://archive.nyu.edu/bitstream/2451/14958/2/USEDBOOK16.pdf.

Sprint. 2006. "Sprint Mobile Broadband: Enhancing Productivity in the Insurance Industry and Beyond." http://www.sprint.com/business/resources/065455-insurancecs-1g.pdf.

Stanford, D. 2002. "Telework, Better Productivity Cited as Broadband Benefits." *Washington Internet Daily,* March 26.

Strategic Networks Group. 2003. "Economic Impact Study of the South Dundas Township Fiber Network." Prepared for the U.K. Department of Trade and Industry, Ontario. http://www.berr.gov.uk/files/file13262.pdf.

University of California Berkeley. 2006. "New Wireless Networking System Brings Eye Care to Thousands in India." Berkeley, CA. http://www.berkeley.edu/news/media/releases/2006/06/06_telemedicine.shtml.

UNCTAD (United Nations Conference on Trade and Development). 2006a. *Information Economy Report 2006: The Development Perspective.* New York: UNCTAD.

———. 2006b. "Using ICTs to Achieve Growth and Development." Background paper for the Expert Meeting in Support of the Implementation and Follow-up of WSIS, December 4–5, Geneva.

Van Welsum, Desirée, and Graham Vickery. 2007. "Broadband and the Economy." Background paper for the 2008 Information, Computer, and Communications Policy Conference on "The Future of the Internet Economy," October 4–5, Ottawa, Canada.

Varian, Hal, Robert E. Litan, Andrew Elder, and Jay Shutter. 2002. "The Net Impact Study." http://www.netimpactstudy.com/Net Impact_Study_Report.pdf.

von Hippel, Eric. 2005. *Democratizing Innovation.* Cambridge, MA: MIT Press.

Waverman, Leonard, Meloria Meschl, and Melvyn Fuss. 2005. "The Impact of Telecoms on Economic Growth in Developing Countries." Vodafone Policy Paper 2, Vodafone, Berkshire, U.K.

World Bank. 2008a. *Global Economic Prospects: Technology Diffusion in the Developing World.* Washington, DC: World Bank.

———. 2008b. World Development Indicators Online Database. Washington, DC.

Zilber, Julie, David Schneier, and Philip Djwa. 2005. "You Snooze, You Lose: The Economic Impact of Broadband in the Peace River and South Similkameen Regions." Prepared for Industry Canada, Ottawa.

Chapter 4

Advancing the Development of Backbone Networks in Sub-Saharan Africa

Mark D. J. Williams

Expanding access to advanced information and communication technology (ICT) services will be a key factor in sub-Saharan Africa's economic and social development. Cross-country data show that ICT investment fosters higher long-term economic growth (Roller and Waverman 2001). Small businesses with access to mobile phones can generate sustained increases in the incomes of poor people in developing countries (Jensen 2007). The impact of broadband is harder to quantify because less data are available,[1] but emerging evidence suggests that access to advanced ICT services—such as those that require broadband for delivery—can also have positive economic and social effects (Goyal 2008).

As understanding of the benefits of ICT has grown, African governments have begun to give priority to it and to focus on providing affordable ICT services to as many people as possible. For example, in the introduction to Rwanda's 2006 ICT strategy, President Paul Kagame wrote: "We have high expectations of ICT and its transformative effects in all areas of the economy and society. Communications technology has fundamentally changed the way people live, work, and interact socially, and we in Rwanda have no intention of being left behind or standing still as the rest of the globe moves forward at an ever increasing pace" (Government of Rwanda 2006, foreword).

In response to the dramatic success of policy reforms in expanding access to mobile phone services in sub-Saharan Africa, policy makers and investors are exploring more advanced ICT services (Balancing Act 2007; Global Insight 2007; Telegeography 2008). Indeed, many policy makers in the region consider access to broadband a key driver of economic and social development. Yet broadband connectivity remains lower than in other parts of the world, and prices are high. For example, a basic DSL (digital subscriber line) package costs an average of $366 a month in sub-Saharan Africa,[2] compared with $6–$44 in India (ITU 2007; OECD 2006).[3] The average price of entry-level broadband in the OECD is $22 per month.[4]

The limited availability of low-cost backbone network capacity is one of the factors constraining sub-Saharan Africa's development of broadband connectivity. Backbone networks are the high-capacity links that carry communications traffic between fixed points in the networks and form a crucial component in the communications supply chain. This chapter explains why backbone networks are important for delivering broadband connectivity and describes the current pattern of backbone infrastructure development in sub-Saharan Africa and the market dynamics underlying it. This analysis provides the basis for the policy recommendations outlined at the end of the chapter. The potential benefits of broadband are analyzed in chapter 3 and so are not discussed here.

The Significance of Backbone Networks

The technical and economic characteristics of backbone networks place them at the heart of communications infrastructure and strongly affect the commercial viability of communications services, particularly broadband connectivity.

The Role of Backbone Networks in Delivering Telecommunications Services

The process of supplying communications services can be thought of as a supply chain (figure 4.1). At the top of the chain is the international connectivity that provides links to the rest of the world. At the second and third levels are the regional and domestic backbone networks that carry traffic from international communications infrastructure and within countries. The fourth level is the "intelligence" in the networks that route traffic. Below this are the access networks that link core networks to customers. Finally, there is a suite of retail services, including customer acquisition, billing, and customer care, that allows providers to function. The hierarchical nature of networks means that the volume of traffic carried by backbone networks can be relatively high even if the customer base is small.

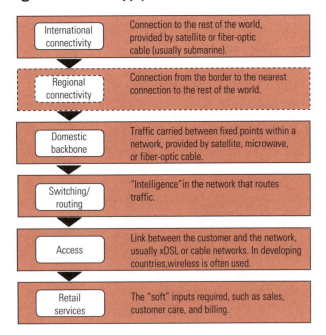

Figure 4.1 The Supply Chain for Communications

Source: Author.

The Economic Impact of Backbone Networks

Backbone networks have a major impact on the commercial viability of ICT services, particularly broadband. In a typical mobile phone network, the backbone network accounts for 10–15 percent of total network costs.[5] The cost of backbone networks is much higher for operators providing broadband connectivity, particularly in small towns and rural areas. If an area does not have a backbone network offering low-cost network services, broadband connectivity is unlikely to be commercially viable.

Backbone networks have high fixed costs and low variable costs, which means that the average cost of capacity falls as traffic volumes increase. Figure 4.2a shows how average costs fall as traffic volume increases, while figure 4.2b shows how spreading traffic across more than one network raises average costs.

The picture presented in figure 4.2 is a static one that does not take into account the dynamic effects of competition. The cost advantage of aggregating traffic onto a single high-capacity network needs to be offset against the inefficiencies created by the lack of infrastructure competition. An illustration of this is the high price typically charged for backbone network services by incumbent operators, even where they have a monopoly in this market segment.

Competition among operators does not necessarily require each operator to have its own backbone network. Network interconnection enables one operator to use the backbone network of another, provided that it can access it on reasonable terms. This is achieved either through a competitive wholesale market for backbone services or through regulatory controls that allow open access to networks.

In a fully liberalized ICT market, the upstream elements (that is, the higher levels of the supply chain shown in figure 4.1) are typically consolidated into a few large companies with high-capacity networks, while the downstream components tend to be smaller and more geographically disaggregated. In the United States, for example, this vertical disaggregation results in a three-tier industry structure. The first tier is made up of Internet service providers (ISPs) with extensive international and domestic communications infrastructure. Second-tier ISPs are large national companies, also often with their own infrastructure, that have interconnection arrangements with ISPs at other tiers. Third-level ISPs provide services directly to users.

Thus, the economic significance of backbone networks is determined by two factors. The first is the reduction in

Figure 4.2 Economic Impact of Backbone Networks

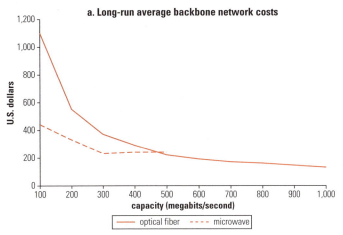
a. Long-run average backbone network costs

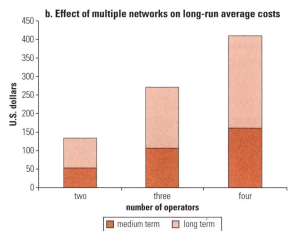
b. Effect of multiple networks on long-run average costs

Source: Ingénieurs Conseil et Economistes Associés 2008.

overall costs that occurs when traffic is channeled through high-capacity networks with lower average costs. The second is the opportunity for smaller players to enter the market by purchasing low-cost backbone network services without having to build their own network. These two factors are interrelated. By aggregating the traffic generated by smaller players onto higher-capacity backbone networks, average costs are reduced.

The Dynamics of Backbone Network Development

This section assesses the backbone network infrastructure in sub-Saharan Africa and describes the dynamics of the markets and regulatory systems that have influenced it. It focuses on three issues: the adequacy of backbone network infrastructure, network ownership, and geographic patterns of network development.

The Adequacy of Backbone Network Infrastructure

Contrary to common assumptions, there is extensive backbone network coverage in sub-Saharan Africa, with about 508,000 kilometers of terrestrial backbone infrastructure (microwave and fiber-optic cables) serving around three-quarters of communications users.[6] The remaining one-quarter of users connect to networks using backbone infrastructure based on satellite links.[7]

About a third of the terrestrial backbone in sub-Saharan Africa is owned by fixed operators, including both formerly and currently state-owned incumbents and new entrants. The other two-thirds of terrestrial backbone infrastructure and almost all satellite-based backbone infrastructure are owned by mobile operators. This setup is the opposite of that in Western Europe and North America, where mobile operators often focus on the wireless access layer of network infrastructure and lease backbone services from fixed network operators (Hanna and Ramarao 2006).

Most backbone infrastructure in sub-Saharan Africa is low-capacity wireless networks. Only 12 percent of terrestrial infrastructure in the region is fiber-optic cable; the rest is microwave. The share of fiber optics is even smaller when satellite-based infrastructure is taken into account.

The mix of wireline and wireless infrastructure varies considerably by type of operator. Among mobile operators in the region, some 99 percent of the length of backbone networks is made up of microwave technology; just 1 percent is fiber. Fixed operators have far more fiber in their networks—about 40 percent. Satellite capacity is generally used for transmission in thinly populated areas, between parts of networks where coverage is not contiguous, and during the early stages of network rollouts. This situation is also the opposite of that in more advanced markets, where fiber-optic backbone networks dominate and wireless technologies are used as backbone infrastructure primarily in remote and inaccessible areas.

Detailed technical information on the capacity of backbone networks in sub-Saharan Africa is confidential. But choices of basic network technology indicate likely capacity limits in the region. As shown in figure 4.2a, for a given length of network, capacity requirements determine the optimal choice of backbone network technology. Microwave networks are the cheapest option for low volumes of traffic, while fiber-optic networks are preferable for higher traffic. Satellites are the cheapest technology for backbone links connecting points that are far apart, but they typically carry low volumes of traffic. Thus, the predominance of wireless technologies—both microwave and satellite—indicates that backbone networks in sub-Saharan Africa are low capacity and generally incapable of carrying the large volumes of traffic generated by mass market broadband connectivity. Most of the backbone network infrastructure in the region was designed to carry voice traffic, which requires much lower bandwidth than the services offered to broadband customers. This is one reason the networks were built mainly using wireless technologies.

The cost structures of different technologies are another reason for the predominance of wireless technologies in the region's backbone networks. Between 60 percent and 80 percent of the costs of fiber-optic networks come from the civil works associated with laying fibers (Hanna and Ramarao 2006). These fixed costs do not vary with the volume of traffic that a network carries. In fact, the only costs in fiber-optic networks affected by capacity are the costs of transmission equipment, which typically account for less than 10 percent of total network costs. The cost structure of wireless backbone networks is very different. A much smaller share of total costs is fixed relative to network capacity, so total costs are more directly affected by the volume of traffic carried. Thus, the initial cost of wireless networks is much lower, while the marginal cost of increasing network capacity is higher. This is an important reason why, in an uncertain market during the early stages of network development, operators are more likely to invest in wireless backbone networks than fiber-optic networks—even if, in retrospect, it might have been cheaper to use fiber in the long run.

A consequence of this preference for wireless networks is that operators are less likely to have excess backbone network capacity than might have been the case had they invested in fiber-optic networks.[8] This has implications for the market in backbone services. Operators that have a fiber backbone network with spare capacity have a strong commercial incentive to sell that capacity and, because its marginal cost is low, competition among operators could be expected to lower prices. By contrast, an operator with a predominantly microwave backbone network is likely to install the amount of capacity that it needs to meet its own requirements. If it were to decide to sell backbone capacity wholesale, additional capacity would have to be installed. Thus, a microwave-based operator has less incentive to enter this market and, if it did, competition with other operators would be less likely to drive down prices as quickly or as far.

Backbone Network Ownership Structure

Telecommunications markets in most countries in sub-Saharan Africa have developed as a series of vertically integrated businesses operating in parallel. Backbone networks are generally part of these vertically integrated businesses, and there is little wholesale trading of backbone services. In addition, there are few examples of joint ventures to build and operate terrestrial backbone networks and there is little sharing of backbone network facilities. This situation stands in contrast to that in countries with more advanced telecommunications markets, where there is extensive vertical disaggregation of networks and network operators can choose to own only certain parts of the network supply chain and buy network components from other operators.

The few sub-Saharan African countries that have encouraged full infrastructure competition at the backbone level provide an instructive contrast to this general assessment of the situation in the region. Kenya and Nigeria, for example, have both allowed carrier networks to enter the market, while Uganda and Zambia have allowed their electricity transmission companies to operate as wholesale backbone network operators.

Regulatory frameworks often help maintain vertically integrated networks and discourage the development of wholesale markets for backbone services. For example, countries such as Burkina Faso have allowed mobile phone operators to build backbone networks to provide services to their retail customers but not to other operators on a wholesale basis. Such restrictions limit opportunities to exploit economies of scale in network infrastructure and reduce incentives to invest in high-capacity backbone networks.

Moreover, some countries—such as Botswana before its recent revision of sector legislation (Ovum 2005)—have given incumbent operators legal monopolies on backbone network services. Such regulations do force the market into vertical separation because competing operators are prevented from building their own backbone networks. But they also prevent the development of a market in backbone network services and so limit overall investment, often resulting in low capacity and poor quality of service.

The development of wholesale markets for backbone services is also constrained by the dynamics of markets in their early stages of development. When operators are competing to roll out networks in a country, they may not have an incentive to provide backbone services to their competitors because doing so could reduce their competitive advantage. In Uganda, the ISP Infocom was unable to negotiate interconnection agreements to use the backbone networks of MTN and UTL, the country's two biggest network operators. Although Infocom does not offer mobile voice services, the offering by data service providers of voice-over-Internet protocol (VoIP) services and the presence of MTN and UTL in the data services market mean that these operators may have considered Infocom a competitor and so could not reach an agreement to sell backbone services. Infocom, however, was ultimately able to reach an agreement to buy capacity on the electricity transmission network, which operates as a wholesale backbone services provider in Uganda.

Geographic Patterns of Backbone Network Infrastructure

Fiber-optic backbone networks in sub-Saharan Africa have mainly been developed in and between major urban areas and on international routes. Fixed operator backbone networks, which account for most high-capacity fiber-optic networks in the region, cover only about a fifth of the population.[9] The focus of backbone network infrastructure in specific parts of the country is further concentrated by infrastructure competition in the few countries with fully liberalized markets. Entrants in these countries have focused their backbone network construction in areas where incumbent operators already have networks. This is illustrated for four sub-Saharan African countries in figure 4.3.

This pattern of network development shares some features with that of countries in other regions. A 2004 review of the U.K. leased line market (the market for capacity on backbone networks) by the regulator, the Office of Communications (Ofcom), found that the backbone market was highly competitive on intra- and interurban routes, particularly for high-bandwidth services. In areas with competition among networks, the former state-owned monopoly operator, BT, retained around three-quarters of the market share for low-bandwidth leased line services (64 kilobits per second to 8 megabits per second) but less than 10 percent of the market for high-bandwidth

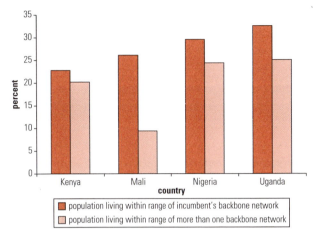

Figure 4.3 Population Covered by Incumbent and Competing Networks in Four African Countries, 2007

Source: Hamilton 2007.

services (155 megabits per second and faster). In India, market liberalization has resulted in multiple backbone network operators entering the market and competing across the full range of services, but only on a limited set of routes. Reliance, for example, has 67,000 kilometers of fiber-optic network and competes with more than 10 other fixed-line operators, primarily on major interurban routes (TRAI 2006).

The concentration of backbone networks in urban areas and on interurban routes reflects the demand for and cost of providing services. In urban areas, residents have higher incomes and there are more businesses; both of these characteristics generate demand for advanced ICT services—and so there is more demand for backbone network capacity. At the same time, the fixed costs of networks mean that the average cost of providing services to people in urban areas is lower than in rural areas. Thus, there is a strong commercial incentive for networks to focus on urban areas and high-traffic routes between these areas.

There is one significant difference between the geographic pattern of backbone network development in sub-Saharan Africa and that in many other regions. Outside the region, incumbent operators (either currently or formerly state-owned) generally play a key role in providing backbone infrastructure in areas where infrastructure competition is absent. In the United Kingdom, BT is the sole supplier of backbone services in half of the market (Ofcom 2004). In India, despite Reliance's extensive investment in backbone network infrastructure, its fiber-optic network is still only 15 percent of the length of the incumbent operator's, which is more than 450,000 kilometers long. Such situations are not the case in most sub-Saharan African countries, where incumbent operators often have limited network coverage and do not provide backbone services of sufficient quality or affordability. Thus, these operators are not in a position to develop backbone network infrastructure in less profitable areas and be the "backbone provider of last resort."

Cross-border connectivity is another emerging feature of the geographic pattern of backbone infrastructure development in sub-Saharan Africa. Many communications networks in the region traditionally evolved as stand-alone networks without direct cross-border interconnection. Most international traffic has been carried by satellite, even where the destination was a neighboring country. But recently, cross-border terrestrial backbone infrastructure has started to develop. For example, fiber-optic networks on both sides of the Kenya-Uganda border are interconnected, and a link is being built across the Rwanda-Uganda border. Similar cross-border connections are being constructed across the region. Extensive cross-border connectivity is also occurring across Western Europe, where many pan-European networks connecting major urban areas have emerged since market liberalization.

Commercial factors are helping drive this emergence of cross-border connectivity in sub-Saharan Africa. First, a lot of international communications traffic is intraregional, since personal and business links are often within a region. One market response to this has been the development of retail packages in which customers pay local call charges when roaming within the region (Global Insight 2006). Such offers are likely to stimulate intraregional traffic, strengthening operators' incentives to interconnect their networks.

Second, demand for the Internet is becoming a major driver of network development. Most Internet traffic generated by sub-Saharan African customers is international because most content is hosted outside the region. Operators can route this traffic between countries where they operate to exploit economies of scale in international gateways. This effect will increase significantly when international submarine fiber-optic connectivity improves in the region. Such cables land at specific locations along the coast, and all traffic carried on them must be routed through these locations. Thus, operators in landlocked countries wishing to use submarine fiber-optic cables need regional connectivity to access them. Even operators in countries with direct access to such cables may wish to develop regional backbone networks to provide access to alternative locations for the cables as backup. Cross-border connectivity will become more profitable as broadband expands in the region and the data traffic carried by submarine fiber-optic cables grows. Thus, cross-border backbone network development will likely continue to develop.

Policies to Improve the Development of Backbone Networks

Sub-Saharan Africa has widespread but low-capacity backbone networks operating in parallel. Higher-capacity fiber-optic networks are concentrated in urban areas, between cities and, increasingly, on cross-border routes. These patterns have emerged because operators initially designed their networks to carry voice traffic, which requires lower-capacity backbone networks. Where high-capacity networks have been built, they have focused on the most profitable and populated areas. In addition, regulatory restrictions on infrastructure competition have often limited the development of high-capacity backbone networks, as have the considerable political and commercial risks associated with investing in fixed wireline infrastructure in the region.

Policy makers face two main challenges in developing backbone network infrastructure. The first is establishing and encouraging competitive markets in backbone infrastructure. The second is providing some form of financial support to encourage the development of high-capacity networks in commercially unattractive areas. Addressing these challenges will require a twin-track approach:

- Create an enabling environment for competition in infrastructure and services by fully liberalizing markets to encourage infrastructure competition and allow aggregation of traffic onto higher-capacity networks.

- Stimulate rollout in underserved areas, especially rural areas and small towns.[10]

Several policy options can be used to tackle these issues. These are summarized in table 4.1 and explained in more detail in the sections that follow.

Table 4.1 Policy Options for Expanding Backbone Networks

Create an enabling environment for competition in infrastructure and services	Stimulate rollout in underserved areas
Remove regulatory obstacles to investment and competition Remove limits on the number of network licenses Encourage the entry of alternative infrastructure providers Remove constraints on the market for backbone services Improve regulation of backbone networks **Reduce investment costs** Facilitate access to passive infrastructure Promote infrastructure sharing **Reduce political and commercial risks** Provide risk guarantees and political risk insurance Aggregate demand **Promote competition in the downstream market** Promote downstream competition through effective regulation	**Share infrastructure** Give operators incentives to cooperate in developing backbone infrastructure in areas where infrastructure competition is not commercially viable **Provide competitive subsidies** Give operators subsidies to build and operate backbone networks in underserved areas, with services provided on a nondiscriminatory basis **Reduce taxes and levies** Give operators incentives to build networks in underserved areas by lowering sector levies or contributions to universal service funds

Source: Author.

Create an Enabling Environment for Competition in Infrastructure and Services

Many sub-Saharan African countries do not provide incentives for private investment and competition in backbone networks—and, in some cases, discourage or obstruct it. Promoting private investment and competition among backbone networks allows market forces to aggregate traffic onto higher-capacity networks, lowering costs and stimulating downstream investment and competition among ISPs and other providers. Several policy initiatives are needed to create an enabling environment for infrastructure competition; they fall into four groups.

1. Remove regulatory obstacles to investment and competition

Remove limits on the number of network licenses. Many sub-Saharan African countries that have nominally liberalized their network markets still have formal or informal limits on the number of licenses that they issue (World Bank 2008). There is little economic justification for such limits.

Encourage the entry of alternative infrastructure providers in the backbone network market. Electricity transmission networks, oil and gas pipelines, and railway networks can provide major cost advantages in the development of fiber-optic backbone networks. By encouraging these (usually state-owned) networks to establish operating companies that run fiber-optic assets and by licensing them, countries can bring them into the telecommunications market as providers of backbone capacity. This practice has been successful in other regions (box 4.1) and in sub-Saharan African countries such as Uganda and Zambia (though not in others, such as Ghana).

Lift constraints on the market for backbone services. Many sub-Saharan African countries impose constraints on operators with backbone networks and those that use them. These constraints include restrictions on the sale of network services and requirements to buy backbone network services from specific operators—usually state-owned incumbent operators. Removing these restrictions would allow operators to buy and sell backbone services to and from whatever operator they wished. Such an environment would consolidate traffic, providing an incentive to upgrade backbone networks to fiber-optic technology and so lowering average costs.

Improve regulation of backbone networks. Difficulties in enforcing contracts and service agreements have been a major constraint to the development of markets for backbone network services. In the short term, the ability to enforce legal contracts in commercial courts is unlikely to improve much in most sub-Saharan African countries. Still, regulators can improve the situation by establishing clear regulations on interconnection and access to backbone networks, amending licenses (if necessary) to increase the enforceability of such rules, establishing effective quality controls and clear procedures for resolving disputes, collecting accurate data

> **Box 4.1** Alternative Infrastructure Providers in Morocco
>
> Morocco has three main backbone network operators: the incumbent Maroc Telecom, Meditel (a major mobile phone provider), and Maroc Connect (an Internet service provider recently awarded a general telecommunications license). In 2005, Meditel was given a license to develop a fixed-line network (including backbone) and Maroc Connect was given a global network and service license. However, rather than building full backbone networks, both operators obtain some backbone network capacity from two alternative infrastructure operators: the Office National des Chemins de Fer (the national railway carrier, which has a nationwide backbone network infrastructure of about 1,100 kilometers) and the Office National d'Electricité (the national power company, which has a nationwide infrastructure of aerial fibers of about 4,000 kilometers). In addition, a company called Marais entered the Moroccan backbone network market in 2007. The market liberalization and the presence of alternative network operators have allowed the new entrants to decide whether to build their own backbone networks or purchase backbone services from other operators. The entry of Marais into the backbone network market indicates that there is further scope for network development and competition.
>
> **Source:** Ingénieurs Conseil et Economistes Associés 2008.

on service quality, and sharing knowledge and experiences with regulators from other countries. Regional approaches to regulating backbone network infrastructure, through participating in regional organizations, may also provide a way of improving the quality of regulation.

2. Reduce investment costs

Facilitate access to passive infrastructure. Civil works account for most of the cost of constructing fiber-optic cable networks (Hanna and Ramarao 2006). These major fixed and sunk costs increase the risks for network investors. By lowering these costs and the associated risks, governments can significantly increase incentives for private investment. Such changes can be made in a number of ways—for example, by providing open access to existing infrastructure and including passive communications infrastructure in the design of other forms of public infrastructure (such as roads, railways, and electricity transmission lines; box 4.2).

Promote infrastructure sharing where it does not undermine competition. By sharing backbone network infrastructure, builders of backbone networks can reduce costs and so make such investments more commercially viable. This is particularly relevant for fiber-optic networks in urban areas, where the costs of laying new fibers can be high, or in rural areas, where the revenues generated by such networks are low. However, caution may be needed in taking this approach. Infrastructure sharing arrangements are hard to enforce if the parties involved are not willing to do them on a commercial basis. Moreover, the sharing of facilities may help sustain collusive arrangements between competing operators (box 4.3).

3. Reduce political and commercial risks

Cut political and regulatory risks through risk guarantees and insurance. In uncertain political and regulatory environments, operators are likely to favor investments in scalable wireless networks instead of fiber-optic networks (which have high fixed, sunk costs). This uncertainty limits the extent to which operators are willing to invest in high-capacity infrastructure that could then be used to consolidate traffic and reduce average costs. These risks can be reduced by building confidence in the regulatory process, and mitigated by using instruments such as partial risk guarantees and political risk insurance (World Bank 2002).

Reduce commercial risk by aggregating demand. Governments can lower commercial risks and transactions costs for operators by acting as a central purchaser of services on behalf of all public institutions—including those at lower levels of government (such as schools, health centers, and local governments; box 4.4). But while there are potential advantages to this approach of demand aggregation, companies in sub-Saharan Africa often have a hard time

> **Box 4.2** Spain's Provision of Passive Infrastructure for Fiber-Optic Networks
>
> In 2003, the government of Spain passed legislation requiring that the design and construction of new buildings include common communications passive infrastructure in elements such as ducting, building risers, and access points. Building managers are required to make this infrastructure available to any operator seeking to provide household access to fiber-optic networks.
>
> This law directly affects the establishment of household access to fiber-optic networks and the construction of privately developed buildings. The same principle can be applied to the development of backbone networks in public infrastructure such as roads and railways.
>
> **Source:** Ministerio de Ciencia y Tecnología 2003.

> **Box 4.3** Sharing Network Infrastructure in Uganda
>
> Uganda's telecommunications market has five mobile operators and several Internet service providers. MTN and UTL are the two national wireline operators and are also significant mobile operators. Both companies also have operations in Rwanda. Thus, they have a common interest in establishing a communications link across the border from Uganda to Rwanda. In 2007, MTN constructed a fiber-optic cable from Kigali, Rwanda, to the border with Uganda. The company also recently announced a deal with its competitor UTL to jointly develop the fiber-optic network on the Uganda side—a good example of competing operators forming a cooperative arrangement to lower the costs of developing fiber-optic networks outside major urban areas. However, such an arrangement raises concerns for the market in Rwanda since the only fiber-optic connection to the country will be jointly controlled by the only two network operators in Rwanda. Such concerns may ease as more licenses are issued in Rwanda and competition develops.
>
> **Source:** Author.

collecting revenues from public institutions, even for utility services such as water and electricity. Thus, an issue to consider for this policy approach is the extent to which the credit risk associated with the public sector as a customer offsets the commercial advantages of the bulk purchase of backbone services. This credit risk can be reduced by using prepayment and escrow mechanisms.

4. Promote competition in downstream markets

Promote competition among downstream network operators and service providers. Network operators and service providers wishing to enter downstream markets—that is, those interested in building access networks and offering services to customers—will need either to build their own backbone networks or to access those of other companies. Governments can stimulate the development of backbone networks by promoting downstream competition and ensuring that operators have access to upstream backbone network infrastructure.

Stimulate Rollout in Underserved Areas

Incumbent operators in sub-Saharan Africa have found it difficult to build and operate backbone networks that meet the needs of the market. Relying on these operators to provide backbone networks outside the main urban areas—a model that has been adopted in other regions—would be difficult to replicate in sub-Saharan Africa. Instead, a partnership with the private sector is more likely to ensure that networks are built and operated efficiently. Three types of such partnerships are discussed here: shared infrastructure, competitive subsidies, and other incentive-based private sector models. While other models

> **Box 4.4** Developing Infrastructure by Aggregating Demand in the Republic of Korea
>
> Korea's government provided financing to develop the country's broadband infrastructure in the form of a prepayment for the future provision of broadband services to public institutions. Between 1995 and 1997, the government provided $200 million toward the $2.2 billion cost of building a fiber-optic network. The remainder of the funding was provided by the private sector, mainly Korea Telecom. The second phase, between 1998 and 2000, focused on the access network, for which the government contributed $300 million of a total investment of $7.3 billion. The final phase, between 2001 and 2005, involved upgrading the entire network, and the government contributed $400 million of a total investment of $24 billion. In exchange for these upfront payments, operators were required to provide broadband services to public institutions over an extended period. Thus, the government's financing can be thought of as prepayment for services that, although representing only a small portion of total investment costs, provided the private sector with sufficient incentive to develop the networks and contribute their own resources. This initiative was undertaken in the context of an overall policy of promoting broadband that included full market liberalization.
>
> **Source:** Author.

are available, these three basic models are representative of the broad scope of policy options, with their respective advantages and disadvantages. Hybrids of the models discussed here are also possible.

1. Shared infrastructure model

Under a shared infrastructure approach, existing private operators would form a consortium to build and operate backbone networks in underserved areas. The government would provide public resources to ensure that the network meets public policy goals such as focusing investment on underserved areas, achieving cost-oriented wholesale prices, and ensuring nondiscrimination between buyers of services. This regulatory protection can be written into the consortium structure through the leverage obtained by public support for the investment. But these policy goals run counter to the commercial interests of the consortium members who would benefit from charging above-cost prices and discrimination against users that are not consortium members. Thus, such an arrangement would require ongoing regulatory oversight.

The *advantages* of the shared infrastructure model are the following:

- The backbone network would be built and operated by private companies that already operate facilities in the country and so have experience likely to improve the chances of their success in operating the network.

- The operators would partially finance the network, reducing the cost to the government and ensuring that the operators have a financial stake in its success.

- The companies operating the network would also be its main customers, giving them an incentive to ensure that it is run efficiently and effectively.

The *disadvantages* are the following:

- A consortium in an otherwise competitive market could allow operators to collude and reduce competition.

- Any consortium is unlikely to include all players in the market, particularly as the market develops and new companies enter. Thus, members of the consortium have an incentive to raise prices and discriminate against nonmembers.

- Because this model does not have a competitive bidding process, it is difficult to assess the level of subsidy required for the network.

The shared infrastructure model has been used to develop the Eastern Africa Submarine Cable System (EASSy), a submarine cable project established by a consortium of operators from the region and partially financed by a group of international financial institutions. The involvement of the international financial institutions was used to establish an open access model to ensure that access to the cable is available to all operators in the region, regardless of whether they are members of the consortium (box 4.5).

2. Competitive subsidy model

A competitive subsidy approach uses a competitive process to award a license to build and operate a backbone network based on government specifications. The government provides resources to the licensee through in-kind or cash payments. The contract specifies the terms under which backbone network services are provided, including the type, quality, and price. These are key aspects of the contractual design because they determine the network's impact on downstream users and have a major impact on how much investors are willing to pay to obtain a license.

This model has a number of variations, based on the ownership structure of the network. At one end of the spectrum of options is a network entirely owned by a private company that receives a government subsidy to build a network that meets the government's policy goals. At the other end is one where the public and private sectors are joint owners of the backbone network. In all cases, the contract to build and operate the network, as well as the associated license, is awarded competitively through a minimum-subsidy auction (Wellenius, Foster, and Malmberg-Calvo 2004).

The *advantages* of the competitive subsidy model are the following:

- The government achieves its goals while leveraging the private sector's skills, expertise, and investment resources.

- The private operator has a commercial interest in operating the network as efficiently and effectively as possible.

- This approach is simpler than the consortium approach because fewer parties are involved. If it does not succeed initially, there is recourse to alternative operators or alternative models.

- Similar approaches have been used to promote the rollout of rural access networks in sub-Saharan Africa. There is also relevant experience with similar structures from other sectors that could provide useful benchmarks.

The *disadvantages* are the following:

- Government support to specific operators may undermine competition.

- It can be difficult to obtain accurate information on the performance of licensees and to impose penalties for failure to deliver.

Box 4.5 A Shared Model for Backbone Infrastructure Development in East Africa

EASSy is a submarine fiber-optic cable from South Africa to Sudan, with connections to 10 countries along its route. The system's termination points will connect to the global communications network. The project has been developed by a consortium of more than 20 telecommunications operators, mostly from East and Southern Africa, with support from the International Finance Corporation, European Investment Bank, African Development Bank, Agence Française de Développement, and Kreditanstalt für Wiederaufbau.

The system has been designed to minimize the problems associated with the absence of effective competition and regulation. This is done through a special purpose vehicle that is a member of the consortium and owned by a group of smaller operators from the region. This special purpose vehicle is allowed to sell network capacity in any market in the region on an open-access, nondiscriminatory basis—providing competition to other members of the consortium. The agreements that established the special purpose vehicle require it to pass through to customers any cost savings arising from increased traffic volumes. These mechanisms for competition and pass-through of cost reductions are intended to lower prices and increase access.

Source: Author.

- If the backbone operator has any financial connection to downstream operators, it will have an incentive to discriminate in favor of them.

- It can be politically difficult to justify large public subsidies to private companies in which the government does not maintain an equity stake.

France provides an example of this type of public-private partnership used to develop backbone infrastructure (box 4.6).

3. Other incentive-based private sector models

All countries require operators to pay taxes and levies that typically consist of general taxes—applicable to all companies in the economy—and sector-specific taxes or levies. One common levy is a contribution to universal service or access funds. Such contributions are usually calculated as a percentage of revenues and are collected annually from operators. In most cases, these funds are intended to be used to subsidize access to services in rural areas. But in many countries they are not used effectively, often remaining undisbursed by the government and sometimes diverted for other uses.

Governments can give operators an incentive to develop backbone networks in commercially unattractive areas by offering to reduce these levies in exchange for the operators meeting specific targets. This can be done on a competitive basis—a limited number of companies are awarded the levy reduction and they have to compete for it, or it could be available to all. Such "pay-or-play" schemes are not common in the telecommunications sector but have recently been receiving increasing attention (box 4.7).

The *advantages* of other incentive-based private sector models are the following:

- Private companies own and operate the networks, increasing the likelihood that they will be managed efficiently and effectively.

- The government can specify the type of network that it requires and the terms on which services are sold.

- No cash changes hands between operators and the government.

- Government retains the option of penalizing any failure to meet obligations by removing the financial incentive (that is, making them pay, instead of play).

The *disadvantages* are the following:

- Any network built under such a scheme would be privately owned by operators competing in the market, and these operators would have strong incentives to discriminate against competitors. Thus, this option would require strong monitoring and regulation.

Box 4.6 A Public-Private Partnership for Backbone Infrastructure in France

Limousin is a rural region in central France with limited broadband services. To raise access to urban levels, the government launched the DORSAL project to develop a backbone network capable of delivering access to high-speed Internet. The project is structured as a public-private partnership with a 20-year concession to build and operate a backbone network and to construct a broadband wireless network using worldwide interoperability for microwave access (WiMAX) technology and capable of supporting high-speed, value added services. The project will cost 85 million euros, split between the public (45 percent) and private (55 percent) sectors. The fiber-optic backbone network was completed in mid-2007 and downstream competition has developed. Customers in the project area now have access to third-party service providers offering a wide range of broadband services, such as Internet protocol television (IPTV), voice-over-Internet protocol (VoIP), and high-speed data services, in competition with France Telecom.

Source: Ingénieurs Conseil et Economistes Associés 2008.

> **Box 4.7** Sweden's Incentive-Based Mechanisms for Developing Backbone Networks
>
> Since it launched its first broadband policy in 1999, the Swedish government has provided subsidies for broadband rollout through several programs, including tax incentives for operators building networks in rural areas and grants to municipalities to build fiber-optic networks. The total value of these subsidies is an estimated $820 million. This policy has been quite successful and a government survey in 2007 found that, taking into account both wireless and wireline access, Sweden was coming close to 100 percent coverage of broadband. However, a government-appointed committee in 2008 determined that 145,000 people and 39,000 businesses still did not have access to wireline broadband (i.e., fiber, DSL, or cable) and recommended that government spend another $500 million on grants to municipalities and operators to invest in high-speed networks.
>
> The financial incentives for infrastructure development provided by the government have been part of an overall package of policy measures used to promote broadband that includes stimulating competition, subsidizing network rollout in high-cost areas, encouraging municipalities to develop operator-neutral backbone networks, and promoting the use of state-owned businesses to develop fiber-optic infrastructure.
>
> **Source:** Atkinson, Correa, and Hedlund. 2008.

- Pay-or-play schemes may sometimes limit competition in a particular area since the winner will be operating with a government subsidy.
- It can sometimes be difficult to ascertain precisely the amount of financial support that the government is implicitly giving operators through a pay-or-play scheme.

Conclusion

As the pace of broadband development accelerates globally and economies adapt to better and more widespread connectivity, the importance of broadband connectivity will continue to grow. Thus, the widening gap between sub-Saharan Africa and other developing countries is a major policy issue for many countries in the region. Most incumbent operators in sub-Saharan African countries are not strong enough to be an effective backbone network of last resort. Thus, the model of market liberalization and regulation of access to the incumbent's network—which has been successful in the European Union, North America, and increasingly in Asia and Latin America—is not directly relevant in the region. The main challenges facing policy makers in the region are ensuring that entrants have access to existing infrastructure developed by private operators and that networks are built in areas where commercial operators are not currently willing to invest. Both objectives have to be achieved without discouraging the private sector from investing in network infrastructure.

This chapter has outlined a market-based approach to policy for the development of backbone networks in sub-Saharan Africa. This approach harnesses both the investment resources and operational expertise of the private sector to help meet public policy goals, reducing financial and operational burdens on the public sector. It also builds on the model of infrastructure competition that has been very successful in other segments of the communications market in sub-Saharan Africa. Thus this chapter is consistent with the general approach to ICT that has been adopted in most countries in the region.

Still, the detailed design and implementation of the public-private partnership models discussed here will require innovation by governments and regulators in the region. There are few clear, off-the-shelf examples from other parts of the world that can be directly transposed into the sub-Saharan African context. But this dearth of ready-made examples can be considered an opportunity rather than a problem. It provides policy makers with an opportunity to tailor policy solutions suited to their specific challenges.

Notes

1. Definitions of broadband vary, and no single definition is universally accepted. This chapter defines broadband as an Internet connection that is always on and provides a download speed of at least 256 kilobits per second.
2. Population-weighted average price of cheapest broadband package. World Bank staff calculations based on data from ITU (2007).
3. India's regulatory authority reported that the average cost of broadband was $12–$18 a month (assuming usage of three hours a day; TRAI 2006).
4. Simple average of monthly subscription prices in U.S. dollars in all OECD countries as of October 2007.
5. This estimate is based on discussions with operators.
6. This figure for terrestrial infrastructure is an underestimate because data were not available for some operators. In addition, the data in this section are for 47 sub-Saharan African countries but exclude South Africa because its backbone network infrastructure is highly developed and unrepresentative of countries in the region.
7. The key metrics used to measure the adequacy of terrestrial backbone network infrastructure are length (in kilometers) and capacity (in megabits per second). Satellite links are also measured in terms of capacity, but the distance between two nodes on the network is irrelevant.
8. This is changing, however. The commercial success of mobile operators in sub-Saharan Africa, increases in traffic arising from a growing customer base, and a shift in strategy toward more data services have led more operators to consider investing in fiber-optic networks that once were considered too financially risky.
9. Defined as the population living within 10 kilometers of a backbone network node.
10. This approach to backbone policy is analogous to the standard approach for analyzing telecommunications access in rural areas, where policies are designed to narrow two "gaps"—the market efficiency gap and the market access gap. The first gap is addressed by improving the functioning of the market; the second requires external financial support.

References

Atkinson, Robert D., Daniel K. Correa, and Julie A. Hedlund. 2008. "Explaining International Broadband Leadership." ITIF (Information Technology and Innovation Foundation), Washington, DC. http://www.itif.org/files/ExplainingBBLeadership.pdf.

Balancing Act. 2007. *Balancing Act News Update*, Issue 374. http://www.balancingact-africa.com/news/back/balancing-act_374.html.

Global Insight. 2006. "Sub-Saharan Africa: Celtel Launches 'One Network' in East Africa."

———. 2007. "The Shifting Geography of Broadband in Sub-Saharan Africa."

Government of Rwanda. 2006. "An Integrated ICT-Led Socio-Economic Development Plan for Rwanda 2006–2010. The NICI-2010 Plan." Kigali.

Goyal, Aparajita. 2008 "Information Technology and Rural Markets: Theory and Evidence from a Unique Intervention in Central India" Working paper. University of Maryland, College Park, MD.

Hamilton, Paul. 2007. "Analysis of the Extent of Development of Transmission Backbone Networks in Sub-Saharan Africa." Study commissioned by World Bank, Washington, DC.

Hanna, Milad, and Balaji Ramarao. 2006. "Cost Optimization for Transmission and Backhaul Technologies." Accenture. http://www.accenture.com/NR/rdonlyres/0000aae5/nwhotfppdswdgywuynxzagosezkmkgtr/acs_cost_opt_pov.pdf.

Ingénieurs Conseil et Economistes Associés (ICEA). 2008. "Strategies for the Promotion of Backbone Communications Networks in Sub-Saharan Africa." Study commissioned by World Bank, Washington, DC.

ITU (International Telecommunication Union). 2007. "Telecommunications/ICT Markets and Trends in Africa 2007." Geneva. http://www.itu.int/ITU-D/ict/statistics/material/af_report07.pdf.

Jensen, Robert. 2007. "The Digital Provide: Information (Technology), Market Performance and Welfare in the South Indian Fisheries Sector." *The Quarterly Journal of Economics* 122(3): 879–924.

Ministerio de Ciencia y Tecnología. 2003. "Por el que se aprueba el reglamento regulador de las infraestructuras comunes de telecomunicaciones para el acceso a los servicios de telecomunicación en el interior de los edificios y de la actividad de instalación de equipos y sistemas de telecomunicaciones." Madrid.

OECD (Organisation for Economic Co-operation and Development). 2006. "Internet Traffic Exchange: Market Developments and Measurement of Growth." Paris. http://www.oecd.org/dataoecd/25/54/36462170.pdf.

———. 2007. *OECD Broadband Portal*. Paris. http://www.oecd.org/sti/ict/broadband.

Ofcom (Office of Communications). 2004. "Review of the Retail Leased Lines, Symmetric Broadband Origination and Wholesale Trunk Segments Markets." London.

Ovum. 2005. "Recommendations on Further Liberalization of the Telecommunications Industry of Botswana: A Final Report to the Botswana Telecommunications Authority." London.

Roller, Lars-Hendrik, and Leonard Waverman. 2001. "Telecommunications Infrastructure and Economic Development: A Simultaneous Approach." *The American Economic Review* 91 (4): 909–23.

Telegeography. 2008. *Global Internet Geography.* http://www.telegeography.com/products/gig/.

TRAI (Telecom Regulatory Authority of India). 2006. "Study Paper on Analysis of Internet and Broadband Tariffs in India." Delhi.

Wellenius, Björn, Vivien Foster, and Christina Malmberg-Calvo. 2004. "Private Provision of Rural Infrastructure Services: Competing for Subsidies." Policy Research Working Paper 3365, World Bank, Washington, DC.

World Bank. 2002. "The World Bank Guarantee: Leveraging Private Finance for Emerging Markets." Project Finance and Guarantees Department Washington, DC.

———. 2008. *Africa Infrastructure Country Diagnostic for Sub-Saharan Africa, 2008.* Washington, DC.

Chapter 5

How Do Manual and E-Government Services Compare? Experiences from India

Deepak Bhatia, Subhash C. Bhatnagar, and Jiro Tominaga

Governments in both developed and developing countries are taking active steps to leverage the potential of information and communication technology (ICT) to improve the efficiency, effectiveness, and accountability of public sector organizations. In many countries, harnessing the power of ICT is a critical element of achieving goals to improve governance. As of 2003, 70–90 countries had national e-strategies, with e-government the most common area of focus (World Bank 2006).[1]

E-strategies generally define e-government as the provision of services and information by electronic means through an integrated conduit. A number of developing countries have adopted e-government, and while there are several success stories, a high failure rate—more than 50 percent—has also been reported (Heeks 2003b). In fact, ensuring robust performance by large-scale information systems has proven to be a challenge even for countries with highly sophisticated technical skills.

Given such a high level of risks and the substantial financial resources often required, investment decisions about e-government should be based on realistic estimations of the expected value and costs involved. But measuring the value of e-government is not a simple task. While the returns of ICT investments in the private sector can be measured by their contributions to the firms' profits, the value expected from ICT in the public sector accrues to stakeholders in many different ways and cannot always be measured in monetary terms. Thus, a first step toward more strategic ICT investment planning is to develop a framework for assessing the value derived from implementing e-government projects.

This chapter reviews recent efforts by various governments and researchers to define the value generated by ICT use in the public sector primarily in developed countries. Building on this understanding, the chapter presents an assessment framework that can be used to estimate the value delivered to different stakeholders from an e-government project. This framework of ex post assessment is then applied to five state- or municipal-level projects in India that used ICT to improve the quality of public service delivery, using data from a survey commissioned by the World Bank (Bhatnagar 2007). The assessment analyzes positive and negative changes reported by citizens on the costs of accessing public services and the perceptions of service and governance quality that came with the shift from manual to electronic service delivery in the five projects.

The chapter has three sections. The first reviews several existing frameworks for assessing ICT investment in public organizations and presents the argument for the framework proposed in this chapter. The second section assesses the five projects in India using the proposed framework. The final section summarizes findings, analyzes some of the reasons for the perceived impacts, and explores the application of the proposed framework to ex ante assessments.

Research Review: Assessing Public Sector ICT Projects

A significant number of academic and policy studies have been conducted on effective methods to evaluate public sector ICT projects. For example, United Nations Online Network in Public Administration and Finance (UNPAN 2005) and Brown University (2007) provide macro-level estimates of e-government activity using appraisal indexes focused on supply-side, quantifiable measures such as government Web presence, network coverage, institutional and regulatory support, and human capital provision. These assessments tend to focus almost exclusively on measuring physical access to certain types of ICT without incorporating issues such as affordability, appropriateness, ICT capacity and training, and the regulatory and macroeconomic environment (Bridges.org 2005).

A second group of studies, which have occurred in an unsystematic manner and are largely anecdotal, provide project-level evaluations with little prospect for synthesis from past approaches. This group includes a considerable number of case studies of ICT projects, often commissioned by international development institutions, research institutes, nongovernmental organizations (NGOs), and government agencies. A small number of studies in this group have focused on project impacts in terms of how they affect citizens and government agencies, though their findings have seldom been linked to broader national, regional, or global trends—for example, in terms of changes in the approach to development and governance policy (Grant 2005; Heeks 2003a; Madon and Kiran 2002).

Recent research on information systems and management has focused on the underlying change processes involved in introducing ICT. Of importance here is the evolving historical and institutional background in which ICT projects are implemented, as the organizational background has been shown to influence the process by which resources allocated for ICT projects lead to successful or unsuccessful systems (Avgerou and Madon 2004). The growing body of literature on enterprise information systems focuses on the process of change within projects. The information systems success model presented by Delone and McLean (1992) has been widely cited. This model is based on an analysis of 180 papers and proposes six major, interlinked impact variables: systems quality, information quality, information systems use, user satisfaction, individual impact, and organizational impact. Impact on individuals was later broken down into economic benefits and social benefits (Seddon and Kiew 1996).

The goals pursued by a public sector information technology (IT) system tend to be more diverse than those of a private sector information system. Public sector agencies are expected to achieve various objectives. Some of these are the same as the private sector goals—for example, administrative efficiency and operational effectiveness. But pursuit of other public sector goals, such as fairness, transparency, and support for public goods, adds complexity. Measuring the impact of ICT investments in the public sector, therefore, has been a difficult task.

A fundamental difficulty arises from the fact that for many of the goals pursued and costs incurred, it is difficult to apply market prices. In addition, measurement is hindered by the need for government agencies to deliver services to a wide range of constituents and by the complexities arising from cross-agency contributions to final service delivery (eGEP 2006). Thus, efforts to develop a framework for assessing the effect of ICT applications in the public sector need to encompass the benefits to multiple stakeholders, including both users and implementing agencies.

Policy makers around the world recognize these challenges. However, as ICT is penetrating the public sector rapidly, they need to find practical solutions that will help them make appropriate IT investment decisions and demonstrate reasonable returns on investment to constituents.

Diverse approaches have been taken thus far. For example, MAREVA (méthode d'analyse et de remontée de la valeur, or method of analysis of and value enhancement) developed by France's Agence pour le Développement de l'Administration Electronique (Electronic Administration Development Agency) with help from BearingPoint, relies on return on investment calculations to measure project impact. The WiBe Economic Efficiency Assessment methodology used by Germany's federal administration calculates economic efficiency in monetary terms, taking into account several aspects of IT project costs and benefits: project urgency, qualitative and strategic importance, and qualitative effects on external customers. Under the WiBe methodology, a measure is considered economically efficient if a positive capital value is achieved over the calculation period, usually five years for IT projects (Germany, Federal Ministry of the Interior 2004).

The U.S. government's Performance Reference Model builds on value chain and program logic models to reflect how value is created as inputs (such as technology) and are used to create outputs (through processes and activities), which in turn affect outcomes (such as mission, business, and customer results). And in Australia, the link between user demand and value generation is a key underlying principle of the Demand Assessment Methodology and Value Assessment Methodology. The methodologies are based on the realization that consideration of demand forces enables the government to assess services from the perspective of users.

All these government methodologies broadly focus on the same dimensions. They offer two levels of assessment: first, in terms of how projects provide a business case justification for expenditure and whether they meet the targets set for them, and second, in terms of how projects meet the goals of the agencies concerned and, in turn, how that helps achieve wider government strategies. These assessment frameworks are guided by the strategic outcomes pursued by governments, representing broad policy priorities that drive government activities.

The framework used in this chapter builds on these recent studies and government attempts to assess ICT projects in the public sector. The recent studies described above indicate the diverse nature of benefits pursued by the public sector through ICT investments. They also suggest the difficulty of quantifying the benefits of public sector services. There is a significant emphasis on the links between ICT investments and overall policy objectives. To address these challenges and needs, this chapter defines key stakeholders as users and service providers, specifies the intended benefits to them, and measures the direction of effects using surveys.

In defining expected benefits for users, reductions in costs of accessing services and improvements in the overall quality of services and governance—both of which are based on the perspectives of users—are key measures of performance. For agencies implementing projects (that is, service providers), increases in transactions and revenues and reductions in costs of providing services are key elements of better performance. The main indicators for these performance measures are summarized in table 5.1.

In addition, the design of the survey and the methodology for measurement add some unique features for e-government in developing countries. These features are:

- Services are delivered in public centers by operators across computerized counters connected to back-end databases. Because citizens do not access services on their own through portals, Internet surveys are not an option for collecting data on their experiences with the system. Thus, face-to-face surveys were conducted.

- Reducing corruption is an important goal of many e-government projects, so survey instruments incorporated ways of measuring payments of bribes in addition to other elements of performance.

- Since the quality of record keeping and management information systems is generally poor in the manual systems used by government agencies, a relatively high reliance on perceived changes (rather than disaggregated data) on costs, transaction volumes, and revenues is required.

Survey respondents were asked to compare current services with their recollection of manually delivered services. This, however, is evidently not the ideal way of measuring changes. The survey used in this chapter adopted this approach because the baseline data prior to the introduction of e-government were lacking. Some data on transactions and revenues before computerization were available from implementing agencies. However, information on operational costs was not fully available.

This assessment therefore extrapolated changes in operational efficiency using an additional survey of supervisors who make day-to-day decisions on the operations of the five projects. These supervisors were asked to provide their perceptions of general trends in matters such as employee workloads and the number of workers involved in delivery services.[2] The latter is particularly important because labor tends to be the largest cost in service-oriented activities.

Analysis of Five E-Government Projects in India

The following sections apply the framework described above to analyze five projects in India that aimed to improve the quality of public service delivery through e-government. The services provided by these projects vary—from issuance of land records for farmers, to registration of property sale and purchase deeds, to payment for public utility services, to payment of taxes. A common feature of these projects is that they are intended to improve public services through computerization of service delivery points. Because these are locations where citizens have direct contact with public

Table 5.1 Performance Measures and Indicators for ICT Projects for Users and Service Providers

Stakeholder	Performance measure	Key indicator
Users	Costs of accessing services	Number of trips required to complete the service
		Travel costs, based on distance to service delivery site
		Waiting times at service delivery site
		Payments of bribes to receive satisfactory services
	Quality of service and governance	User perceptions of service quality, transparency, and accountability (five-point scale)
Implementing agencies	Transaction volume and revenue	Growth in transaction volumes
		Growth in revenues
	Operating cost	Supervisor perceptions of cost impact

Source: Data from surveys.

sector organizations, they are quite significant in shaping citizen perceptions of public services. The five projects assessed in this chapter are as follows:

- *Bhoomi*. This project has been operational since 2001. At the time of the survey in 2006, 203 kiosks in the state of Karnataka were delivering two online services: issuance of records of rights, tenancy, and crop inspection register (RTC) and filing of requests for mutation (changes in land ownership) for affecting changes in land records.

- *Karnataka Valuation and E-Registration (KAVERI)*. KAVERI has been operational since 2003. By 2006, 201 offices of the subregistrar in Karnataka were delivering three key services: registration of property sale and purchase deeds, issuance of nonencumbrance certificates, and issuance of copies of previously registered deeds.

- *Computer-aided Administration of Registration Department (CARD)*. CARD was launched in 1998. By 2006, 387 offices of the subregistrar in Andhra Pradesh were delivering the same three online services provided by KAVERI.

- *eSeva*. In 2002, 45 eSeva centers became operational in Hyderabad, the capital of Andhra Pradesh. These one-stop service centers now deliver 135 services from central, state, and local governments and public utilities, and are used by 3.1 million people per month at 275 locations in 190 towns.

- *Ahmedabad Municipal Corporation (AMC) Civic Centers*. Launched in 2002, this project uses 16 civic centers to deliver three important services: annual payment of property taxes, issuance of birth and death certificates, and issuance of shop licenses.

The framework presented in the previous section was used to assess whether these e-government projects have improved key measures of performance for public services. The assessment is based on data collected through surveys of randomly selected users and project supervisors in the implementing agencies conducted between June and September 2006.

While ICT is an important component of these sample projects, associated efforts such as changes in administrative processes and introduction of more client-oriented public services are also integral parts of the projects. As mentioned earlier, this assessment largely relies on the survey of stakeholder perceptions of public services before and after these projects. It is not an evaluation of investments in narrowly defined ICT, which typically includes hardware, software, and communication equipment. Rather, it is a review of the outcomes of packaged efforts involved in the transition from manual services to those using ICT as a key component.

User Evaluations of Manual and E-Government Services

This section presents the data collected from the survey of users. Data from 240–250 randomly selected users in eight service centers were collected for each project. Questions were asked to encapsulate their experiences using the computerized service delivery systems as well as the previous manual systems. (Annex table 5A.1 provides a profile of the respondents.) Respondents were asked to answer about 120 questions related to the following:

- Costs of accessing services, including the number and cost of trips required to complete a service, the amount of bribes paid to complete the service satisfactorily, and the waiting time at the service center

- Overall quality of service, rated by a five-point scale, also including questions regarding likely attributes to the quality of service, for example, responsiveness of staff, convenience of office location, office hours, and facilities at the service center

- Quality of overall governance, for example, transparency, corruption, fairness of treatment, quality of feedback, and levels of accountability.

Respondents were also asked to rate the manual and computerized systems based on a common set of 18 attributes covering cost of access, convenience, and quality of service delivery and governance.

Cost of accessing services. Access costs are a key measure of the performance and accessibility of public services. In most cases, computerized services are perceived as improvements over manual services. The number of trips required to complete a service fell in all five projects, with an average reduction of 0.75 trip per user. CARD users claimed a decrease of 1.38 trips—the largest reduction among the five projects (figure 5.1).

E-government helped achieve these improvements in several ways. With manual services, users often were not sure when the service they requested would be delivered. As a result, they had to make several visits to the respective government office to see if the requested services had been completed. Computerization has made service delivery more predictable. E-government systemizes the work flow for delivering services, and automation makes each step more efficient. For example, in CARD, the process of making a manual copy of a 30–40-page registry document has been replaced by scanning, making it possible for the agency to adhere to promised delivery times. Moreover, delays at any stage can be easily monitored and corrective action taken.

Savings in travel costs depend on the number of trips taken and the distance traveled per trip. For example, in the Bhoomi project, the travel costs increased for many users after computerization even though the number of trips was reduced (figure 5.2). This is mainly because service delivery sites were moved from village offices to *taluka* (an administrative division consisting of a number of villages, towns, and possibly a city) headquarters. Consequently, many users must travel farther to receive services, substantially increasing travel costs.

Waiting times at service sites decreased in all five projects. For projects where the reduction in waiting time was statistically significant (only Bhoomi RTC was not statistically significant), the reductions ranged from 16.2 minutes (AMC) to 96.2 minutes (CARD; figure 5.3). With manual systems, users often had to wait in long lines to receive services. The e-government system has increased the efficiency of document processing.

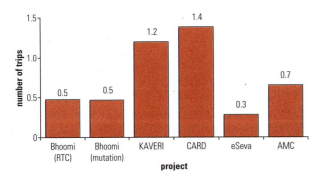

Figure 5.1 User Report of Number of Trips Saved, by Project

Source: Data are from surveys.

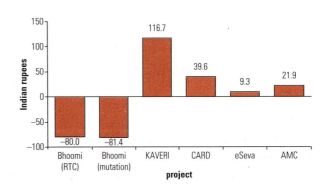

Figure 5.2 User Reports of Changes in Travel Costs, by Project

Source: Data are from surveys.

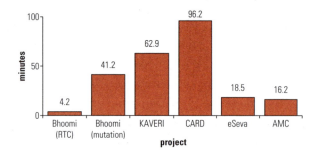

Figure 5.3 User Reports of Reductions in Waiting Times, by Project

Source: Data are from surveys.

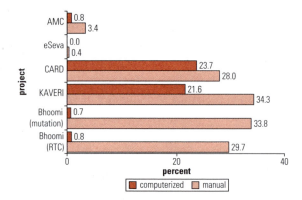

Figure 5.4 User Reports of Reductions in Bribe Payments, by Project

Source: Data are from surveys.

Bribes paid to officials are another cost for users, because bribes are often required to accelerate service delivery. Under e-government projects, fewer users need to pay bribes than under manual systems. The decline in bribery is more substantial in some projects than others (figure 5.4). For example, in the Bhoomi project, about 30 percent of users paid bribes under the manual system—while less than 1 percent pay bribes in the e-government system.

The share of respondents among KAVERI and CARD users who had to pay bribes continues to be high, despite computerization, at 22 and 24 percent, respectively (see figure 5.4). A more detailed study of one of KAVERI's delivery sites suggests that bribery tends to occur when the e-government system experiences a breakdown. In the specific case that was studied, there was a surge in property transactions in an area being developed as a residential estate. The e-government center had the capacity to handle 15–20 transactions a day, yet at peak load hundreds of clients a day were seeking to register their property—leading to a complete breakdown in e-government services and a surge in bribes for accelerated transactions. Computerized scheduling of property registration may have led to an orderly queue, but this feature was not part of the system design.

It is difficult to ascertain all the reasons for the sharp reduction in bribery in the Bhoomi project. But it appears that the introduction of a first-in-first-out system for handling service applications and an increase in transparency contributed to this success (box 5.1).

Though the frequency of bribe payments has declined, the average amount paid increased under every project (table 5.2). General inflation could be part of the reason, but higher bribes also suggest that the complexity of the favors requested through bribes may have increased under e-government.

A qualitative study of each project would provide better explanations for the variations in bribery across different projects. It should be noted here that estimates of bribes are less accurate than other results, because the sample of respondents paying bribes was small, resulting in higher standard errors in the bribe estimates.

The survey results presented in this section suggest that e-government services have generally lowered access costs for users. But the size of changes varies by project, and the differences are partly caused by project design. For example, travel costs increased significantly for users under the Bhoomi project, due to the change in location of service centers. This design was adopted in the initial phase until a reliable communication infrastructure could be built to provide services in rural areas—underscoring the binding nature of infrastructure limitations on the impact of electronic service delivery.

Quality of services. Users were asked to rate manual and computerized services on a five-point scale. They consistently rated the computerized service higher than manual service in overall quality, and the higher ratings were found to be statistically significant (figure 5.5). eSeva exhibited particularly significant improvement in service quality. Despite significant reductions in the number of trips, travel costs, and waiting times, KAVERI and CARD showed smaller improvement in perceived service quality.

Users were also asked to rate, using the same five-point scale, 18 common attributes encompassing broad aspects of service delivery for each project. These attributes included elements such as service delivery location, accessibility, convenience, cost, transparency, and service orientation.[3]

Box 5.1 Explaining the Bhoomi Project's Success in Reducing Corruption

The Bhoomi project recognized the existence of corruption (see Bhatnagar and Chawla 2007) and sought to reduce it by reengineering many processes. All mutation requests must be logged into the system by an operator, and such requests may subsequently be rejected after formally recording a reason. A first-in–first-out system was also introduced to handle mutation applications in the computerized system, thus removing the opportunity for officers to give an application priority that was available under the manual system. To minimize resistance from village accountants who stood to lose income from bribes, a fresh set of recruits was hired to operate the kiosks.

Bhoomi provides a complete audit trail—including biometric login—for any changes made to the database, to ensure that actions can be traced. Anyone can get the copy of the records of rights, tenancy, and crop inspection register (RTC) by paying a fee. Thus records become transparent, as it is easy for landowners to verify data on the landholdings of neighbors. In fact, farmers request nearly a quarter of RTCs, to verify the correctness of records for land they own. At some locations touch-screen terminals can be used to view records.

The Bhoomi project had a vision for how the scale and scope of services will expand over time. Although in the first phase farmers had to go to a specific Bhoomi kiosk to get the RTC, 11 centers have since been interconnected, and the data are also mirrored on a central server. In addition, the RTC can be issued by private kiosks authorized to connect to the central database. Bhoomi also exchanges information with KAVERI, making it mandatory for citizens who buy or sell rural land to get land records modified. It is expected that the enhanced data exchange will further increase transparency and accountability.

Source: Bhatnagar and Chawla 2007.

Table 5.2 User Reports of Bribes Paid to Functionaries and Intermediaries/Agents, by Project

System	Bhoomi RTC	Bhoomi Mutation	KAVERI	CARD	eSeva	AMC
Amount of bribes paid to officials (rupees)						
Manual	48.7	39.6	215.1	726.8	200.0	83.3
Sample size	71	46	81	65	1	8
Standard error	5.13	1.95	26.96	137.98	n.a.	18.32
Computerized	50	200	576	1,094	0	150
Sample size	2	1	51	55	n.a.	2
Standard error	n.a.	n.a.	78.149	224.380	n.a.	50.000
Other amounts paid to intermediaries/agents (rupees)						
Manual	39.4	46.1	91.9	256.4	0	77.3
Sample size	62	53	24	78	n.a.	11
Standard error	1.71	3.58	11.43	23.76	n.a.	7.87
Computerized	78.8	50.0	200.8	398.6	0	60.0
Sample size	4	1	8	72	n.a.	1
Standard error	41.25	n.a.	74.72	40.79	n.a.	n.a.

Source: Data are from surveys.
Note: n.a = not applicable.

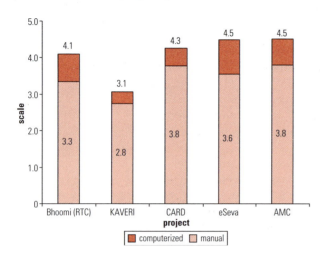

Figure 5.5 User Perceptions of Increased Service Quality, by Project

Source: Data are from surveys.

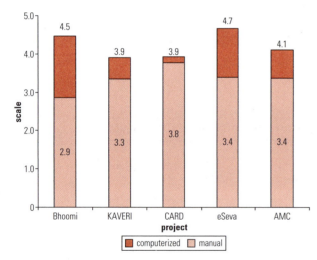

Figure 5.6 Changes in User Perceptions of Overall Composite Score, by Project

Source: Data are from surveys.

A weight was assigned to each attribute based on the proportion of respondents who selected the attribute, and weighted scores were calculated for each attribute as the product of the weight and the difference between the average scores for the attributes under manual and computerized systems. The sum of these differences denotes the improvement in composite scores of computerized service delivery.

The results indicate that the Bhoomi project shows the greatest improvement, despite its manual version having the lowest score (figure 5.6). eSeva also shows a significant improvement, as its computerized service centers are rated close to "very good" in the composite score. KAVERI and CARD indicate only marginal improvements over manual systems, consistent with the low ratings they received for overall service quality.

Differences in user satisfaction among projects may be explained by a variety of internal and external factors. In some projects IT-enabled improvements were closely aligned with the attributes of public service delivery considered important by users. For example, the survey asked respondents to list the three factors they considered most important for the respective services. The most frequently cited attributes were related to transactional efficiency, such as increase in convenience, reductions in visits and waiting times, better governance (e.g., less corruption), and better quality of service (e.g., reduced error rates and increased convenience) (table 5.3).

Users of KAVERI showed a strong preference for less corruption, increased transparency, error-free transactions, and shorter waiting times. CARD users also showed a preference for less corruption as well as for increased convenience, shorter waiting times, and fair treatment by officials providing services. But the design of the KAVERI and CARD projects provided little focus on reducing corruption. Thus, it is not surprising that perceived improvements in service quality were small in these two projects.

In addition to differences among projects, the survey results suggest variations in performance within projects. Earlier analyses of client impacts were based on the total number of respondents, typically surveyed in eight locations. Data were analyzed to understand if there were differences by location of service facility on a few key dimensions across all projects. For improvements in composite scores, there is significant variation among the KAVERI, AMC, and Bhoomi projects. Variations in the CARD project are small because improvements were marginal. In eSeva, the overall improvement was significant, but variations among service delivery sites are small.

Variations in service quality are largest among Bhoomi service kiosks. The key indicators of surveyed Bhoomi service kiosks are shown in table 5.4. Kiosk 4 had the largest deterioration in composite score, with users indicating 0.58 point deterioration in overall performance relative to manual systems.

Table 5.3 Users' Top-Four Desired Features of Services, by Project

Project	Features
Bhoomi	Error-free transactions, reduced delays in transactions, shorter waiting times, and fewer visits
KAVERI	Less corruption, increased transparency, error-free transactions, and shorter waiting times
CARD	Increased convenience, shorter waiting times, less corruption, and fair treatment
eSeva	Increased convenience, shorter waiting times, more convenient time schedules, and fair treatment
AMC	Increased convenience, less corruption, greater transparency, and good complaint handling system

Source: Data are from surveys.

Table 5.4 Variations in Service Quality across Seven Bhoomi Project Kiosks

Indicator	Location							Average
	Kiosk 1	Kiosk 2	Kiosk 3	Kiosk 4	Kiosk 5	Kiosk 6	Kiosk 7	
Number of mutations per year	4,860	6,880	9,888	15,564	54,132	11,491	32,050	16,861
Average distance of users to kiosk (kilometers)	21	15	23	21	12	15	12	17
Composite score difference (five-point scale)	0.14	1.17	−0.01	−0.58	2.14	2.68	1.92	1.60
Travel cost saved (rupees)	−148.9	−119.1	−140.0	−460.0	−79.6	−38.8	−33.8	−81.4
Waiting time saved (minutes)	−51.8	−30.0	45.0	−60.0	51.3	116.0	54.1	41.2
Preference for computerization (percent)	69.7	62.1	66.7	40.0	98.3	100.0	100.0	79.3
Power supply breakdown								
All the time	0	0	0	0	0	0	0	0
Often	42.9	0	50.0	33.3	0	0	0	20.8
Sometimes	28.6	100.0	50.0	66.7	100.0	75.0	100.0	66.7
Rarely	28.6	0	0	0	0	25.0	0	12.5
Never	0	0	0	0	0	0	0	0

Source: Data are from surveys.

Note: Data for kiosk 5 were collected from two centers located in the same district and a sample of 60 (twice that of the other kiosks).

The last row in table 5.4 shows the incidence of power supply breakdowns based on responses from three to four operators at each kiosk (half of all operators working at the kiosk). In kiosks 1, 3, and 4, which rate relatively poorly on composite scores and preference over manual systems, more operators reported that power supply breakdowns occur often. This suggests that variations in power supply contributed to the composite score rating and preference for e-government under the Bhoomi project.

Recognizing the significance of the quality of available infrastructure to users' perceptions of services, service delivery site operators were surveyed to understand the quality of infrastructure in these projects. Operators were asked about the frequency of system breakdowns, such as failures in power supply, connectivity, hardware, and software. The results are predictable, with urban projects such as AMC and eSeva showing better infrastructure than the Bhoomi project where centers are in semiurban areas (figure 5.7). This implies that KAVERI and CARD, which were rated relatively low on all the aspects assessed, are not especially prone to system failures. The Bhoomi project, which substantially reduced the incidence of bribery and had the largest improvements in user perceptions, experienced more technical problems than KAVERI and CARD. Although the quality of infrastructure is important, its impact on user perceptions of service delivery was largely contained.

Implementing Agencies of E-Government Projects

For implementing agencies, a major expected outcome of e-government is increased operational efficiency. This chapter examines trends in the amount of fees and taxes collected, the number of transactions, and operating costs. Tracing accurate cost and revenue data before the computerization turned out to be an extremely difficult exercise, since only limited efforts had been made by implementing agencies to produce such data systematically. In addition, because computerization projects are implemented as part of larger portfolios of public services delivered by implementing agencies, it was often difficult for the agencies to separate costs of individual initiatives. This lack of baseline data made it impossible to accurately identify status before computerization.

Despite the limitations, the information available at the time of this study is summarized in table 5.5. It suggests that both the number of transactions and the revenues from taxes and fees increased quite substantially in most cases. The data for KAVERI and CARD, however, were more complete with respect to the situation before computerization. In both projects, considerable increases were recorded for the year when computerized services started. Though some of the transaction and revenue increases might be due to e-government, it was not possible to establish an unambiguous relationship. For example, eSeva's integrated service centers provide multiple services from different government departments at one location. This makes it more convenient for users to pay utility bills, which is why the number of transactions increased so quickly. Manual systems required citizens to visit different service centers operated by individual departments. Similarly, the increased revenues collected by CARD and KAVERI could be the result of improvements in the collection of stamp duties and the introduction of more transparent, rule-based processes for valuing property.

Given the lack of comprehensive data, the study tried to gauge the effects of computerization on implementing agencies through a survey of 85 supervisors of the five projects. These supervisors manage day-to-day operations of the service centers and are in a good position to track how e-government affects operational efficiency. The survey asked supervisors their perceptions of changes in costs, governance, and work efforts, using a five-point scale. The results suggest that the supervisors surveyed did not perceive a significant decrease in key components of operating costs, namely, those related to labor, facilities, and communications. More than half of the supervisors said that labor costs either remained the same or increased (figure 5.8)[4] and that labor is the major component of operating costs for these services.

A separate review of KAVERI and CARD, however, reveals a different picture, suggesting that operational efficiency has improved in both projects. Total costs as a percentage of fee revenue fell from 6.5 percent to 4.2 percent between 2003/04

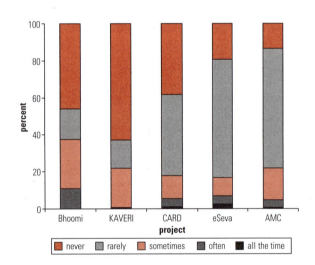

Figure 5.7 System Breakdowns, according to Service Delivery Site Operators, by Project

Source: Data are from surveys.

Table 5.5 Impact of Computerization on Agency, by Project (percentage of annual growth)

Indicator	Bhoomi		KAVERI	CARD	eSeva	AMC
	RTC	Mutation				
Tax revenues	n.a.		28.7	24.0	n.a.	44.7
Transaction fees	22.2	87.8	11.6	24.7	120.2	17.5
Number of transactions	22.2	87.8	10.6	16.8	87.7	38.0

Source: Data are from surveys.

Note: The years for current and base figures for tax revenues, transaction fees, and number of transactions vary for each project according to project starting date and data availability. Growth rates are annualized. n.a.= Not applicable.

Figure 5.8 Computerization-Driven Changes in Operating Costs for Implementing Agencies

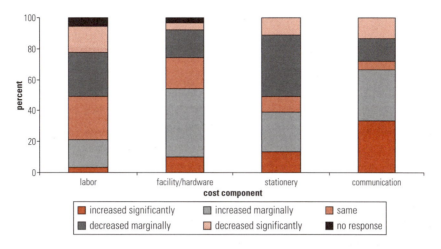

Source: Data are from surveys.

and 2005/06 for KAVERI, and from 5.2 percent to 3.7 percent between 2002/03 and 2004/05 for CARD. But available data are not sufficient to ascertain whether similar improvements occurred in the other projects.

Part of the ambiguous results in efficiency gain may be explained by the level of process changes implemented in these projects. Most supervisors reported that the extent of reengineering and integration of services was moderate during the computerization process (figure 5.9). Business process change is a challenging task for any organization, particularly when it involves workforce reductions. Moreover, public organizations generally have fewer incentives for cost efficiency than do private firms.

Still, process change is a key driver for increasing efficiency in any institution. It forces agencies to review workflow and approval procedures, with the goal of implementing better processes enabled by technology that was not available when existing processes were developed. Automating arcane, inefficient processes with IT would have little impact. The mixed efforts in business process change and the lack of systematic tracking of cost data may be rooted in the low priority that implementing agencies place on increasing operational efficiency.

Supervisor perceptions on changes in rent-seeking behavior corroborate user perceptions on bribery. An overwhelming proportion of supervisors think that corruption has decreased since computerization. They sense that discretionary power—through means such as denying services to citizens—has narrowed (figure 5.10). They are also more

Figure 5.9 Impact of Computerization on Business Process Change

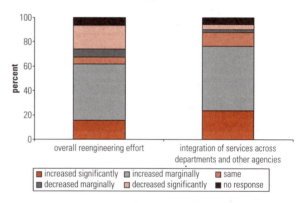

Source: Data are from surveys.
Note: For "overall reengineering effort," "increased significantly" implies significant reengineering and "decreased significantly" implies insignificant reengineering.

Figure 5.10 Impact of Computerization on Corruption and Service Denial

Source: Data are from surveys.

How Do Manual and E-Government Services Compare? Experiences from India 77

aware of the need to comply with service standards specified in citizen charters. Their perception of the increase in data accuracy is consistent with the reduction in error rates reported in the user survey.

Limitations of This Assessment and Areas for Further Research

This chapter assesses five e-government projects primarily through surveys of service users and providers. Though the study generated a rich data set providing insights on how these projects are functioning, there are limitations. First, the study of impacts on users relied on assessments of manual and computerized service delivery systems through a survey of users familiar with both systems. These users, however, had to rely on their memories when assessing the manual systems. In cases where computerized systems were introduced many years ago, memories could be inaccurate. And since four of the five projects have replaced manual delivery with computerized processes, most current users have no option of accessing services manually. Moreover, a lack of reliable baseline data on operational performance from implementing agencies prevented the study from identifying changes in the operational efficiency of the service providers. In addition, the surveys would have yielded more robust results had they used larger samples.

Second, a more detailed analysis of the overall enabling environment, which could influence the performance of the implementing agencies and delivery systems on the indicators used in this assessment, would have helped separate the effect of computerization from certain factors. These include the existence and quality of government strategies for e-government, the amount of political and popular support for improving public sector organizations, the overall macroeconomic situation, and access to telecommunication and IT services.

While recognizing these limitations, the study was able to establish that a very large number of users prefer computerized service delivery systems and provide some explanations for this preference. The five projects analyzed succeeded in reducing some key costs for users in accessing public services, including the number of trips to and length of waiting times at service delivery locations (table 5.6; see annex table 5A.2 for more detailed results). The frequency of paying bribes to service officials has also fallen. But in some cases, travel costs have increased substantially because the number of service delivery points fell after computerization.

Users appear to recognize the improvement of computerized over manual services, as they consistently rated the computerized services to be of higher quality. While users' desired attributes of services related to the five projects differed, the attributes cited most often were increased transactional efficiency and improved governance and quality. Although the revenues of implementing agencies increased, available information makes it uncertain whether this change is due to computerization. And based on the survey of agency supervisors, computerization has had an ambiguous effect on operational efficiency. Supervisor observations that room for discretion and corruption decreased since computerization corroborate the findings on user behavior—that is, fewer bribe payments.

Some of these benefits of computerization were achieved across all five projects, which were implemented independently in diverse environments. Although the results of the study should be interpreted with caution, they point to the potential of ICT use in improving public service delivery. For further confirmation of this potential, the costs and benefit of ICT investments need to be studied in more projects and in different countries.

Given the experiences and lessons from the assessment in this chapter, it is suggested that future attempts to measure

Table 5.6 Summary Results of User Survey, by Project

Indicator	Bhoomi RTC	Bhoomi Mutation	KAVERI	CARD	eSeva	AMC
Number of trips saved	0.48	0.47	1.20	1.38	0.29	0.65
Travel costs saved (rupees)	−79.96	−81.38	116.68	39.63	9.34	21.85
Waiting times saved (minutes)	4.23	41.21	62.92	96.24	18.50	16.16
Difference in overall service quality score (five-point scale)	0.76	0.95	0.32	0.48	0.95	0.70

Source: Data are from surveys.

the impact of e-government explicitly incorporate two additional dimensions. First, they should include more comprehensive analysis of the enabling environment and the external factors that have potential effects on outcomes; and second, they should provide a thorough examination of project components to clarify interventions that generate positive outcomes. The proposed conceptual framework, summarized in table 5.7, is quite similar to the hierarchical layers used in the measurement frameworks developed by some governments presented in the research review of this chapter.

Conclusion

This chapter has tried to ascertain the size and direction of impacts of selected electronic service delivery projects in India. The literature review of impact assessments revealed that frameworks used by several governments should focus more on the impact that ICT can bring to implementing agencies. Yet the beneficiaries of electronic service projects are primarily citizens and businesses that use them. Thus, this chapter proposes a framework that reflects the benefits and costs for users as well as implementing agencies.

The assessment framework was used to undertake ex post assessment of ongoing projects. The basic concept should be applicable for ex ante evaluation, which aims to arrive at realistic estimates of the nature and scale of improvements anticipated from planned projects. Ex ante evaluation would require that certain assumptions be made. For example, the extent and adoption rate of user demand need to be estimated. Project designs and investment levels must be commensurate with anticipated benefits to users. In addition, the quality of the enabling environment should be assessed to gauge the expected impacts of planned projects (see table 5.6). Detailed surveys at the initial stages of project planning, to establish the preproject costs of accessing services and the quality of services, would eventually help establish a baseline from which improvements can be projected and future monitoring processes designed.

This chapter has identified several factors that could be critical for successful e-government projects in developing countries. First, it showed that service centers played an important role in determining success in the sample projects, due to the fact that most service users in the projects analyzed do not have Internet access at home. Thus, service delivery through computerized service centers is essential to reach

Table 5.7 An Assessment Framework for E-Government

Goal	Example of performance indicators	Key feature of the enabling environment
Outcomes		
Increased efficiency	Financial and time savings in government activities	Overall e-government strategies
Increased transparency and accountability	Public perceptions, such as user satisfaction and score cards	Political and popular support for cross-agency coordination and public sector reform
Higher-quality public services	Financial and time savings for citizens	Telecommunications infrastructure and cost structures for increasing ICT access
Better access to services	Increased public service timeliness and responsiveness	Supportive legal and regulatory frameworks
	Reduced errors	Balances with competing priorities (such as roads and education)
	Financial saving per transaction	Macroeconomic changes
Outputs		
Reengineered processes	Comparisons of old and new business processes	
New ICT systems	Technical reviews of IT infrastructure, applications, and performance	
Increased service coverage	Variety of available services	
	IT support capacity	
	Service training	

Source: Data are from surveys.

Annex: Information from User Surveys on E-Government Projects in India

Table 5A.1 Profile of Respondents to User Surveys on E-Government Projects

	Bhoomi	KAVERI	CARD	eSeva	AMC
Number of respondents	242	237	232	253	239
Nature of clients	Farmers	Property owners	Property owners	Urban dwellers	Urban dwellers
Education					
Illiterate	36.78	27.00	9.91	3.95	2.09
Schooled	52.07	55.70	58.62	57.71	63.18
Graduate	11.16	17.30	31.47	38.34	34.73
Profession					
Workers	90.91	69.20	38.79	33.99	17.15
Business	4.13	12.24	34.91	27.27	45.19
White collar	1.65	6.75	10.34	22.53	17.15
Supervisor	3.31	11.81	15.95	16.21	20.50
Average annual income (rupees)					
Less than 5,000	92.15	71.73	43.10	40.71	15.07
5,000–10,000	5.79	19.83	44.40	42.69	65.75
More than 10,000	2.07	8.44	12.50	16.60	19.18
Location					
Urban	28.93	32.49	70.26	100.00	100.00
Rural	71.07	67.51	29.74	0.00	0.00

Source: Data are from surveys.

Table 5A.2 Summary of Findings of User Surveys on E-Government Projects

| Indicator | Bhoomi | | KAVERI | CARD | eSeva | AMC |
	RTC	Mutation				
Number of trips saved	0.479	0.473	1.200	1.384	0.285	0.654
Standard error	0.139	0.152	0.119	0.115	0.089	0.145
Significance	***	***	***	***	***	***
Travel cost saved (rupees)	−79.958	−81.381	116.684	39.632	9.342	21.853
Standard error	5.083	11.305	18.103	6.317	2.228	5.256
Waiting time saved (minutes)	4.230	41.206	62.915	96.240	18.498	16.164
Standard error	9.433	9.149	7.003	7.950	1.642	1.579
Significance	Not significant	***	***	***	***	***
Difference in overall service quality score (five-point scale)	0.756	0.946	0.316	0.475	0.947	0.700
Standard error	0.067	0.078	0.037	0.046	0.044	0.049
Significance	***	***	***	***	***	***

Source: Data are from surveys and authors' analysis.
Note: *** Significant at the 99 percent confidence level.

target audiences. This situation is similar in many other developing countries; how the service centers in the sample projects are operated might be a model that other countries should consider. India's government is implementing a National E-Government Program at an estimated cost of $5.5 billion to address the access issue (Seshadri 2007). Part of this ambitious plan is to create 100,000 citizen service centers in rural areas by expanding networking infrastructure.

This chapter's assessment also suggests that in order to ensure public support, e-government project design incorporate attributes considered important by users. Thus, efforts to analyze user needs and demand should precede e-government project design. The quality of supporting infrastructure—such as power supply—is also important to the success of e-government projects. In the five Indian projects studied, service centers with more frequent power outages were rated lower by users. Power supply capacity is often taken as a given by e-government project planners because it is not possible to dramatically improve it in a short period. Additionally, the level and quality of Internet service are fixed in many areas. Project design should be adjusted to such given conditions. If satisfactory outcomes from electronic services cannot be expected due to limited capacity of supporting infrastructure, alternative methods based on manual procedures should be explored.

Finally, data from users and service operators indicate that the five Indian e-government projects studied likely helped reduce corruption in service delivery centers. There can be many reasons for this outcome, but reductions in discretionary power appear to make a major contribution. Further studies on the extent of reengineering, improvements in supervision, and other environmental factors can help identify features that need to be built into the design of e-government systems to reduce bribery.

Notes

1. According to World Bank (2006), the United Nations Development Programme estimates the number to be 90, while McConnell International puts it at 70.
2. Directly accounting operating costs before and after computerization was the preferred approach for measuring cost reductions, though there was no reliable way to collect cost data before computerization. It was also difficult to properly account for the costs of different services.
3. Common attributes included convenience of the location of service delivery centers or offices; convenience of working hours; simplicity of the design and layout of forms; clarity of language used in forms and other written communication; clarity and simplicity of processes and procedures; time and effort required to receive service; courtesy, helpfulness, and knowledge of staff; cost of receiving service; incidence of error-free transactions; level of corruption; fair treatment; level of transparency; system for handling complaints; accountability of officers; quality of service area facilities; efficiency of queuing systems; and protection of confidential data.
4. About one-third said that labor costs have decreased, while a quarter said that they have increased. The remainder said that labor costs have not changed.

References and Other Resources

Accenture. 2002. "2002 eGovernment Leadership—Realizing the Vision." Third Annual eGovernment Benchmarking Report, Accenture, Dublin.

Ahuja, A., and A. P. Singh. 2006. "Evaluation of Computerization of Land Records in Karnataka." *Economic and Political Weekly* (January 7): 69–71.

Andhra Pradesh IT and Communications Department. 2004. "Profiles of IT Projects." Andhra Pradesh IT and Communications Department, Hyderabad.

Avgerou, C., and S. Madon. 2004. "Framing IS Studies: Understanding the Social Context of IS Innovation." In *The Social Study of Information and Communication Technology: Innovation, Actors, and Contexts*, eds. C. Avgerou, C. Ciborra, and F. Land, 162–84. Oxford, U.K.: Oxford University Press.

Bhatnagar, Subhash. 2006. "Electronic Delivery of Citizen Services: Andhra Pradesh's E-Seva Model." In *Reinventing Public Service Delivery in India*, ed. Vikram K. Chand, 95–124. New Delhi: Sage Publications.

———. 2007. "Impact Assessment Study of Computerized Service Delivery Projects from India and Chile." Working paper, World Bank, Washington, DC.

Bhatnagar, Subhash, and Rajeev Chawla. 2007. "Online Delivery of Land Titles to Rural Farmers in Karnataka." In *Ending Poverty in South Asia: Ideas That Work*, eds. Deepa Narayan and Elena Glinskaya, 219–43. Washington, DC: World Bank.

Bridges.org. 2005. "E-ready for What? E-readiness in Developing Countries: Current Status and Prospects Toward the Millennium Development Goals." http://www.bridges.org/publications/18.

Brown University. 2007. "Global E-Government 2007." Taubman Center for Public Policy, Brown University, Providence, RI.

Centre for Electronic Governance. 2004. "Evaluations of E-Governance Projects." IIMA (Indian Institute of Management), Ahmedabad. http://www.iimahd.ernet.in/egov/documents/evaluations-of-egovernance-projects-af.pdf.

Commonwealth of Australia. 2004. "Demand and Value Assessment Methodology for Better Government Services." Information Management Office, Canberra.

Delone, W. H., and E. R. McLean. 1992. "Information Systems Success: The Quest for the Dependent Variables." *Information Systems Research* 3: 60–95.

eGEP (eGovernment Economics Project). 2006. "Measurement Framework Final Version." Prepared for the eGovernment Unit, Directorate General, Information Society and Media, European Commission, Brussels.

Germany, Federal Ministry of the Interior. 2004. "WiBe 4.0—Recommendations on Economic Efficiency Assessments in the German Federal Administration, in Particular with Regard to the Use of Information Technology." Federal Ministry of Interior, Berlin, Germany.

Grant, G. 2005. "Realizing the Promise of Electronic Government." *Journal of Global Information Management* 13 (1): i–iv.

Harmon, E., S. C. Hensel, and T. E. Lukes. 2006. "Measuring Performance in Services." *The McKinsey Quarterly* 1: 1–6. http://www.mckinseyquarterly.com/Measuring_performance_in_services_1730_abstract.

Heeks, R. 2003a. "Most e-Government-for-Development Projects Fail: How Can Risks Be Reduced?" www.sed.manchester.ac.uk/idpm/research/publications/wp/igovernment/documents (accessed November 26, 2007).

———. 2003b. "Success and Failure Rates of E-Government in Developing/Transitional Countries: Overview." University of Manchester, IDPM (Institute of Development Policy and Management), U.K. http://www.egov4dev.org/sfoverview.htm.

Lanvin, B. 2005. "METER e-Strategies Monitoring and Evaluation Toolkit." In *E-Development: From Excitement to Effectiveness*, ed. Robert Schware. Prepared for the World Summit on the Information Society, Tunis, 16–18 November.

Lau, E. 2005. "Electronic Government and the Drive for Growth and Equity." Paris, Organisation for Economic Co-operation and Development.

Lobo, A., and S. Balakrishnan. 2002. "Draft Report Card on Service of Bhoomi Kiosks." Public Affairs Centre, Bangalore, India.

Madon, S., and G. R. Kiran. 2002. "Information Technology for Citizen-Government Interface: A Study of FRIENDS Project in Kerala." Department of Information Technology, Trivandrum, India.

OECD (Organisation for Economic Co-operation and Development). 2003. "The e-Government Imperative." OECD E-Government Studies, OECD, Paris.

Panneervel, P. 2005. "Urban e-Governance." In *E-Governance: A Change Management Tool*, 131–59. Jaipur, India: Rawat Publications.

Rama Rao, T. P., V. Rao, S. C. Venkata, Bhatnagar, and J. Satyanarayana. 2004. "E-Governance Assessment Frameworks." Center for Electronic Governance, Ahmenabad, and National Institute for Smart Governance, Hyderabad. http://www.mit.gov.in/download/NISG_EAF_18-05-04.pdf.

Satyanarayana, J. 2000. "Computer-aided Registration of Deeds and Stamp Duties." In *Information and Communication Technology in Rural Development*, eds. Subhash Bhatnagar and Robert Schware, 48–65. Washington, DC: World Bank.

Seddon, P. B., and M. Y. Kiew. 1996. "A Partial Test and Development of Delone and McLean's Model of IS Success." *Australian Journal of Information Systems* 1: 90–105.

Seshadri, T. K. 2007. "e-Governance: NeGP" (National E-Governance Action Plan). Department of Information Technology, Delhi. http://www.mit.gov.in/default.aspx?id=115.

3i Infotech Limited. "Summary Assessment Report of e-Governance Project: Bhoomi, Karnataka." Department of Information Technology, New Delhi, India.

Tripathy, R. K., and Dilip Mahajan. 2004. "City Civic Centres and E-Governance." Ahmedabad Municipal Corporation, Ahmedabad, India.

UNPAN (United Nations Online Network in Public Administration and Finance). 2005. *E-Government Readiness Report*. New York: United Nations.

World Bank. 2006. *Information and Communications for Development: Global Trends and Policies*. Washington, DC: World Bank.

Chapter 6

National E-Government Institutions: Functions, Models, and Trends

Nagy K. Hanna and Christine Zhen-Wei Qiang
with Kaoru Kimura and Siou Chew Kuek

The term *e-government* (electronic government) refers to the use of information and communication technology (ICT) to enhance the range and quality of public services to citizens and businesses while making government more efficient, accountable, and transparent (Schware 2005). E-government goals may include improving the following:

- Management of public finances, human resources, and service delivery

- Access to and quality of public services, particularly for poor people (World Bank 2003)

- Investment climates, including lowering regulatory burdens and business-to-government transaction costs

- Government transparency and accountability.

Different e-government programs give priority to some goals over others, in line with national development priorities.

This chapter focuses on national institutions responsible for leading, promoting, coordinating, and facilitating e-government programs. Given the many challenges of moving e-government programs from aspirations to development results in view of the fast-changing nature of technology and the need to constantly adapt to the changing expectations of different stakeholders, this chapter identifies the functions, models, and trends of e-government institutions responsible for translating vision into reality. It provides a survey of current e-government institutional practices and takes steps toward systematically assessing institutional options and innovations.

Context matters in institutional development (North 1990), and no single institutional model will fit all countries—developing or developed. Although governments share common challenges, they are starting from very different points in e-readiness and administrative development. Thus, they need solutions adapted to different circumstances. In addition, evidence about the effectiveness and impact of alternative institutional arrangements is emerging only slowly. This chapter focuses on identifying basic institutional models, and their strengths and weaknesses, rather than on prescriptions for best practices. Much can be learned from good examples around the world, emerging trends, and systematic assessment of available options.

This chapter first highlights the importance of institutional development for e-government programs. It then identifies strategic institutional design issues in leading and coordinating e-government. To help analyze trends in institutional arrangements, the chapter outlines key functions of effective e-government institutions. Next, the chapter identifies four basic models that countries have used to fulfill these functions and compares the models' strengths and weaknesses. It reports on emerging trends in the adoption of institutional models and the use of institution-building methods. The chapter concludes by emphasizing both the importance of best fit based on country circumstances and the need for continued institutional innovation.

Why Is Institutional Development So Important for E-Government?

ICT is a useful tool that can enable public agencies to change from routine-based, command-and-control organizations that are inwardly focused on administration to knowledge-based, networked, learning organizations that are externally focused on service (OECD 2005). This shift requires changes not only in front-end transactions and delivery of services to clients but also in integration and reengineering of back-end and core business processes in and across government agencies.

That ICT can assist in such development efforts does not mean that it will inevitably do so, or that it is easy to realize the potential benefits. ICT is expensive and complex. It is also a "disruptive technology." It changes the ways bureaucracies organize and work, power is distributed or controlled, and information is shared or protected. E-government projects have a mixed record (Heeks 2003). The main barriers are institutional—lack of leadership and the capabilities needed to leverage ICT for development strategies and to integrate ICT investments with organizational, process, and skill changes.

Moving to e-government is a major transformational and change management exercise. It entails a managerial revolution and an institutional and political reform process facilitated by technology (Rubino-Hallman and Hanna 2006). Competent leaders and empowered institutions are needed to overcome resistance to process and organizational changes, prioritize and manage complex investments, change skills and mind-sets, coordinate across multiple agencies and project portfolios, avoid duplicate efforts, leverage economies of scale, and maintain a long-term vision of transformation while insisting on concrete short-term results.

Many countries have made unsuccessful attempts to deliver e-government programs. This is largely because they lacked adequate institutional mechanisms for the programs' creative design, effective implementation, objective evaluation, and continual adaptation (Schware 2005). Even though institutions play a decisive role in the formulation and implementation of e-government strategies and programs, they are often treated as an afterthought. Some countries have ignored the need to create umbrella agencies to coordinate highly interdependent e-government activities. Others have lacked a clear division of responsibilities between various government branches and agencies, creating political and bureaucratic obstacles for e-government and inhibiting the proper allocation of resources and policy coordination across government. Yet others have overcentralized e-government management under a single agency or ministry, contributing to a separation between ICT policy and investment decisions and mainstream development issues.

Governments, supported by donors, have often resorted to creating project implementation units to carry out new investment programs, including e-government. The underlying assumption is that e-government development is a one-off project or a blueprint that can be designed by international consultants and implemented by a temporary project unit created specifically to follow the accountability and governance requirements of the donor. Lacking a vision of the leadership and institutional capabilities required for sustainable development, such project implementation units often reduce or crowd out (rather than complement) already weak state capacity (Fukuyama 2004). Different donors may work with different ministries and place their project implementation units within those ministries—reinforcing isolation, fragmentation, and duplication of e-government networks and applications.

Developing e-government, however, is a process, not a product or a blueprint. It is a continuous process of policy development, investment planning, innovation, learning, and change management (Fountain 2001; Ramsey 2004). This process must fit with and respond to a dynamic development strategy that supports evolving national goals and creates sustained institutional reforms and public service improvements. The challenge is to build effective governance and institutional frameworks for ICT-enabled public sector modernization and make the new competencies part of the country's human and institutional resources.

Moreover, the institutional culture and governance frameworks of the public sector sometimes do not fit well with the aims of innovation and transformation and the modalities of an integrated approach to e-government. Current arrangements often emphasize stability, a silo mentality, an inwardly focused bureaucracy, separation between the public and private sectors, and the isolation of technology managers from mainstream public policy leadership. These gaps need to be bridged.

At the same time, institutional development is path dependent (North 1990). Countries must deal with their institutional legacies while adapting and innovating new ones. The design of e-government institutions should be guided by a deep understanding of the political economy of reform and modernization. This is bound to be a long-term process that involves experimentation, learning, and adaptation (Rodrik 2004).

Thus, institutional changes and innovations are needed to manage the cross-cutting nature of e-government activities in fundamental, unprecedented ways. Strong leadership, governance, and organization make it possible for economic and social systems to function effectively during periods of change and transformation. They provide the strategies, implementation methods, coordination tools, and monitoring and evaluation mechanisms that enable innovative efforts to be undertaken and scaled-up programs to be successful. Specialized institutions and new competencies are required to create, acquire, adapt, and diffuse technologies and to synchronize them with associated policy reforms, intangible investments, managerial innovations, and organizational changes.

What Strategic Issues Arise When Designing E-Government Institutions?

Countries have created various institutional arrangements to cope with the governance issues and coordination challenges posed by e-government. These include shifting from one model to another, experimenting with hybrids, and developing entirely new models. Still, countries share the same basic choices and considerations:

- *Integration with development.* What kinds of institutional arrangements are needed to integrate e-government with a country's development strategy and state modernization? What role should central ministries (finance, planning, or economy) play in the process? Which policy makers should decide on e-government investments that are congruent with national development policies and goals?

- *Synergies between e-government and the rest of e-development.*[1] What kinds of institutional leadership and networks are needed to tap the synergies among e-government, telecommunications infrastructure, ICT literacy and human resources, ICT as a sector or core competency, and ICT as an enabler or productivity driver for all sectors of the economy?

- *Coordination across e-government components.* How should governments coordinate and balance their ICT-enabled transformation? How can the technological imperatives of building a common enterprise architecture be reconciled with the need to empower agencies and ministries to articulate their service priorities, implement their ICT-enabled service transformations, and integrate ICT with their sector strategies? How can public leaders achieve client-centered public services that span agencies and ministries? Beyond coordination, what incentives and institutional frameworks could encourage collaboration?

- *Degree of centralization.* How much should governments centralize or decentralize planning and decision making in e-government investments? Which elements of e-government are amenable to central direction and coordination, and which are best left to bottom-up initiatives and decentralized innovation? What institutional arrangements are needed to promote both bottom-up innovation and top-down reforms, and to enable scaling up of successful local e-government initiatives? How can e-government institutions enforce this optimal level?

- *Fit with institutional architecture and capabilities.* How should new e-government institutions and capabilities be designed to fit with—or perhaps transform—a country's political culture and institutional structures? For example, what kinds of institutional arrangements and capabilities would be most conducive to building effective partnerships among government, the private sector, academia, and civil society? What role should be played by the ministry currently responsible for ICT? How much authority and autonomy should be given to a central coordinating ICT agency?

What Are the Key Functions of Effective E-Government Institutions?

The analysis in this section is based on a review of national approaches to e-government leadership (box 6.1) in 30 developing and developed countries (see annex 6A

> **Box 6.1** The Functions of E-Government Institutions
>
> The key functions of central leadership institutions are grouped into three areas: strategy and policy making, governance and coordination, and facilitation of e-government implementation.
>
> The area of *strategy and policy making* aims to ensure that e-government goals, policies, and strategies are consistent with the country's overall development and state modernization objectives. Having an interministerial committee chaired by the head of state indicates that high priority and leadership are given to e-government strategy and policy making.
>
> *Governance and coordination* functions include the following:
>
> - Developing governmentwide information infrastructure, shared networks, data centers, business processes, and one-stop service delivery centers
>
> - Formulating e-laws and frameworks for IT (information technology) governance
>
> - Mobilizing, prioritizing, and allocating resources for e-government infrastructure and services
>
> - Monitoring, evaluating, and communicating lessons of experience, providing feedback, and ensuring accountability.
>
> *Facilitation of e-government implementation* is handled in most countries by a single ministry responsible for e-sector (vertical e-government) applications. This study does not track implementation of these single-sector e-government applications. However, ministries involved with vertical or sectoral applications often need technical support to implement their e-services. Common support services—such as IT human resources development, public-private partnerships, and IT procurement and contract management—may be provided by a specialized central agency. In some countries, ICT agencies or councils of chief information officers (CIOs) help share experiences and lessons across ministries.
>
> **Source:** Authors' analysis.

for a list of the countries). The review was shaped by three questions:

- What is the country's arrangement for e-government strategy and policy making?

- What is the country's approach to e-government governance and coordination?

- How does the country facilitate e-government implementation?

Strategy and Policy Making

In more than two-thirds of the countries studied, e-government strategy and policy are coordinated by an interministerial committee—often led by the head of state and part of his or her (or the cabinet's) office. Examples include China's State Council Information Leading Group, the Republic of Korea's Presidential Committee on Government Innovation and Decentralization, Kenya's Directorate of e-Government, Mexico's President's Office for Government Innovation (box 6.2), Pakistan's National E-Government Council, Tunisia's E-Government Ministerial Committee, and the United Kingdom's Office of the e-Envoy and subsequently e-Government Unit.

These committees formulate e-government strategy and policy and direct their implementation across ministries and agencies. Though these entities are rarely granted executive powers, they act as independent bodies for strategic oversight and policy coordination for a range of ministries. Other institutions remain responsible for implementing specific components of the national e-government plan.

To develop e-government strategies and policies, countries must rigorously analyze their development and state modernization priorities and encourage active participation by all major stakeholders. E-government is a highly dynamic process, with constant innovations in technologies, applications, products, and processes. It cannot be pushed or defined solely by government. Institutional frameworks should provide opportunities for all major stakeholders—government, the private sector, academia, and civil society—to build mutual understanding and

> **Box 6.2 Kenya's and Mexico's Experiences with Formulating E-Government Strategy Using Interministerial Steering Committees**
>
> Kenya makes e-government a priority in its development agenda. The Directorate of e-Government, located in the Cabinet Office under the Office of the President, formulates the country's e-government strategy. This directorate is chaired by the Secretary for e-Government and is the secretariat to two interministerial committees: the Cabinet Standing Committee and the Permanent Secretaries' Committee.
>
> The Cabinet Standing Committee is chaired by the Minister of Public Service and provides the political leadership needed for the e-government program. The Permanent Secretaries' Committee is made up of permanent secretaries from all ministries and provides the institutional ownership and support needed to marshal staff and resources to expedite e-government implementation.
>
> In Mexico, the President's Office for Government Innovation sets the direction for e-government and coordinates e-government activities within the framework of the country's Good Government Agenda. The office provides political support and leadership for e-government, which includes establishing an e-government network and ensuring broad participation across agencies.
>
> In late 2005, an interministerial commission—drawing on 50 ministries and departments—was created for e-government development. The commission serves as a governance and regulatory framework in support of the national e-government strategy. The Executive Council under the commission sets policies and standards and is responsible for coordinating, implementing, and operating horizontal initiatives (such as gateways and citizen portals).
>
> **Sources:** Authors' analysis; Evalueserve 2007.

provide input into e-government strategies and policies. This is especially important given the size and interdependencies of e-government investments, innovation efforts, and spillover effects for major stakeholders.

E-government evolves along with a country's needs and implementation capabilities. Therefore, to ensure continuity as well as adaptation, the strategy formulation process must be institutionalized. Institutionalization is also needed to secure ownership and commitment to the strategy adopted and to translate shared visions and strategy documents into actions. Links to development can be forged only when the e-government strategy process is driven by institutional mechanisms that engage and coordinate potential e-government users from all sectors of the economy.

Governance and Coordination

Organizing e-government involves assigning responsibilities for governing, coordinating, prioritizing, and monitoring e-government programs and activities. Given that e-government has emerged relatively recently as a national issue and given its pervasive impact, many countries have made e-government a specific portfolio in order to ensure that shared infrastructure is in place, e-government applications are prioritized, adequate resources allocated to agencies, interoperability promoted through common standards, and outputs and outcomes monitored and evaluated.

The fact that national e-government portfolios (where they exist) reside in several different ministries and involve various administrative arrangements implies that e-government does not have a natural home for governance and coordination. Regardless, e-government institutions should be able to perform several governance and coordination functions, as described below.

Developing governmentwide information infrastructure, shared networks, data centers, common business processes, and one-stop service delivery centers. Governments need to reform, reengineer, and connect systems and processes that have resulted from decades of inwardly focused operating strategies. Ministries and agencies often have independent ICT programs, and some operational independence is needed. But when e-government funds are mainly invested autonomously or coordination is limited to single applications or donor portfolios, it results in duplication, interoperability problems, and substantial waste of resources.

One role of central e-government institutions is to promote, develop, and support common information infrastructure and applications, including governmentwide networks, government portals, data centers, and common business processes (for example, for financial and accounting systems, payment systems, human resource management, and public procurement systems). They also need to coordinate or integrate service delivery channels and thus move government agencies from fragmented, multiple, discrete channels to a networked, multichannel approach to service delivery (OECD 2005).

Identifying and standardizing common functions across government address the challenges arising from the silo structure of public administration. It reduces duplication of systems and processes, captures process innovations and reusage solutions across government, focuses on improving core activities and outsourcing secondary ones, consolidates ICT expertise, and promotes interoperability and administrative simplification. These efforts imply that central e-government institutions should have resources under their control to invest in shared information infrastructure, induce collaboration across agencies to develop standardized business processes and shared networks, and push for governmentwide interoperability.

Formulating e-laws and frameworks for IT governance. To set and enforce common laws, regulations, and IT governance in support of e-government development and operation, governments should create institutions responsible for, among other things, developing e-government policies and legal and regulatory frameworks for issues such as e-transactions, e-security and privacy, and access to information. E-laws are likely to affect many stakeholders. Thus, their formulation and enforcement involve more than the ministry of justice or ICT. A central agency or institutional mechanism should therefore lead and coordinate the process of designing and adapting such laws and of monitoring and evaluating their impact. Such an agency should also harmonize country-specific e-laws with international conventions and best practices.

A number of countries—including developing ones such as Jordan, Morocco, Romania, and Vietnam—have made adopting and promoting governmentwide ICT architecture frameworks, approaches, and technology standards an integral part of their e-government strategies. A key requisite for achieving compatibility and interoperability in government departments is the establishment of institutional mechanisms and organizational processes for enforcement and compliance. Equally important is the institutional setup for maintaining and updating ICT architecture and standards. ICT architecture and governance frameworks should be dynamic and reflect fast-changing technologies and innovation possibilities.

Several countries have opted for a centralized institutional structure to facilitate interoperability. For example, the United Kingdom has established a centralized accreditation authority to implement its e-Government Interoperability Framework. A similar role is played by Canada's Treasury Board and Singapore's Infocomm Development Authority.

Mobilizing, prioritizing, and allocating resources for e-government infrastructure and services. Most developing countries suffer from huge deficits in the reach and quality of public services. Thus, there is often a temptation to do everything at once, and political pressures, growing expectations, and interest groups often encourage new ICT agencies or e-government institutional arrangements to take on too many projects and spread resources across too many initiatives. Although many governments invest heavily in ICT and e-government programs, investment levels are seldom a good gauge of progress or results. In fact, even substantial investments in e-government often fail to bring about the results they are intended to achieve (Fountain 2001; Heeks 2003).

Moreover, in developing countries, public resources for e-government are likely to be scarce. Absorptive capacity, change management capabilities and leadership, and project management and technical skills are also binding constraints on e-government. As a result, governments often rely on strategic analysis to survey and prioritize public services, develop sequenced investments, and mobilize resources outside the public budget for such investments. Donor agencies have encouraged such planning and prioritization. The goal is to identify and allocate resources to high-impact services.

But such efforts have often been treated as one-time events, driven by ad hoc institutional arrangements or donors. Yet technology, service priorities, and infrastructure need to sustain change over time. This requires that e-government institutions develop new and rigorous frameworks to maximize the impact of their investments and ensure that the resulting outputs are affordable, scalable, and sustainable.

E-government institutions must have a strong influence over ICT resource management, particularly through ICT budgeting and procurement. Budgeting and procurement are key to translating the prioritizing and sequencing of investments in services into reality. Given the scarcity of public funds and skills, innovative financing schemes and partnerships with the private sector and civil society are needed.

Monitoring, evaluating, and communicating lessons of experience; providing feedback; and ensuring accountability. Evaluating e-government programs is challenging. Even most developed countries have done only limited assessments of how well ICT investments have been used. Governments need to develop systematic monitoring and evaluation mechanisms that can serve as tools for improving program management, answering questions from stakeholders, meeting official reporting requirements, increasing the understanding of program strategies and goals, and promoting interest in and support for e-government programs and activities (see chapter 5 on existing frameworks for monitoring and evaluating e-government). Furthermore, information from monitoring and evaluation must be used to redesign, change direction, and implement new strategies where necessary.

Monitoring and evaluation is often confused with cost-benefit analysis. The latter is an administrative practice in efficiency-focused investment choices, whereas the former is about realizing public value in service offerings as part of the business strategy practice. E-government programs are concerned with creating public value and achieving development results. Canada provides a good example of best practices in using monitoring and evaluation for timely feedback and accountability with its series of studies on the use of e-government services, and with its use of the findings to reshape its e-government strategy and investments.[2]

E-government development often neglects strategic communication of visions shared, progress made, impacts measured, and lessons learned to all concerned stakeholders (Hanna 2007a, 2007b). Yet without such awareness and communication, e-government cannot be broadly owned or sustained or integrated with the overall development agenda. As a demanding transformational task, e-government requires mobilizing policy makers to lead policy reforms and institutional changes and mobilizing potential communities of ICT users to innovate and press for change (box 6.3).

The ICT governance and coordination functions described above lie at the heart of mandates for e-government institutions. They could make the difference between e-government success and failure; between exploiting economies of scale and suffering substantial duplication of investments; and between focused, coherent investment portfolios and diffuse, poorly planned resource allocations. These functions are key to creating a vibrant e-government ecosystem and an enabling legal and regulatory environment for e-services. Establishing a shared vision of modern, ICT-enabled government and developing the needed monitoring and evaluation systems further ensure an adaptive and learning strategy for public sector modernization and service innovation.

Facilitating E-Government Implementation

Coordination is important for minimizing redundancy and duplication, but it is insufficient for redesigning and reforming business processes, facilitating collaboration and knowledge sharing, and implementing a user-focused approach to delivering services. Most government agencies are unlikely to have in-house expertise that can simultaneously define a country's ICT requirements, cost-effectively procure IT hardware and software, engage in business process reengineering and change management of services, institute public-private partnerships and service-level agreements, and establish timely access to best practices in adopting new technology or knowledge of trends transforming governments around the world.

Thus, it is necessary to have institutions with the expertise, budgets, and other means to facilitate implementation of e-government strategies and ensure that key stakeholders are engaged at all levels. Given the scarcity of ICT and change management skills in the public sector and the potential economies of scale involved in e-government development, facilitation and technical support functions are often shared across agencies or provided by the private sector. Moreover, implementation problems change over time, presenting novel situations that demand innovation and peer support. E-government institutions should facilitate partnerships in e-government investments and operations and promote collaboration among government agencies for process innovation and integrated solutions.

> **Box 6.3** The Need to Build Strong Demand for E-Government Services and Institutions
>
> Most successful institutional reforms have occurred when societies have generated strong domestic demand for institutions. In developing countries, insufficient demand for institutions is the most important obstacle to institutional development (Fukuyama 2004).
>
> Effective demand for e-government institutions can be created by building business and civil society pressure for better public services. It can be nurtured by raising awareness among societal leaders and exposing them to international best practices. Citizens should be made owners of e-government programs. They should be engaged—through political leaders and e-government institutions—in shaping the kind of government, information society, and knowledge economy they would like to have and in realizing their shared vision (Stiglitz, Orszag, and Orszag 2000). The media can play a critical role here, as it did in the Republic of Korea and several other East Asian countries (Jeong 2006).
>
> Demand for new institutions or for reforms to existing ones is often time sensitive. When such demand emerges, it is usually the product of crisis or a major change in the political environment. There are serious limitations to the ability of external partners or donors to create demand for institutions and so to transfer knowledge about building new institutions. Thus, such windows of opportunity should be anticipated and quickly captured.
>
> **Source:** Authors' analysis.

A key function of e-government institutions is human resource development and capacity building to help government agencies absorb and manage ICT. The central e-government ministry or agency may take the lead in the professional development of ICT specialists and chief information officers (CIOs) in government, including defining roles and career structures and certifying education programs for minimal qualifications and professional development. E-government institutions can also support the development of communities of practices and knowledge sharing systems for public ICT professionals, and raise awareness among policy makers.

What Are the Institutional Models for E-Government?

This section describes four models that governments have used to create a national institutional framework to lead the e-government agenda and fulfill the key functions of strategy and policy making, governance and coordination, and facilitation of implementation (table 6.1). The actual arrangements are more diverse than suggested by these groupings and do not fit neatly into these simplifying models. Moreover, countries' institutional arrangements shift over time— often from one model to another. Thus, the countries here are classified based on their most distinguishable structural features for the most recent period or longest duration, with the understanding that these institutional models are used only for comparative analysis and for detecting patterns and trends in a rich, complex institutional reality.

Model 1: Policy and Investment Coordination

Working from the ministry of finance (or treasury, economy, budget, or planning) gives the entity responsible for governing and coordinating e-government activities direct access to the funding it needs. In addition, it enables easy control over funds required by other ministries in pursuing e-government goals set for them. It also facilitates integration of the e-government agenda with the country's overall economic development agenda. This model seems to have worked relatively well in countries that have a powerful central agency with cross-cutting mandates (examples of such organizations include the Treasury Board of Canada and the U.S. Office of Management and Budget). It enforces policies and priorities through the budget process, yet allows effective decentralization of implementation.

Most countries using this model started adopting e-government initiatives early on and made sustained

Table 6.1 Models for E-Government Institutions in Various Countries

Model	Countries	Benefits	Drawbacks
Policy and investment coordination (cross-cutting ministry such as finance, treasury, economy, budget, or planning)	Australia, Brazil, Canada, Chile, China, Finland, France, Ireland, Israel, Japan, Rwanda, Sri Lanka, United Kingdom, United States	Has direct control over funds required by other ministries to implement e-government. Helps integrate e-government with overall economic management.	May lack the focus and technical expertise needed to coordinate e-government and facilitate implementation.
Administrative coordination (ministry of public administration, services, affairs, interior, state, or administrative reform)	Bulgaria, Arab Republic of Egypt, Germany, Republic of Korea, Mexico, Slovenia, South Africa	Facilitates integration of administrative simplification and reforms into e-government.	May lack the technical expertise required to coordinate e-government or the financial and economic knowledge to set priorities.
Technical coordination (ministry of ICT, science and technology, or industry)	Ghana, India, Jordan, Kenya, Pakistan, Romania, Singapore, Thailand, Vietnam	Ensures that technical staff is available; eases access to nongovernmental stakeholders (firms, NGOs, and academia).	May be too focused on technology or industry and disconnected from administrative reform.
Shared or no coordination	Russian Federation, Sweden, Tunisia	Least demanding and with little political sensitivity (does not challenge the existing institutional framework and responsibilities of ministries).	May lead to rivalries among ministries. No cross-cutting perspective. Fails to exploit shared services and infrastructure and economies of scale.

Source: Hanna and Qiang 2009 (chapter 6 in this volume).

commitments. Today, Canada is a leader in terms of e-government strategies; its government laid out a clear, specific, comprehensive, actionable strategy at an early stage. The strategy has been effectively rolled out to and implemented by government departments. Other countries, such as the United States, have had similar success attributable to early beginnings and sustained commitments. These countries also tend to spend more government funds per capita on ICT than do most other countries (Booz Allen Hamilton 2002). A drawback is the lack of focus and technical expertise of the coordinating body.

Model 2: Administrative Coordination

Countries that adopt a model of e-government led by the ministry of public administration (or services, affairs, interior, state, or administrative reform) coordinate e-government within the framework of their good governance agendas. This model facilitates the integration of e-government efforts with administrative reform, simplification, and decentralization. It raises the visibility of the e-government agenda and encourages broad participation across agencies. Moreover, increasing government efficiency and transforming public services are the ultimate goals of any e-government initiative, making this model outcome-oriented rather than technology-driven. But a potential downside is that the leading ministry may lack the technical expertise and budget mechanisms required to ensure technical coordination.

Model 3: Technical Coordination

Governing and coordinating e-government activities under a technical ministry such as the ministry of ICT (or science and technology or industry) ensures that specialized technical staff are available to address ICT issues. This approach may be a natural evolution of the traditional role of the ministry of telecommunications—typically when the approach to e-government is focused on infrastructure. It may also have the advantage of involving the private sector and other nongovernmental stakeholders more effectively in the e-government process and thus allow for innovative public-private partnerships. But if the technical ministry has limited leadership competencies, the e-government agenda remains outside broad public sector reform efforts and the core development agenda. Accordingly, strong financial mechanisms with well-defined carrots and sticks must be in place to ensure compliance and cooperation.

Model 4: Shared or No Coordination

In this model, e-government development and implementation functions are distributed among existing ministries. Thus, each ministry is responsible for the part of the e-government strategy that falls within its field of expertise. This model does not involve any new coordination mechanisms and is the least politically demanding, making it the easiest to adopt for the short term. Funding for e-government activities comes from the ministries' budgets. However, agencies set up their own information systems—and in some cases, proprietary communications networks—leading to duplication and impairing information sharing. This approach is likely to result in uneven development across ministries and missed opportunities to leverage economies of scale in shared infrastructure, applications, and support services.

The choice of institutional location for e-government governance and coordination may reflect more general tendencies or legacies: faced with a new challenge, a government may have a preference about where it locates responsibility. While administrative control can be wielded to ensure cross-agency coordination, placement of e-government responsibility under each model has clear advantages and disadvantages that should be borne in mind—and perhaps complemented by capacity building and cross-agency policy and strategy mechanisms.

Alternative Models

The four institutional models just described focus on the leading or central institution for e-government strategy and policy making, and governance and coordination. But governments have increasingly created and experimented with new arrangements outside the ministerial structure—including ICT agencies and councils of CIOs—to overcome sectoral silos and civil service constraints and to create a new capability to engage various stakeholders and agencies in facilitating implementation.

ICT agency. When implementing ICT strategies, governments inevitably compete with the private sector for scarce ICT talent. For example, the first CIOs (or their equivalents) in Italy, the United Kingdom, and the United States were recruited from the private sector. Such competition is no longer local: it is global. This highlights the challenges involved in hiring, training, and retaining skilled staff for e-government institutions and ICT programs in ministries and agencies. This challenge goes beyond ICT specialists, and includes people with a broad understanding of and talent in public sector reform, ICT-enabled business process reengineering, service innovation, supply chain management, public-private partnerships, change management, knowledge management, and transformational leadership.

Several countries have created dedicated executive ICT agencies in their civil services but have given them special autonomy and salary structures to attract and motivate the best technical talent. Such agencies prioritize investments and coordinate and monitor implementation of e-government, often under the supervision of an interministerial committee that sets policies and strategies. The chiefs of these agencies sometimes serve as national CIOs. Moreover, such agencies are often charged with developing mechanisms that encourage all stakeholders to become involved in e-government and the exchange of information, experience, and best practices through focus groups, workshops, seminars, and online tools. Bulgaria, Ireland, the Republic of Korea, and Singapore have adopted variations of such central ICT agencies (box 6.4).

Given the innate conservatism of public agencies and the transformative nature of e-government, it is not surprising when government leaders turn to bodies outside standard ministerial structures. Having a focal ICT agency also makes it easier and more effective to focus on e-government goals. The creation of such an agency typically involves adopting comprehensive approaches to integrate e-government with broader development strategies.

However, a new entity may struggle to obtain needed political weight and resources. An ICT agency's impact on e-government thus depends on institutional links to the leadership of the line ministries responsible for process transformation and sectoral (vertical) applications, as well as strong ties to powerful ministries such as finance or public administration. The viability of an ICT agency also critically depends on the authorizing environment, and whether political leaders are committed to giving the agency the autonomy needed to act in an agile manner and avoid political interference in staffing and day-to-day management.

A variation of the ICT agency described above is a public-private partnership or quasi-public enterprise. Private sector participation in public sector ICT policy and strategy formulation, as well as rigorous public-private partnership frameworks for investing in and implementing e-government programs, are more common in developed than in

> **Box 6.4** Singapore: Pioneering a Centrally Driven Public ICT Agency
>
> Singapore developed one of the world's first national ICT plans in 1980. Successive plans and e-government institutions have become increasingly broadened, deepened, and decentralized.
>
> The InfoComm Development Authority was created in 1999 with the merger of the National Computer Board and the Telecommunications Authority of Singapore.[a] The Authority operates under the Ministry of Information, Communications, and the Arts. As the government's chief information office, it drives the implementation of Singapore's e-government action plan and provides the technical expertise for various e-government programs, under the guidance and oversight of the interministerial E-government Policy Committee and the Ministry of Finance's E-government Office.
>
> Since 2002, the InfoComm Development Authority has focused on creating relevant content, promoting public use of e-government services, and ensuring universal access to such services. Government agencies were required to survey their customers' needs and launch marketing campaigns to promote e-government services through the single window, eCitizen. To secure ownership of e-government by civil servants, the InfoComm Development Authority has been empowering public officers with training and resources for ICT-enabled innovation and knowledge sharing. The InfoComm Education Program was launched to equip officers with needed ICT skills, and the Knowledge Management Experimentation Program provided seed funding to encourage public agencies to pioneer knowledge management projects that nurture knowledge sharing.
>
> Some of the technical expertise developed in the public sector under the InfoComm Development Authority was subsequently transferred to semipublic enterprises such as National Computer Services to deliver e-government advisory services beyond Singapore. More recently, Singapore has been positioning itself to go global, sourcing talent from and partnering with other Asian countries and leveraging infrastructure and capital to become a knowledge services hub for the world economy.
>
> **Source:** Authors' analysis.
>
> a. See http://www.ida.gov.sg.

developing countries. But in recent years such partnerships and business influence on government ICT use have been increasing in developing countries—particularly where public sector performance suffers from civil service constraints and the private sector's technological know-how is relatively advanced.

Under this institutional innovation, an ICT agency would be semipublic, operating like a business but ultimately answerable to a country's political leadership (Hanna 2008). Such an agency typically has a government-appointed board of directors, the chair of which reports directly to the head of state, and is composed of representatives of key stakeholders from the private sector and civil society. The agency's responsibilities may cover only the central leadership of e-government or extend to the entire range of the e-development agenda. The national chief information officer may be the chief executive of such an agency. Sri Lanka is currently experimenting with this institutional model (box 6.5). In India, the National Institute for Smart Government was created as a public-private partnership, with joint financing from the government and the National Association of Software Services Companies, to advise on e-government progress.

To succeed, these public-private partnerships should be staffed by experienced development strategists, ICT professionals in various disciplines, and project coordination specialists who can liaise between the public and private sectors. These staff members could be hired from the public or the private sector, as available. A hybrid staff will reflect the diversity of skills and experiences needed to cut across the public, private, and civil society sectors, and be able to understand and partner with diverse groups of beneficiaries. The staff must strike the right balance of business culture and public values and accountability.

> **Box 6.5** Sri Lanka: Pursuing Institutional Innovation in a Turbulent Political Environment
>
> Sri Lanka's Information and Communication Technology Agency (ICTA) represents one form of public-private governance, an agency under the head of state and governed by the Companies Act. The agency is mandated to operate in a businesslike fashion, following local commercial practices. It is managed by a board of directors made up of representatives from the public and private sectors, academia, and civil society, and is representative of minorities. The chairman of the board answers to the parliament and its committees through the presidential secretariat, provides guidance to the chief operating officer and leadership team of the agency, and approves strategic decisions.
>
> Preliminary assessment indicates that this public-private model has helped promote partnerships and inject a new work ethic and project management practices in an otherwise weak civil service. It has allowed for an action-oriented, "can-do" culture. Freeing ICTA from civil service constraints has been critical to its relative agility and performance—staff are recruited from the private sector, government, civil society, academia, and even the Sri Lankan diaspora.
>
> The agency promises an institutional arrangement that will lead to public sector modernization and, more broadly, ICT-enabled development. However, the agency's high-performance, high-reward business culture may have at times created tensions with government agencies' hierarchical, unmotivated, overstaffed, turf-bound bureaucracy. Moreover, this model raises issues concerning financial sustainability: its viability depends on the fiscal space and autonomy, as well as institutional stability, provided to it by the political leadership of the country.
>
> **Source:** Hanna 2008.

The main advantages of the public-private partnership model are that the ICT agency is free from government bureaucratic requirements and has the flexibility to react swiftly to changing demands. In addition, the agency can more easily hire the required cutting-edge staff at competitive wages. It also has the freedom to provide shared technical services (such as for network infrastructure) to the government or to contract out to the private sector. By pursuing public-private partnerships and extensive outsourcing, the agency can remain lean, focused, and agile. Finally, active private sector participation would help the agency operate in a businesslike way and make the best use of scarce resources. This model fits well with—and compensates for—the weak civil service environment in many developing countries.

One disadvantage of a public-private partnership in an e-government institution is that it may not receive the political and financial support it needs if it is not directly linked to a powerful ministry or the prime minister's office. In addition, the public sector bureaucracy may reassert control over the agency, and political interference may reduce the effectiveness of agency staff and undermine its businesslike culture. A comparative study of public-private partnership innovations in different sociopolitical contexts is warranted to reach more generalized and robust conclusions about the merits of this model.

Council of chief information officers. About one-third of the countries in this study's analysis are instituting or experimenting with national councils of CIOs, supported by CIOs in ministries and agencies. This approach combines centralized governance and coordination with decentralized implementation and ownership. The role of such councils has evolved and become increasingly critical to e-government development. These councils vary in mandate but often involve addressing common CIO concerns and challenges, such as investment planning, IT procurement practices, information security policies, and IT human resource development (box 6.6). They also have been engaged in CIO capacity development by providing inputs into defining core competencies, accrediting CIO education and training programs, and sharing information

Box 6.6 Chief Information Officer Councils in Various Countries

Australia. Significant e-government matters affecting all jurisdictions are processed by the Online and Communications Council. The council includes a cross-jurisdiction CIO committee chaired by the Australian government's CIO—who also chairs the Australian Government Information Management Office (AGIMO)—and the Chief Information Officer Committee, which investigates ICT issues, endorses solutions, and undertakes strategic ICT projects. The AGIMO and Chief Information Officer Committee also collaborate with the Business Process Transformation Committee, which coordinates reform of agencies' business processes (ICA-IT 2006a, pp. 1–6).

Canada. To promote interjurisdictional collaboration, the Public Sector Chief Information Officer Council and the Public Sector Service Delivery Council bring together various levels of CIOs and leading service officials to exchange best practices, conduct joint research, and evaluate and pursue opportunities to adopt common practice and collaborate on integrated service delivery.[a]

Singapore. The ICT Committee aims to share experiences, promote integration across agencies, streamline processes, and share data (Tan 2007). The CIO Forum, comprised of CIOs from key agencies, was created in 2004 to promote interagency sharing of best practices and systems as well as consultancy on and review of central systems and investments, increasing opportunities for collaboration. The forum also provides a venue for giving feedback to central authorities on servicewide e-government initiatives (ICA-IT 2006b, pp. 3–5; Infocomm Development Authority of Singapore 2005, p. 4).

South Africa. The Government Information Technology Officers Council serves as a coordination and oversight unit, involved in the development of IT security policy, e-government policy and strategy, IT procurement guidelines, and project coordination.[b]

United Kingdom. The CIO Council was created to support the Cabinet Office's E-government Unit on research, monitoring of major government IT projects and investment decisions, management of and career development for government IT professionals, and management and analysis of relationships with strategic government ICT suppliers. The council also enables partnerships between IT professionals in various areas of government.[c]

United States. The Federal CIO Council's role includes developing recommendations for IT management policies, procedures, and standards; identifying opportunities to share information resources; and assessing and addressing the federal government's IT workforce needs. It also addresses cross-cutting issues—such as financial management and procurement—with other federal executive agencies.[d]

a. See http://www.tbs-sct.gc.ca/organisation/ciob-ddpi_e.asp and http://www.servicecanada.gc.ca/en/about/index.shtml.
b. See http://www.southafrica.info/public_services/citizens/services_gov/sagovtonline.htm and http://www.dpsa.gov.za/egov_documents.asp.
c. See http://ec.europa.eu/idabc/servlets/Doc?id=21032 and http://www.cio.gov.uk/about _the_council/the_cio_council.asp.
d. See http://www.cio.gov/index.cfm?function=aboutthecouncil.

and best practices among CIOs. CIO councils are expected to play an increasing role in consensus building, vertical and horizontal communication, team-based problem solving, and knowledge sharing.

Many countries view e-government as a catalyst for developing indigenous ICT industries and local technological capabilities. Furthermore, there is growing awareness that e-government depends on other elements of e-development, including IT literacy among citizens and small enterprises, countrywide connectivity and access, public policies on e-commerce, and the availability of local skills to adapt and support information systems. National ICT agencies or CIO councils may help tap synergies and coordinate investments across all the key elements of e-development.

What Are the Broad Trends in the Evolution of E-Government Leadership Institutions

The promises of e-government have been slow to come to fruition because deep transformation of how government works and relates to citizens and businesses is difficult and time consuming. Transformation takes sustained leadership and targeted incentives to reshape relationships and create networked, adaptive, ICT-enabled government agencies. It also requires building coalitions, aligning e-government programs with political goals, and achieving effective coordination across agencies, including effective implementation and learning.

The models of e-government institutions presented in this chapter are used for comparative analysis and for detecting broad trends in a complex reality of rich institutional innovation and learning. They can serve as starting points or options for governments interested in advancing their institutional frameworks for e-government. Hybrids of these models are increasingly common, tailored to a country's needs and conditions. Governments can choose from and build on these basic approaches, understanding the advantages and disadvantages of each.

The country studies suggest some broad trends in the evolution of e-government institutions:

- *Countries have moved from ad hoc responses, informal processes, and temporary relationships to institutionalized structures that respond to the challenges of developing e-government.* At the outset of the ICT revolution, when awareness of e-government's potential was nascent, governments convened task forces, commissions, and panels to advise them on directions to take. These ad hoc entities typically made their recommendations to relevant ministers or the head of state. Among the countries that turned to such task forces were Singapore in 1992 and the United States in 1993, followed by China, Japan, and the Republic of Korea, among others (Wilson 2004). At that time, the central message was to raise awareness about the enabling role of ICT across the bureaucracy and society. Over time, these temporary bodies and ad hoc processes were transformed into permanent institutions and formal coordination mechanisms. The ad hoc processes were often used to reach out to key leaders and constituencies beyond government and to identify potential leaders and stakeholders for the subsequent institutions. In many countries, the institutionalized structures were given legislative mandates to enhance their influence and authority, often covering issues of ICT budgeting, procurement, and data and technology architecture.

- *There has been a move toward direct, institutionalized engagement of the head of state or an interministerial steering committee to formulate national e-government strategies and policies.* This process occurs as part of the search for an overarching strategic framework for e-government development in the knowledge economy, placing the capacity for orchestration and policy coordination under the highest authority. A common trend for e-government leadership is to place a coordinating unit within the office of the president or to establish a policy coordinating committee chaired by the prime minister or head of state. The head of the coordinating unit or committee becomes the visible e-leader, using e-government as a core component of his or her public management reform agenda and, more broadly, as a key to transforming the country to a knowledge-based, innovation-driven economy.

- *Emphasis is shifting from computerization and technology management to public sector reform, service transformation, process innovation, and cross-agency integration.* As e-government programs mature, countries move beyond concern about front-end electronic delivery of services. Instead, they start to rationalize and integrate back-office processes and the entire value chain and to fully integrate e-government with the governance framework and activity of each sector and agency. There is also a shift in mind-set from an inside-out, agency-bound perspective to an outside-in, client-oriented perspective of service delivery. In the process, the role of central agencies also changes, from providing top-down solutions to playing catalytic roles for service innovation and cross-agency coordination. The aim is to facilitate public service innovation at all levels of government, institutionalize and scale up process innovation, promote collaboration across boundaries, engage more stakeholders, and disseminate best practices—and thus achieve deeper transformation and sustainable improvements in public sector performance.

- *As a further evolution, many countries are opting to create strong, independent national e-government agencies.* These agencies tend to focus on policy development, governance mechanisms, integrated government approaches to public interaction, enterprise architecture, and strategic investments that cut across many agencies. In some countries, such as Canada, Ireland, and Singapore, these relatively independent agencies tend to coordinate all components of e-development, of which e-government is key. They often operate under a special act or civil service framework that allows them to provide competitive compensation and attractive career paths and to operate in a businesslike manner, yet enjoy the legitimacy and authority of top political leadership and retain alignment with public service value creation. The shift to this model is driven by growing recognition that e-government development is a cross-sectoral, cross-agency, cross-hierarchical process. It is a major transformation that requires political leadership, a holistic view of government, and the ability to partner with nongovernmental actors. These needs are more likely to be achieved by an agile, independent agency, a semipublic enterprise, or a powerful coordinating ministry such as finance or economy.

- *E-government institutions are taking on increasing responsibility for promoting and managing private-public partnerships.* A key competency of e-government institutions is the capacity to identify, procure, and manage private-public partnerships on behalf of the entire government. They should also be able to establish the policy and legal frameworks to support the sound procurement and management of such partnerships by individual government agencies—consistent with the politically acceptable role of the state, allocation of risks to parties most likely to mitigate them, and relative competencies of the country's public and private sectors. E-government institutions are expected to ensure that private-public partnerships are priority projects. A central cross-government pool of expertise in private-public partnerships is likely to be needed to supplement any nascent capacity in line agencies that contract for them. The degree of centralization of this function will vary. It may be limited to sharing information and broad guidance, promoting the use of private-public partnerships to accelerate e-government financing and implementation, and developing the legal and regulatory framework for such partnerships. Or the role may extend to approving private-public partnerships entered into by line ministries, understanding and monitoring the fiscal costs of the partnerships, and directly establishing and executing complex private-public partnerships on behalf of all government agencies.

- *Broadly, the nature and priorities of e-government are changing, and the institutional models adopted tend to evolve with the maturity of a country's e-government programs and its changing development priorities.* In recent years, a number of countries have shifted responsibility for their e-government portfolios. Each change reflects the countries' needs, given the point they reached in developing e-government. These changes should be viewed as responses to strategic policy needs and issues as they develop and implement solutions, rather than absolute illustrations of right or wrong approaches. For example, some countries are shifting from political or ad hoc e-government programs to more systematic administrative control in order to institutionalize e-government and lock in the gains they have achieved (Mexico and Portugal). Other changes have been driven by an increased focus on the use of e-services following a rapid increase in online services (Canada and the United Kingdom). In terms of tie-in with related policy areas, some countries have separated their e-government and information society portfolios (Australia and the United Kingdom), while others have consolidated their leadership of these portfolios (Norway and Sri Lanka). Many countries are currently engaged in internal discussions about the impact of e-government on the public sector in general and the consequences that this should have in how initiatives should be structured.

- *The structures and functions of central e-government institutions evolve in response to the growing decentralization of government services to the state and city levels.* Subnational economies—particularly cities—are playing a central role in economic growth, competitiveness, and globalization. Leading states and cities have greater agility to pilot e-government services and seize opportunities in rapidly changing environments (Lanvin and Lewin 2006). Accordingly, e-government program success will depend on institutional arrangements at the state and city levels, where most government services are delivered, pilots and innovations are conducted, and partnerships with central governments are forged.[3] This movement to decentralize government functions tends to favor the administrative coordination model of e-government, where e-government functions are assigned to the ministry of public administration and local government (or services, affairs, interior, state, or administrative reform). Central e-government institutions then become engaged in disseminating best practices across states and cities, providing matching funds for innovation in local e-government services, addressing common human and infrastructure constraints to local e-government efforts, and leveraging economies of scale across local jurisdictions, among other activities.[4]

Conclusion

The basic e-government institutional functions and models identified in this review suggest the wide range of possibilities open to governments. Governments have moved from ad hoc responses to institutionalized structures to lead and manage e-government programs. They have put increasing emphasis on engaging top political leadership in their e-government programs and have devoted increasing attention to ICT-enabled process innovation and institutional reform. Moreover, some governments have changed their institutional arrangements and developed new models for e-government in response to institutional learning, technological progress, and new phases in e-government.

Today's knowledge and evaluation research does not enable definitive prescriptions for the best e-government institutional model, especially given the diverse conditions facing both developing and developed countries. But understanding available options, current trends, and the core capabilities that e-government institutions must possess is critical to building effective institutions that can achieve ICT-enabled transformation. Identifying appropriate institutional functions and capabilities should guide institutional development and capacity building efforts for better governance and coordination of e-government programs.

Although there is no one-size-fits-all institutional model for all countries, the strengths and weaknesses of the models described above suggest reasonable approaches for countries at different phases of e-government development. The appropriate level of centralization and decentralization is a key consideration in the design of national e-government institutions. The balance is often determined by a country's general political and institutional architecture and the availability and distribution of local capacity.

Another key institutional design issue is the balance between, on the one hand, technological leadership to invest in sound technologies and manage complex systems development projects and, on the other hand, business and institutional leadership to ensure general management ownership and true business process and service transformation. Businesses have been struggling much longer with various governance and institutional arrangements to get this balance right (Weill and Ross 2004). The models described in this chapter present alternative emphases in striking this balance. Shifts to recent and hybrid models suggest that e-government programs have evolved from computerization and online front-end delivery of services to organizational transformation. Governments are experimenting with and learning to manage this new paradigm—but it has taken about a decade for leading governments to appreciate this paradigm shift.

Another key lesson is the importance of cross-sector partnerships. Adopting a national e-government strategy will always demand a comprehensive policy approach. The cross-cutting nature of ICT makes it highly challenging to use traditional institutional arrangements that designate the entire agenda to a single ministry. E-government requires strong coordination of activities among various government agencies. Public leadership styles need to change from silo thinking and turf protection to management through collaboration and partnerships across agencies.[5] Equally important is the need to build partnerships among government, the private sector, and civil society to account for the needs and capabilities of the private sector and civil society.

Top-down leadership and institutional coordination must be complemented by bottom-up collaboration and local initiative. Centrally driven coordination alone will not be sufficient for e-government to mature and lead to continuous innovation in governance, service provision, and citizen participation. It must be complemented by bottom-up initiatives, knowledge sharing, and incentives for collaboration across bureaucratic boundaries. Implementation facilitation and peer-to-peer coordination are essential complements to central coordination. National e-government institutions and CIO councils are increasingly used to support their counterpart state- and municipal-level e-government institutions and councils. Countries continue to innovate and experiment with institutional arrangements to maintain an appropriate balance. These innovations should be identified, evaluated, and disseminated.

Further research is urgently needed to understand the governance and institutional mechanisms needed to guide e-government. This survey, which focuses on central institutions at the national level, is only a start, as much of e-government's potential for decentralization and much of the rich institutional experience at subnational levels remain untapped. The various models and trends of e-government institutions at the state and city levels need to be examined, perhaps starting from the typology of models identified here at the national level. It would also be useful to investigate the implications of decentralized e-government strategies and of virtual vertical integration of public services across all levels of government, the division of functions between central and local e-government institutions, and the role of central institutions in scaling up local successes and supporting the development of common capabilities among local e-government institutions.

Research is also required to further understand the institutional implications of different priorities being assigned to e-government program goals. Would an emphasis on increasing efficiency in public sector management and e-government programs imply more centralized e-government institutional mechanisms and adoption of the technical coordination model? Would a goal of transparency and anticorruption highlight the governance agenda and point to the advantage of the administrative coordination model? Would an emphasis on making public services work for poor people lead to relatively broad mandates for national ICT agencies to overcome access barriers and the underlying causes of the digital divide?

Similarly, research is needed on the supply and demand of CIO capacity and professional development programs, and on the kinds of networks and support services required to help public CIOs in developing countries break out of their isolation and increase access to peer groups in developed countries.

Finally, this survey represents only a snapshot of institutional arrangements for e-government development—a field that is fast changing as countries are continuously adapting and replacing their institutional models. Thus, a regular mechanism for monitoring, updating, and evaluating countries' institutional arrangements is needed.

Annex: Characteristics of E-Government Institutions in Selected Countries

Annex table 6A.1 describes the characteristics of e-government institutions for 30 countries. Because of its length, the table can be found at the end of this chapter on pages 101–102.

Notes

1. For a treatment of e-leadership institutions in the broader context of e-development, see Hanna (2007b).
2. See http://www.iccs-isac.org.
3. The same arguments can be made for other knowledge economy institutions (Hanna 2007c). Much of the experimentation and many of the support services and partnerships must be forged at the regional, city, and cluster levels, where cooperation, competition, and institutional partnerships occur.
4. India provides an example of the services supplied by the central Department of IT at the federal level. The department diffuses and scales up successful priority e-government applications at the state level and adapts and matches central support to local state priorities.
5. This is a major human resource management challenge for the public sector because government employment practices and incentives often lead to turf protection, one way of doing business, ossification of business processes and practices, and reluctance to collaborate across sectors.

References

Booz Allen Hamilton. 2002. *International e-Economy Benchmarking: The World's Most Effective Policies for the e-Economy.* London.

Cohen, Yizak. 2002. "A Three-Year Master Plan 'E-Government' Initiative 2003–2005." Israeli Ministry of Finance, General Accountant Office, Jerusalem.

Evalueserve. 2007. "Country Assessments for Identifying Potential Public Private Partnerships in e-Government: Mexico." Country assessment report for the World Bank.

Fountain, Jane E. 2001. *Building the Virtual State.* Washington, DC: Brookings Institution.

Fukuyama, Francis. 2004. *State-Building: Governance and World Order in the 21st Century.* Ithaca, NY: Cornell University Press.

Hanna, Nagy. 2004. "Why National Strategies Are Needed for ICT-Enabled Development." Information Solution Group Working Paper, World Bank, Washington, DC.

———. 2007a. *From Envisioning to Designing e-Development: The Experience of Sri Lanka.* Washington, DC: World Bank.

———. 2007b. "e-Leadership Institutions for the Knowledge Economy." World Bank Institute Working Paper, World Bank, Washington, DC.

———. 2008. *Transforming Government and Empowering Communities: The Experience of Sri Lanka.* Washington, DC: World Bank.

Heeks, Richard, ed. 2003. *Reinventing Government in the Information Age: International Practice in IT-Enabled Public Sector Reform.* London: Routledge.

ICA-IT (International Council for Information Technology in Government Administration). 2006a. "Country Reports—Australia." Paper presented at the ICA 40th Conference, Guadalajara, Mexico, 12–14 September. http://www.ica-it.org/conf40/docs/Conf40_country_report_Australia.pdf.

———. 2006b. "Country Reports—Singapore." Paper presented at the ICA 40th Conference, Guadalajara, Mexico, 12–14 September. http://www.ica-it.org/conf40/docs/Conf40_country_report_Singapore.pdf.

Infocomm Development Authority of Singapore. 2005. *Report on Singapore e-Government.* http://www.igov.gov.sg/NR/rdonlyres/C586E52F-176A-44B6-B21E-2DB7E4FA45D1/11228/2005ReportonSporeeGov.pdf.

Jeong, Kuk-Hwan. 2006. *E-Government: The Road to Innovation: Principles and Experiences in Korea.* Seoul: Gil-Job-E Media.

Lanvin, Bruno, and Anat Lewin. 2006. "The Next Frontiers of E-Government: Local Governments May Hold the Key to Global Competitiveness." In *Global Information Technology Report 2006-07,* 51–68. Geneva: World Economic Forum.

North, Douglass C. 1990. *Institutions, Institutional Change and Economic Performance.* Cambridge, U.K.: Cambridge University Press.

OECD (Organisation for Economic Co-operation and Development). 2005. *e-Government for Better Government.* Paris: OECD.

Ramsey, Todd. 2004. *On Demand Government: Continuing the E-government Journey.* Lewisville, Texas: IBM Press.

Rodrik, Dani. 2004. "Getting Institutions Right." *CESifo DICE Report.* University of Munich, Center for Economic Studies, and Ifo Institute for Economic Research.

Rubino-Hallman, Silvana, and Nagy K. Hanna. 2006. "New Technologies for Public Sector Transformation: A Critical Analysis of e-Government Initiatives in Latin America and the Caribbean." *Journal for e-Government* 3 (3): 3–39.

Schware, Robert, ed. 2005. *E-development: From Excitement to Effectiveness.* Washington, DC: World Bank.

Stiglitz, Joseph, Peter R. Orszag, and Jonathan M. Orszag. 2000. "The Role of Government in a Digital Age." Study commissioned by the Computer and Communications Industry Association, Washington, DC. http://unpan1.un.org/intradoc/groups/public/documents/APCITY/UNPAN002055.pdf.

Tan, Pauline. 2007. "National e-Government Strategies: Designing for Success." e-Development Thematic Group Global e-Government Dialogue Series, World Bank, Washington, DC. http://go.worldbank.org/U2MOGH2U30.

Weill, Peter, and Jeanne W. Ross. 2004. *IT Governance: How Top Performers Manage IT.* Cambridge, MA: Harvard Business School.

Wilson, Ernest J., III. 2004. *The Information Revolution and Developing Countries.* Cambridge, MA: MIT Press.

World Bank. 2003. *World Development Report: Making Services Work for Poor People.* Washington, DC: World Bank and Oxford University Press.

Yong, James S. L. 2005. *e-Government in Asia.* Singapore: Times Media Publishing.

Table 6A.1 Characteristics of E-Government Institutions in Selected Countries

Country	Strategy formulation		E-government program coordination			Facilitator of implementation				
	No steering committee	Steering/inter-ministerial committee/office	Policy and investment coordination (ministry of finance, treasury, economy, budget, or planning)	Administrative coordination (ministry of public administration, services, affairs, interior, state, or administrative reform)	Technical coordination (ministry of ICT)	No facilitator	Ministry of ICT	ICT agency under other ministries (e.g., public services, finance)	CIO council	Public-private partnership/ quasi-public ICT agency
Australia		X	X[abcd]					X[e]	X	
Brazil		X		X[bd]			X	X		
Bulgaria		X		X[acd]			X	X		X
Canada		X	X[abcd]	X[ce]				X[ef]	X	
Chile		X[g]	X[ab]				X	X[f]		
China		X[g]	X[acd]				X			
Egypt, Arab Rep. of		X		X			X[f]			
Finland		X[g]	X[a]					X		
Ghana	X				X[ab]		X		X	
India		X[g]	X[c]		X[abcd]					X
Ireland		X[g]	X[abc]					X[ef]		X
Israel		X	X[abc]				X		X	
Japan		X[g]	X[abc]						X[f]	
Jordan	X				X[bcd]		X			
Kenya		X		X[abcd]				X[ef]		
Korea, Rep. of		X[g]		X[ab]			X		X	X
Mexico		X		X[abd]			X		X[a]	
Pakistan		X[g]			X[abcd]		X			
Romania		X[g]			X[a]		X			
Russia		X	X		X	X				
Rwanda		X	X[ab]				X			

(Table continues on the next page.)

Table 6A.1 *continued*

Country	Strategy formulation		E-government program coordination			Facilitator of implementation			
	No steering committee	Steering/inter-ministerial committee/office	Policy and investment coordination (ministry of finance, treasury, economy, budget, or planning)	Administrative coordination (ministry of public administration, services, affairs, interior, state, or administrative reform)	Technical coordination (ministry of ICT)	No facilitator	Ministry of ICT	ICT agency under other ministries (e.g., public services, finance)	CIO council	Public-private partnership/ quasi-public ICT agency
Singapore		X	X[c]		X[ab]		X[ef]		X[f]	
Slovenia	X			X[abcd]				X[f]	X	
South Africa	X			X[abd]						X[f]
Sri Lanka	X		X[abc]						X	X[f]
Thailand		X[g]	X		X[ab]		X			
Tunisia		X[g]	X[g]					X		
United Kingdom		X	X						X	
United States		X	X[cd]						X[ef]	
Vietnam		X			X[bd]		X			

Source: Authors' analysis.

Note: CIO = chief information officer.

a. Develop governmentwide information infrastructures, shared networks, data centers, and common business processes.
b. Formulate e-laws and frameworks for IT governance.
c. Mobilize, prioritize, and allocate resources for e-government infrastructure and services.
d. Monitore, evaluate, and communicate lessons of experience, providing feedback, and ensuring accountability.
e. Handle ICT procurement.
f. Process reengineering.
g. Chaired by the head of state, head of government, or cabinet office.

Chapter 7

Realizing the Opportunities Presented by the Global Trade in IT-Based Services

Philippe Dongier and Randeep Sudan

Advances in information technology (IT) and global connectivity, combined with waves of economic liberalization, have given rise to a new dimension of globalization: cross-border trade in services. The service sector has been growing steadily and already accounts for 70 percent of employment and 73 percent of gross domestic product (GDP) in developed countries and for 35 percent of employment and 51 percent of GDP in developing countries (UNCTAD 2008). As infrastructure and skills improve in developing countries, cross-border trade in services is expected to continue to expand.

This chapter aims to help policy makers take advantage of the opportunities presented by increased cross-border trade in IT services and IT-enabled services (ITES). It begins by defining the two industries and estimating the potential global market opportunities for trade in each.[1] Then it discusses economic and other benefits for countries that succeed in these areas. It also analyzes factors crucial to the competitiveness of a country or location—including skills, cost advantages, infrastructure, and a hospitable business environment—and examines the potential competitiveness of small countries and of least developed countries. The chapter concludes by discussing policy options for enabling growth in the IT services and ITES industries.

Much of the analysis and policy advice presented here is based on inputs from consultants, policy experts, and industry leaders, including work conducted by McKinsey & Company under a recent consulting engagement with the World Bank and the Information for Development Program (*info*Dev). Analysis based on expert knowledge was found to be more useful than efforts to conduct quantitative analysis of various policy options. The large number of potential explanatory variables would require extensive data that are not yet available given the limited number of countries with significant experience in the IT services and ITES industries.

Large Markets and Growing Opportunities

IT services typically include IT applications and engineering services, while ITES involve a wide range of services delivered over electronic networks (table 7.1). These are two broad segments, however, and the sophistication of the services in each varies considerably.

Estimating the market size for trade in IT services and ITES is difficult given definitional issues and the relative novelty of the field. Official statistics are often not available or not reliable, and calculations based on balance of payments and trade in services do not accurately isolate IT services and ITES. As a result, much of the data on the size of the current market comes from private surveys, consulting firms, and anecdotal evidence. According to McKinsey estimates, the annual addressable market for IT services and

Table 7.1 Types of IT Services and IT-Enabled Services

IT services		IT-enabled services
Application services	Engineering services	Business process services
Application development and maintenance	**Manufacturing engineering**	**Horizontal processes**
Application development	Upstream product engineering	Customer interaction and support (including call centers)
Application development integration and testing	Concept design	Human resource management
	Simulation	
Application maintenance	Design engineering	Finance and administration
System integration	Downstream product engineering	Supply chain (procurement logistics management)
Analysis	Computer-aided design, manufacture, and engineering	**Vertical processes**
Design		Banking
Development	Embedded software	Insurance
Integration and testing	Localization	Travel
Package implementation	Plant and process engineering	Manufacturing
IT infrastructure services	**Software product development**	Telecommunications
Help desks	Product development	Pharmaceuticals
Desktop support	System testing	Other
Data center services	Porting[a]/variants	**Knowledge process outsourcing**
Mainframe support	Localization	Business and financial research
Network operations	Maintenance and support	Animation
Consulting	Gaming	Data analytics
IT consulting		Legal process and patent research
Network consulting		Other high-end processes

Sources: Adapted from BPAP (2007) and data from McKinsey & Company.

a. *Porting* is the process of adapting software to run on a different computer and/or operating system.

ITES is approximately $475 billion. Less than 15 percent of that market, however, is being exploited—about $65 billion in 2007 (figure 7.1).

Among the various segments of IT application services, opportunities are large in traditional services (about $100 billion),[2] system integration ($50 billion), application development and maintenance ($43 billion), and consulting ($6 billion). For IT engineering services, opportunities are significant in mechanical design and production (about $45 billion), embedded software ($40 billion), and plant engineering ($35 billion).

Estimates of the market size of ITES vary significantly. According to Gartner Research (2008a), the global market is expected to grow from $171 billion in 2008 to $239 billion in 2012. But estimates by NASSCOM (National Association of Software and Services Companies)-Everest (2008) are more than three times that amount, at $700 billion–800 billion by 2012, out of a total cost base of $17 trillion for key industry verticals in source markets.[3] Most estimates of the addressable ITES market are derived by estimating spending on a range of business process functions and evaluating the potential for delivering such functions remotely. Figure 7.2 shows the relative importance of various vertical and horizontal functions in India's ITES industry.

Despite the variation in estimates, it is clear that the demand for IT services and ITES is very large, and that only a small percentage of the potential has been realized. The limiting factors appear to be on the supply side. Countries that meet the requirements of the untapped market are likely to experience rapid growth in their IT and ITES sectors.

Economic Impacts of Developing IT and ITES Industries

India is the global leader in the provision of both IT services and ITES (figure 7.3). Two developed countries—Canada and

Figure 7.1 Global Opportunities for IT Services and IT-Enabled Services
(US$ billion)

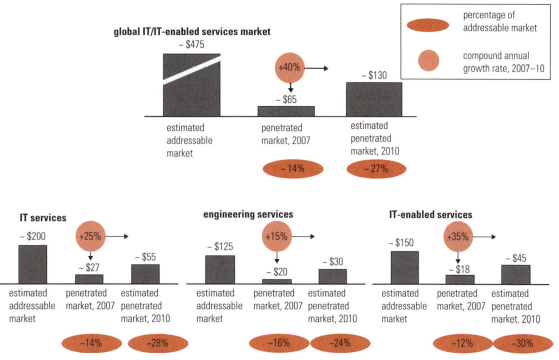

Source: McKinsey & Company.
Note: According to NASSCOM (2009), the addressable market for global sourcing of IT services and ITES was $500 billion in 2008.

Figure 7.2 India's Addressable Market for Vertical and Horizontal IT-Enabled Service Functions

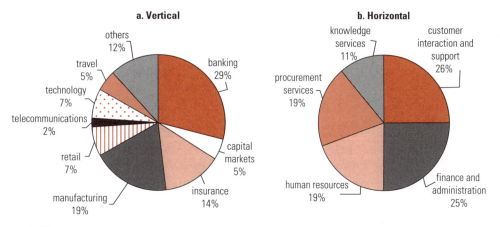

Source: NASSCOM-Everest 2008.

Ireland—have also done particularly well in the industries, as have a few developing countries, notably China, Mexico, and the Philippines. Several countries in Central and Eastern Europe (the Czech Republic, Hungary, Poland, Romania, and Russia) have also developed their capacity in IT services and ITES, though on a much smaller scale. The expansion of IT services and ITES has provided these countries with a wide range of economic and social benefits. In India, the Philippines, and Ireland, for example, the industries have created jobs, raised incomes, and increased exports and GDP.

Realizing the Opportunities Presented by the Global Trade in IT-Based Services

Figure 7.3 Global Distribution of Offshore IT Services and IT-Enabled Service Markets

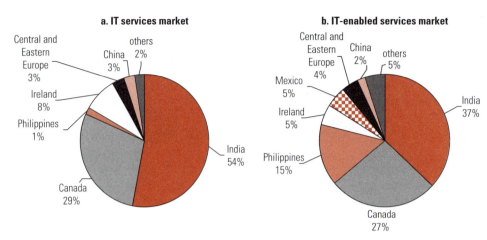

Sources: McKinsey & Company 2008; NASSCOM-Everest 2008; Tholons 2006.

India

The best-known IT services and ITES success story is India. In 2007–08,[4] total exports of IT services and ITES from India stood at $40.4 billion ($23.1 billion in IT application services, $6.4 billion in engineering and research and development (R&D) services, and $10.9 billion in ITES). The IT services and ITES industries contributed one-quarter of the country's total exports and nearly half of service exports in 2007.[5] In addition to the exports, $11.6 billion of domestic software services were also produced. In sum, IT services and ITES represent 5.5 percent of India's GDP and together grew at a remarkable rate of 33.7 percent in 2007 (NASSCOM 2008a). Going forward, India's IT services and ITES exports are forecasted to reach $60 billion by 2010, when the sector is expected to represent almost 7 percent of GDP (NASSCOM-McKinsey 2005).

A study on the output linkages of India's IT services and ITES sectors conducted by Credit Rating Information Services of India Limited (CRISIL) concluded that the total turnover of $30.3 billion for the sectors in 2005–06 implied spending of $14.3 billion in the domestic economy, which in turn generated additional output of $14 billion in sectors linked to IT services and ITES (NASSCOM-CRISIL 2007).[6]

The IT services and ITES industries have an important impact on the labor market in India. The industries directly employ 2.01 million people in jobs that pay 50 to 100 percent more than comparable service sector jobs. Nearly 80 percent of these jobs (1.56 million) cater to exports of IT services and ITES, while another 0.45 million serve the domestic market. In addition, the sectors create indirect employment opportunities in industries such as construction, retail, transport,[7] and telecommunications, as well as induced employment due to higher spending on goods and services such as food, transportation, entertainment, health, and medical services. McKinsey estimates that each new job in IT services and ITES in India has led to the creation of between three and four new jobs in other sectors (NASSCOM-McKinsey 2005).[8] Other estimates put the number of new non-IT/ITES jobs at four for each job created in IT services and ITES (NASSCOM-CRISIL 2007). Altogether, an estimated 8 million to 10 million employees directly or indirectly support the IT services and ITES industry in India.

The Philippines

The Philippines is another important beneficiary of international trade in IT services and ITES, as it is now one of the top destinations for IT services and ITES companies in the world. Growth of the sector in the Philippines has been impressive: total IT services and ITES revenues reached $6 billion in 2008, up from $100 million in 2001.[9] As of mid-2008, the industries employed 345,000 people, up from 100,000 in 2004. Moreover, in the Philippines, as in India, workers in this sector are typically paid 50 to 100 percent more than in other service jobs and tend to fall into the top income quintile (Roxas-Chua 2008).

The BPAP projects that it is possible for the IT services and ITES industries in the Philippines to continue their rapid growth, doubling their combined share of the global market from 5 percent to 10 percent and producing revenues of about $13 billion and direct employment for close to 1 million people by the end of 2010. Employment of this scale means that the sector would account for 27 percent of all new jobs created in the country by 2010.[10]

The BPAP estimates that each new job created in IT services and ITES in the Philippines results in two to three new jobs in other sectors. An increase in direct employment of 600,000 people by 2010 would therefore create 1.2 million–1.8 million additional new jobs indirectly as employees consume housing, food, transport, and consumer goods and employers invest in telecommunications, building rentals, water, and other core services. By 2010, the IT services and ITES industries could represent as much as 8.5 percent of GDP (BPAP 2007).

Ireland

Ireland has built an IT services and ITES sector that is widely regarded as essential to the country's rapid economic growth. Until the late 1980s, Ireland was the poorest country in Western Europe and suffered from deteriorating infrastructure, high unemployment (20 percent), and a well-documented brain drain to the United States, the United Kingdom, and elsewhere.

In the years since, directed efforts by the Industrial Development Agency (IDA) to build the country into an IT services and ITES destination using corporate tax incentives, enterprise zones, and other incentives, along with European Union (EU) aid and successful marketing efforts, resulted in a high-tech industry that employed 80,000 people by 2000. The call center program introduced by the IDA in 1992 was particularly successful: by mid-1998, around 50 call centers employed 6,000 people, twice as many as the original plan (Barry and van Welsum 2005).

The direct economic impact of the growth of IT services and ITES in Ireland has been mainly from activity in financial and other ITES services. Following the establishment of the International Financial Services Center (IFSC) in 1987, almost 450 international financial institutions operate in Dublin, including half of the world's top-50 banks and half of the top-20 insurance companies. The IFSC focuses on international wholesale banking and treasury, securitization, fund management, fund administration, and insurance (Economist Intelligence Unit 2008). Financial services companies employ 16,000 people and pay an estimated 15 percent of all corporate tax in Ireland.

Other Impacts: Social Benefits, Policy Reforms, and Country Brand

Success in the IT services and ITES sector engenders a number of other positive impacts. An important one is the positive impact on the status of women. Women account for about 65 percent of the total professional and technical workers in IT services and ITES in the Philippines. In India, women make up 30 percent of the IT services and ITES workforce—a much higher rate of female participation than in the service sector in general—and the share is expected to grow to 45 percent by 2010. In Ireland, 70 percent of call center employees are women. In all these cases, women represent a greater number of high-paying jobs than in most other sectors of the economy.

In addition to direct economic and social benefits, a focus on developing the IT services and ITES sector can catalyze fiscal, regulatory, and legal reforms. Policy reforms are often easier to enact when a "new" export-oriented sector such as IT services and ITES is targeted, since entrenched special interests are less directly affected than when reforming other sectors. This appears to have been the case in several states in India, where IT services and ITES companies have been exempted from many of the regulations that make doing business in India a slow and uncertain process. As the value of more efficient fiscal, regulatory, and legal regimes becomes increasingly appreciated, innovations and reforms can be extended to other sectors of the economy.

Finally, success in IT services and ITES presents opportunities for repositioning the image of a country, a "branding" effect that can have profound implications. In India, the positive impact of industry leaders such as Genpact, Wipro, TCS, and Infosys points to this effect. As one commentator put it, "More importantly, [the IT sector's] impact was psychological. It signaled to the world that India was much more than its old historical stereotypes. It suddenly . . . made the world think that every Indian was smart and could fix [its] computers. That helped entrepreneurs in India from all industry segments because it gave them a more receptive environment in which to do business" (Masani 2008).

Assessing Potential Competitiveness

Governments that wish to take advantage of global opportunities in IT services and ITES can benefit from a structured assessment of the strengths and weaknesses of their countries' locations. In recent years, a number of consulting firms have developed benchmarking frameworks, locational indices, and rating criteria for determining the readiness and attractiveness of different locations for IT services and ITES industries. Among these studies, there is broad agreement that several key factors determine locational competitiveness: availability of employable skills (including IT skills), competitive costs, quality of public infrastructure relevant to the IT services and ITES industries, and an overall environment that is conducive to business. Table 7.2 provides a more detailed list of factors in each of these categories.

Table 7.2 Frameworks for Assessment of Locations for IT Services and IT-Enabled Services

AT Kearney's Global Services Location Index	Gartner's 10 criteria	Hewitt's International Benchmarking Model	McKinsey's Locational Readiness Index
People and skills availability	**Infrastructure**	**Infrastructure**	**Quality of infrastructure**
Remote service sector experience and quality ratings	Power	Real estate	Telecom and network
	Telecommunications	Telecom	Real estate
Labor force availability	Transport	Power	Transportation
Education and language	**Labor pool**	**Connectivity**	Power
Attrition risk	Quality	**Talent**	**Talent**
Financial attractiveness	Quantity	Availability	Availability
Compensation costs	Scalability	Quality	Suitability
Infrastructure costs	Work conditions	Cost	Willingness
Tax and regulatory costs	**Educational system**	**General demographics**	Accessibility
Business environment	Quality	**Environment**	Trainability
Country environment	Number of institutions	Macroeconomic	**Cost**
Infrastructure	New grads in IT	Business environment	Labor cost
Cultural exposure	**Cost**	Geopolitical environment	Infrastructure cost
Security of intellectual property	Labor	**Clusters**	Corporate tax
	Real estate	**Incumbent IT/ITES industry**	**Market maturity**
	Infrastructure		IT/ITES employees as percentage of total service sector employment
	Telecom		IT/ITES as percentage of services GDP
	Political and economic environment		Presence of industry association
	Stability of government		**Risk profile**
	Corruption		Regulatory risks
	Geopolitical risks		Country investment risks
	Financial stability		Data protection
	Language		**Other incentives**
	Government support		**Environment**
	Promotional		Government support
	Institutional		Business and living environment
	Education		Accessibility
	Cultural compatibility		Living environment
	Cultural attributes		
	Adaptability		
	Proximity		
	Ease of travel		
	Global and legal maturity		
	Data and intellectual property security and privacy		

Source: Authors' summary.

Talent Pool

Together with the existence of competitive broadband telecommunication markets, the availability of employee skills is the single most important factor in the growth of the IT services and ITES sectors. In fact, their growth has created a situation in which skills scarcity creates opportunities for countries new to the industries to offer and develop strong local talent pools. India, which has about 30 percent of the global supply of low-wage labor for the IT services and ITES industries (McKinsey Global Institute 2005), is likely to have a talent shortfall of 0.8 million–1.2 million by 2012 (NASSCOM-Everest 2008). In 2007, India's top-five IT companies alone hired 120,000 new employees, and many Indian companies have begun recruiting from international talent pools. TCS, for example, now employs more than 10,000 non-Indians, who make up 9.1 percent of its staff (Wadhwa, de Vitton, and Gereffi 2008). Just five years ago, the company employed fewer than 100 non-Indians.

A typical assessment of the university graduate talent pool by an IT services or ITES company considers a number of aspects, including the following:

- Suitability for employment—that is, meeting a quality standard for work in the industry and having the necessary language (not necessarily English) skills. In a study conducted by the McKinsey Global Institute, an assessment of the available talent pool across 28 developing countries found that, on average, only about 13 percent of generalist graduates had the necessary qualifications (including language) for being employed in the sector (Farrell 2007). Education content is rarely aligned with industry needs.

- Willingness to work in the industry—a function of both the stature of the industry and other job options available.

- Accessibility—that is, proximity to a proposed IT/ITES site or a willingness to relocate.

- Trainability—that is, of the nonemployable cohort, the number who could be employable following short-term training courses.

An important consideration for many large companies is the scalability of the suitable talent pool, that is, whether it is large and growing so that firms can scale up their businesses without having to look for talent in another location. In addition to the above factors, companies considering investment in IT services and ITES also look at parameters such as average retention and turnover rates, maximum number of hours in a work week, average premium for overtime work, minimum wage, conditions of employment mandated by legislation, regulations on severance and termination of employees, restrictions on expatriates working in the country, and ease of travel clearances for visiting executives (Sutherland Global Services 2008).

Cost

Primary cost considerations, from the point of view of a company making an investment decision, include the cost of labor (from entry-level employees to seasoned managers); infrastructure costs; selling, general, and administrative expenses (SG&A); and facilities costs. Table 7.3 presents an illustrative example of the share of different cost components for IT services and ITES businesses and suggests that the most important cost elements are wages, physical infrastructure, and training.

The evaluation of cost also reflects fiscal or other incentives provided by the government to encourage investment, as well as tariff or trade restrictions regarding imports and exports, corporate tax rates, regulations on profit remittances and repatriation of capital, capital gains on assets and other property transfers, and special incentives and tax holidays. Companies assess the different business taxes (value added tax [VAT], withholding tax, excise duties, stock transaction taxes, capital gains tax, documentary stamp tax, customs duty, and local taxes), and also seek information on tax treaties and their effects on tax rates (Sutherland Global Services 2008).

Table 7.3 Relative Percentage of Components in the Total Cost of Offshoring

Cost component	IT services firm	ITES firm
Wage rate	46	42
Physical infrastructure and support	18	17
Training and productivity	9	10
Transition and governance	7	8
Communications	3	5
Disaster recovery and business continuity	3	5
Resource redeployment	4	3
Travel costs	3	3
Advisory services	4	2
Exchange rate changes	3	2
Resource redundancy	1	2

Source: Vashistha and Vashistha 2006.

Given the cost advantage of most developing countries compared to developed countries, tax incentives may not be required to enhance the attractiveness of developing country locations. However, they often signal the importance that governments attach to the IT services and ITES sectors and demonstrate governments' commitment to nurturing a conducive business environment for the industries.

Infrastructure

Infrastructure consists of availability, quality, and reliability of services such as telecommunications (including broadband), power, and transportation, along with availability of suitable real estate. Competitive broadband telecommunications markets are a particularly critical factor for the growth of trade in IT services and ITES. In addition to cost and quality, most IT services and ITES companies require redundancy in terms of telecommunication links. It is important, therefore, to ensure that more than one international carrier is available, and that there is more than one international gateway and one international cable linking the location to competitive global communication networks.

In countries with unreliable public infrastructure, companies look for customized facilities such as IT parks—with modern office space, high-speed broadband links, reliable power supply (including backup supply), security services, and ancillary infrastructure including banks, travel desks, restaurants, transportation systems, and hotel accommodation for visiting executives. They also look for availability of land and business-friendly procedures such as quick building clearances for real estate development. The availability of international airports with good flight connections near IT/ITES locations is also an important factor.

Business and Living Environment

The general business and living environment of a country—government policies toward foreign direct investment (FDI), incidence of corruption, labor laws, ease of travel to and from the country, and general quality of life—is also important in a company's decision about whether to invest there. Many of these factors can be more easily controlled by focusing on a discrete sector like IT services and ITES, and later expanding efforts to the broader market environment. There are numerous cases of countries offering special status for IT/ITES investors to speed them through the formalities and insulate them from the more difficult aspects of doing business locally. The Agency to Promote and Facilitate Investments in Remote Services and Technology (APFIRST) in Andhra Pradesh in India and the IDA in Ireland, for example, cut through red tape to help IT services and ITES companies start local operations, while the broader business environment strengthened more slowly.

The living environment also influences companies' decisions about where to locate in terms of availability of healthcare facilities, international schools and other high-quality academic institutions, entertainment facilities, civic infrastructure, public safety, and hygiene.

Country risk relates to stability and transparency of law, macroeconomic stability, treatment of foreign capital, and data and intellectual property law protection, to name a few. Potential investors weigh these risks according to their own history and the risk-taking profile of their management. Companies make decisions about locating IT services and ITES investments based on their assessment of the judicial system, the average duration to resolve disputes, the legal framework for contract enforcement, average time to resolve contractual conflicts, opportunities to arbitrate locally, the legal framework for intellectual property protection, and antitrust laws, among other factors (Sutherland Global Services 2008).

In summary, the elements that make a country an attractive IT services and ITES investment destination are a combination of depth and quality of the talent pool, cost advantage, availability and quality of infrastructure, and other factors that facilitate the smooth and predictable day-to-day running of a business. These factors, with different weightings based on different approaches, tend to be common to the indices and tools used by industry analysts and consulting firms (such as the ones summarized in table 7.2) active in the IT services and ITES industries.

Relative Competitiveness of Small Countries

Given the large addressable market for IT services and ITES, there is an opportunity for many countries to participate and benefit. In recent years, an increasing number of countries have begun to develop their IT services and ITES industries, viewing them as important potential sources of economic growth. South Africa, for example, is emerging as an attractive ITES destination, benefiting from English language abilities (South Africa Online 2005). Similarly, the Arab Republic of Egypt, Morocco, Senegal, and Tunisia are also developing a range of ITES operations, including call and contact centers, and Israel is starting to emerge as a location for packaged application development.

An important question is whether the opportunities presented by the IT services and ITES sector are possible only for countries with a large talent pool, or whether small economies and least developed countries can benefit as well. Scalability is an important success factor, as many companies prefer locations where scaling up is possible. This is particularly true for large, "commodity" market segments that require a large number of workers with comparable skills, such as telemarketing and consumer support call centers, and providers of standard back-office functions such as accounting and IT support. Countries with large and growing employable labor forces thus may have a competitive advantage in capturing a share of the global IT services and ITES markets.

For a number of niche segments, however, the basis for differentiation may center on factors other than scalability. In R&D, for example, skills quality appears to be a much more important differentiating factor. In addition to the case of Ireland discussed earlier in this chapter, the following country examples illustrate potential growth opportunities in niche segments.

Mauritius. With an area of 2,040 square kilometers and a population of 1.27 million, Mauritius employed approximately 7,000 people in IT services and ITES in 2007, compared to only 2,000 in 2003. Mauritius used the competitive advantage of its historical and cultural ties with India to establish Ebene Cybercity, an IT park. A soft credit from India of $100 million in 2001, along with a bilateral agreement facilitating travel between India and Mauritius, made it possible to attract investment from a number of Indian companies, including Infosys, which has set up a disaster recovery center on the island (CNET News.com 2003; ITU 2004). Mauritius has successfully attracted leading international players such as Accenture, TNT Group, Teleforma, Ceridian, and EURO CRM (Burton 2007), and has begun transitioning to higher value–added activities such as advisory, design, and legal services (MBOI 2007). As another indication of Mauritius's success, A.T. Kearney (2007) ranked the country ahead of other locations well known for their attractiveness, in spite of Mauritius's smaller size, higher costs, and direct competition with the other countries in the francophone market.

Malta. Malta, with an area of 316 square kilometers and a population of 0.4 million, is an even smaller country than Mauritius. The World Economic Forum's *Global Information Technology Report 2007–2008* ranks Malta third, after Singapore and Tunisia, in terms of government success in promoting information and communication technology (ICT). Between 2001 and 2004, Malta embarked on the e-Malta strategy. In 2008, it launched a new national ICT strategy that aims to make Malta a "smart island" (IDABC 2008). The country has successfully attracted specialized software firms like Crimsonwing, Uniblue, GFI, Anvil, 2i, and RS2, in addition to leading IT players such as Oracle, Microsoft, Hewlett-Packard, and SAP (Malta Enterprise 2008). The call center that HSBC established in Malta for its U.K. operations has grown to over 450 employees (*Times of Malta* 2008). Malta has also become an attractive destination for remote gaming: it now hosts 10 percent of all remote gaming companies in the world, including Betfair, Expekt, Unibet, Interwetten, and CBM Bookmakers (Malta Enterprise 2008).

In 2007, the government embarked on the SmartCity Malta project with a $300 million investment from Dubai Internet City's Tecom. Encompassing an area of 360,000 square meters, SmartCity Malta will consist of office, residential, and retail space focused on attracting ICT and media companies (SmartCity Malta 2008). The first phase of the project was inaugurated in June 2008. SmartCity Malta is the largest foreign investment in the island country and the single largest job creation initiative in Malta's history, committed to creating 5,600 jobs over eight years (OANA 2008).

The examples of Ireland, Mauritius, and Malta suggest that size is not a binding constraint in the potential for countries to benefit from global IT services and ITES opportunities. Small countries can aim at specific niches: leveraging language skills, as in the case of Malta; building on historical and cultural ties, as in Mauritius; using advantages such as state-of-the-art infrastructure, a skilled workforce, security, and a strong legal system, as in Singapore; or operating at the high end of the value chain to compete on quality, as in Ireland.

Relative Competitiveness of Least Developed Countries

The question of whether least developed countries have the potential to become players in the global IT services and ITES industries is an engaging issue. Industry experts and current trends suggest that countries with severe constraints in infrastructure, a small employable labor force, and no clear competitive advantage enabling differentiation in high-end markets may not be immediately attractive to investors and companies looking to establish IT services and ITES operations. Such countries may, however, recognize and plan for the longer-term potential that the industries represent, assuming these constraints can be addressed over time.

Deliberate investment in human resource and infrastructure development, in a manner that is geared to meeting the skill requirements of the IT services and ITES industries globally, is likely to be a sound policy for least developed countries. In a context where companies around the world are learning to operate in different geographies, it can be expected that locations equipped with employable skills, decent infrastructure, and a stable and conducive business environment will be able to take advantage of the opportunities presented by the IT services and ITES industries.

It is important to note that a relatively small investment and a small number of jobs in the IT services and ITES industries can have a considerable impact on the economy of a country. While the IT services and ITES industries currently employ less than 1 percent of the labor force in India, for example, the sector is responsible for one-quarter of the country's exports. Although India may be viewed as a unique case, evidence suggests that the IT services and ITES sectors may contribute just as much to other, smaller economies. The percentage of population employed in the IT services and ITES industries in Mauritius was nearly four times as high as in India in 2007. Least developed countries may see the case of Mauritius as an example and invest in human resources, infrastructure, and the general business environment in their own countries in order to position themselves for success in the medium and long terms.

Policy Options to Enhance Competitiveness

A fundamental question faced by governments is whether to focus on industry-specific policy, such as the development of the IT services and ITES industries, in addition to working to improve the broader business environment.

Policies Targeted at the IT Services and ITES Industries

Opponents of industry-specific policy point to the dismal record of governments in supporting specific sectors, emphasizing that the task is best left to markets. Governments should, in their view, focus on macroeconomic stability, ensure property rights and contract enforcement, and improve the general environment for doing business.

Proponents of targeted industry support point out that (1) countries that have succeeded have seen their governments making deliberate interventions to catalyze growth of the sector; (2) many of the policy enablers needed by the IT services and ITES industries involve "no-regret" interventions that also benefit the rest of the economy; and (3) a broader approach to policy, aimed at the overall business environment and not at the IT services and ITES industries specifically, is likely to miss key interventions and be out of sync with the dynamic needs of these industries.

Countries successful in adopting IT/ITES industry policy. Although proactive policies may not be a sufficient condition for building successful IT services and ITES industries in an individual country, all the success cases reviewed here have involved active government support—albeit support not necessarily focused on the needs of the IT services or ITES industries.

In India, long-term investment in world-class technology institutes produced a critical mass of technology leaders able to compete globally. In the state of Andhra Pradesh, education policies in the late 1990s and early 2000s liberalized entry of private technology institutes in the tertiary education market, multiplying the number of engineering graduates available for the IT sector in only a few years. The Software Technology Parks of India (STPI) initiative that was launched by the government in 1991 to overcome infrastructural and procedural constraints by providing data communication facilities, office space, and "single window" statutory services was extremely beneficial. The technology parks proved essential to the growth of the industry given the broader context of deficient infrastructure and bureaucratic red tape. India's telecommunications policies of 1994 and 1999 allowed private sector investments into the sector and cleared the path for establishment of alternative international gateways that were also critical to development of the IT services and ITES industries.

In Canada, the government offered special incentives to IT and ITES companies that would develop a significant volume of contact center operations in the Atlantic provinces. Ireland's emergence as an IT and international financial services center is widely recognized as partly the result of proactive government policies that encouraged investment in these industries. In the Philippines, the Board of Investments has actively targeted the IT services and ITES sectors and has been credited by the local industry association for its key role in supporting the rapid growth of the sector.

No-regret interventions. Investments that enable the IT services and ITES industries include those in education, infrastructure, and regulatory reform. All of these in turn contribute to improving the broader business environment. While a less targeted approach may not prioritize these actions in the timeline demanded by the industries, most of the reforms and investments required to develop the IT services and ITES industries can be seen as "no regret" actions. For example, a critical mass of low-cost labor with English language skills, problem-solving abilities, and basic IT proficiency is likely to be useful to other industries in the event that the IT services and ITES industries do not develop. In addition, high-capacity telecommunications infrastructure and modern power infrastructure are likely to benefit other industries. Ancillary investment in IT parks, where the bulk of the development investment is made by private developers, does not represent a large, risky public expenditure (other than the land, which is often provided as equity by governments). In this sense, government support for the IT services and ITES sector is a low-risk strategy and is consistent with the argument that public interventions should create positive externalities.

Institutions and Leadership Following an Adaptive and Engaged Approach to Policy

Locations successful in attracting IT services and ITES companies do more than rely on highly structured strategies—rather, they have leadership and institutions that follow an adaptive and engaged approach to policy. Given the fast-moving nature of the industries globally, the domain of policy and investment promotion is a constantly moving target. Unless the institutional framework is agile, it will be difficult to adapt to changing market conditions and achieve and sustain success. The institutional structures for promoting IT services and ITES ideally should include a level of government that is sufficiently high to have cross-cutting oversight and should promote close engagement between the public and private sectors in order to adapt policy to evolving opportunities and sources of competitive advantage. A number of examples bear out the efficacy of such an institutional approach.

Ireland's IDA, a government-sponsored development agency funded primarily through government grant-in-aid, has achieved significant success in attracting IT services and ITES investments to the country. The inward investment program launched by the IDA has been a major driving force behind the growth of the Irish economy, contributing to 35 percent of GDP and over 85 percent of manufactured exports (IDA 2006). By its own account, "IDA is a full service national development agency, a so-called 'one-stop shop.' It deals with all aspects of inward investment—the planning, promoting, marketing, negotiating and processing of investment proposals; provision of financial incentives and property solutions; helping new investors get started and working with them to maximize their contribution to the Irish economy" (IDA 2006). And 9 of IDA's 13 board members are from the private sector.

Another example of a successful development agency in the context of IT services and ITES is APFIRST (renamed and reconstituted as APInvest in 2005) in Andhra Pradesh. Established in 2001 to promote investment and development in key sectors including offshore IT and business process outsourcing (BPO), APFIRST was set up as a one-stop contact agency, with a dedicated budget for marketing and promotional activities, authorization to grant incentives such as single-window clearances in order to attract investors, and a dedicated account manager for key investors. The agency has been a resounding success.

From 2001 to 2005, when the global ITES industry grew at a cumulative average growth rate of 49 percent, the ITES sector grew at more than twice this rate in Andhra Pradesh (110 percent). Starting from a low base of $14 million in 1995, Andhra Pradesh's exports of IT services and ITES grew to $450 million in 2001 and to more than $6 billion in 2007 (*The Hindu* 2008).

The success of APFIRST illustrates the importance of an investment promotion institution that has cross-cutting oversight. When APFIRST was trying to attract Microsoft to establish a campus in Hyderabad, for example, it had to work with a number of government departments to clinch the investment. APFIRST negotiated with the Indian School of Business (ISB) to provide part of its land to Microsoft (the ISB was compensated with additional land), facilitated funding of roads to the campus, and arranged for an alternative source of electricity at the site. The 54-acre Hyderabad facility[11] is the second-largest Microsoft campus in the world, after the company's headquarters in Redmond, Washington. When courting Dell to make an investment decision, APFIRST worked with the Andhra Pradesh State Council of Higher Education to train students who could be hired by Dell for the company's Hyderabad call center. It also worked with telecommunications companies to provide high-speed bandwidth with redundant links for the Dell facility. Since each major company had its own set of requirements, APFIRST's ability to coordinate across government and existing business institutions was critical in successfully attracting new companies.

A holistic marketing and business development focus was another factor contributing to the success of APFIRST. The organization hired a leading management consultancy to obtain market intelligence and identify competitive advantages while simultaneously leveraging the firm (in addition to its own and Andhra Pradesh's leadership) to reach out to key decision makers in top global companies.

Development institutions in Chile and Malaysia have pursued similar relationship building initiatives. Chile's economic development agency, CORFO (Corporación de Fomento de la Producción de Chile), established a partnership with the Thunderbird School of Global Management in Glendale, Arizona, to undertake market research and establish contacts with key organizations such as the American Electronics Association and the San Jose Business Incubator (Nelson 2007). Malaysia set up an International Advisory Committee for its Multimedia Super Corridor to facilitate engagement with leading global players.

In addition to government investment promotion institutions, industry associations can also be effective in carrying out branding and industry promotion initiatives. NASSCOM, for example, has not only played an important advocacy role with policy makers for the IT services and ITES industries, but also has successfully created an India brand that is now recognized internationally (World Bank 2008). NASSCOM's success in a branding initiative has been emulated by agencies in other countries, notably the Brazilian Association of Information Technology and Communication Companies (BRASSCOM), the Bulgarian Association of Software Companies (BASSCOM), and the Ghana Association of Software and IT Services Companies (GASSCOM).

Governments need to be proactive in attracting strategic anchor investors in order to gain a critical mass of investors. This critical mass can generate dynamic cluster effects and help raise visibility as a potential destination for IT services and ITES. When Andhra Pradesh succeeded in getting Microsoft to locate a software development center in Hyderabad, it became much easier to attract follow-on investments from other high profile companies such as Oracle, IBM, and Accenture, which in turn triggered a cluster effect that encouraged investment from many more companies.

Policy Options for Nurturing and Expanding the Talent Pool

After access to high-bandwidth telecommunications infrastructure, the availability of employable talent is the single most important determinant for the growth of the IT services and ITES industries in the long term. As mentioned above, public education content is too often divorced from the needs of industry. When examining policies related to the talent pool, institutional mechanisms for aligning skills development with the needs and requirements of the industries are in our view the most important factor for success. In this regard, the government of Mexico facilitated a new organization in 2008, MexicoFIRST, founded by Cámara Nacional de la Industria Electrónica de Telecomunicaciones e Informática (CANIETI) and Asociación Nacional de Instituciones de Educación en Informática (ANIEI) to work closely with the Asociación Mexicana de la Industria de Tecnologías de Información (AMITI) and ProSoft, a government agency tasked with promoting the IT services and ITES industries. MexicoFIRST will closely interface with IT and ITES companies on the one hand and Mexican universities on the other to identify the training needs of the companies and to facilitate training programs at the universities to meet those needs.

Another important policy intervention is to improve the quality of education in order to develop generic skills that are relevant to a broad spectrum of industries. An example of this approach is the NASSCOM assessment of competency (NAC) framework, which the organization developed in consultation with a large number of ITES players. The NAC has emerged as India's national standard for generic skills and recruitment of entry-level talent for the ITES industry (NASSCOM 2007), and NASSCOM has rolled out the framework in partnership with a number of state governments in India. The skill testing themes under NAC are shown in table 7.4. The test scores indicate areas for improvement, allowing customization of further training. NASSCOM has subsequently developed a NAC-Tech certification (NASSCOM 2008b) that is focused on benchmarking engineering skills for the IT industry. This too is being rolled out in partnership with state governments. Applying and enforcing common industry certification not only helps to align skills with industry requirements, but also provide IT services and ITES companies with a more accurate estimate of the talent pool available and reduces their recruitment costs.

Still another important policy intervention aimed at nurturing the talent pool is the establishment of mechanisms to allow just-in-time training for IT services and ITES. A number of countries are providing training grants for this purpose. South Africa offers a training and skills support grant toward the cost of company-specific training up to 12,000 rand (approximately $1,700) per agent. Under its ICT Capacity Building Program, Sri Lanka offers grants to fund a portion of the training costs of IT services and ITES companies. Sri Lanka also offers grants up to $10,000 to bring in a specialized trainer from abroad under a "train the trainer" program. In November 2007, the president of the Philippines directed the Technical Education and Skills Development Agency (TESDA) to allocate 350 million pesos (approximately $8 million) to provide scholarships for training 70,000 call center agents (TESDA 2008). Singapore has a national Skills Development Fund for upgrading worker skills and has launched the Initiatives in New Technology scheme to establish new capabilities within companies or industries by encouraging manpower development in the application of new technologies, industrial R&D, and professional know-how (SEDB 2008).

Given the significant shortage of skills, many large IT services and ITES companies are taking up skills development initiatives, building dedicated training centers, and employing hundreds of training staff. Infosys's new Global Education Center in Mysore, India, for example, has more than 300 full-time faculty and is able to train 13,500 employees at a time. The company invested more than $120 million in this 335-acre, 2-million-square-foot facility. Satyam, after establishing a 240,000-square-foot School of Leadership in India, has announced that it will build a Satyam Technology and Learning Center within Deakin University at Geelong in the State of Victoria, in Australia (Gartner Research 2008b).[12] Given the need to address the shortage of skills, a number of IT services and ITES companies are collaborating with academic institutions. Some examples in India are the following (Wadhwa, de Vitton, and Gereffi 2008):

- Accenture-Xavier Labor Research Institute (XLRI) Academy

- VLSI (very-large-scale integration) Finishing School, established as a partnership between Cadence and the University of California Extension at Santa Cruz

- NIIT Institute of Process Excellence, a joint venture between NIIT and Genpact

- NIIT Institute of Finance, Banking and Insurance, a joint venture between NIIT and ICICI Bank

- Infosys's "Campus Connect" program, which brings faculty members from 470 engineering colleges to its training institute for a two-week residential training program

- Satyam's effort to help 103 universities with faculty training, course design, and implementation of e-learning infrastructure

- 24/7's partnership with 200 colleges and even with high schools to prepare students for the BPO industry.

In addition to IT services and ITES companies' efforts to create incentives for training potential employees, some governments and universities have used public funding and public-private partnerships to nurture and expand the talent pool. These initiatives have been designed to expand existing university infrastructure and faculty,

Table 7.4 NASSCOM IT-Enabled Services Skill Competence Testing Themes

Test	Competencies assessed
Keyboard skills	Typing speed and accuracy
Spoken English	Voice clarity, fluency, vocabulary, grammar/sentence construction, accent, and situation comprehension
Writing ability (multiple choice and essay)	Message clarity and comprehension
Listening	Comprehension and accent comprehension
Numerical and analytical	Numerical ability and logical reasoning

Source: NASSCOM 2007.

develop competencies that are benchmarked globally, and forge linkages for skills development with private sector and best-in-class institutions (see box 7.1).

Partnerships with leading standards organizations, industry associations, universities, and companies can prove highly advantageous for developing globally benchmarked skills. Universities in the Philippines, for example, offer courses in finance and accounting similar to those in the United States because accounting principles used in the Philippines are modeled after the U.S. generally accepted accounting principles (GAAP). This has made the Philippines a natural choice for U.S. banks and financial institutions seeking to offshore parts of their operations. Similarly, Sri Lanka has a large number of qualified accountants. (The Chartered Institute of Management Accountants [CIMA], one of the world's largest professional accounting bodies, has its second-largest number of management accountants in Sri Lanka, after the United Kingdom.) Consequently, companies engaged in investment research find Sri Lanka to be an attractive offshoring destination.

Examples of other global skills providers related to IT services and ITES include the Customer Operations Performance Center Inc. (COPC), the world's leading authority on operations management and performance improvement for contact centers. Carnegie Mellon's Software Engineering Institute is a world leader in standards such as Capability Maturity Model integration (CMMI) and has developed a range of programs including those relating to improvement of personal and team software processes. Similarly, the Project Management Institute's Project Management Professional credential program is recognized globally.

Policy Options for Reducing Costs

The biggest component of cost in the IT services and ITES industries is labor. While labor costs are typically difficult for a government to influence, some labor market distortions, such as minimum wage laws, severance requirements, restrictions on women working, or restrictions on nighttime work, have the potential to be addressed. The government of the state of Goa in India, for example, exempted the IT industry from the Minimum Wages Act of 1948 because wages in the industry were much higher than minimum wage and IT companies were averse to being subjected to frequent inspections, rent seeking, and bureaucratic requirements surrounding compliance with the act.

This chapter does not advocate indiscriminate use of tax incentives and subsidies, as they may allow inefficient firms and sectors to persist, may result in decreased tax revenue while firms might have invested without the subsidy, and are difficult to withdraw once given (McKinsey Global Institute [2003] elaborates on the convenience and value of tax incentives). Targeted fiscal and other government incentives to catalyze growth of the IT services and ITES industries can, however, be helpful. Toward this end, several governments have decreased net costs experienced by individual firms or lowered the income tax rate for a specific sector. Examples include a reduced income tax rate of 10 percent for key software enterprises identified by the government (China), an income tax holiday on profit from exports (India and Singapore), 100 percent tax exemption for qualifying companies for 10 years (Malaysia) or 7 years (Republic of Korea), 100 percent tax exemption for pioneer status companies (Singapore), and fiscal subsidies linked to the number of jobs created (India). In 2002, the government of India revised Section 10A of the Income Tax Act to allow for accelerated depreciation of up to 60 percent for hardware and other equipment in the first year after purchase for IT services and ITES companies.

Other fiscal benefits include adjustment of capital expenses and VAT, as well as duty waivers for IT equipment. Malaysia, for example, offers a 100 percent deduction on capital expenditures. China imposes no customs duty or import VAT for software companies importing capital equipment. India exempts software from customs duty, allows duty-free imports into IT parks, makes computer systems freely importable, and exempts second-hand computers donated to state schools from customs duties. Korea exempts companies set up with foreign investment from custom duties, VAT, and special excise tax.[13] It also offers a 100 percent exemption from dividend withholding tax for foreign investment in technology.

Policy Options to Address Infrastructure Barriers

Broadband connectivity at globally competitive prices is a necessary condition for a successful IT services and ITES sector. Governments need to create an enabling environment for establishing competitive and effective markets in order to attract investment, extend infrastructure access, and

Box 7.1 Government and University Initiatives in Skills Development for IT Services and IT-Enabled Services

Public funding initiatives. Ireland presents several good examples of publicly funded initiatives to expand existing infrastructure, faculty, and IT curricula at universities, colleges, and schools in order to ultimately expand the IT talent pool. By end-2001, Ireland had invested IR£40 million (approximately $79 million) in its "Schools IT 2000" initiative, which provided IT equipment, infrastructure and training, and curriculum resources to schools. The University of Limerick has established a College of Informatics and Electronics that brings together the disciplines of mathematics, software, computing, communications, and electronics. The Dublin City University is focused on the development of skills in the areas of business, science and electronics, computer technology, communications, and languages. Business and IT skills curricula were also introduced in other universities. Partly as a result of these efforts, Ireland now has among the highest proportion of science and engineering graduates as a percentage of all university graduates (31.9 percent) in the European Union (Eurostat 2004).

Partnerships with private sector and best-in-class institutions. Various governments have played a critical role in encouraging ICT-related partnerships with the private sector and academic institutions. Singapore has been one of the most proactive in this regard, starting with the creation of the Industrial Training Board (ITB) in 1973. The ITB established an extensive system of training advisory committees with industry participation, introduced industry-based training schemes in partnership with companies, and established arrangements for keeping training staff abreast of the latest technological developments. The last of these was done, for example, through memorandums of understanding with companies including Mitsubishi Electric Asia, Robert Bosch (SEA), Siemens, IBM, Cisco, and Sun Microsystems (Lee and others 2008). The SEDB also began working with large companies to set up specialized training facilities, such as the ones for Tata Group's precision engineering plant in Singapore. The InfoComm Development Agency of Singapore has been active in forging global partnerships to improve ICT sector skills. For instance, in 2006 it partnered with Carnegie Mellon University's Entertainment Technology Center and the National University of Singapore's School of Computing in order to develop a degree program in interactive digital media (CMU 2006).

In Malaysia, the Penang Skills Development Centre (PSDC) is a joint partnership between the government, academia, and industry.[a] Established in 1989, it has a membership of 141 companies and is led by the private sector. The Chittagong Skills Development Center[b] in Bangladesh is a similar public-private partnership focused on skills development for the ICT, manufacturing, and services sectors. The Center was established in 2006 in partnership with the PSDC, government agencies, industry associations, and ICT companies such as Alcatel, Ericsson, Huawei, and ZTE.

The government of Andhra Pradesh in India is yet another interesting example of proactive promotion of public-private partnerships (IIIT 2007). The International Institute of Information Technology (IIIT) in Hyderabad was started in 1998 with initial support of buildings and seed funding by the state government. Over time, the IIIT has become an autonomous, self-supporting institution and has developed active relationships with major IT companies including IBM, Signal Tree, Motorola, Oracle, and Satyam, all of which have set up corporate schools on the campus.[c] Andhra Pradesh has also partnered with Dell and GE to offer company-specific training courses in colleges to prepare students for eventual recruitment by those companies.

a. See http://www.psdc.com.my/index.cfm (accessed August 3, 2008).
b. See http://www.csdc.com.bd/ (accessed August 3, 2008).
c. Recently there has been a move to transition these schools into partnerships focusing on high-end research.

improve service quality. Some form of public-private partnerships may be used to encourage the development of broadband networks in commercially unattractive areas; such partnerships have been used for underserved areas in India, Malaysia, Spain, Uganda, and elsewhere (chapter 4 contains a detailed discussion on policies for promoting telecommunications backbone networks).

Korea is a well-known leader in broadband, as a result of policies including full liberalization of the telecommunications market; leveraging of private investment for rapid rollout of broadband infrastructure; and provision of public funding to facilitate uptake of broadband services by citizens, businesses, and public institutions. Rapid deployment of broadband provided important opportunities for Korea's IT industry, and the sector is growing three times faster than the rest of the economy. Particularly fast-growing subsectors include development of search engines and local content. In addition, Korea has developed a competitive advantage in niches of the IT industry, such as online gaming, where Korean companies are the biggest global players (see the analysis in chapter 3 on the economic impact of broadband). In February 2009, Korea announced that it was upgrading its network to boost broadband speeds for citizens to 1 gigabit per second by 2012.

In countries where overall infrastructure is underdeveloped, practical second-best solutions such as IT parks may be justified in order to cluster businesses and thus ease the provision of efficient, high-quality infrastructure services required for development of the IT services and ITES industries. Such an approach may also be helpful in forming a critical mass of investors and attracting a group of support services.

The success of the Stanford Industrial Park (later Stanford Research Park) in California in the United States, which morphed into what is now Silicon Valley, has inspired some governments to establish or facilitate the setting up of IT parks with ambitions beyond provision of basic infrastructure. While there are numerous examples of this, the Malaysian government's development of the Multimedia Super Corridor (MSC) has been one of the more prominent initiatives. Conceptualized in 1996, the MSC aspired to make Malaysia a global IT hub. It has generated revenues of more than RM 13 billion (approximately $4 billion) and 63,000 knowledge-based jobs (MSC Malaysia 2008). More recently, the Dalian Tiandi Software Hub in China is being developed as the "world's largest software, IT service hub" (Livemint.com 2007). The hub will have an area of over 26.5 square kilometers and is funded by private sector investments expected to exceed $2.5 billion (*China Economic Review* 2008).

In 2005, the government of Morocco built the CasaShore zone with world class infrastructure and services and rental costs in line with the most competitive destinations. Building this IT park not only provided the resources that companies need to do business successfully in Morocco, but also clearly signaled the government's commitment to developing the IT services and ITES industries. Likewise, the government of Kenya announced in 2008 that it will develop a 5,000-seat BPO technology park, with a budget of K Sh 900 million (approximately $13 million) already allocated for financing the initial part of the project (Kenya ICT Board 2008). In another case, that of Hitec City in Hyderabad, the government of Andhra Pradesh contributed land as 11 percent equity into the project and provided ancillary infrastructure such as roads, electricity feeders, water, and sewage systems, while Larsen & Toubro was responsible for all other investment in the park.

Competitive incentive packages are often offered for companies to locate to these parks, such as subsidies on the costs of telecommunications (such as in Kenya) and other utilities (such as in the state of Orissa in India). One-stop support services at IT parks range from administration and training to legal and financial services.

Policy Options to Improve the Broader Business Environment

Beyond general policies addressing the broader business environment, policy options include freeing parts of the IT services and ITES industries from burdensome regulation and in some cases providing support from a state agency that has the mandate and the authority to guide businesses through the bureaucratic labyrinth that remains.

The bureaucratic burden may be decreased by removing some licensing requirements and providing expedited approvals for qualifying companies on remaining requirements. Industrial licensing was abolished in India for the electronics sector except for manufacturing electronic aerospace and defense equipment. ITES was declared an "essential services industry" in some of the states in India, allowing "365 × 24 × 7" operations otherwise prohibited by law. In some states in India, a "deemed approval" system that provides automatic approvals if government agencies did not respond to a company request within a stipulated number of days was initiated, and a self-certification option allowed for qualifying companies to self-certify compliance with legal and statutory requirements.

A number of online connectivity and privacy issues are also important elements of the broader business environment that may need to be addressed. Chief among them are the legal validity of online transactions, data security and data privacy protection, Internet protocol (IP) protection, and safeguards against misuse of computing infrastructure (cyber-crime). The enabling environment for legal recognition of online transactions is essential for the IT services and ITES industries. Examples include China's Electronic Signature Law 2005, the formation of a cyber appellate court and digital certification under India's IT Act 2000, and the Malaysian Communications and Multimedia Commission Act. Malaysia's Digital Signature Act and Computer Crimes Act, Singapore's Computer Misuse Act and Electronic Transaction Act, enacted in 1998, and Korea's Protection of Information Infrastructure Act 2001 are examples of attempts to provide assurance against the misuse of computers and computing infrastructure. With regard to intellectual property rights, countries can start by bringing patent, copyright, and trademark laws in line with international conventions such as Trade-Related Aspects of Intellectual Property Rights (TRIPS), as China, India, Korea, Malaysia, and Singapore have done, or the World Intellectual Property Organization (WIPO) Copyright Treaty, which China, Korea, Peru, Senegal, and Singapore (to name a few) have signed onto. Raising awareness about these issues in the legal community and among police, prosecutors, and judges is also key.

Movement of capital can be made freer and double taxation can be avoided by permitting 100 percent FDI into IT services and ITES companies and IT parks (as China, India, Malaysia, and Singapore have done), working to form tax treaties with jurisdictions to which earnings would be repatriated (as China, India, Korea, and Singapore have done), and by establishing export agencies such as the Malaysia External Trade Development Corporation (MATRADE) to facilitate trade between local producers and foreign buyers.

Conclusions

The global market for IT services and ITES is large and growing despite fears of a temporary setback because of the global financial crisis. Limitations to growth are mostly on the supply side, in particular in terms of employees with skill sets that meet the requirements of the market. The globalizing market for skills, however, allows developing countries to take advantage of their cost advantage in terms of labor and to make investments in expanding the skills of their labor forces in order to make them suitable for employment in the fast-growing global IT and ITES industries. Successful participation in the industries has been shown to have a positive impact on job creation, exports, economic growth, and social development.

Locations with comparatively large talent pools will have an advantage in attracting IT services and ITES companies because large companies prefer to source services from locations where scalability is feasible. This is particularly true for "commodity" services such as contact centers and standard back-office IT and accounting functions. Recent successes of small countries show that opportunities exist in a range of niche and higher value-added segments where small countries may be able to compete successfully. The timing and scale of gains differ, however, according to a country's skill endowments, infrastructure, cost advantages, and business environment. Countries that are severely constrained in terms of infrastructure and skills may need to focus on longer-gestation programs to develop their talent pools and basic infrastructure, and thus will take longer to realize the benefits of hosting IT services and ITES companies.

In countries that have succeeded in the IT services and ITES industries, governments have adopted a proactive role in promoting the sector. Such support can often be provided with low levels of public funding by leveraging private sector investments. Most of the public interventions to promote the industries—improving education, providing broadband infrastructure, or streamlining government interfaces with business—are essentially "no-regret" moves that carry little risk.

Locations that have successfully developed IT services and ITES appear to have empowered industry development institutions to follow adaptive and engaged approaches to policy. Winning policy development efforts are characterized by adaptation to the rapidly evolving needs of the industries and by ongoing engagement between government leaders and the industries. The private sector can provide governments with invaluable information and insights on available opportunities, market trends, and future skill requirements, and engagement between private and public sectors can also help overcome investment constraints in key areas of infrastructure and human resource development. Given the importance of skills as a driver of growth of the IT services and ITES industries, a focus on quality of education in close alignment with local and global industry needs is essential.

Finally, the importance of leadership for promoting the IT services and ITES industries must be underscored. Extensive commitment and support from the highest echelons of government are essential to make rapid and deliberate policy choices, to implement them effectively, and to overcome bureaucratic resistance.

Notes

1. According to NASSCOM (National Association of Software and Services Companies, India), the global financial crisis is expected to result in reduced technology-related spending for the first two to three quarters of 2009, but it is expected to pick up in 2010; and "greater focus on cost and operational efficiencies in the recessionary environment is expected to enhance global sourcing" (NASSCOM 2009).
2. Traditional services include hardware and software maintenance, network administration, and help desk services.
3. Verticals refer to industries such as banking, insurance, and telecommunications. Horizontals refer to functions common across industries, such as human resource management, finance and administration, and marketing. Verticals account for 60–65 percent of the addressable ITES market, while horizontals account for 35–40 percent.
4. India's financial year is April 1 to March 31.
5. According to the *Economic Survey 2007–08* conducted by the Ministry of Finance, India's total exports in 2006–07 were $128 billion, of which $76.2 billion were service exports. (See http://indiabudget.nic.in/es2007-08/esmain.htm, accessed on August 2, 2008.) Total IT services and ITES exports during 2006–07 were $31.3 billion, which represents 41 percent of India's service exports.
6. Twenty-six percent of gross income spent by employees was housing related, followed by food items, durable goods, and vacation/leisure. In addition, IT services and ITES firms contributed to increased nonwage spending on construction, transportation, communications, and a host of other sectors.
7. According to the NASSCOM-McKinsey Report 2005, the most important employment generation opportunities will occur in construction (an estimated 1.4 million construction site workers will be employed in FY 2010 to meet the demand to develop additional commercial and residential real estate), retail (1.5 million–1.75 million employees in FY 2010), and transport (650,000–700,000 drivers and assistants will be required to meet industry requirements in FY 2010).
8. According to the report, the "two industries (IT and BPO) directly employ nearly 700,000 people and provide indirect employment to approximately 2.5 million workers" (page 15).
9. Despite the global financial crisis, the Philippines was reportedly still on track to achieve its Roadmap 2010 targets, including capturing 10 percent of the global market for ITES (BPAP 2009).
10. Some commentators find this overly optimistic, for example, Magtibay-Ramos, Estrada, and Felipe (2007).
11. The facility consists of the India Development Center, Microsoft IT-India, and the Microsoft Global Services Center.
12. It is unclear if these plans will go ahead in view of revelations in January 2009 that Satyam was involved in the overstating of revenues and profits for a number of years. Tech Mahindra, an Indian niche IT services company, recently won a bid to buy Satyam.
13. This exemption is applicable to high-technology and large-scale manufacturing industries.

References

A.T. Kearney. 2007. *Global Services Location Index*. http://atkearney.com/main.taf?p=5,4,1,132.

Barry, Frank, and Desiree van Welsum. 2005. "Services FDI and Offshoring into Ireland." Organisation for Economic Co-operation and Development, Paris. http://www.oecd.org/dataoecd/14/0/35032060.pdf.

BPAP (Business Processing Association of the Philippines). 2007. *Offshoring and Outsourcing, Philippines Roadmap 2010*. Makati City. http://www.bpap.org/bpap/index.asp?roadmap.

———. 2009. *The Philippine Business Process Outsourcing Newsletter*. http://www.bpap.org/bpap/publications/Breakthroughs%20March%202009_for%20website.pdf.

Burton, Newraj. 2007. "ICT and Business Processing Outsourcing (BPO) in Mauritius." United Nations Economic Commission for Africa, Addis Ababa. http://www.uneca.org/codi/codi5/content/CODI-V-ICT_BPO_Mauritius-Burton-EN.ppt.

China Economic Review. 2008. "Dalian Aims to Be Silicon Valley in China." July 7. http://www.chinaeconomicreview.com/industrial-zones/2008/07/07/dalian-aims-to-be-silicon-valley-in-china.html.

CMU (Carnegie Mellon University). 2006. "Carnegie Mellon Collaborates with National University of Singapore to Create Concurrent Digital Media Degree." Press Release, CMU, Pittsburgh, PA. http://www.cmu.edu/news/archive/2006/november/nov.-9—-etc,-singapore-join-forces.shtml.

CNET News.com. 2003. "Mauritius: We Want Your Call Centers." http://news.cnet.com/Mauritius-We-want-your-callcenters/2030-1011_3-5086029.html.

Criscuolo, Alberto, and Vincent Palmade. 2008. "Reform Teams: How the Most Successful Reformers Organized Themselves." *Public Policy for the Private Sector* 318: 1–3.

CSDC (Chittagong Skills Development Center). *CSDC Home*. Agrabad C/A, Chittagong. http://www.csdc.com.bd/.

Economist Intelligence Unit. 2008. "Country Profile: Ireland." EIU, London.

Engardio, Pete. 2008. "Mom-and-Pop Multinationals: Improved Software and Services Allow the Smallest Businesses to Outsource Work around the Globe." *BusinessWeek*, July 3. http://www.businessweek.com/magazine/content/08_28/b4092077027296.htm?chan=top+news_top+news+index_small+business.

Eurostat. 2004. "Highest Proportion of Graduates in Science & Engineering in Sweden, Ireland and France." Press release, Eurostat Press Office, Luxembourg. http://www.lex.unict.it/cde/documenti/vari/2004/040119laureati_en.pdf.

Expert Group on Future Skill Needs. 2008. "Future Requirement for High-Level ICT Skills in the ICT Sector." http://www.forfas.ie/publications/forfas080623/egfsn080623_future_ict_skills.pdf.

Farrell, Diana, ed. 2007. *Offshoring: Understanding the Emerging Global Labor Market*. Boston, MA: Harvard Business School Press.

Gartner Research. 2008a. "Gartner on Outsourcing, 2008–2009." Stamford, CT. http://www.gartner.com/resources/164200/164206/gartner_on_outsourcing_20082_164206.pdf.

———. 2008b. "Satyam Initiative to Tap Local Skills in Mature Economy." Stamford, CT. http://www.gartner.com/DisplayDocument?id=650508.

Hewitt Associates. 2006. "Improving Business Competitiveness and Increasing Economic Growth in Ghana: The Role of ICT and IT-Enabled Services." infoDev, World Bank, Washington, DC. http://www.infodev.org/en/Publications.170.html.

Houghton, Alistair. 2008. "Global Search for Software Experts Knows No Bounds." *Liverpool Daily Post*, July 9. http://www.liverpooldailypost.co.uk/business/business-local/2008/07/09/global-search-for-software-experts-knows-no-bounds-64375-2130 8137/.

IDA (Industrial Development Agency) Ireland. 2006. "Guide to IDA Ireland's Legislation, Structure, Functions, Rules, Practices, Procedures and Records." IDA Ireland, Dublin. http://www.idaireland.com/uploads/documents/IDA_Publications/FOI_Manual_November_06_2.pdf.

IDABC (Interoperable Delivery of European eGovernment Services to Public Administrations, Business and Citizens). 2008. "e-Government in Malta." http://ec.europa.eu/idabc/servlets/Doc?id=31512.

International Institute of Information Technology (IIIT). 2007. "Welcome to IIIT-Hyderabad." Hyderabad, Andhra Pradesh. http://www.iiit.ac.in/institute/institute.php.

ITU (International Telecommunication Union). 2004. "The Fifth Pillar: Republic of Mauritius ICT Case Study." http://www.itu.int/itudoc/gs/promo/bdt/cast_int/86187.pdf.

Kenya ICT Board. 2008. "BPO Park Receives KSH900 in the 2008/9 Budget." http://www.ict.go.ke/inner.php?cat=news&sid=106.

Lee, Sing Kong, Goh Chor Boon, Birger Fredriksen, and Tan Jee Peng. 2008. *Toward a Better Future: Education and Training for Economic Development in Singapore Since 1965*. Washington, DC and Singapore: World Bank and National Institute of Education.

Livemint.com. 2007. "China Starts Building World's Largest Software, IT Service Hub." http://www.livemint.com/2007/09/18152056/China-starts-building-world8.html.

Magtibay-Ramos, Nedelyn, Gemma Estrada, and Jesus Felipe. 2007. "An Analysis of the Philippine Business Process Outsourcing Industry." ERD Working Paper No. 93, Asian Development Bank, Manila. http://www.adb.org/Documents/ERD/Working_Papers/WP093.pdf.

Malta Enterprise. 2008. "Key Industry Profiles: ICT and Electronics." http://www.maltaenterprise.com/ict.aspx.

Masani, Zareer. 2008. "India Still Asia's Reluctant Tiger." BBC Radio 4, London. http://news.bbc.co.uk/2/hi/business/7267315.stm.

MBOI (Mauritius Board of Investments). 2007. "BPO Flash (October 2006 to March 2007)." http://www.investmauritius.com/download/BPO%20flashMarch2007.pdf.

McKinsey & Company. 2008. Development of IT and ITES Industries—Impacts, Trends, Opportunities, and Lessons Learned for Developing Countries: Exhibits to Economic Impact Discussion." Presentation to the World Bank in June.

McKinsey Global Institute. 2003. "New Horizons: Multinational Company Investment in Developing Economies." http://www.mckinsey.com/mgi/publications/newhorizons/index.asp.

———. 2005. "The Emerging Global Labor Market: Part II—The Supply of Offshore Talent in Services." http://www.mckinsey.com/mgi/publications/emergingglobrallabormarket/part2/index.asp.

Ministry of Finance, Government of India. "Union Budget and Economic Survey." National Informatics Center, New Delhi. http://indiabudget.nic.in/es2007-08/esmain.htm.

MSC (Multimedia Super Corridor) Malaysia. 2008. "Did You Know? Facts and Figures March 2008." http://www.mscmalaysia.my/topic/12066956321774.

NASSCOM (National Association of Software and Services Companies). 2007. "NAC—NASSCOM Assessment of Competence." http://www.nasscom.in/Nasscom/templates/NormalPage.aspx?id=15539.

———. 2008a. "Indian IT-BPO Industry Factsheet." http://www.nasscom.in/Nasscom/templates/NormalPage.aspx?id=2374.

———. 2008b. "NASSCOM Rolls Out NAC-TECH across India." http://www.nasscom.in/Nasscom/templates/NormalPage.aspx?id=53326.

———. 2009. "Indian IT-BPO Industry Factsheet." http://www.nasscom.org/Nasscom/templates/NormalPage.aspx?id=53615.

NASSCOM-CRISIL. 2007. "The Rising Tide: Employment and Output Linkages of IT-ITES." http://www.nasscom.in/upload/51269/NASSCOM_CRISIL.pdf.

NASSCOM-Everest. 2008. "Roadmap 2012—Capitalizing on the Expanding BPO Landscape." http://www.nasscom.in/Nasscom/templates/NormalPage.aspx?id=53361.

NASSCOM-McKinsey. 2005. "NASSCOM-McKinsey Report 2005: Extending India's Leadership of the Global IT and BPO Industries." http://www.mckinsey.com/locations/india/mckinseyonindia/pdf/NASSCOM_McKinsey_Report_2005.pdf.

Nelson, Roy C. 2007. "Transnational Strategic Networks and Policymaking in Chile: CORFO's High Technology Investment Promotion Program." *Latin American Politics and Society* 49 (2): 149–81.

OANA (Organisation of Asia-Pacific News Agencies). 2008. "Phase 1 of SmartCity Malta Launched." http://www.oananews.org/view.php?id=2220&ch=AST.

PSDC (Penang Skills Development Center). *PSDC Homepage*. Penang, Malaysia. http://www.psdc.com.my/index.cfm.

Rodrik, Dani. 2007. *One Economics, Many Recipes: Globalization, Institutions, and Economic Growth*. Princeton, NJ: Princeton University Press.

Roxas-Chua, Ray Anthony. 2008. "The Philippine Offshoring and Outsourcing Industry." Presented at the World Bank conference "The Potential of Global Sourcing of Services for Achieving Sustainable and Inclusive Growth" Washington, DC, April 16. http://go.worldbank.org/JPF64G90L0.

SEDB (Singapore Economic Development Board). 2008. "Initiatives in New Technology (INTECH)." SEDB, Singapore. http://www.edb.gov.sg/etc/medialib/downloads/forms_2008.Par.0012.File.tmp/INTECT.pdf.

SmartCity Malta. 2008. "Master Plan." http://www.smartcity.ae/malta/township.html.

South Africa Online. 2005. "South African Contact Centre Industry Leads the World." http://www.southafrica.co.za/the_good_news_72_7.html.

Sutherland Global Services. 2008. "Country Selection Template." Unpublished document, Sutherland Global Services.

TESDA (Technical Education and Skills Development Authority). 2008. "PGMA Grants More Funds for Training Call Center Agents and Other BPO Workers." TESDA, Manila. http://www.tesda.gov.ph/page.asp?rootID=3&sID=244&pID=3.

The Hindu. 2008. "State Nets Rs 26,000 Crore from IT Exports," May 31. http://www.thehindu.com/2008/05/31/stories/2008053153000400.htm.

Tholons. 2006. "Emergence of Centers of Excellence." Unpublished report.

Times of Malta. 2008. "HSBC Inaugurates Call Centre Extension," February 15. http://www.timesofmalta.com/articles/view/20080215/local/hsbc-inaugurates-call-centre-extension.

UNCTAD (United Nations Conference on Trade and Development). 2008. *Globalization for Development: The International Trade Perspective*. New York: United Nations. http://www.unctad.org/en/docs/ditc20071_en.pdf.

Vashistha, Atul, and Avinash Vashistha. 2006. *The Offshore Nation: Strategies for Success in Global Outsourcing and Offshoring*. New York: McGraw Hill.

Wadhwa, Vivek, Una Kim de Vitton, and Gary Gereffi. 2008. "How the Disciple Became the Guru." http://www.globalizationresearch.com/.

WEF (World Economic Forum). 2008. *Global Information Technology Report 2007–2008*. Geneva: WEF. http://www.insead.edu/v1/gitr/wef/main/analysis/showcountrydetails.cfm.

World Bank. 2006. "Project Appraisal Document for eGhana." World Bank, Washington, DC. http://www-wds.worldbank.org/external/default/WDSContentServer/WDSP/IB/2006/07/14/000012009_20060714140033/Rendered/PDF/366720rev0pdf.pdf.

———. 2008. "e-Development Thematic Group Seminar." Session at the seminar "India's Emergence as a Global IT Player: The Role of NASSCOM," Washington, DC, April 15. http://go.worldbank.org/QM4LL9C2C0.

Part II

Key Trends in ICT Development

David A. Cieslikowski, Naomi J. Halewood, Kaoru Kimura, and Christine Zhen-Wei Qiang

Part I of this report shows that information and communication technology (ICT) is playing a vital role in advancing economic growth and improving governance. The potential impact that ICT can have on individuals, businesses, and governments depends largely on how policies are formulated and technology and markets evolve. Thus, it is important for countries to possess timely ICT data and benchmarks to facilitate policy making that extends the reach of ICT and increases its development impact.

ICT development is multifaceted, ranging from the rollout of telecommunications infrastructure—providing voice, data, and media services—and information applications tailored to specific sectors and functions (for example, banking and finance, land management, education, and health) to the implementation of electronic government (e-government) and the development of information technology (IT) and IT-enabled industries (including IT goods and services). There are many opportunities for developing countries to advance development through the innovative use of ICT. Banking services and job search text messaging services delivered through mobile phones and portable devices used by farmers and fishermen to track crop prices and market demand are a few examples that are changing the lives of people in developing countries.

Part II of the report includes this chapter describing the key trends in ICT development; sections titled ICT Performance Measures: Methodology and Findings, with its annex on progress in measuring ICT, and User's Guide to ICT At-a-Glance Country Tables; and the ICT at-a-glance (AAG) tables for 150 economies. The tables contain 29 indicators on ICT sector structure, capacity, and performance.

The present chapter uses the data from the AAG tables to demonstrate the progress that many developing countries have made in recent years in improving ICT access, use, quality, affordability, trade, and applications, and to show how that progress relates to enabling policies and regulations.

This ICT trends analysis and the ICT at-a-glance tables use standard World Bank income and region classifications. For income classification, every economy is classified as low income, middle income (subdivided into lower middle income and upper middle income), or high income based on gross national income (GNI) per capita. Regional groupings include only low- and middle-income economies and are based on the World Bank's operational regions, which may differ from common geographic designations. For further details on income and region classifications, see the section titled User's Guide to ICT At-a-Glance Country Tables.

Mobile Phones Have Narrowed the Gaps in Voice Communications Worldwide

The overall trends in voice communications in developing countries are positive. At the end of 2007 there were about 1.1 billion fixed telephone lines and 3.3 billion mobile phone

Figure 1 Mobile Phone Subscriptions in Developing and Developed Countries, 2000–07

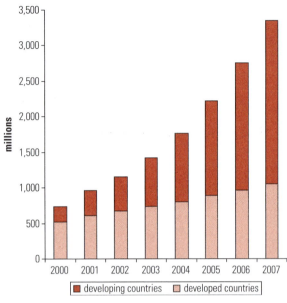

Source: International Telecommunication Union (ITU), World Telecommunication/ICT Indicators Database.

Figure 2 Status of Competition in Fixed and Mobile Telephony in Developing and Developed Countries, 2007

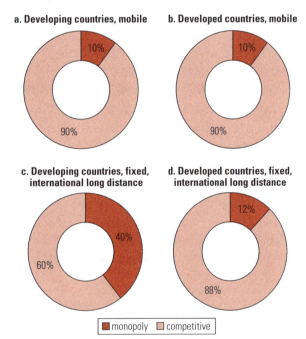

Source: ITU, World Telecommunication Regulatory Database; based on World Bank classification of economies.
Note: For the developing-country panel at the top, a sample of 127 countries was used; for that at the bottom, 134 countries. For the developed-country panels, a sample of 49 countries was used.

Figure 3 Mobile Telephony Penetration before and after the Introduction of Competition

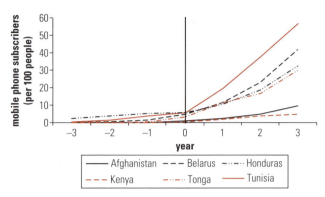

Source: ITU, World Telecommunication/ICT Indicators Database.
Note: Year 0 in the figure indicates the year of entry of a second mobile operator.

subscriptions worldwide. The proportion of mobile phone subscriptions in developing countries increased from about 30 percent of the world total in 2000 to more than 50 percent in 2004—and to almost 70 percent in 2007 (figure 1).

There are two principal reasons that the uptake of mobile telephony in developing countries has overtaken fixed-line service. First, wireless technology can be deployed more quickly because it requires less upfront investment in infrastructure than do fixed telephone systems. This translates to lower prices and hence stronger customer demand. Second, liberalization of fixed-line markets, which were often dominated by state-owned monopolies, started later, while mobile phone markets were generally opened to one or more new entrants from the start (figure 2).

The introduction of competition in the mobile telephony market has often led to an immediate growth in voice services (figure 3). Countries that have taken decisive steps to establish independent regulators and foster competition have seen notable improvements in sector performance. In some cases, the announcement of a plan to issue a new license has been effective in triggering growth, encouraging the existing mobile phone operator to improve service, reduce prices, and increase market penetration before the new entrant started operations.

Access to Advanced ICT Services Is the Natural Next Step in Development

Although Internet use took off more recently and has not grown as rapidly as voice communications, the worldwide

Figure 4 Number of Internet Users by Region, 2000 and 2007

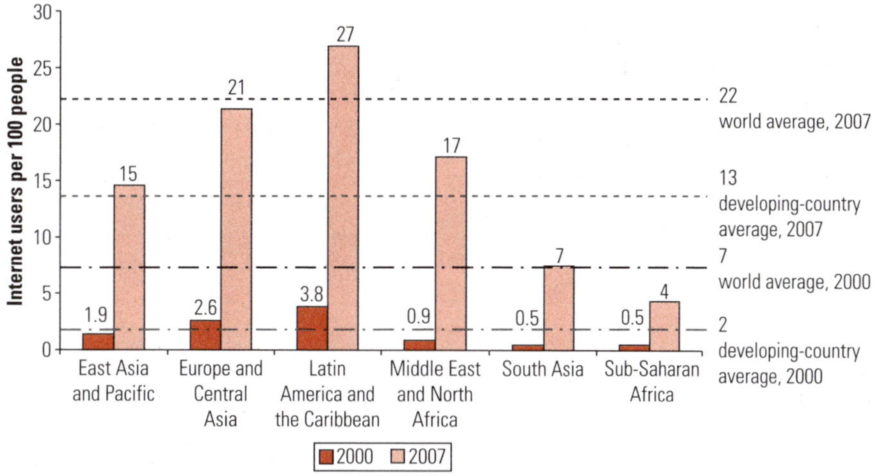

[Bar chart showing Internet users per 100 people for 2000 and 2007:
- East Asia and Pacific: 1.9 (2000), 15 (2007)
- Europe and Central Asia: 2.6 (2000), 21 (2007)
- Latin America and the Caribbean: 3.8 (2000), 27 (2007)
- Middle East and North Africa: 0.9 (2000), 17 (2007)
- South Asia: 0.5 (2000), 7 (2007)
- Sub-Saharan Africa: 0.5 (2000), 4 (2007)

Reference lines: world average 2007 = 22; developing-country average 2007 = 13; world average 2000 = 7; developing-country average 2000 = 2]

Source: ITU, World Telecommunication/ICT Indicators Database.

number of Internet users more than tripled between 2000 and 2007. In developing countries, the number of Internet users jumped about 10 times over the same period, from 76 million to 726 million. But the disparity in Internet penetration among developing countries and regions remains high, ranging from about 27 users per 100 people in Latin America and the Caribbean and 21 users per 100 people in Europe and Central Asia to 4 per 100 people in Sub-Saharan Africa in 2007 (figure 4). Unlike mobile phones, personal computers are often too costly for many households in developing countries. Public Internet access points, therefore, have played an important role in introducing the Internet to people in rural areas and low-income countries. Moreover, as wireless technology evolves and markets expand, more people in both developed and developing countries are using mobile phones to access the Internet.

The demand for always-on, high-capacity Internet services in both developed and developing countries is increasing. Advanced Internet service—i.e., beyond what can be achieved through dial-up connections—has become more important as the demand for data and value-added services grows. Broadband allows for large volumes of data to be transmitted and facilitates cheaper voice communications (for example, by routing calls over the Internet). It is also enabling voice, data, and media services to be transmitted over the same network. Such convergence could have an enormous impact on economic and social development—increasing productivity, lowering transaction costs, facilitating trade, and increasing retail sales and tax revenues. Where broadband has been introduced in rural areas of developing countries, villagers and farmers have gained better access to market prices of crops, training, and job opportunities. However, in 2007, average broadband penetration in low-income economies was just 2 percent of the population and was concentrated in urban centers. Clearly, the benefits of broadband are not yet available to most people in developing economies.

Although the capacity of broadband service is measured by the advertised speed available to consumers, speed may be constrained by the availability of bandwidth, which is increasing faster in developed countries with robust infrastructure than in developing countries. In high-income economies, average per capita international bandwidth increased from 586 bits per second (bps) in 2000 to 18,240 bps in 2007. Among developing regions, Europe and Central Asia and Latin America and the Caribbean have the greatest capacity. Between 2000 and 2007, bandwidth per capita increased from 12 bps to 1,114 bps in Europe and Central Asia and from 8 bps to 1,126 bps in Latin America and the Caribbean. With improved fiber-optic connectivity, some countries in South Asia are seeing a rise in international bandwidth, yet in terms of international bandwidth per capita, South Asia and Sub-Saharan Africa are still well behind other regions (figure 5).

Broadband penetration is closely associated with per capita income (figure 6). The Republic of Korea, however, through ambitious policies and support for broadband infrastructure investments, now has one of the world's highest rates of broadband subscribers—well above that of many other economies with higher per capita incomes. Estonia has also made great progress through a national strategy to build an ICT-enabled economy. It is experiencing a surge in Internet and broadband use, triggered in part by the development of an electronic environment for its banking sector. Both countries have leveraged public-private partnerships to facilitate ICT development and aim to use ICT to increase connectivity between citizens, businesses, and government.

Affordability Unleashes the Potential Impact of ICT Services on Economic Growth

The price of ICT access continues to fall due to technological advances, market growth, and increased competition, a trend that is especially important in allowing people in developing countries to take full advantage of ICT services. In recent years, steep price reductions have contributed to the rapid expansion in mobile phone use in many countries (figure 1, figure 7). Increased use of prepaid service allows mobile customers to make payments in small amounts instead of having to commit to fixed monthly subscriptions. Such cards enable even low-income consumers to have access to mobile communications, leading to higher penetration rates in poor and rural areas.

Pricing for Internet access has also been falling in many countries, including some in sub-Saharan Africa (figure 8). Still, the average price for Africa as a whole continues to be well above the world average. The gap in affordability is even more stark: in 2006, the Internet price basket for sub-Saharan Africa was about 62 percent of average monthly per capita income, while it was about 12 percent in South Asia, and less than 9 percent in all other developing regions. In high-income economies, Internet service costs less than 1 percent of average monthly income.

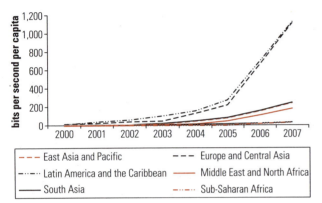

Figure 5 International Bandwidth in Developing Regions, 2000–07

Sources: ITU, World Telecommunication/ICT Indicators Database; World Bank.

Figure 6 Broadband Penetration and Gross National Income in Various Economies, 2007

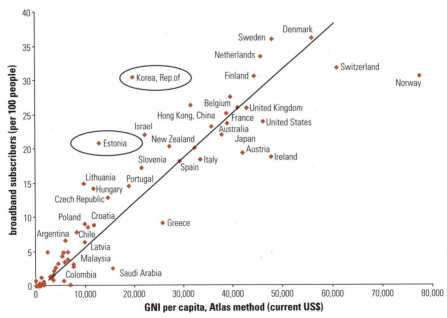

Sources: ITU, World Telecommunication/ICT Indicators Database; World Bank, World Development Indicators Database.

Figure 7 Average Annual Change in Price of Mobile Phone Services in Various Countries, 2004–06

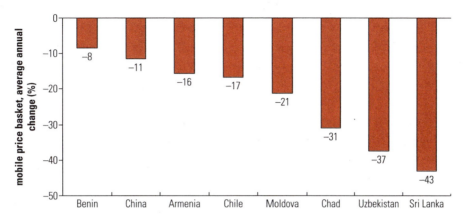

Source: ITU, World Telecommunication/ICT Indicators Database.
Note: Mobile price basket is based on the Organisation for Economic Co-operation and Development (OECD) low user definition and is calculated based on the prepaid price for 25 calls per month spread over the same mobile network, other mobile networks, and mobile to fixed calls and during peak, off-peak, and weekend times. It also includes 30 text messages per month. Countries that have experienced significant price reductions in mobile phone service prices do not necessarily have the lowest prices.

Figure 8 Monthly Price of Internet Services in Various Sub-Saharan African Countries, 2005–07

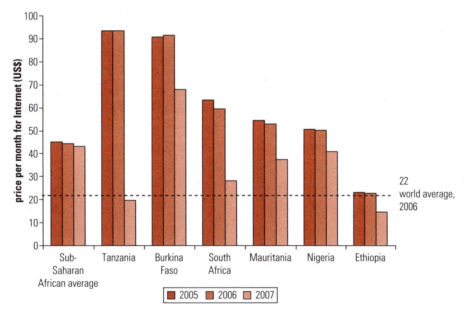

Source: ITU, World Telecommunication/ICT Indicators Database.
Note: Price basket for Internet service is calculated based on the cheapest available tariff for accessing the Internet for 20 hours a month (10 hours peak and 10 hours off-peak). The basket does not include the telephone line rental but does include any telephone usage charges.

Developing Countries Are Benefiting from ICT Exports

Although the level of exports of ICT goods and services does not necessarily reflect high rates of ICT use in a country, it does indicate the importance of a country's ICT sector and its international competitiveness. As barriers to trade in ICT goods and services are removed, opportunities for developing countries to benefit from such exports will likely grow.

Some developing countries have already become key exporters of ICT goods and services. The top-five exporters of ICT goods in 2006 were China ($299 billion); the United

States ($169 billion); Hong Kong, China ($136 billion); Japan ($125 billion); and Singapore ($124 billion). In terms of the share of ICT goods exports in total goods exports, economies in the East Asia and Pacific Region were leaders: the Philippines (56 percent); Singapore (46 percent); Malaysia (45 percent); Hong Kong, China (42 percent); and China (31 percent).

Trade in ICT services includes communications services (telecommunications, business network services, teleconferencing, support services, and postal services) and computer and information services (databases, data processing, software design and development, maintenance and repair, and news agency services). India's software exports jumped from about $1 billion in 1995 to $22 billion in 2006, and generated employment of about 1.6 million people. India leads all other developing countries in exports of communication, computer, and information services as a share of total service exports, at 42 percent in 2006/07 (figure 9). Other developing countries on the top-10 list include Costa Rica, Guyana, the Republic of Yemen, Romania, and Senegal.

ICT Applications Are Transforming Government and Commerce

Between 2000 and 2007, the ICT sector accounted for 3 to 7 percent of GDP, regardless of the country's level of income. With robust ICT infrastructure as a foundation, countries are using ICT applications to further develop specific sectors. And developing countries are well positioned to benefit from adopting new ICT applications because they often do not have the added expenses of maintaining and transitioning from legacy IT systems to newer technologies.

Governments are increasingly important users of ICT, particularly in the context of e-government, making them a major actor in fostering ICT uptake and setting IT standards. E-government initiatives usually aim to make public administration more efficient, increase government accountability and transparency, and improve delivery of public services to citizens and businesses.

The United Nations Web Measure Index (table 1) evaluates the availability and sophistication of governments' Web presence and use. A look at the five indicators that make up the index shows not only that governments in developed countries are much higher up the ICT adoption ladder but that they are at a more advanced stage where ICT has been embedded in daily workflows—easing transactions, recordkeeping, and sharing of information among government agencies and between government, citizens, and businesses. Many developing countries are still at an earlier stage, focused on physical implementation of IT systems and networks.

In terms of advancement of e-commerce, developing countries have less than one-hundredth the number of secure servers as developed countries (table 1). A secure, reliable business-enabling environment is a key element of successful e-commerce. Privacy and security concerns

Figure 9 ICT Service Exports as a Percentage of Total Service Exports for the Top-Five Countries, 2000–06/07

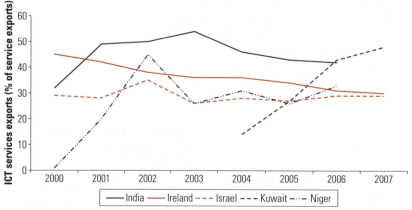

Source: International Monetary Fund (IMF), Balance of Payments Statistics Yearbook Database.
Note: The most recent data available for India and Niger are from 2006.

Table 1 Measures of E-Government and E-Commerce in Developing and Developed Countries

Country group	Web measure index, 2007	E-Commerce (secure Internet servers per 1 million people), December 2008
Developed	0.61	662.6
Developing	0.27	5.2
Ratio of developed to developing	2.2	128.1

Sources: UNPAN 2008; Netcraft Secure Server Survey; based on World Bank classification of economies.

about the transmission of personal or financial information over the Internet are major issues for both consumers and firms and may explain why they may be reluctant to use the Internet to make transactions. The number of secure servers indicates how many companies are conducting encrypted transactions over the Internet. Differences in this measure are stark. While developed countries had about 662 such servers per 1 million people in December 2008, developing countries had about 5. Canada alone, with 30,200 secure servers, had more than all developing countries combined, with 28,399.

Conclusion

Countries that have taken steps to create a competitive market environment for ICT generally have a larger share of people using ICT services than those that have not. One important outcome of competition is that it lowers prices for ICT services. But while prices are falling rapidly, ICT services are still unaffordable for many people in low-income countries. As a result, a large share of the population in these countries has yet to realize the potential of ICT for economic and social development.

Among ICT services, voice communications, particularly mobile telephony, has led the way. Access to new services such as broadband remains limited in developing countries. Yet the expansion of broadband networks plays a catalytic role in the development of trade and e-government. In some developing countries, trade in ICT goods and services has sparked export-led growth and job creation. ICT applications are also transforming how governments deliver public services to citizens and businesses.

References

IMF (International Monetary Fund). Balance of Payments Statistics Yearbook Database. Washington, DC: IMF.

UNPAN (United Nations Public Administration Network). United Nations eGovernment Knowledge Readiness Knowledge Base. (accessed August 2008). http://www2.unpan.org/egovkb/

ICT Performance Measures: Methodology and Findings

Over the past few years, attempts have been made by many organizations to measure the relative level of development of the information and communication technology (ICT) sector in individual economies. The multitude of efforts is evidence of the importance of ICT in a country's economic and social transformation and as a cohesive force for integrating a country into the global economy. This report introduces a new approach—the World Bank's country ICT performance measures—and this section describes the methodology behind the measures and summarizes the findings.

The country ICT performance measures aim to provide a quick and effective way for policy makers to compare their countries' ICT performance against that of other countries and to benchmark progress over time. Rather than creating an aggregate ICT index, the World Bank's performance measures group indicators into various dimensions of ICT development, which allows users to see how a country has progressed within these dimensions relative to its peers, to use the measures as a tool to analyze relative ICT performance, and to understand the interrelations among various aspects of ICT development. More important, the measures are intended to stimulate policy dialogue about how to improve the scores through targeted sector policies and investments.

Methodology

A total of 150 developing and developed economies with a population greater than 1 million are included in the calculation of the country ICT performance measures, based on internationally recognized indicators for measuring ICT (see the annex to this section). These measures are calculated using the *average percentile method* and assess ICT capacities along three dimensions: (1) access to ICT, (2) affordability of ICT, and (3) adoption of ICT applications in government and business.

Specifically, the method for calculating scores for each dimension is as follows:

- *Indicator selection and data collection*. Between two and five indicators are selected that best represent and measure the dimension, subject to the availability of country-level data from international organizations, supplemented by World Bank staff estimates based on official national sources (table 2).

- The *access dimension* features five penetration indicators covering a range of ICT services and products. The *affordability dimension* consists of three indicators measuring a monthly basket of fixed-line telephone, mobile phone, and Internet service charges, expressed as a percentage of per capita income. The *applications dimension* consists of indicators measuring ICT adoption including the United Nations Web Measure Index and the number of secure Internet servers per 1 million inhabitants, a proxy for the availability of e-business services in an economy.

Table 2 Indicators for the Country ICT Performance Measures

Dimension	Indicator	Source
Access	A1. Telephone lines (per 100 people)	ITU
	A2. Mobile cellular subscriptions (per 100 people)	
	A3. Internet users (per 100 people)	
	A4. Personal computers (per 100 people)	
	A5. Households with a television set (%)	
Affordability	P1. Price basket for residential fixed line (% of monthly GNI per capita)	ITU and World Bank
	P2. Price basket for mobile call (% of monthly GNI per capita)	ITU and World Bank
	P3. Price basket for Internet (% of monthly GNI per capita)	ITU
Applications	AP1. United Nations Web Measure Index	UNPAN
	AP2. Secure Internet servers (per 1 million people)	Netcraft

Table 3 Example of How a Country ICT Performance Measure Is Calculated (Mauritius)

Dimension	Indicator	Data	Percentile	Score
Access	A1. Telephone lines (per 100 people)	28.51	73.83	
	A2. Mobile cellular subscriptions (per 100 people)	74.13	57.33	
	A3. Internet users (per 100 people)	10.97	73.15	
	A4. Personal computers (per 100 people)	17.55	74.32	
	A5. Households with a television set (%)	95.68	70.27	
	Average		69.78	
	Percentile of the average (aggregate percentile)		70.67	8
Affordability	P1. Price basket for residential fixed line (% of monthly GNI per capita)	1.18	75.00	
	P2. Price basket for mobile call (% of monthly GNI per capita)	0.91	81.21	
	P3. Price basket for Internet (% of monthly GNI per capita)	3.58	61.64	
	Average		72.62	
	Percentile of the average (aggregate percentile)		74.67	8
Applications	AP1. United Nations Web Measure Index	0.47	63.01	
	AP2. Secure Internet servers (per 1 million people)	44.81	70.92	
	Average		66.97	
	Percentile of the average (aggregate percentile)		70.67	8

Source: World Bank, World Development Indicators Database.

- *Aggregation.* The percentile of each indicator is calculated, and a simple average of the percentiles is taken. Then the percentile of the average is calculated.

- *Calculation of measures.* Economies are scored on a scale from 1 to 10 based on the aggregate percentile values, with 10 given to the highest performance decile and 1 to the lowest decile.

Where possible, indicators used for the access and affordability dimensions reflect 2007 data. Where data for 2007 are not available, earlier years have been used. Indicators for the adoption of ICT applications dimension use 2008 data. Table 3 uses Mauritius as an example of how the country ICT performance measures are calculated.

To avoid distortion of the results, the methodology does not "pre-judge" an economy's ICT performance according to its income level, sector structure, or knowledge capacity of its population. The country ICT performance measures, however, can easily be compared to income data (table 4) in order to examine the impact of economic development, level of competition, independent regulation, and literacy levels on ICT take-up and use, and vice versa.

This methodology has several advantages:

- It is robust. More than 95 percent of the scores using the average percentile method are identical to those calculated using other methodologies, including the matching percentile method and the unobserved component method.

- It is flexible enough to incorporate additional indicators for each dimension, or to include other core areas as new dimensions of ICT development unfold. This will allow the framework to evolve over time while ensuring comparability of the scores.

- It is straightforward, easy to explain, and transparent.

- It avoids subjective manipulation of data based on underlying assumptions about the data (for example, assigning weights). Subjective manipulation of data is particularly problematic for the ICT sector, where ongoing technical and market evolution make predictions difficult, and where weighting might express the prevailing view of a particular country or region.

- It deals with missing data relatively well, by averaging across indicators.

Findings

The results of the country ICT performance measure calculations are summarized in table 5 (see pages 140–43).

Overview

Overall, as would be expected, there is a close relationship between the country ICT performance measures and income levels (see table 4). The leading economies in ICT performance are mainly developed economies, attesting their high levels of accessibility, affordability, and ICT usage. Canada, Denmark, the Republic of Korea, the Netherlands, Norway, Sweden, and the United Kingdom score 10 (the maximum) on all three measures.

The highest-scoring developing economies are all upper middle income and mostly countries in Central and Eastern

Table 4 Average Country ICT Performance Measures, by Income Level

Income	Access to ICT services	Affordability of ICT services	Adoption of ICT applications in government and businesses
High	9.06	9.08	9.00
Upper middle	6.79	6.82	6.57
Lower middle	4.98	5.07	4.93
Low	2.27	2.14	2.50

Source: World Bank staff.

Europe (Croatia, Latvia, Lithuania, Poland, Serbia, and Turkey) and Latin America (Chile, Costa Rica, and Mexico), but they also include Malaysia and Mauritius. Most of these countries have effective regulation, competitive telecommunications markets, and strong governmental support for ICT. In the case of the Central and Eastern European countries, convergence of domestic legislation with the European Union (EU) telecommunications framework has led to more liberalized ICT industries. These countries have also benefited from the establishment and strengthening of robust institutions to enforce fair competition rules.

Among the high-scoring Latin American countries, results show that the early ICT sector reform practiced by Chile and Mexico is paying dividends. While Costa Rica historically has had a relatively uncompetitive sector, the government has dedicated resources to telecommunications, particularly fixed telephony, throughout the country. In addition, Costa Rica's comparatively high level of literacy and efforts to develop an export-oriented, IT-enabled services industry contribute to its high applications score. Similarly, Malaysia and Mauritius have pursued outward-looking export strategies for ICT goods and services and liberalized their telecommunications sectors to attract infrastructure investment.

Within the group of lower-middle-income countries, China, Jordan, Thailand, and Ukraine score highest, while Vietnam, Uzbekistan, and Pakistan have the highest performance measures in the low-income bracket.

At the bottom of the overall measures are seven countries with a score of 1 (the minimum) in all three dimensions, all of which are in sub-Saharan Africa. In each of these seven countries, access to telephony is below 5 percent of the population, access to the Internet and personal computers below 0.1 percent, and less than 10 percent of the households have a television. Combined tariffs on fixed and mobile phone services and Internet services in these countries are two to five times the average income, and there are extremely low levels of ICT application use. Although weak ICT performance of these countries is partly the result of low average incomes, it is mainly due to weak regulation, limited competition, lack of private investment, and lack of supporting infrastructure such as electricity. In many other countries in the region, significant improvements have taken place, particularly in the access and affordability dimensions. In fact, sub-Saharan Africa has experienced the highest mobile phone growth rate of all regions since market reform began in about 2000.

Some developing countries—such as Serbia, Croatia, Ukraine, Macedonia, Syria, Jordan, Vietnam, and Moldova in terms of access and Malaysia, Jordan, Peru, Guatemala, India, and Mongolia in terms of adoption of ICT applications—stand out as better ICT performers than their incomes would suggest. However, some countries are not doing as well in ICT performance as they could be given their income level. Countries in this category include upper-middle-income economies such as Botswana, Cuba, Gabon, and Libya (figure 10).

The country ICT performance measures also illustrate different levels of achievement for the various dimensions in each developing region (see figure 11). Europe and Central Asia scores highest in the access measure, whereas the Middle East and North Africa leads in affordability and Latin America and the Caribbean leads in adoption of ICT applications. The developing region with the lowest measures is Sub-Saharan Africa, scoring less than 3 on the 1–10 scale in all three dimensions.

Access

Telecommunications infrastructure development has witnessed explosive growth over the past decade, due in part to the liberalization of the sector and the introduction of mobile technology. Hong Kong, China, leads in the access measure. This is a reflection of both the high level of competition in its ICT sector, particularly in proportion to its market size, and a population eager to adopt ICT technology.[1]

Croatia and Serbia are the top-scoring developing countries in terms of access. They both have a relatively high level of fixed-line access (more than 40 per 100 people), their inhabitants each have more than one cell phone on average, and the penetration of Internet subscribers is at par with countries in the Organisation for Economic Co-operation

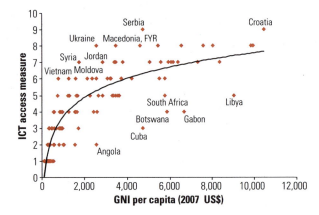
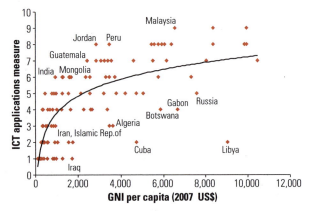

Figure 10 Relation between the Country ICT Performance Measures (for Access and Applications) and Income per Capita, Developing Countries

Sources: World Bank staff.

and Development (OECD). This is partly the result of low tariffs (2 percent of per capita income) on fixed and mobile phone service and Internet service.

Most countries have achieved a penetration of five telephone lines (fixed or mobile) per 100 people. Telecommunications equipment prices have fallen sharply, and wireless technologies allow quicker installation compared to the past. Remaining access gaps are therefore no longer primarily due to engineering challenges or capacity limitations. Supply constraints continue to exist in some developing countries because of market structures that inhibit access to ICT products and services and high costs of infrastructure development in remote and geographically challenging areas. Poor affordability (especially for more advanced ICT services and products such as Internet and personal computers) and low digital literacy are also constraining factors to enhanced access.

Figure 11 Average Country ICT Performance Measures, by Region

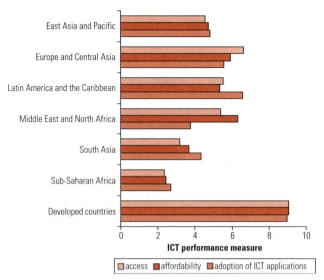

Sources: World Bank staff.

Affordability

Affordability of ICT services around the world varies widely. Some 60 percent of the 150 economies included in the ICT performance measures have monthly fixed-line tariffs less than 5 percent of per capita income. Denmark, the United Arab Emirates, Singapore, the United States, and Norway—countries that have the most affordable ICT services in the world—have price baskets for ICT services equivalent to 0.5 percent or less of average income. However, close to 30 percent of the countries analyzed have monthly fixed-line tariffs that amount to more than 10 percent of income. The highest variation is in Internet pricing; here, more than one-third of the economies have monthly price baskets more than 25 percent of per capita income.

Interpreting the affordability measure is difficult because low tariffs increase access, but they also have the potential to reduce the likelihood of financial sustainability of the ICT sector. Several developing countries that do well in terms of ICT affordability (those with low tariffs relative to their income levels) are those that have not undertaken extensive sector liberalization. For example, Iraq, Libya, and Oman have telecommunications tariffs that make up a relatively low percentage of per capita income, yet until recently, they also had monopolistic telecommunications markets. One reason that tariff rebalancing has been limited is lack of political will, keeping tariffs on fixed lines (and consequently dial-up Internet) relatively low. The downside of low tariffs is that relatively little capital for investment is generated, which also influences mobile pricing in these countries, since mobile operators must compete to some degree with low fixed-line tariffs. The challenge for developing countries is to strike a balance between affordability of ICT services to the users and sustainability of sector growth.

It is important to note that the affordability measure takes into account both the prices of ICT services and the income levels to reflect the financial ability to pay on the demand side. Therefore, countries with relatively low per capita income and relatively high tariffs (and therefore lower affordability) tend to have a bigger access gap beyond what a competitive market can deliver, which might require more public interventions.

Applications

Governments and businesses are increasingly exploring the potential of high-speed networks to improve internal communications, increase the efficiency of transactions, and deliver better services to customers. Although most economies have established some basic level of e-government, in many developing economies, this is still often limited to a Web site displaying information. A growing number of countries also have some form of e-commerce, but at a rudimentary stage.

The United States has the highest score in the ICT applications measure, leading the world in the number of secure Internet servers per 1 million people. It also ranks high in the United Nations Web Measure Index, behind only Denmark and Sweden.

Among developing economies, Latin American countries perform relatively well in the applications area. Argentina, Brazil, Chile, Costa Rica, Mexico, Peru, and Uruguay, for example, all score 8 or higher out of 10 on applications; one reason is that common culture and heritage allow economies of scale and the ability to leverage the experience of neighboring nations. Another factor is the region's relatively high literacy rate and school enrollment, both of which are the highest among developing regions.

Challenges

The task of capturing a country's ICT development in an aggregated measure remains a challenge. Aspects that are worth revisiting in the future are the inclusion of measures of impact as well as a number of data-related issues and their influence on selecting indicators.

Impact Measures

Currently, the country ICT performance measures focus mainly on output indicators. No impact metrics (for example, the impact of ICT development on economic growth, labor productivity, and employment opportunities) are incorporated in the framework.

Reliable impact measures (and evidence of causality) of ICT use are still being developed. In addition, the data required for impact indicators are usually collected through surveys, implying high collection costs in both developed and developing countries. Scarcity of data and inconsistent methodologies continue to make cross-country comparison difficult. Nevertheless, this is an important area for future research given the growing interest in understanding the role ICT plays in economic and social development.

Data Limitations

Due to data limitations, the country ICT performance measures have incorporated only a core set of indicators for which reliable figures are readily available for the majority of the 150 economies included in this report.

Perhaps the most significant data limitation is to access indicators for emerging ICT technologies. Mobile broadband subscribers, voice-over-Internet protocol (VoIP) subscribers, and pay TV subscribers are such examples. Data for these indicators are available from several sources, though they have limited country coverage. In some cases, such as mobile broadband, limited coverage may suggest that the service is not yet widely used. Yet these cases may represent future trends of a dynamic and ever-evolving sector. A data consolidation effort has thus been initiated (see the annex) in anticipation that data for these important indicators will become more widely available in the near future.

Ongoing data collection efforts also include indicators for the usage and quality dimensions of ICT sector development. International voice traffic (minutes per person per month), mobile telephone usage (minutes per user per month), short message service (SMS) usage (messages per user per month), and Internet users (per 100 people) are grouped under the usage dimension. Telephone faults (per 100 main lines per year), broadband subscribers (percentage of total Internet subscribers), and international Internet bandwidth (bits per person) are selected to reflect quality and capacity of ICT services. As the data availability of these indicators reaches a high threshold, usage and quality can be added as new dimensions to the country ICT performance measures.

Lack of data also limits the understanding of adoption of ICT applications across different sectors of an economy. Data on ICT expenditures—which reflect the purchase of software and hardware by businesses and consumers and hence their propensity to use it—are not available beyond developed countries and some large developing countries. Indicators that might measure take-up of ICT in the education and health sectors, for example, such as percentage of schools connected to the Internet and percentage of clinics with personal computers or Internet connections, are also not widely available for many developing countries. Relevant indicators for these areas have thus not yet been included in this edition of the country ICT performance measures.

The two indicators included in the applications measure, namely the United Nations Web Measure Index and the number of secure servers, might not be representative of the wider applications dimension, as they focus on the adoption of ICT applications in government and business. The World Bank and the United Nations are involved in ongoing discussions regarding the development of improved measures for e-government, such as the number of government services available online and the percentage of uptake of online service delivery. Statistics of business use of ICT are generally collected from special ICT business surveys or as a module in business surveys. Most OECD and EU countries already collect these data annually, and other countries are beginning to collect indicators such as percentage of businesses using computers, Internet, or with a Web presence and business use of Internet by type of activities with similar frequency (see annex tables A1 and A4).

As the availability, quality, and comparability of ICT data improve over time (see the annex about the global Partnership on Measuring ICT for Development), the country ICT performance measures will be able to incorporate additional indicators in each dimension, or even add new dimensions. The intention is to update the performance measures annually.

Conclusions

ICT is a multifaceted sector. Measurement of its different aspects is necessary to give a diverse perspective across the spectrum of ICT capabilities. The country ICT performance measures cover access, affordability, and applications as key dimensions in assessing ICT development across economies. A transparent and objective average percentile method has been applied for 150 economies.

Despite the challenges of creating such performance measures, benchmarking, if used wisely, can provide useful information and meaningful analysis for policy purposes. Although benchmarking often cannot establish the causal link, comparisons with better-performing economies can nevertheless be a useful input in developing policy, drawing lessons for improvement and progress, and ultimately, promoting more effective ICT development. The categorization of indicators into different dimensions, for its part, helps countries recognize both the areas in which they are doing well and those in which they have room for improvement, thus allowing policy makers to focus resources on the appropriate dimension of ICT.

(Section continues on the following page.)

Table 5 Country ICT Performance Measures, by Income Level and Economy, 2007

Income level/economy	Access to ICT services	Affordability of ICT services	Adoption of ICT applications in government and business
High income			
Australia	10	9	10
Austria	10	9	10
Belgium	9	9	9
Canada	10	10	10
Czech Republic	9	8	9
Denmark	10	10	10
Estonia	10	8	10
Finland	9	10	9
France	9	10	10
Germany	10	10	9
Greece	9	9	7
Hong Kong, China	10	10	9
Hungary	9	8	9
Ireland	10	9	10
Israel	8	9	9
Italy	9	9	8
Japan	9	9	10
Korea, Rep. of	10	10	10
Kuwait	8	9	7
Netherlands	10	10	10
New Zealand	9	9	10
Norway	10	10	10
Oman	6	9	7
Portugal	9	7	9
Puerto Rico	6	5	8
Saudi Arabia	8	9	6
Singapore	10	10	9
Slovak Republic	8	8	8
Slovenia	9	9	8
Spain	9	8	9
Sweden	10	10	10
Switzerland	10	10	9
Trinidad and Tobago	7	8	7
United Arab Emirates	8	10	9
United Kingdom	10	10	10
United States	9	10	10
Average	**9.06**	**9.08**	**9.00**
Upper middle income			
Argentina	7	7	8
Belarus	6	7	5

Table 5 continued			
Income level/economy	Access to ICT services	Affordability of ICT services	Adoption of ICT applications in government and business
Botswana	4	6	4
Brazil	7	5	8
Bulgaria	8	6	7
Chile	7	7	8
Costa Rica	7	8	8
Croatia	9	7	7
Cuba	3	5	2
Gabon	4	6	4
Jamaica	6	5	6
Kazakhstan	7	7	5
Latvia	8	7	8
Lebanon	6	6	6
Libya	5	9	2
Lithuania	8	8	9
Malaysia	8	8	9
Mauritius	8	8	8
Mexico	7	7	9
Panama	6	6	7
Poland	8	8	8
Romania	7	7	7
Russia	8	7	5
Serbia	9	8	5
South Africa	5	5	8
Turkey	8	7	7
Uruguay	7	6	8
Venezuela, R.B. de	7	8	6
Average	*6.79*	*6.82*	*6.57*
Lower middle income			
Albania	5	6	6
Algeria	5	6	3
Angola	2	4	5
Armenia	5	5	4
Azerbaijan	6	4	5
Bolivia	4	4	6
Bosnia and Herzegovina	7	7	5
Cameroon	2	3	2
China	6	8	6
Colombia	6	6	7
Congo, Rep. of	2	3	2
Dominican Republic	5	5	7
Ecuador	6	4	7

(*Table continues on the following page.*)

Table 5 continued

Income level/economy	Access to ICT services	Affordability of ICT services	Adoption of ICT applications in government and business
Lower middle income continued			
Egypt, Arab Rep. of	6	6	6
El Salvador	5	7	7
Georgia	6	4	5
Guatemala	4	4	7
Honduras	4	4	6
India	4	5	6
Indonesia	4	5	4
Iran, Islamic Rep. of	7	9	3
Iraq	3	8	1
Jordan	7	6	8
Lesotho	2	3	4
Macedonia, FYR	8	5	6
Moldova	6	3	5
Mongolia	5	5	6
Morocco	5	4	4
Namibia	5	4	4
Nicaragua	4	3	5
Paraguay	5	5	6
Peru	5	4	8
Philippines	5	5	6
Sri Lanka	4	6	5
Sudan	3	3	1
Swaziland	4	5	4
Syrian Arab Rep.	7	5	2
Thailand	7	7	7
Tunisia	6	7	5
Turkmenistan	4	4	1
Ukraine	8	6	6
West Bank and Gaza	5	6	4
Average	*4.98*	*5.07*	*4.93*
Low income			
Afghanistan	2	1	3
Bangladesh	3	3	3
Benin	2	2	2
Burkina Faso	1	1	2
Burundi	1	1	1
Cambodia	2	2	3
Central African Republic	1	1	1
Chad	1	1	1
Congo, Dem. Rep. of	1	1	1

Table 5 *continued*

Income level/economy	Access to ICT services	Affordability of ICT services	Adoption of ICT applications in government and business
Côte d'Ivoire	3	2	2
Eritrea	1	1	1
Ethiopia	1	2	1
Gambia, The	3	2	4
Ghana	2	3	4
Guinea	1	3	1
Guinea-Bissau	1	1	1
Haiti	4	2	2
Kenya	3	1	4
Kyrgyzstan	4	3	4
Laos	3	3	2
Madagascar	2	1	3
Malawi	2	2	2
Mali	1	2	3
Mauritania	4	2	3
Mozambique	2	1	3
Myanmar	1	2	2
Nepal	2	3	4
Niger	1	1	1
Nigeria	3	3	3
Pakistan	4	4	5
Papua New Guinea	3	3	3
Rwanda	1	2	3
Senegal	3	3	4
Sierra Leone	1	2	2
Somalia	2	1	1
Tajikistan	3	4	5
Tanzania	2	1	2
Togo	3	2	3
Uganda	1	1	3
Uzbekistan	5	6	3
Vietnam	6	4	5
Yemen	3	4	1
Zambia	2	2	1
Zimbabwe	3	2	2
Average	**2.27**	**2.14**	**2.50**

Source: World Bank staff.

Note: Low-income economies are those with a GNI per capita of $935 or less in 2007. Middle-income economies are those with a GNI per capita of more than $936 but less than $11,456. Lower-middle-income and upper-middle-income economies are separated at a GNI per capita of $3,705. High-income economies are those with a GNI per capita of $11,456 or more. Although the measures are presented in separate tables by income group, they were calculated using the full sample of 150 economies.

Annex: Progress in Measuring ICT

Measuring the impact of ICT on development and evaluating the outputs of ICT interventions are essential for making ICT strategies relevant and holding governments accountable for their implementation. This was highlighted at the World Summits on the Information Society (WSIS) held in Geneva in 2003 and Tunis in 2005 and attended by 50 heads of state, prime ministers, and vice presidents and 80 ministers and vice ministers from 180 countries. Recognizing that comparable ICT statistics are important for formulating policies and making informed decisions concerning ICT, the Partnership on Measuring ICT for Development was launched in 2004. Partnership members include the United Nations Conference on Trade and Development (UNCTAD), the International Telecommunication Union (ITU), the Organisation for Economic Co-operation and Development (OECD), the United Nations Educational, Scientific, and Cultural Organization (UNESCO), Institute for Statistics (UIS), the United Nations Economic Commission for Latin America and the Caribbean (UNECLAC), the United Nations Economic Commission for Western Asia (UNESCWA), the United Nations Economic Commission for Asia and the Pacific (UNESCAP), the United Nations Economic Commission for Africa (UNECA), Eurostat, and the World Bank.

The partnership's main objectives are to do the following:

- Continue to raise awareness among policy makers on the importance of statistical indicators for monitoring ICT policies and carrying out impact analysis

- Expand the core list of indicators to other areas of interest such as ICT in education, government, and health, building on the original core ICT list of access and use by individuals, households, and businesses, and production and trade in ICT goods and services

- Conduct technical workshops at the regional level to exchange national experiences and discuss methodologies, definitions, survey vehicles, and data collection efforts

- Assist statistical agencies in developing economies in their ICT data collection and dissemination efforts, including the development of national databases to store and analyze survey results

- Develop a global database of ICT indicators and make it available on the World Wide Web.

In May 2008, the Partnership on Measuring ICT for Development published *The Global Information Society: A Statistical View*, which included some of the 41 core ICT indicators for participating countries. The core ICT indicators were organized in the following main sections (see http://measuring-ict.unctad.org for a complete list):

- ICT infrastructure and access
- Access and use of ICT by households and individuals
- Access and use of ICT by businesses
- ICT sector and trade in ICT goods.

The data on access and use by individuals and households were drawn from survey data. The OECD and Eurostat have collected such data from their member countries since 2002, and ITU collects a more limited subset of these data from all countries that report these data. Tables A1–A4 (see pages 145–52) present a subset of the data from *The Global Information Society*. Table A5 (see page 152) includes several new measures on ICT occupations and use, as well as emerging ICT technologies. Data for these ICT indicators are drawn from several sources and have limited country coverage.

Statistics of business use of ICT are generally collected from special ICT business surveys or as a module in business surveys. Most OECD and European Union (EU) countries collect these data annually, and other countries

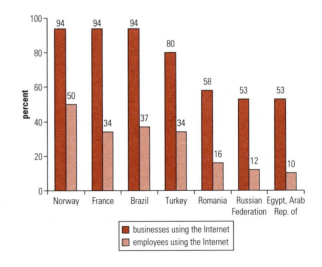

Figure A1 Internet Use by Businesses and Employees, Selected Countries, 2005 and 2006

Source: Partnership on Measuring ICT for Development 2008.

are beginning to collect indicators of business use of ICT, often based on OECD and Eurostat methodological recommendations. For developing countries, UNCTAD provides a framework for data collection and collects available data annually from developing countries. Few developing countries collect data on business use of ICT, but they are increasingly using the core list of indicators and applying the partnership-sponsored standards. Where data for developing countries are available, they reveal a fairly low use of ICT by businesses and employees (see figure A1).

Selected Core ICT Indicators on Household and Business Users of ICT

Table A1 Use of the Internet by Individuals and Businesses, 2005–07
(percent)

Economy	Individuals using the Internet	Employees using computers	Business using computers	Businesses using the Internet	Businesses with a Web presence	Businesses receiving orders over the Internet	Businesses placing orders over the Internet
Afghanistan	21	—	—	—	—	—	—
Argentina	—	40	100	96	74	46	45
Australia	69	—	96	87	52	21	55
Austria	69	53	98	98	80	18	53
Azerbaijan	10	9	38	9	33	—	—
Belarus	—	—	84	38	27	—	—
Belgium	69	57	97	95	72	15	46
Bermuda	80[a]	62	80	71	56	14	41
Bhutan	10[a]	—	—	—	—	—	—
Botswana	4[b]	—	—	—	—	—	—
Brazil	21	48	99	94	50	50	52
Bulgaria	34	21	90	75	44	5	8
Canada	72	—	—	95	71	13	65
Chile	37	—	60	49	39	4	7
China	12	—	—	47	24	12	10
Hong Kong, China	61	58	88	83	52	3	22
Macao, China	46	—	76[a]	53[a]	26[a]	16[a]	21[a]
Costa Rica	22	—	—	—	—	—	—
Cuba	24	59	95	71	24	1	4
Cyprus	41	43	95	86	50	7	25
Czech Republic	52	39	97	95	74	9	28
Denmark	85	68	98	98	85	36	61
Dominican Republic	16	—	—	—	—	—	—
Ecuador	7	—	—	—	—	—	—
Egypt, Arab. Rep.of	—	18	100	53	71	35	21
Estonia	66	38	94	92	63	16	27
Finland	81	67	99	99	81	12	56
France	66	63	99	94	65	16	26
Germany	75	56	96	95	77	20	57
Greece	36	37	97	94	64	8	15
Honduras	15	—	—	—	—	—	—
Hungary	53	31	89	80	53	13	15
Iceland	91	58	99	97	72	30	61
Ireland	61	54	97	94	67	25	59
Israel	42[b]	—	—	—	—	—	—

(*Table continues on the following page.*)

Table A1 continued

Economy	Individuals using the Internet	Employees using computers	Businesses using computers	Businesses using the Internet	Businesses with a Web presence	Businesses receiving orders over the Internet	Businesses placing orders over the Internet
Italy	41	40	96	93	61	4	29
Japan	68	—	—	98	86	16	21
Korea, Rep. of	80	—	97	96	59	8	34
Latvia	59	25	92	80	43	4	16
Lithuania	50	27	92	88	47	17	25
Luxembourg	79	49	98	93	65	—	—
Macedonia, FYR	29	—	—	—	—	—	—
Mauritius	17	—	94	87	46	33	35
Mexico	20	—	—	—	—	—	—
Morocco	46	—	—	—	—	—	—
Netherlands	86	61	100	97	81	29	47
New Zealand	69	—	96	95	63	37	60
Norway	87	59	97	94	76	26	70
Oman	6	—	—	—	—	—	—
Panama	22	32	90	80	—	39	44
Paraguay	8	—	—	—	—	—	—
Peru	12[a]	—	—	—	—	—	—
Poland	49	38	93	89	60	8	26
Portugal	42	35	95	83	42	6	24
Qatar	—	—	84	68	99	51	41
Romania	28	22	77	58	41	4	11
Russian Federation	—	30	91	53	28	24	31
Serbia	33	—	—	—	—	—	—
Singapore	60	—	93	91	75	15	34
Slovak Republic	62	39	97	93	65	—	—
Slovenia	57	48	97	96	65	12	23
Spain	55	49	98	93	50	9	17
Sweden	82	66	96	96	90	24	73
Switzerland	64	57	99	98	92	23	58
Thailand	14	—	88	70	51	11	14
Trinidad and Tobago	27	—	—	—	—	—	—
Turkey	15	41	88	80	60	—	—
United Kingdom	75	51	96	93	81	20	67
United States	68	—	—	—	—	—	—
Uruguay	29	—	—	—	—	—	—
West Bank and Gaza	36	—	—	—	—	—	—

Source: Partnership on Measuring ICT for Development 2008.

Note: Data are for the most recent year in the period. — = not available.

a. Data refer to 2003.

b. Data refer to 2004.

Notes for Table A1

The source for all seven indicators in this table is Partnership on Measuring ICT for Development (2008).

Individuals using the Internet is calculated by dividing the total number of age-specific individuals who used the Internet (from any location) in the past 12 months by the total number of age-specific individuals.

Employees using computers is calculated by dividing the number of employees using computers (in all businesses within the scope of the survey) by the total number of employees (in all businesses within the scope of the survey). *Employees* refers to all persons working for the business, not only those working in clerical jobs. They include working proprietors and partners as well as employees.

Businesses using computers/the Internet is calculated by dividing the number of businesses within the scope of the survey using computers/the Internet during the 12-month reference period by the total number of businesses within the scope of the survey. A *computer* includes a desktop, portable or handheld computer (for example, a personal digital assistant), minicomputer, and mainframe. A computer does not include equipment with some embedded computing abilities, such as mobile phones or TV sets, nor does it include computer-controlled machinery or electronic tills. The *Internet* refers to Internet protocol (IP)–based networks: the World Wide Web, an extranet over the Internet, electronic data interchange (EDI) over the Internet, Internet accessed by mobile phones, and Internet-based e-mail.

Businesses with a Web presence is calculated by dividing the number of businesses within the scope of the survey with a Web presence by the total number of businesses within the scope of the survey. A *Web presence* includes a Web site, home page, or presence on another entity's Web site (including a related business). It excludes inclusion in an online directory and any other Web pages where the business does not have substantial control over the content of the page.

Businesses receiving orders over the Internet is calculated by dividing the number of businesses within the scope of the survey receiving orders over the Internet by the total number of businesses within the scope of the survey. *Receiving orders* includes orders received via the Internet, whether or not payment was made online. It includes orders received via Web sites, specialized Internet marketplaces, extranets, EDI over the Internet, Internet-enabled mobile phones, and e-mail. It also includes orders received on behalf of other organizations—and orders received by other organizations on behalf of the business. *Receiving orders* excludes orders that were cancelled or not completed.

Businesses placing orders over the Internet is calculated by dividing the number of businesses within the scope of the survey placing orders over the Internet by the total number of businesses within the scope of the survey. *Placing orders* includes orders placed via the Internet, whether or not payment was made online. It includes orders placed via Web sites, specialized Internet marketplaces, extranets, EDI over the Internet, Internet-enabled mobile phones, and e-mail. *Placing orders* excludes orders that were cancelled or not completed.

Table A2 Location of Internet Use by Individuals

	Percentage of individuals using the Internet at						
Economy	Home, 2003–07[a]	Work, 2005–07[a]	Place of education, 2005–07[a]	Another person's home, 2005–07[a]	Community Internet access facility, 2005–07[a]	Commercial Internet access facility, 2005–07[a]	Other places, 2005–07[a]
Australia	88	45	23	38	—	—	—
Austria	82	48	10	8	n.a.[a]	n.a.[a]	5[a]
Azerbaijan	69	38	17	11	1	44	11
Belgium	89	34	10	8	n.a.[a]	n.a.[a]	5[a]
Brazil	50	40	26	31	10	22	—
Bulgaria	71	38	12	6	n.a.[a]	n.a.[a]	16[a]
Canada	61	26	12	—	10	—	20
Chile	40	19	35	—	2	28	5
China	76	33	13	—	—	32	1
Hong Kong, China	91	42	14	—	2	1	4
Macao, China	86	26	12	—	—	—	8
Costa Rica	32	27	20	5	0	46	1
Cyprus	70	51	16	15	n.a.[a]	n.a.[a]	9[a]
Czech Republic	76	42	19	15	n.a.[a]	n.a.[a]	6[a]

(*Table continues on the following page.*)

Table A2 continued

Economy	Percentage of individuals using the Internet at						
	Home, 2003–07[a]	Work, 2005–07[a]	Place of education, 2005–07[a]	Another person's home, 2005–07[a]	Community Internet access facility, 2005–07[a]	Commercial Internet access facility, 2005–07[a]	Other places, 2005–07[a]
Denmark	95	53	13	17	n.a.[a]	n.a.[a]	16[a]
Dominican Republic	20	32	34	28	8	41	2
Estonia	83	43	18	15	n.a.[a]	n.a.[a]	7[a]
Finland	89	49	21	35	n.a.[a]	n.a.[a]	20[a]
France	—	40	8	36	—	—	—
Germany	89	42	10	18	n.a.[a]	n.a.[a]	10[a]
Greece	62	44	11	—	n.a.[a]	n.a.[a]	17[a]
Honduras	10	11	7	—	0	81	1
Hungary	74	40	21	23	n.a.[a]	n.a.[a]	11[a]
Iceland	93	63	30	48	n.a.[a]	n.a.[a]	30[a]
Ireland	77	39	11	5	n.a.[a]	n.a.[a]	9[a]
Italy	78	48	13	22	n.a.[a]	n.a.[a]	16[a]
Japan	83	34	12	—	4	5	—
Korea, Rep. of	95	32	17	7	4	21	17
Latvia	77	40	19	15	n.a.[a]	n.a.[a]	12[a]
Lithuania	80	40	24	23	n.a.[a]	n.a.[a]	13[a]
Luxembourg	92	44	11	11	n.a.[a]	n.a.[a]	3[a]
Macedonia, FYR	32	17	19	9	n.a.[a]	n.a.[a]	54[a]
Mauritius	73	28	23	2	2	9	0
Mexico	34	24	16	2	1	42	—
Morocco	28	7	3	3	1	71	—
Netherlands	97	50	13	16	n.a.[a]	[a]	5[a]
New Zealand	88	36	16	24	9	11	0
Norway	92	56	15	18	n.a.[a]	n.a.[a]	13[a]
Paraguay	21	25	15	2	—	51	1
Poland	74	33	23	23	n.a.[a]	n.a.[a]	13[a]
Portugal	68	43	21	32	n.a.[a]	n.a.[a]	20[a]
Romania	67	34	21	12	n.a.[a]	n.a.[a]	9[a]
Serbia	76	32	13	18	n.a.[a]	n.a.[a]	6[a]
Singapore	82	50	25	13	6	5	—
Slovak Republic	60	51	21	20	n.a.[a]	n.a.[a]	15[a]
Slovenia	85	53	18	25	n.a.[a]	n.a.[a]	16[a]
Spain	74	45	13	25	n.a.[a]	n.a.[a]	21[a]
Sweden	91	52	14	22	n.a.[a]	n.a.[a]	12[a]
Thailand	33	28	46	17	2	—	—
United Kingdom	87	45	13	19	n.a.[a]	n.a.[a]	11[a]
United States	80	36	23	—	—	—	—
Uruguay	41	26	14	11	3	52	—

Source: Partnership on Measuring ICT for Development 2008.

Note: Data are for the most recent year in the period. — = not available; n.a. = not applicable.

a. Community and commercial access facilities are included in other places.

Notes for Table A2
Percentage of individuals using the Internet at home, work, place of education, another person's home, community/commercial Internet access facility, and other places is, for international comparability, presented as the proportion of age-specific individuals using the Internet at each location. When a person's workplace is located at his or her home, then he or she would answer yes to the home category only. Note that respondents could report use at multiple locations. A community Internet access facility includes public libraries, publicly provided Internet kiosks, digital community centers, and other government agencies; access is typically free or low cost and is available to the general public. Commercial Internet access facilities include cyber cafés, hotels, airports, and the like; although the venue is commercial, the cost is not necessarily at full market price. (Partnership on Measuring ICT for Development 2008)

Table A3 Frequency of Internet Use by Individuals, 2005–07

	Percentage of individuals using the Internet			
Economy	At least once a day	At least once a week but not every day	At least once a month but not every week	Less than once a month
Australia	50	41	8	1
Austria	67	25	6	2
Azerbaijan	41	51	5	3
Belgium	73	21	4	1
Brazil	36	47	12	3
Bulgaria	64	28	6	2
Canada	64	26	5	2
Costa Rica	34	38	24	5
Cyprus	61	30	7	2
Czech Republic	50	36	13	1
Denmark	81	12	4	1
Estonia	67	25	7	1
Finland	78	17	4	1
France	65	25	7	3
Germany	63	25	8	3
Greece	56	27	13	4
Hong Kong, China	72	19	5	4
Hungary	72	22	5	1
Iceland	82	14	3	1
Ireland	56	32	9	2
Italy	82	5	9	4
Korea, Rep. of	71	21	2	5
Latvia	67	27	5	1
Lithuania	61	32	6	1
Luxembourg	72	20	6	2
Macedonia, FYR	44	40	13	3
Mauritius	33	47	15	5
Mexico	20	68	10	2
Morocco	55	34	8	3
Netherlands	79	17	3	1
New Zealand	58	30	6	5
Norway	77	17	4	1

(*Table continues on the following page.*)

Table A3 continued

Economy	Percentage of individuals using the Internet			
	At least once a day	At least once a week but not every day	At least once a month but not every week	Less than once a month
Poland	61	27	10	2
Portugal	67	21	8	3
Romania	49	41	9	1
Serbia	50	37	8	4
Singapore	70	22	8[a]	—
Slovak Republic	58	32	8	1
Slovenia	72	20	6	2
Spain	57	28	11	4
Sweden	73	22	4	1
Thailand	23	60	17	1
United Kingdom	67	24	6	3
Uruguay	37	48	12	3
West Bank and Gaza	49	40	10	0

Source: Partnership on Measuring ICT for Development 2008.

Note: Data are for the most recent year in the period. — = not available.

a. Includes individuals who used the Internet at least once a month and those who used it less than once a month.

Notes for Table A3
Percentage of individuals using the Internet at least once a day, at least once a week but not every day, at least once a month but not every week, and less than once a month is calculated by dividing the total number of age-specific individuals who used the Internet (from any location) in the last 12 months by the total number of age-specific individuals. (Partnership on Measuring ICT for Development 2008)

Table A4 Proportion of Businesses Using the Internet, by Type of Activity, 2005–06

Economy	Percentage of businesses using the Internet for								
	Sending or receiving e-mail	Getting information about goods or services	Getting information from government	Other information searches or research	Internet banking or financial services	Transacting with public authorities	Providing customer services	Delivering products online	Other types of activity
Argentina	97	88	75	40	84	57	43	6	55
Australia	—	—	—	—	—	50	—	—	—
Austria	—	—	58	—	88	77	69	—	—
Azerbaijan	—	—	26	—	26	26	—	—	—
Belgium	—	—	56	—	88	46	68	—	—
Brazil	98	78	59	82	80	84	31	14	—
Bulgaria	—	—	57	—	53	48	42	—	—
Canada	98	—	—	—	—	—	—	—	—
Chile	99	—	—	—	—	—	—	—	—
China	80	65	46	39	—	37	35	11	—
Hong Kong, China	97	96	73	—	42	—	23	43	—
Macao, China[a]	89	—	20	69	—	—	15	—	4

Table A4 continued

	Percentage of businesses using the Internet for								
Economy	Sending or receiving e-mail	Getting information about goods or services	Getting information from government	Other information searches or research	Internet banking or financial services	Transacting with public authorities	Providing customer services	Delivering products online	Other types of activity
Cyprus	—	—	51	—	57	39	48	—	—
Czech Republic	—	—	75	—	91	69	72	—	—
Denmark	—	—	83	—	94	82	81	—	—
Egypt, Arab Rep. of	93	59	—	59	27	6	36	—	0
Estonia	—	—	72	—	98	69	63	—	—
Finland	—	—	87	—	93	90	75	—	—
France	—	—	—	—	77	66	—	—	—
Germany	—	—	38	—	77	44	75	—	—
Greece	—	—	76	—	74	72	64	—	—
Hungary	—	—	54	—	68	53	45	—	—
Iceland	—	—	86	—	94	80	71	—	—
Ireland	—	—	79	—	86	82	63	—	—
Italy	—	—	81	—	81	79	59	—	—
Korea, Rep. of	89	61	54	78	67	43	35	13	2
Latvia	—	—	46	—	91	44	40	—	—
Lithuania	—	—	77	—	94	83	46	—	—
Luxembourg	—	—	77	—	76	84	52	—	—
Netherlands	—	—	65	—	76	66	59	—	—
New Zealand	—	—	68	—	87	77	30	—	—
Norway	—	—	73	—	92	73	75	—	—
Panama	97	81	68	61	70	36	39	—	70
Poland	—	—	56	—	75	53	57	—	—
Portugal	—	—	64	—	75	64	40	—	—
Romania	—	—	65	—	52	58	35	—	—
Russian Federation	92	55	43	—	15	—	5	5	—
Singapore	93	93	—	—	64	—	—	42	—
Slovak Republic	—	—	73	—	84	74	53	—	—
Slovenia	—	—	74	—	93	68	61	—	—
Spain	—	—	57	—	85	58	49	—	—
Sweden	—	—	81	—	92	81	84	—	—
Switzerland	—	98	—	60	85	57	21	22	—
Thailand	81	—	—	65	10	—	24	21	14
Turkey	—	—	56	—	75	63	16	38	—
United Kingdom	—	—	55	—	73	52	76	—	—

Source: Partnership on Measuring ICT for Development 2008.

Note: Data are for the most recent year in the period. — = not available.

a. Data refer to 2003.

Notes for Table A4
Percentage of businesses using the Internet for sending or receiving e-mail, getting information about goods or from government, other information searches, Internet banking or financial services, transacting with public authorities, providing customer services, delivering products, and other types of activities is, for international comparability, presented as the proportion of businesses within the scope of the survey undertaking each activity. Interacting with government organizations includes downloading or requesting forms, completing or lodging forms online, making online payments, and purchasing from or selling to government organizations. Customer services include providing online or e-mailed product catalogues or price lists, product specification or configuration, post-sales support, and order tracking. Delivering products online refers to products delivered over the Internet in digitized form, for example, reports, software, music, videos, and computer games; and online services such as computer-related services, information services, travel bookings, and financial services. (*Partnership on Measuring ICT for Development 2008*)

Table A5 Supplemental ICT Data, 2007

Economy	ICT-related occupations in the total economy, OECD narrow definition (percent)	ICT-related occupations in the total economy, OECD broad definition (percent)	Mobile broadband subscribers, 4th quarter (number)[a]	Broadband price (US$/megabits/sec)	VoIP subscribers (thousands)	Female Internet users (% females in survey age group)	Pay TV subscribers (thousands)
Argentina	—	—	153,851	149.6	—	—	—
Australia	3.6	20.8	7,085,182	22.3	—	68	—
Austria	3.0	20.5	1,751,553	19.7	—	55	3,000
Belgium	2.9	21.7	366,240	5.0	—	59	4,750
Brazil	—	—	2,237,337	20.0	—	32	5,200
Bulgaria	—	—	661,587	8.2	—	28	—
Canada	5.4	27.3	671,191	5.8	1,871	—	10,600
China	—	—	0	15.4	—	—	160,000
Hong Kong, China	—	—	2,752,949	8.3	—	62	1,518
Croatia	—	—	560,149	12.0	—	—	700
Cyprus	—	—	96,782	35.1	—	—	75
Czech Republic	4.5	22.4	666,086	8.6	—	39	1,280
Denmark	4.0	27.2	360,731	15.2	—	74	2,211
Egypt, Arab Rep. of	—	—	159,398	64.9	—	—	2,700
Estonia	2.6	21.8	20,289	8.0	—	64	414
Finland	4.4	24.9	811,694	2.8	—	73	1,232
France	2.6	20.1	8,448,387	2.0	—	54	12,857
Germany	3.1	21.6	9,323,490	3.0	—	58	36,500
Greece	2.2	14.9	3,195,106	7.6	—	23	200
Hungary	2.7	22.6	789,703	6.7	—	47	3,000
Iceland	3.1	22.5	0	8.4	—	—	64
India	—	—	0	18.3	—	—	70,000
Indonesia	—	—	2,323,183	99.7	—	—	—
Ireland	2.4	20.9	1,671,323	22.1	—	47	1,081
Italy	2.8	22.2	23,751,802	3.4	2,405	28	6,782
Japan	—	—	75,029,396	0.7	16,766	71[b]	37,218
Korea, Rep. of	—	—	25,167,406	0.9	—	70	15,000

Table A5 *continued*

Economy	ICT-related occupations in the total economy, OECD narrow definition (percent)	ICT-related occupations in the total economy, OECD broad definition (percent)	Mobile broadband subscribers, 4th quarter (number)[a]	Broadband price (US$/ megabits/sec)	VoIP subscribers (thousands)	Female Internet users (% females in survey age group)	Pay TV subscribers (thousands)
Latvia	—	—	46,418	164.7	—	51	550
Lithuania	—	—	43,657	7.8	—	41	430
Luxembourg	3.2	30.6	201,432	6.5	—	—	—
Morocco	—	—	107,309	5.0	—	—	2,637
Netherlands	3.9	23.4	2,206,473	1.4	2,172	77	7,056
Norway	4.8	23.8	1,027,639	4.3	459	78	1,500
Peru	—	—	0	68.3	—	26	—
Philippines	—	—	3,052,600	12.7	—	—	—
Poland	2.8	17.9	2,395,011	8.5	—	37	6,543
Portugal	2.8	14.3	4,148,841	3.1	—	31	1,964
Romania	—	—	1,712,621	6.5	—	20	—
Russian Federation	—	—	45,277	28.1	—	21	14,500
Senegal	—	—	0	73.0	—	—	20
Singapore	—	—	1,757,701	2.6	—	—	526
Slovak Republic	3.5	19.1	337,591	10.9	—	49	1,150
Slovenia	2.8	22.0	232,449	4.0	—	48	470
Spain	2.9	18.6	9,829,357	25.8	—	40	3,678
Sweden	4.9	24.6	2,643,214	1.8	623	72	2,979
Switzerland	5.2	23.0	1,250,082	2.9	—	—	3,227
Thailand	—	—	58,680	30.0	—	15[b]	2,000
Turkey	1.7	11.8	0	16.6	—	7[c]	—
United Kingdom	3.2	28.0	16,616,643	3.7	2,000	61	12,618
United States	3.7	20.2	52,476,426	12.4	12,404	67[b]	96,800
Venezuela, R. B. de	—	—	1,414,476	81.4	—	—	—

Sources: Eurostat, International Telecommunication Union, national statistical/telecommunications offices and service providers, OECD forthcoming, Telecommunications Management Group, and Wireless Intelligence.

Note: — = not available.

a. Those countries showing zero have yet to license or deploy technologies that can provide high-speed mobile broadband service (see definition in Notes for Table A5).

b. Data refer to 2006.

c. Data refer to 2005.

Notes for Table A5

ICT-related occupations in the total economy (narrow and broad OECD definitions)—there is no commonly adopted definition of ICT-skilled employment, but three categories of ICT competencies are distinguished:

- *ICT specialists,* who have the ability to develop, operate, and maintain ICT systems. ICTs constitute the main part of their job—they develop and put in place ICT tools for others.

- *Advanced users,* who are competent users of advanced, and often sector-specific, software tools. ICT is not the main job but a tool.

- *Basic users,* who are competent users of generic tools (for example, Microsoft Word, Excel, Outlook, and PowerPoint) needed for the information society, e-government, and working life. Here too, ICTs are a tool, not the main job.

The first category is the *narrow* measure of ICT-skilled employment, and the sum of all three categories is the *broad* measure. The broad measure mainly includes knowledge workers, managerial workers, and data workers. Data across countries are not strictly comparable since the classifications and the selections of occupations are not harmonized. (OECD forthcoming)

Mobile broadband subscribers is the number of subscribers to cellular mobile networks with access to data communications (for example, the Internet) at broadband speeds. There is no standard definition of the threshold speed for broadband, but for the purpose of statistical collection, high-speed broadband is defined as speeds greater than or equal to 256 kilobits per second in one or both directions. This includes mobile broadband technologies such as WCDMA, WCDMA HSDPA, CDMA2000 1xEV-DO, and CDMA 2000 1xEV-DO Rev. A. These services are typically referred to as 3G or 3.5G. Subscriber data are drawn from the Wireless Intelligence database; data presented in this table include technologies with speeds greater than or equal to 256 kilobits per second. (Wireless Intelligence)

Broadband price (U.S. dollars per megabits per second [Mbps]) is calculated by dividing the monthly subscription charge in dollars by the theoretical download speed. This figure is calculated for each recorded sample, and the lowest monthly cost per Mbps is given. The prices included are those advertised and may or may not include Internet service provider (ISP) charges. (International Telecommunication Union)

VoIP (voice-over-Internet protocol) subscribers is the number of subscribers to telephone service over a broadband Internet connection that provides the user with their own telephone number and ability to place and receive calls to or from other telephone subscribers. This differs from personal-computer-based Internet telephony services such as Skype. Data for most countries are based on the top providers of VoIP service, and therefore may understate the total number of VoIP subscribers in a country. (National telecommunication authorities/service providers and World Bank staff estimates)

Female Internet users (percentage of females in survey age group) is a percentage of all females of a specific age group. Age ranges and frequency of use (for example, once a month, once in the last three months, at least once in the last year, and so on) vary by country so comparisons must be treated with caution. (Eurostat and national statistical offices)

Pay TV subscribers is the number of customers paying for a subscription-based television service. Some countries report all cable and direct TV to home satellite connections regardless of whether they require a subscription. In some countries, the number of subscriptions can exceed the number of households due to multiple subscriptions (for example, subscribing to multiple pay television services delivered over different platforms such as cable, satellite, or broadband Internet). (Telecommunications Management Group)

Note

1. There are 5 mobile network operators, 10 local fixed telecommunication service network operators, and 169 Internet service providers in Hong Kong, China, as of January 2009 (see http://www.ofta.gov.hk) for a population of less than 7 million.

References

OECD (Organisation for Economic Co-operation and Development). Forthcoming. *OECD Information Technology Outlook 2008.* Paris: OECD.

Partnership on Measuring ICT for Development. 2008. *The Global Information Society: A Statistical View.* Santiago, Chile: United Nations.

User's Guide to ICT At-a-Glance Country Tables

The World Bank ICT at-a-glance country tables present the most recent country-specific ICT data from many sources available in one place. They offer a snapshot of the economic and social context and the structure and performance of the ICT sector in each of the 150 countries covered in the report.

Overview

Tables

Economies are presented alphabetically. Data are shown for 150 economies with populations of more than 1 million for which timely and reliable information exists. The table of Key ICT Indicators for Other Economies presents 59 additional economies—those with sparse data, smaller economies with populations of between 30,000 and 1 million, and others that are members of the World Bank Group.

The data in the tables are categorized into four sections:

- *Economic and social context,* which provides a snapshot of the country's macroeconomic and social environment

- *Sector structure,* which provides an overview of regulatory and policy status in the telecommunications sector

- *Sector efficiency and capacity,* which includes information on investment and revenue and employees per subscriber in the telecommunications sector

- *Sector performance,* which provides statistical data on the ICT sector with indicators for access, usage, quality, affordability, trade, and ICT applications.

Aggregate Measures for Income Groups and Regions

The aggregate measures for income groups include 209 economies (those economies listed in the at-a-glance country tables plus those in the Other Economies table) wherever data are available.

The aggregate measures for regions include only low- and middle-income economies (note that these measures include developing economies with populations of less than 1 million, including those listed in the Other Economies table). The country composition of regions is based on the World Bank's analytical regions and may differ from common geographic usage (see the section on the classification of economies at the end of the User's Guide).

Values for the indicators under *ICT sector structure* that are nonnumerical cannot be aggregated into income and regional groups.

Charts

The GNI per Capita, Atlas Method, chart shows country and regional averages from 2000 to 2007.

The ICT MDG Indicators chart is based on the three ICT indicators selected for measuring the Millennium

Development Goals (MDGs). The three ICT indicators are telephone lines (per 100 people), mobile cellular subscriptions (per 100 people), and Internet users (per 100 people). For more information, visit the World Bank's MDG Web site: http://www.developmentgoals.org/.

The Price Basket for Internet Service chart (based on the OECD low user mobile price basket) tracks the price of Internet service based on the cheapest available tariff for accessing the Internet for 20 hours a month (10 hours peak and 10 hours off-peak). The basket does not include the telephone line rental but does include any telephone usage charges. Country and region information is presented when available.

The ICT Service Exports chart shows ICT exports as a percentage of total service exports for 2000–07 for the country and the region in which the country is located. ICT service exports include computer and communications and information services.

Data Consistency and Reliability

Considerable effort has been made to standardize the data collected. However, full comparability of data among countries cannot be ensured and care must be taken in interpreting the indicators.

Many factors affect availability, comparability, and reliability: statistical systems in some developing countries are weak; statistical methods, coverage, practices, and definitions differ widely among countries; and cross-country and intertemporal comparisons involve complex technical and conceptual problems that cannot be unequivocally resolved. Data coverage may not be complete because of special circumstances or because economies are experiencing problems (such as those stemming from conflicts) that affect the collection and reporting of data. For these reasons, although data are drawn from the sources thought to be most authoritative, they should be construed only as indicating trends and characterizing major differences among economies rather than offering precise quantitative measures of those differences.

Data Sources

Data are drawn from Global Insight, the International Monetary Fund (IMF), the International Telecommunication Union (ITU), Netcraft, TeleGeography, the United Nations Commodity Trade Statistics Database (UN Comtrade), the United Nations Department of Economic and Social Affairs (UNDESA), the United Nations Educational, Scientific Organization (UNESCO) Institute for Statistics (UIS), the United Nations Public Administration Network (UNPAN), Wireless Intelligence, the World Information Technology and Services Alliance (WITSA), and the World Bank.

Classification of Economies

For operational and analytical purposes, the World Bank's main criterion for classifying economies is GNI per capita. Every economy is classified as low income, middle income (these are subdivided into lower middle and upper middle), or high income. Note that classification by incomes does not necessarily reflect development status. Because GNI per capita changes over time, the country composition of income groups may change, but one consistent classification, based on GNI per capita in 2007, is used in Part II of this publication.

Low-income economies are those with a GNI per capita of $935 or less in 2007. Middle-income economies are those with a GNI per capita of more than $936 but less than $11,456. Lower-middle-income and upper-middle-income economies are separated at a GNI per capita of $3,705. High-income economies are those with a GNI per capita of $11,456 or more.

For more information on these classifications, see table 2 below and the World Bank's country classification Web site: http://www.worldbank.org/data/countryclass/countryclass.html.

Symbols

The following symbols are used throughout the at-a-glance tables:

—

The symbol — means that data are not available or that aggregates cannot be calculated because of missing data in the year shown.

0 or 0.0

0 or 0.0 means zero or less than half the unit shown.

$ refers to U.S. dollars.

Table 6 Classification of Economies by Income and Region, FY2009

Low and middle income
East Asia and Pacific
  American Samoa (UMC)
  Cambodia (LIC)
  China (LMC)
  Fiji (UMC)
  Indonesia (LMC)
  Kiribati (LMC)
  Korea, Democratic Rep. of (LIC)
  Lao PDR (LIC)
  Malaysia (UMC)
  Marshall Islands (LMC)
  Micronesia, Federated States of (LMC)
  Mongolia (LMC)
  Myanmar (LIC)
  Palau (UMC)
  Papua New Guinea (LIC)
  Philippines (LMC)
  Samoa (LMC)
  Solomon Islands (LIC)
  Thailand (LMC)
  Timor-Leste (LMC)
  Tonga (LMC)
  Vanuatu (LMC)
  Vietnam (LIC)

Europe and Central Asia
  Albania (LMC)
  Armenia (LMC)
  Azerbaijan (LMC)
  Belarus (UMC)
  Bosnia and Herzegovina (LMC)
  Bulgaria (UMC)
  Croatia (UMC)
  Georgia (LMC)
  Kazakhstan (UMC)
  Kyrgyz Republic (LIC)
  Latvia (UMC)
  Lithuania (UMC)
  Macedonia, FYR (LMC)
  Moldova (LMC)
  Montenegro (UMC)
  Poland (UMC)
  Romania (UMC)
  Russian Federation (UMC)
  Serbia (UMC)
  Tajikistan (LIC)
  Turkey (UMC)
  Turkmenistan (LMC)
  Ukraine (LMC)
  Uzbekistan (LIC)

Latin America and the Caribbean
  Argentina (UMC)
  Belize (UMC)
  Bolivia (LMC)
  Brazil (UMC)
  Chile (UMC)
  Colombia (LMC)
  Costa Rica (UMC)
  Cuba (UMC)
  Dominica (UMC)
  Dominican Republic (LMC)
  Ecuador (LMC)
  El Salvador (LMC)
  Grenada (UMC)
  Guatemala (LMC)
  Guyana (LMC)
  Haiti (LIC)
  Honduras (LMC)
  Jamaica (UMC)
  Mexico (UMC)
  Nicaragua (LMC)
  Panama (UMC)
  Paraguay (LMC)
  Peru (LMC)
  St. Kitts and Nevis (UMC)
  St. Lucia (UMC)
  St. Vincent (UMC)
  Suriname (UMC)
  Uruguay (UMC)
  Venezuela, R. B. de (UMC)

Middle East and North Africa
  Algeria (LMC)
  Djibouti (LMC)
  Egypt, Arab Rep. of (LMC)
  Iran, Islamic Rep. of (LMC)
  Iraq (LMC)
  Jordan (LMC)
  Lebanon (UMC)
  Libya (UMC)
  Morocco (LMC)
  Syrian Arab Rep. (LMC)
  Tunisia (LMC)
  West Bank and Gaza (LMC)
  Yemen, Republic of (LIC)

South Asia
  Afghanistan (LIC)
  Bangladesh (LIC)
  Bhutan (LMC)
  India (LMC)
  Maldives (LMC)
  Nepal (LIC)
  Pakistan (LIC)
  Sri Lanka (LMC)

Sub-Saharan Africa
  Angola (LMC)
  Benin (LIC)
  Botswana (UMC)
  Burkina Faso (LIC)
  Burundi (LIC)
  Cameroon (LMC)
  Cape Verde (LMC)
  Central African Republic (LIC)
  Chad (LIC)
  Comoros (LIC)
  Congo, Dem. Rep. of (LIC)
  Congo, Rep. of (LMC)
  Côte d'Ivoire (LIC)
  Eritrea (LIC)
  Ethiopia (LIC)
  Gabon (UMC)
  Gambia, The (LIC)
  Ghana (LIC)
  Guinea (LIC)
  Guinea-Bissau (LIC)
  Kenya (LIC)
  Lesotho (LMC)
  Liberia (LIC)
  Madagascar (LIC)
  Malawi (LIC)
  Mali (LIC)
  Mauritania (LIC)
  Mauritius (UMC)
  Mayotte (UMC)
  Mozambique (LIC)
  Namibia (LMC)
  Niger (LIC)
  Nigeria (LIC)
  Rwanda (LIC)
  São Tomé and Principe (LIC)
  Senegal (LIC)
  Seychelles (UMC)
  Sierra Leone (LIC)
  Somalia (LIC)
  South Africa (UMC)
  Sudan (LMC)
  Swaziland (LMC)
  Tanzania (LIC)
  Togo (LIC)
  Uganda (LIC)
  Zambia (LIC)
  Zimbabwe (LIC)

High-income OECD
Australia
Austria
Belgium
Canada
Czech Republic
Denmark
Finland
France
Germany
Greece
Hungary
Iceland
Ireland
Italy
Japan
Korea, Rep. of
Luxembourg
Netherlands
New Zealand
Norway
Portugal

(Table continues on the following page.)

Table 6 *continued*

Slovak Republic	Cayman Islands	Monaco
Spain	Channel Islands	Netherlands Antilles
Sweden	Cyprus	New Caledonia
Switzerland	Equatorial Guinea	Northern Mariana Islands
United Kingdom	Estonia	Oman
United States	Faeroe Islands	Puerto Rico
	French Polynesia	Qatar
Other high income	Greenland	San Marino
Andorra	Guam	Saudi Arabia
Antigua and Barbuda	Hong Kong, China	Singapore
Aruba	Isle of Man	Slovenia
Bahamas, The	Israel	Taiwan, China
Bahrain	Kuwait	Trinidad and Tobago
Barbados	Liechtenstein	United Arab Emirates
Bermuda	Macao, China	Virgin Islands (U.S.)
Brunei Darussalam	Malta	

Source: World Bank.

Note: This table classifies all World Bank member economies and all other economies with populations of more than 30,000. Economies are divided among income groups according to 2007 GNI per capita, calculated using the World Bank Atlas method. LIC = low-income country, $935 or less; LMC = lower-middle-income country, $936–$3,705; UMC = upper-middle-income country, $3,706–$11,455. A high-income country is at $11,456 or more GNI per capita.

Definitions and Data Sources

This section provides definitions and sources of the indicators used in the World Bank ICT at-a-glance country tables.

Economic and Social Context

Population is based on the de facto definition of *total population*, which counts all residents regardless of legal status or citizenship—except for refugees not permanently settled in the country of asylum, who are generally considered part of the population of their country of origin. The values shown are mid-year estimates. (World Bank)

Urban population is the mid-year population of areas defined as urban in each country and reported to the United Nations. (United Nations)

GNI per capita (World Bank Atlas method) is gross national income converted to U.S. dollars using the World Bank Atlas method, divided by the mid-year population. GNI is the sum of value added by all resident producers plus any product taxes (less subsidies) not included in the valuation of output plus net receipts of primary income (compensation of employees and property income) from abroad. (World Bank)

GDP growth shows the annual percentage rate of growth of gross domestic product at market prices based on constant price local currency. GDP is the sum of gross value added by all resident producers plus any product taxes (less subsidies) not included in the valuation of output. (OECD and World Bank)

Adult literacy rate is the percentage of people ages 15 and older who can, with understanding, read and write a short, simple statement about their everyday life. (UNESCO)

Gross primary, secondary, and tertiary school enrollment is the combined number of students enrolled in primary, secondary, and tertiary levels of education, regardless of age, as a percentage of the population of official school age for the three levels. (UNESCO)

Sector Structure

Separate telecommunications regulator refers to whether a separate telecommunications regulator exists. (ITU)

Status of main fixed-line telephone operator shows the status of the incumbent fixed-line operator. *Public* refers to a fully state-owned operator, *private* refers to a fully private operator, and *mixed* refers to a partially private operator. (ITU)

Level of competition: international long distance refers to the level of competition for international long distance telephone calls (M = monopoly, P = partial competition, C = full competition). (ITU)

Level of competition: mobile telephone service refers to the level of competition for digital cellular mobile services (M = monopoly, P = partial competition, C = full competition). (ITU)

Level of competition: Internet service refers to the level of competition for retail Internet access service (M = monopoly, P = partial competition, C = full competition). (ITU and World Bank)

Sector Efficiency and Capacity

Telecommunication revenue is revenue from the provision of telecommunication services such as fixed line, mobile, and data. (ITU)

Mobile and fixed-line subscribers per employee refers to are telephone subscribers (mobile and fixed-line) divided by the total number of telecommunications employees. (ITU)

Telecommunications investment is total telecommunication investment (capital expenditure) as a percentage of telecommunication revenue. (ITU)

Sector Performance

Access

Telephone lines refers to telephone lines connecting subscribers to the telephone network. (ITU)

Mobile cellular subscriptions refers to subscribers to a public mobile telephone service using cellular technology. (ITU)

Internet subscribers includes people who pay for access to the Internet (dial-up, leased line, and fixed broadband). The number of subscribers measures all those who are paying for Internet use, including the so-called "free Internet" used by those who pay via the cost of their telephone call, those who pay in advance for a given amount of time (prepaid), and those who pay for a subscription (either flat-rate or volume-per-usage based). (ITU)

Personal computers is defined as self-contained computers designed to be used by a single individual. (ITU)

Households with a television set refers to the percentage of households with a television set. (ITU)

Usage

International voice traffic is the sum of international incoming and outgoing telephone traffic (in minutes). (ITU)

Mobile telephone usage measures the minutes of use per mobile user per month. (Wireless Intelligence)

Internet users includes subscribers who pay for Internet access (dial-up, leased line, and fixed broadband) and people with access to the worldwide computer network without paying directly, either as the member of a household, or from work or school. Therefore, the number of Internet users will always be much larger than the number of subscribers, typically by a factor of 2–3 in developed countries and more in developing countries. (ITU)

Quality

Population covered by mobile telephony is the percentage of people within range of a mobile cellular signal regardless of whether they are subscribers. (ITU)

Fixed broadband subscribers includes the number of broadband subscribers with a digital subscriber line, cable modem, or other high-speed technologies connected to the Internet. Reporting countries may have different definitions of broadband, so data are not strictly comparable across countries. Data are presented as a percentage of total Internet subscribers. (ITU)

International Internet bandwidth is the contracted capacity of international connections between countries for transmitting Internet traffic. (ITU and TeleGeography)

Affordability

Price basket for residential fixed line is calculated as one-fifth of the installation charge, the monthly subscription charge, and the cost of local calls (15 peak and 15 off-peak calls of three minutes each). (ITU and World Bank)

Price basket for mobile service is calculated based on the prepaid price for 25 calls per month spread over the same mobile network, other mobile networks, and mobile to fixed calls and during peak, off-peak, and weekend times. It also includes 30 text messages per month. The calculation is based on the OECD low user mobile price basket definition. (ITU and World Bank)

Price basket for Internet service is calculated based on the cheapest available tariff for accessing the Internet for 20 hours a month (10 hours peak and 10 hours off-peak). The basket does not include the telephone line rental but does include any telephone usage charges. (ITU and World Bank)

Price of call to United States is the cost of a three-minute, peak-rate, fixed-line call from the country to the United States. (ITU)

Trade

ICT goods exports and ICT goods imports are defined as goods that are either intended to fulfill the function of information processing and communication by electronic means, including transmission and display; or use electronic processing to detect, measure, and/or record physical phenomena or to control a physical process. ICT goods exports and imports include the following broad categories: telecommunications equipment, computer and related equipment, electronic components, audio and video equipment, and other ICT goods. Re-exports (exports of foreign goods in the same state as previously imported) are also included. (UN Comtrade)

ICT service exports includes communications services (telecommunications, business network services, teleconferencing, support services, and postal services) and computer and information services (databases, data processing, software design and development, maintenance and repair, and news agency services). This is the balance of payment definition. (IMF)

Applications

ICT expenditure includes computer hardware (computers, storage devices, printers, and other peripherals); computer software (operating systems, programming tools, utilities, applications, and internal software development); computer services (information technology consulting, computer and network systems integration, Web hosting, data processing services, and other services); and communications services (voice and data communications services) and wired and wireless communications equipment. (Global Insight and WITSA)

E-government Web measure index measures the level of sophistication of a government's online presence based on five stages of e-government evolution: emerging presence, enhanced presence, interactive presence, transactional presence, and networked presence (1 = best). (UNDESA and UNPAN)

Secure Internet servers is the number of servers using encryption technology for Internet transactions. (Netcraft)

// World Bank • ICT at a Glance

Afghanistan

	Afghanistan 2000	Afghanistan 2007	Low-income group 2007	South Asia Region 2007
Economic and social context				
Population (total, million)	—	—	1,296	1,522
Urban population (% of total)	21	24	32	29
GNI per capita, World Bank Atlas method (current US$)	—	—	574	880
GDP growth, 1995–2000 and 2000–07 (avg. annual %)	—	10.7	5.6	7.3
Adult literacy rate (% of ages 15 and older)	28	—	64	63
Gross primary, secondary, tertiary school enrollment (%)	*12*	*43*	*51*	*60*
Sector structure				
Separate telecommunications regulator	No	Yes		
Status of main fixed-line telephone operator	Public	Public		
Level of competition[a]				
International long distance service	C	P		
Mobile telephone service	—	P		
Internet service	—	P		
Sector efficiency and capacity				
Telecommunications revenue (% of GDP)	4.1	5.1	*3.3*	*2.1*
Mobile and fixed-line subscribers per employee	—	861	*301*	660
Telecommunications investment (% of revenue)	—	37.8	—	—
Sector performance				
Access				
Telephone lines (per 100 people)	—	—	4.0	3.2
Mobile cellular subscriptions (per 100 people)	—	—	21.5	22.8
Internet subscribers (per 100 people)	—	—	*0.8*	1.3
Personal computers (per 100 people)	—	—	*1.5*	3.3
Households with a television set (%)	—	62	16	42
Usage				
International voice traffic (minutes/person/month)[b]	—	—	—	—
Mobile telephone usage (minutes/user/month)	—	—	—	364
Internet users (per 100 people)	—	—	5.2	6.6
Quality				
Population covered by mobile cellular network (%)	—	72	*54*	*61*
Fixed broadband subscribers (% of total Internet subscrib.)	—	0.8	3.4	18.9
International Internet bandwidth (bits/second/person)	—	—	26	31
Affordability				
Price basket for residential fixed line (US$/month)	—	17.8	5.7	4.0
Price basket for mobile service (US$/month)	—	5.6	11.2	2.4
Price basket for Internet service (US$/month)	—	24.0	29.2	8.0
Price of call to United States (US$ for 3 minutes)	—	0.39	2.00	2.02
Trade				
ICT goods exports (% of total goods exports)	—	—	*1.4*	*1.2*
ICT goods imports (% of total goods imports)	—	—	*6.7*	*8.1*
ICT service exports (% of total service exports)	—	—	—	39.0
Applications				
ICT expenditure (% of GDP)	—	—	—	5.7
E-government Web measure index[c]	—	0.27	0.11	0.37
Secure Internet servers (per 1 million people, Dec. 2008)	—	—	0.5	1.1

Sources: Economic and social context: UIS and World Bank; Sector structure: ITU; Sector efficiency and capacity: ITU and World Bank; Sector performance: Global Insight/WITSA, IMF, ITU, Netcraft, UN Comtrade, UNDESA, UNPAN, Wireless Intelligence and World Bank. Produced by the Global Information and Communication Technologies Department and the Development Economics Data Group. For complete information, see Definitions and Data Sources.

Notes: Use of italics in the column entries indicates years other than those specified. — Not available. GDP = gross domestic product; GNI = gross national income; ICT = information and communication technology; and MDG = Millennium Development Goal.

a. C = competition; M = monopoly; and P = partial competition. **b.** Outgoing and incoming. **c.** Scale of 0–1, where 1 = highest presence. **d.** Millennium Development Goal indicators 8.14, 8.15, and 8.16.

Albania

	Albania 2000	Albania 2007	Lower-middle-income group 2007	Europe & Central Asia Region 2007
Economic and social context				
Population (total, million)	3	3	3,435	446
Urban population (% of total)	42	46	42	64
GNI per capita, World Bank Atlas method (current US$)	1,170	3,300	1,905	6,052
GDP growth, 1995–2000 and 2000–07 (avg. annual %)	5.2	5.3	8.0	6.1
Adult literacy rate (% of ages 15 and older)	*99*	99	83	98
Gross primary, secondary, tertiary school enrollment (%)	68	*69*	68	82
Sector structure				
Separate telecommunications regulator	Yes	Yes		
Status of main fixed-line telephone operator	Public	Public		
Level of competition[a]				
International long distance service	M	P		
Mobile telephone service	C	P		
Internet service	C	C		
Sector efficiency and capacity				
Telecommunications revenue (% of GDP)	1.8	6.0	*3.1*	2.9
Mobile and fixed-line subscribers per employee	40	*710*	*624*	532
Telecommunications investment (% of revenue)	28.4	10.4	25.3	22.0
Sector performance				
Access				
Telephone lines (per 100 people)	5.0	*8.9*	15.3	25.7
Mobile cellular subscriptions (per 100 people)	1.0	72.3	38.9	95.0
Internet subscribers (per 100 people)	0.2	*0.6*	6.0	13.6
Personal computers (per 100 people)	0.8	*3.8*	*4.6*	10.6
Households with a television set (%)	*90*	*90*	79	96
Usage				
International voice traffic (minutes/person/month)[b]	6.0	10.4	—	—
Mobile telephone usage (minutes/user/month)	—	59	322	154
Internet users (per 100 people)	0.1	*14.9*	12.4	21.4
Quality				
Population covered by mobile cellular network (%)	*84*	97	*80*	92
Fixed broadband subscribers (% of total Internet subscrib.)	0.0	*0.0*	40.4	32.5
International Internet bandwidth (bits/second/person)	4	216	199	1,114
Affordability				
Price basket for residential fixed line (US$/month)	3.0	5.5	*7.2*	5.8
Price basket for mobile service (US$/month)	—	20.9	9.8	11.8
Price basket for Internet service (US$/month)	—	16.3	16.7	12.0
Price of call to United States (US$ for 3 minutes)	4.59	1.34	2.08	1.63
Trade				
ICT goods exports (% of total goods exports)	0.7	1.0	*20.6*	1.8
ICT goods imports (% of total goods imports)	4.3	3.5	*20.2*	7.0
ICT service exports (% of total service exports)	4.8	3.6	*15.6*	5.0
Applications				
ICT expenditure (% of GDP)	—	—	6.5	5.0
E-government Web measure index[c]	—	0.39	0.33	0.36
Secure Internet servers (per 1 million people, Dec. 2008)	*0.3*	*4.7*	1.8	23.9

Sources: Economic and social context: UIS and World Bank; Sector structure: ITU; Sector efficiency and capacity: ITU and World Bank; Sector performance: Global Insight/WITSA, IMF, ITU, Netcraft, UN Comtrade, UNDESA, UNPAN, Wireless Intelligence and World Bank. Produced by the Global Information and Communication Technologies Department and the Development Economics Data Group. For complete information, see Definitions and Data Sources.

Notes: Use of italics in the column entries indicates years other than those specified. — Not available. GDP = gross domestic product; GNI = gross national income; ICT = information and communication technology; and MDG = Millennium Development Goal.

a. C = competition; M = monopoly; and P = partial competition. b. Outgoing and incoming. c. Scale of 0–1, where 1 = highest presence. d. Millennium Development Goal indicators 8.14, 8.15, and 8.16.

World Bank • ICT at a Glance

Algeria

	Algeria 2000	Algeria 2007	Lower-middle-income group 2007	Middle East & North Africa Region 2007
Economic and social context				
Population (total, million)	31	34	3,435	313
Urban population (% of total)	60	65	42	57
GNI per capita, World Bank Atlas method (current US$)	1,610	3,620	1,905	2,820
GDP growth, 1995–2000 and 2000–07 (avg. annual %)	3.2	4.5	8.0	4.4
Adult literacy rate (% of ages 15 and older)	*70*	*75*	83	73
Gross primary, secondary, tertiary school enrollment (%)	68	*74*	68	70
Sector structure				
Separate telecommunications regulator	No	Yes		
Status of main fixed-line telephone operator	Public	Public		
Level of competition[a]				
International long distance service	M	P		
Mobile telephone service	M	P		
Internet service	—	C		
Sector efficiency and capacity				
Telecommunications revenue (% of GDP)	0.6	2.7	*3.1*	*3.1*
Mobile and fixed-line subscribers per employee	103	285	*624*	*691*
Telecommunications investment (% of revenue)	*23.9*	*23.7*	*25.3*	*21.7*
Sector performance				
Access				
Telephone lines (per 100 people)	5.8	9.1	15.3	17.0
Mobile cellular subscriptions (per 100 people)	0.3	81.4	38.9	50.7
Internet subscribers (per 100 people)	0.2	*0.6*	6.0	2.4
Personal computers (per 100 people)	0.7	*1.1*	4.6	6.3
Households with a television set (%)	79	*91*	79	94
Usage				
International voice traffic (minutes/person/month)[b]	2.3	1.5	—	2.7
Mobile telephone usage (minutes/user/month)	—	141	322	—
Internet users (per 100 people)	0.5	10.3	12.4	17.1
Quality				
Population covered by mobile cellular network (%)	40	82	*80*	93
Fixed broadband subscribers (% of total Internet subscrib.)	*0.0*	71.1	40.4	—
International Internet bandwidth (bits/second/person)	0	89	199	186
Affordability				
Price basket for residential fixed line (US$/month)	4.5	28.3	7.2	3.9
Price basket for mobile service (US$/month)	—	7.4	9.8	6.5
Price basket for Internet service (US$/month)	—	5.7	16.7	11.6
Price of call to United States (US$ for 3 minutes)	3.67	*2.08*	*2.08*	1.45
Trade				
ICT goods exports (% of total goods exports)	0.0	*0.0*	20.6	—
ICT goods imports (% of total goods imports)	5.9	*6.9*	20.2	—
ICT service exports (% of total service exports)	—	—	15.6	2.6
Applications				
ICT expenditure (% of GDP)	—	2.5	6.5	4.5
E-government Web measure index[c]	—	0.22	0.33	0.22
Secure Internet servers (per 1 million people, Dec. 2008)	—	0.5	1.8	1.3

Sources: Economic and social context: UIS and World Bank; Sector structure: ITU; Sector efficiency and capacity: ITU and World Bank; Sector performance: Global Insight/WITSA, IMF, ITU, Netcraft, UN Comtrade, UNDESA, UNPAN, Wireless Intelligence and World Bank. Produced by the Global Information and Communication Technologies Department and the Development Economics Data Group. For complete information, see Definitions and Data Sources.

Notes: Use of italics in the column entries indicates years other than those specified. — Not available. GDP = gross domestic product; GNI = gross national income; ICT = information and communication technology; and MDG = Millennium Development Goal.

a. C = competition; M = monopoly; and P = partial competition. **b**. Outgoing and incoming. **c**. Scale of 0–1, where 1 = highest presence. **d**. Millennium Development Goal indicators 8.14, 8.15, and 8.16.

Angola

	Angola 2000	Angola 2007	Lower-middle-income group 2007	Sub-Saharan Africa Region 2007
Economic and social context				
Population (total, million)	14	17	3,435	800
Urban population (% of total)	49	56	42	36
GNI per capita, World Bank Atlas method (current US$)	420	2,540	1,905	951
GDP growth, 1995–2000 and 2000–07 (avg. annual %)	6.3	12.9	8.0	5.1
Adult literacy rate (% of ages 15 and older)	67	—	83	62
Gross primary, secondary, tertiary school enrollment (%)	26	—	68	51
Sector structure				
Separate telecommunications regulator	Yes	Yes		
Status of main fixed-line telephone operator	Public	Public		
Level of competition[a]				
International long distance service	M	C		
Mobile telephone service	P	C		
Internet service	P	C		
Sector efficiency and capacity				
Telecommunications revenue (% of GDP)	1.2	2.0	3.1	4.7
Mobile and fixed-line subscribers per employee	43	586	624	499
Telecommunications investment (% of revenue)	—	25.3	25.3	—
Sector performance				
Access				
Telephone lines (per 100 people)	0.5	0.6	15.3	1.6
Mobile cellular subscriptions (per 100 people)	0.2	29.3	38.9	23.0
Internet subscribers (per 100 people)	0.0	0.3	6.0	1.2
Personal computers (per 100 people)	0.1	0.7	4.6	1.8
Households with a television set (%)	9	9	79	18
Usage				
International voice traffic (minutes/person/month)[b]	0.5	—	—	—
Mobile telephone usage (minutes/user/month)	—	118	322	—
Internet users (per 100 people)	0.1	2.9	12.4	4.4
Quality				
Population covered by mobile cellular network (%)	—	40	80	56
Fixed broadband subscribers (% of total Internet subscrib.)	0.0	0.0	40.4	3.1
International Internet bandwidth (bits/second/person)	0	17	199	36
Affordability				
Price basket for residential fixed line (US$/month)	11.9	20.5	7.2	12.6
Price basket for mobile service (US$/month)	—	11.6	9.8	11.6
Price basket for Internet service (US$/month)	—	63.1	16.7	43.1
Price of call to United States (US$ for 3 minutes)	9.32	3.23	2.08	2.43
Trade				
ICT goods exports (% of total goods exports)	—	—	20.6	1.1
ICT goods imports (% of total goods imports)	—	—	20.2	8.2
ICT service exports (% of total service exports)	—	—	15.6	4.2
Applications				
ICT expenditure (% of GDP)	—	—	6.5	—
E-government Web measure index[c]	—	0.44	0.33	0.16
Secure Internet servers (per 1 million people, Dec. 2008)	—	1.4	1.8	2.9

Sources: Economic and social context: UIS and World Bank; Sector structure: ITU; Sector efficiency and capacity: ITU and World Bank; Sector performance: Global Insight/WITSA, IMF, ITU, Netcraft, UN Comtrade, UNDESA, UNPAN, Wireless Intelligence and World Bank. Produced by the Global Information and Communication Technologies Department and the Development Economics Data Group. For complete information, see Definitions and Data Sources.

Notes: Use of italics in the column entries indicates years other than those specified. — Not available. GDP = gross domestic product; GNI = gross national income; ICT = information and communication technology; and MDG = Millennium Development Goal.

a. C = competition; M = monopoly; and P = partial competition. **b**. Outgoing and incoming. **c**. Scale of 0–1, where 1 = highest presence. **d**. Millennium Development Goal indicators 8.14, 8.15, and 8.16.

World Bank • ICT at a Glance

Argentina

	Argentina 2000	Argentina 2007	Upper-middle-income group 2007	Latin America & the Caribbean Region 2007
Economic and social context				
Population (total, million)	37	40	824	561
Urban population (% of total)	90	92	75	78
GNI per capita, World Bank Atlas method (current US$)	7,470	6,040	7,107	5,801
GDP growth, 1995–2000 and 2000–07 (avg. annual %)	2.7	4.7	4.3	3.6
Adult literacy rate (% of ages 15 and older)	97	98	94	91
Gross primary, secondary, tertiary school enrollment (%)	92	90	82	81
Sector structure				
Separate telecommunications regulator	Yes	Yes		
Status of main fixed-line telephone operator	Private	Private		
Level of competition[a]				
International long distance service	P	C		
Mobile telephone service	P	C		
Internet service	C	C		
Sector efficiency and capacity				
Telecommunications revenue (% of GDP)	3.3	3.1	3.3	3.8
Mobile and fixed-line subscribers per employee	700	1,929	566	530
Telecommunications investment (% of revenue)	20.1	—	—	—
Sector performance				
Access				
Telephone lines (per 100 people)	21.4	24.0	22.6	18.1
Mobile cellular subscriptions (per 100 people)	17.6	102.3	84.1	67.0
Internet subscribers (per 100 people)	3.3	8.6	9.4	4.5
Personal computers (per 100 people)	6.9	9.0	12.4	11.3
Households with a television set (%)	95	95	92	84
Usage				
International voice traffic (minutes/person/month)[b]	0.2	0.3	—	—
Mobile telephone usage (minutes/user/month)	89	90	137	116
Internet users (per 100 people)	7.0	25.9	26.6	26.9
Quality				
Population covered by mobile cellular network (%)	30	94	95	91
Fixed broadband subscribers (% of total Internet subscrib.)	5.7	76.1	47.8	81.7
International Internet bandwidth (bits/second/person)	12	2,320	1,185	1,126
Affordability				
Price basket for residential fixed line (US$/month)	20.0	6.8	10.6	9.5
Price basket for mobile service (US$/month)	—	7.8	10.9	10.4
Price basket for Internet service (US$/month)	—	13.6	16.4	25.7
Price of call to United States (US$ for 3 minutes)	2.77	—	1.55	1.21
Trade				
ICT goods exports (% of total goods exports)	0.7	0.6	13.5	11.4
ICT goods imports (% of total goods imports)	15.3	13.1	16.2	15.9
ICT service exports (% of total service exports)	6.5	7.9	4.6	4.7
Applications				
ICT expenditure (% of GDP)	—	6.0	5.2	4.9
E-government Web measure index[c]	—	0.56	0.37	0.44
Secure Internet servers (per 1 million people, Dec. 2008)	6.4	18.3	26.2	18.2

Sources: Economic and social context: UIS and World Bank; Sector structure: ITU; Sector efficiency and capacity: ITU and World Bank; Sector performance: Global Insight/WITSA, IMF, ITU, Netcraft, UN Comtrade, UNDESA, UNPAN, Wireless Intelligence and World Bank. Produced by the Global Information and Communication Technologies Department and the Development Economics Data Group. For complete information, see Definitions and Data Sources.

Notes: Use of italics in the column entries indicates years other than those specified. — Not available. GDP = gross domestic product; GNI = gross national income; ICT = information and communication technology; and MDG = Millennium Development Goal.

a. C = competition; M = monopoly; and P = partial competition. **b**. Outgoing and incoming. **c**. Scale of 0–1, where 1 = highest presence. **d**. Millennium Development Goal indicators 8.14, 8.15, and 8.16.

Information and Communications for Development 2009

World Bank • ICT at a Glance

Armenia

	Armenia 2000	Armenia 2007	Lower-middle-income group 2007	Europe & Central Asia Region 2007
Economic and social context				
Population (total, million)	3	3	3,435	446
Urban population (% of total)	65	64	42	64
GNI per capita, World Bank Atlas method (current US$)	660	2,630	1,905	6,052
GDP growth, 1995–2000 and 2000–07 (avg. annual %)	5.1	12.7	8.0	6.1
Adult literacy rate (% of ages 15 and older)	*99*	*99*	83	98
Gross primary, secondary, tertiary school enrollment (%)	*72*	*71*	68	82
Sector structure				
Separate telecommunications regulator	—	Yes		
Status of main fixed-line telephone operator	Mixed	Mixed		
Level of competition[a]				
International long distance service	M	M		
Mobile telephone service	M	P		
Internet service	C	C		
Sector efficiency and capacity				
Telecommunications revenue (% of GDP)	3.0	3.0	3.1	2.9
Mobile and fixed-line subscribers per employee	71	173	624	532
Telecommunications investment (% of revenue)	46.3	23.6	25.3	22.0
Sector performance				
Access				
Telephone lines (per 100 people)	17.3	*19.7*	15.3	25.7
Mobile cellular subscriptions (per 100 people)	0.6	62.4	38.9	95.0
Internet subscribers (per 100 people)	0.4	*3.0*	6.0	13.6
Personal computers (per 100 people)	0.8	*9.8*	*4.6*	*10.6*
Households with a television set (%)	89	91	79	96
Usage				
International voice traffic (minutes/person/month)[b]	3.3	10.7	—	—
Mobile telephone usage (minutes/user/month)	—	171	322	154
Internet users (per 100 people)	1.3	*5.7*	12.4	21.4
Quality				
Population covered by mobile cellular network (%)	*38*	*88*	80	92
Fixed broadband subscribers (% of total Internet subscrib.)	0.0	1.1	40.4	32.5
International Internet bandwidth (bits/second/person)	3	22	199	1,114
Affordability				
Price basket for residential fixed line (US$/month)	8.1	3.1	7.2	5.8
Price basket for mobile service (US$/month)	—	8.7	9.8	11.8
Price basket for Internet service (US$/month)	—	56.6	16.7	12.0
Price of call to United States (US$ for 3 minutes)	5.34	2.42	2.08	1.63
Trade				
ICT goods exports (% of total goods exports)	2.1	0.6	20.6	1.8
ICT goods imports (% of total goods imports)	7.0	5.8	20.2	7.0
ICT service exports (% of total service exports)	13.0	14.6	15.6	5.0
Applications				
ICT expenditure (% of GDP)	—	—	6.5	5.0
E-government Web measure index[c]	—	0.27	0.33	0.36
Secure Internet servers (per 1 million people, Dec. 2008)	*0.3*	*4.7*	1.8	23.9

GNI per Capita, Atlas Method, 2000–07
Current US$

ICT MDG[d] Indicators, 2000–07
Number per 100 people
— Telephone lines
— Mobile cellular subscriptions
— Internet users

Price Basket for Internet Service, 2003–07
US$/month
— Armenia
— Europe & Central Asia Region

ICT Service Exports, 2000–07
Percentage of total service exports
— Armenia
— Europe & Central Asia Region

Sources: Economic and social context: UIS and World Bank; Sector structure: ITU; Sector efficiency and capacity: ITU and World Bank; Sector performance: Global Insight/WITSA, IMF, ITU, Netcraft, UN Comtrade, UNDESA, UNPAN, Wireless Intelligence and World Bank. Produced by the Global Information and Communication Technologies Department and the Development Economics Data Group. For complete information, see Definitions and Data Sources.

Notes: Use of italics in the column entries indicates years other than those specified. — Not available. GDP = gross domestic product; GNI = gross national income; ICT = information and communication technology; and MDG = Millennium Development Goal.

a. C = competition; M = monopoly; and P = partial competition. **b.** Outgoing and incoming. **c.** Scale of 0–1, where 1 = highest presence. **d.** Millennium Development Goal indicators 8.14, 8.15, and 8.16.

World Bank • ICT at a Glance

Australia

	Australia 2000	Australia 2007	High-income group 2007
Economic and social context			
Population (total, million)	19	21	1,056
Urban population (% of total)	87	89	78
GNI per capita, World Bank Atlas method (current US$)	20,710	35,760	37,572
GDP growth, 1995–2000 and 2000–07 (avg. annual %)	4.4	3.2	2.4
Adult literacy rate (% of ages 15 and older)	—	—	99
Gross primary, secondary, tertiary school enrollment (%)	113	*113*	*92*
Sector structure			
Separate telecommunications regulator	Yes	Yes	
Status of main fixed-line telephone operator	*Mixed*	*Mixed*	
Level of competition[a]			
International long distance service	C	C	
Mobile telephone service	C	C	
Internet service	C	C	
Sector efficiency and capacity			
Telecommunications revenue (% of GDP)	3.2	3.6	*3.1*
Mobile and fixed-line subscribers per employee	181	310	747
Telecommunications investment (% of revenue)	30.9	18.8	*14.3*
Sector performance			
Access			
Telephone lines (per 100 people)	52.5	46.4	50.0
Mobile cellular subscriptions (per 100 people)	44.7	101.2	100.4
Internet subscribers (per 100 people)	20.5	33.8	25.8
Personal computers (per 100 people)	47.0	*60.3*	67.4
Households with a television set (%)	99	99	98
Usage			
International voice traffic (minutes/person/month)[b]	18.7	—	*14.0*
Mobile telephone usage (minutes/user/month)	98	73	353
Internet users (per 100 people)	46.8	68.1	65.7
Quality			
Population covered by mobile cellular network (%)	96	99	*99*
Fixed broadband subscribers (% of total Internet subscrib.)	2.9	68.0	82.6
International Internet bandwidth (bits/second/person)	128	5,472	18,242
Affordability			
Price basket for residential fixed line (US$/month)	17.4	28.4	*26.1*
Price basket for mobile service (US$/month)	—	24.1	13.0
Price basket for Internet service (US$/month)	—	25.1	22.8
Price of call to United States (US$ for 3 minutes)	0.67	—	*0.81*
Trade			
ICT goods exports (% of total goods exports)	3.2	*1.8*	15.2
ICT goods imports (% of total goods imports)	16.3	*12.8*	14.6
ICT service exports (% of total service exports)	7.0	4.6	7.0
Applications			
ICT expenditure (% of GDP)	—	6.6	6.7
E-government Web measure index[c]	—	0.75	0.60
Secure Internet servers (per 1 million people, Dec. 2008)	*176.3*	993.2	662.6

GNI per Capita, Atlas Method, 2000–07
Current US$

ICT MDG[d] Indicators, 2000–07
Number per 100 people

Price Basket for Internet Service, 2003–07
US$/month

ICT Service Exports, 2000–07
Percentage of total service exports

Sources: Economic and social context: UIS and World Bank; Sector structure: ITU; Sector efficiency and capacity: ITU and World Bank; Sector performance: Global Insight/WITSA, IMF, ITU, Netcraft, UN Comtrade, UNDESA, UNPAN, Wireless Intelligence and World Bank. Produced by the Global Information and Communication Technologies Department and the Development Economics Data Group. For complete information, see Definitions and Data Sources.

Notes: Use of italics in the column entries indicates years other than those specified. — Not available. GDP = gross domestic product; GNI = gross national income; ICT = information and communication technology; and MDG = Millennium Development Goal.

a. C = competition; M = monopoly; and P = partial competition. **b**. Outgoing and incoming. **c**. Scale of 0–1, where 1 = highest presence. **d**. Millennium Development Goal indicators 8.14, 8.15, and 8.16.

Austria

	Austria 2000	Austria 2007	High-income group 2007
Economic and social context			
Population (total, million)	8	8	1,056
Urban population (% of total)	66	67	78
GNI per capita, World Bank Atlas method (current US$)	26,010	41,960	37,572
GDP growth, 1995–2000 and 2000–07 (avg. annual %)	2.9	2.0	2.4
Adult literacy rate (% of ages 15 and older)	—	—	99
Gross primary, secondary, tertiary school enrollment (%)	91	*92*	92
Sector structure			
Separate telecommunications regulator	Yes	Yes	
Status of main fixed-line telephone operator	Mixed	Mixed	
Level of competition[a]			
International long distance service	C	C	
Mobile telephone service	P	P	
Internet service	C	C	
Sector efficiency and capacity			
Telecommunications revenue (% of GDP)	2.3	2.1	*3.1*
Mobile and fixed-line subscribers per employee	545	747	747
Telecommunications investment (% of revenue)	18.8	15.3	*14.3*
Sector performance			
Access			
Telephone lines (per 100 people)	49.9	41.0	50.0
Mobile cellular subscriptions (per 100 people)	76.4	119.2	100.4
Internet subscribers (per 100 people)	13.1	30.3	25.8
Personal computers (per 100 people)	36.2	*60.7*	67.4
Households with a television set (%)	97	98	98
Usage			
International voice traffic (minutes/person/month)[b]	25.2	*22.1*	14.0
Mobile telephone usage (minutes/user/month)	*123*	186	353
Internet users (per 100 people)	33.7	67.4	65.7
Quality			
Population covered by mobile cellular network (%)	98	99	*99*
Fixed broadband subscribers (% of total Internet subscrib.)	18.1	64.3	82.6
International Internet bandwidth (bits/second/person)	1,034	20,288	18,242
Affordability			
Price basket for residential fixed line (US$/month)	23.3	*29.0*	26.1
Price basket for mobile service (US$/month)	—	25.1	13.0
Price basket for Internet service (US$/month)	—	27.2	22.8
Price of call to United States (US$ for 3 minutes)	1.19	0.71	0.81
Trade			
ICT goods exports (% of total goods exports)	7.9	*6.3*	15.2
ICT goods imports (% of total goods imports)	10.9	*8.2*	14.6
ICT service exports (% of total service exports)	3.4	6.3	7.0
Applications			
ICT expenditure (% of GDP)	—	5.6	6.7
E-government Web measure index[c]	—	0.67	0.60
Secure Internet servers (per 1 million people, Dec. 2008)	83.2	480.6	662.6

Sources: Economic and social context: UIS and World Bank; Sector structure: ITU; Sector efficiency and capacity: ITU and World Bank; Sector performance: Global Insight/WITSA, IMF, ITU, Netcraft, UN Comtrade, UNDESA, UNPAN, Wireless Intelligence and World Bank. Produced by the Global Information and Communication Technologies Department and the Development Economics Data Group. For complete information, see Definitions and Data Sources.

Notes: Use of italics in the column entries indicates years other than those specified. — Not available. GDP = gross domestic product; GNI = gross national income; ICT = information and communication technology; and MDG = Millennium Development Goal.

a. C = competition; M = monopoly; and P = partial competition. b. Outgoing and incoming. c. Scale of 0–1, where 1 = highest presence. d. Millennium Development Goal indicators 8.14, 8.15, and 8.16.

World Bank • ICT at a Glance

Azerbaijan

	Azerbaijan 2000	Azerbaijan 2007	Lower-middle-income group 2007	Europe & Central Asia Region 2007
Economic and social context				
Population (total, million)	8	9	3,435	446
Urban population (% of total)	51	52	42	64
GNI per capita, World Bank Atlas method (current US$)	610	2,640	1,905	6,052
GDP growth, 1995–2000 and 2000–07 (avg. annual %)	7.3	17.6	8.0	6.1
Adult literacy rate (% of ages 15 and older)	*99*	100	83	98
Gross primary, secondary, tertiary school enrollment (%)	66	67	*68*	*82*
Sector structure				
Separate telecommunications regulator	—	No		
Status of main fixed-line telephone operator	Public	Public		
Level of competition[a]				
International long distance service	M	P		
Mobile telephone service	C	P		
Internet service	—	C		
Sector efficiency and capacity				
Telecommunications revenue (% of GDP)	1.6	2.6	*3.1*	2.9
Mobile and fixed-line subscribers per employee	123	413	*624*	*532*
Telecommunications investment (% of revenue)	27.9	23.6	*25.3*	22.0
Sector performance				
Access				
Telephone lines (per 100 people)	10.0	14.6	15.3	25.7
Mobile cellular subscriptions (per 100 people)	5.2	52.8	38.9	95.0
Internet subscribers (per 100 people)	0.0	2.5	6.0	13.6
Personal computers (per 100 people)	*0.7*	*2.4*	*4.6*	*10.6*
Households with a television set (%)	99	99	79	96
Usage				
International voice traffic (minutes/person/month)[b]	1.0	2.7	—	—
Mobile telephone usage (minutes/user/month)	—	78	322	154
Internet users (per 100 people)	0.1	10.8	12.4	21.4
Quality				
Population covered by mobile cellular network (%)	94	99	*80*	*92*
Fixed broadband subscribers (% of total Internet subscrib.)	0.0	2.6	40.4	32.5
International Internet bandwidth (bits/second/person)	*0*	701	199	1,114
Affordability				
Price basket for residential fixed line (US$/month)	17.6	*5.3*	7.2	5.8
Price basket for mobile service (US$/month)	—	15.1	9.8	11.8
Price basket for Internet service (US$/month)	—	10.3	16.7	12.0
Price of call to United States (US$ for 3 minutes)	7.10	*4.18*	2.08	1.63
Trade				
ICT goods exports (% of total goods exports)	0.3	*0.1*	20.6	1.8
ICT goods imports (% of total goods imports)	10.6	*6.1*	20.2	7.0
ICT service exports (% of total service exports)	—	3.1	15.6	5.0
Applications				
ICT expenditure (% of GDP)	—	—	6.5	5.0
E-government Web measure index[c]	—	0.39	0.33	0.36
Secure Internet servers (per 1 million people, Dec. 2008)	*0.1*	1.6	1.8	23.9

GNI per Capita, Atlas Method, 2000–07
Current US$

ICT MDG[d] Indicators, 2000–07
Number per 100 people
- Telephone lines
- Mobile cellular subscriptions
- Internet users

Price Basket for Internet Service, 2003–07
US$/month
- Azerbaijan
- Europe & Central Asia Region

ICT Service Exports, 2000–07
Percentage of total service exports
- Azerbaijan
- Europe & Central Asia Region

Sources: Economic and social context: UIS and World Bank; Sector structure: ITU; Sector efficiency and capacity: ITU and World Bank; Sector performance: Global Insight/WITSA, IMF, ITU, Netcraft, UN Comtrade, UNDESA, UNPAN, Wireless Intelligence and World Bank. Produced by the Global Information and Communication Technologies Department and the Development Economics Data Group. For complete information, see Definitions and Data Sources.

Notes: Use of italics in the column entries indicates years other than those specified. — Not available. GDP = gross domestic product; GNI = gross national income; ICT = information and communication technology; and MDG = Millennium Development Goal.

a. C = competition; M = monopoly; and P = partial competition. b. Outgoing and incoming. c. Scale of 0–1, where 1 = highest presence. d. Millennium Development Goal indicators 8.14, 8.15, and 8.16.

Bangladesh

	Bangladesh 2000	Bangladesh 2007	Low-income group 2007	South Asia Region 2007
Economic and social context				
Population (total, million)	139	159	1,296	1,522
Urban population (% of total)	24	27	32	29
GNI per capita, World Bank Atlas method (current US$)	360	470	574	880
GDP growth, 1995–2000 and 2000–07 (avg. annual %)	5.2	5.7	5.6	7.3
Adult literacy rate (% of ages 15 and older)	47	53	64	63
Gross primary, secondary, tertiary school enrollment (%)	58	56	51	60
Sector structure				
Separate telecommunications regulator	No	Yes		
Status of main fixed-line telephone operator	Public	Public		
Level of competition[a]				
International long distance service	M	M		
Mobile telephone service	C	C		
Internet service	—	C		
Sector efficiency and capacity				
Telecommunications revenue (% of GDP)	0.8	1.5	3.3	2.1
Mobile and fixed-line subscribers per employee	48	—	301	660
Telecommunications investment (% of revenue)	25.1	—	—	—
Sector performance				
Access				
Telephone lines (per 100 people)	0.4	0.7	4.0	3.2
Mobile cellular subscriptions (per 100 people)	0.2	21.7	21.5	22.8
Internet subscribers (per 100 people)	0.0	0.1	0.8	1.3
Personal computers (per 100 people)	0.1	2.2	1.5	3.3
Households with a television set (%)	20	48	16	42
Usage				
International voice traffic (minutes/person/month)[b]	0.1	0.5	—	—
Mobile telephone usage (minutes/user/month)	313	249	—	364
Internet users (per 100 people)	0.1	0.3	5.2	6.6
Quality				
Population covered by mobile cellular network (%)	40	90	54	61
Fixed broadband subscribers (% of total Internet subscrib.)	0.0	0.0	3.4	18.9
International Internet bandwidth (bits/second/person)	0	4	26	31
Affordability				
Price basket for residential fixed line (US$/month)	10.7	4.0	5.7	4.0
Price basket for mobile service (US$/month)	—	2.6	11.2	2.4
Price basket for Internet service (US$/month)	—	22.1	29.2	8.0
Price of call to United States (US$ for 3 minutes)	4.14	2.02	2.00	2.02
Trade				
ICT goods exports (% of total goods exports)	0.0	0.1	1.4	1.2
ICT goods imports (% of total goods imports)	3.3	4.4	6.7	8.1
ICT service exports (% of total service exports)	3.0	5.7	—	39.0
Applications				
ICT expenditure (% of GDP)	—	8.0	—	5.7
E-government Web measure index[c]	—	0.35	0.11	0.37
Secure Internet servers (per 1 million people, Dec. 2008)	0.0	0.1	0.5	1.1

Sources: Economic and social context: UIS and World Bank; Sector structure: ITU; Sector efficiency and capacity: ITU and World Bank; Sector performance: Global Insight/WITSA, IMF, ITU, Netcraft, UN Comtrade, UNDESA, UNPAN, Wireless Intelligence and World Bank. Produced by the Global Information and Communication Technologies Department and the Development Economics Data Group. For complete information, see Definitions and Data Sources.

Notes: Use of italics in the column entries indicates years other than those specified. — Not available. GDP = gross domestic product; GNI = gross national income; ICT = information and communication technology; and MDG = Millennium Development Goal.

a. C = competition; M = monopoly; and P = partial competition. **b.** Outgoing and incoming. **c.** Scale of 0–1, where 1 = highest presence. **d.** Millennium Development Goal indicators 8.14, 8.15, and 8.16.

World Bank • ICT at a Glance

Belarus

	Belarus 2000	Belarus 2007	Upper-middle-income group 2007	Europe & Central Asia Region 2007
Economic and social context				
Population (total, million)	10	10	824	446
Urban population (% of total)	70	73	75	64
GNI per capita, World Bank Atlas method (current US$)	1,380	4,220	7,107	6,052
GDP growth, 1995–2000 and 2000–07 (avg. annual %)	6.7	8.3	4.3	6.1
Adult literacy rate (% of ages 15 and older)	*100*	100	94	98
Gross primary, secondary, tertiary school enrollment (%)	87	*89*	82	82
Sector structure				
Separate telecommunications regulator	—	No		
Status of main fixed-line telephone operator	Public	Public		
Level of competition[a]				
International long distance service	M	M		
Mobile telephone service	C	C		
Internet service	—	C		
Sector efficiency and capacity				
Telecommunications revenue (% of GDP)	1.5	*2.1*	3.3	2.9
Mobile and fixed-line subscribers per employee	105	*280*	566	532
Telecommunications investment (% of revenue)	28.8	*41.3*	—	22.0
Sector performance				
Access				
Telephone lines (per 100 people)	27.5	37.8	22.6	25.7
Mobile cellular subscriptions (per 100 people)	0.5	71.7	84.1	95.0
Internet subscribers (per 100 people)	0.0	4.2	9.4	13.6
Personal computers (per 100 people)	—	0.8	12.4	10.6
Households with a television set (%)	87	97	92	96
Usage				
International voice traffic (minutes/person/month)[b]	3.3	5.3	—	—
Mobile telephone usage (minutes/user/month)	—	500	137	154
Internet users (per 100 people)	1.9	29.0	26.6	21.4
Quality				
Population covered by mobile cellular network (%)	*62*	93	95	92
Fixed broadband subscribers (% of total Internet subscrib.)	0.0	2.8	47.8	32.5
International Internet bandwidth (bits/second/person)	2	264	1,185	1,114
Affordability				
Price basket for residential fixed line (US$/month)	1.2	*1.6*	10.6	5.8
Price basket for mobile service (US$/month)	—	*11.8*	10.9	11.8
Price basket for Internet service (US$/month)	—	*10.5*	16.4	12.0
Price of call to United States (US$ for 3 minutes)	3.28	*1.90*	1.55	1.63
Trade				
ICT goods exports (% of total goods exports)	1.9	0.8	*13.5*	1.8
ICT goods imports (% of total goods imports)	2.9	3.0	*16.2*	7.0
ICT service exports (% of total service exports)	5.4	6.8	*4.6*	5.0
Applications				
ICT expenditure (% of GDP)	—	—	5.2	5.0
E-government Web measure index[c]	—	0.33	0.37	0.36
Secure Internet servers (per 1 million people, Dec. 2008)	*0.4*	*2.1*	26.2	23.9

Sources: Economic and social context: UIS and World Bank; Sector structure: ITU; Sector efficiency and capacity: ITU and World Bank; Sector performance: Global Insight/WITSA, IMF, ITU, Netcraft, UN Comtrade, UNDESA, UNPAN, Wireless Intelligence and World Bank. Produced by the Global Information and Communication Technologies Department and the Development Economics Data Group. For complete information, see Definitions and Data Sources.

Notes: Use of italics in the column entries indicates years other than those specified. — Not available. GDP = gross domestic product; GNI = gross national income; ICT = information and communication technology; and MDG = Millennium Development Goal.

a. C = competition; M = monopoly; and P = partial competition. **b.** Outgoing and incoming. **c.** Scale of 0–1, where 1 = highest presence. **d.** Millennium Development Goal indicators 8.14, 8.15, and 8.16.

World Bank • ICT at a Glance

Belgium

	Belgium 2000	Belgium 2007	High-income group 2007
Economic and social context			
Population (total, million)	10	11	1,056
Urban population (% of total)	97	97	78
GNI per capita, World Bank Atlas method (current US$)	25,360	41,110	37,572
GDP growth, 1995–2000 and 2000–07 (avg. annual %)	2.7	2.0	2.4
Adult literacy rate (% of ages 15 and older)	—	—	99
Gross primary, secondary, tertiary school enrollment (%)	107	95	92
Sector structure			
Separate telecommunications regulator	Yes	Yes	
Status of main fixed-line telephone operator	Mixed	Mixed	
Level of competition[a]			
International long distance service	C	C	
Mobile telephone service	P	P	
Internet service	C	C	
Sector efficiency and capacity			
Telecommunications revenue (% of GDP)	1.9	3.0	3.1
Mobile and fixed-line subscribers per employee	374	690	747
Telecommunications investment (% of revenue)	33.7	12.3	14.3
Sector performance			
Access			
Telephone lines (per 100 people)	49.1	43.9	50.0
Mobile cellular subscriptions (per 100 people)	54.9	101.1	100.4
Internet subscribers (per 100 people)	11.2	26.9	25.8
Personal computers (per 100 people)	22.4	41.7	67.4
Households with a television set (%)	99	99	98
Usage			
International voice traffic (minutes/person/month)[b]	26.4	—	14.0
Mobile telephone usage (minutes/user/month)	105	158	353
Internet users (per 100 people)	29.3	65.9	65.7
Quality			
Population covered by mobile cellular network (%)	99	100	99
Fixed broadband subscribers (% of total Internet subscrib.)	12.5	94.9	82.6
International Internet bandwidth (bits/second/person)	1,829	24,945	18,242
Affordability			
Price basket for residential fixed line (US$/month)	22.0	33.1	26.1
Price basket for mobile service (US$/month)	—	18.2	13.0
Price basket for Internet service (US$/month)	—	37.6	22.8
Price of call to United States (US$ for 3 minutes)	1.67	0.75	0.81
Trade			
ICT goods exports (% of total goods exports)	6.2	3.7	15.2
ICT goods imports (% of total goods imports)	7.6	4.8	14.6
ICT service exports (% of total service exports)	7.4	8.7	7.0
Applications			
ICT expenditure (% of GDP)	—	5.8	6.7
E-government Web measure index[c]	—	0.54	0.60
Secure Internet servers (per 1 million people, Dec. 2008)	33.2	250.2	662.6

Sources: Economic and social context: UIS and World Bank; Sector structure: ITU; Sector efficiency and capacity: ITU and World Bank; Sector performance: Global Insight/WITSA, IMF, ITU, Netcraft, UN Comtrade, UNDESA, UNPAN, Wireless Intelligence and World Bank. Produced by the Global Information and Communication Technologies Department and the Development Economics Data Group. For complete information, see Definitions and Data Sources.

Notes: Use of italics in the column entries indicates years other than those specified. — Not available. GDP = gross domestic product; GNI = gross national income; ICT = information and communication technology; and MDG = Millennium Development Goal.

a. C = competition; M = monopoly; and P = partial competition. **b.** Outgoing and incoming. **c.** Scale of 0–1, where 1 = highest presence. **d.** Millennium Development Goal indicators 8.14, 8.15, and 8.16.

World Bank • ICT at a Glance

Benin

	Benin 2000	Benin 2007	Low-income group 2007	Sub-Saharan Africa Region 2007
Economic and social context				
Population (total, million)	7	9	1,296	800
Urban population (% of total)	38	41	32	36
GNI per capita, World Bank Atlas method (current US$)	340	570	574	951
GDP growth, 1995–2000 and 2000–07 (avg. annual %)	5.3	3.8	5.6	5.1
Adult literacy rate (% of ages 15 and older)	*35*	41	64	62
Gross primary, secondary, tertiary school enrollment (%)	40	*51*	*51*	*51*
Sector structure				
Separate telecommunications regulator	No	Yes		
Status of main fixed-line telephone operator	—	Public		
Level of competition[a]				
International long distance service	M	M		
Mobile telephone service	M	C		
Internet service	—	—		
Sector efficiency and capacity				
Telecommunications revenue (% of GDP)	2.4	1.1	*3.3*	*4.7*
Mobile and fixed-line subscribers per employee	87	1,539	*301*	*499*
Telecommunications investment (% of revenue)	54.2	5.2	—	—
Sector performance				
Access				
Telephone lines (per 100 people)	0.7	1.2	4.0	1.6
Mobile cellular subscriptions (per 100 people)	0.8	21.1	21.5	23.0
Internet subscribers (per 100 people)	0.0	0.1	*0.8*	*1.2*
Personal computers (per 100 people)	0.1	0.7	*1.5*	*1.8*
Households with a television set (%)	15	*13*	*16*	*18*
Usage				
International voice traffic (minutes/person/month)[b]	0.4	*0.9*	—	—
Mobile telephone usage (minutes/user/month)	—	—	—	—
Internet users (per 100 people)	0.2	1.7	5.2	4.4
Quality				
Population covered by mobile cellular network (%)	*23*	80	*54*	*56*
Fixed broadband subscribers (% of total Internet subscrib.)	*0.0*	24.8	*3.4*	*3.1*
International Internet bandwidth (bits/second/person)	0	17	26	36
Affordability				
Price basket for residential fixed line (US$/month)	11.3	*5.3*	*5.7*	12.6
Price basket for mobile service (US$/month)	—	10.3	11.2	11.6
Price basket for Internet service (US$/month)	—	43.1	29.2	43.1
Price of call to United States (US$ for 3 minutes)	5.93	*4.80*	2.00	2.43
Trade				
ICT goods exports (% of total goods exports)	0.0	*0.0*	*1.4*	*1.1*
ICT goods imports (% of total goods imports)	3.3	*3.3*	*6.7*	*8.2*
ICT service exports (% of total service exports)	—	5.4	—	4.2
Applications				
ICT expenditure (% of GDP)	—	—	—	—
E-government Web measure index[c]	—	0.12	0.11	0.16
Secure Internet servers (per 1 million people, Dec. 2008)	*0.1*	0.2	0.5	2.9

Sources: Economic and social context: UIS and World Bank; Sector structure: ITU; Sector efficiency and capacity: ITU and World Bank; Sector performance: Global Insight/WITSA, IMF, ITU, Netcraft, UN Comtrade, UNDESA, UNPAN, Wireless Intelligence and World Bank. Produced by the Global Information and Communication Technologies Department and the Development Economics Data Group. For complete information, see Definitions and Data Sources.

Notes: Use of italics in the column entries indicates years other than those specified. — Not available. GDP = gross domestic product; GNI = gross national income; ICT = information and communication technology; and MDG = Millennium Development Goal.

a. C = competition; M = monopoly; and P = partial competition. **b.** Outgoing and incoming. **c.** Scale of 0–1, where 1 = highest presence. **d.** Millennium Development Goal indicators 8.14, 8.15, and 8.16.

Information and Communications for Development 2009

World Bank • ICT at a Glance

Bolivia

	Bolivia 2000	Bolivia 2007	Lower-middle-income group 2007	Latin America & the Caribbean Region 2007
Economic and social context				
Population (total, million)	8	10	3,435	561
Urban population (% of total)	62	65	42	78
GNI per capita, World Bank Atlas method (current US$)	1,000	1,260	1,905	5,801
GDP growth, 1995–2000 and 2000–07 (avg. annual %)	3.5	3.6	8.0	3.6
Adult literacy rate (% of ages 15 and older)	*87*	*91*	83	91
Gross primary, secondary, tertiary school enrollment (%)	84	*86*	68	81
Sector structure				
Separate telecommunications regulator	Yes	Yes		
Status of main fixed-line telephone operator	Mixed	Private		
Level of competition[a]				
International long distance service	M	M		
Mobile telephone service	P	C		
Internet service	C	C		
Sector efficiency and capacity				
Telecommunications revenue (% of GDP)	4.7	6.8	3.1	3.8
Mobile and fixed-line subscribers per employee	451	376	624	530
Telecommunications investment (% of revenue)	43.4	3.8	25.3	—
Sector performance				
Access				
Telephone lines (per 100 people)	6.1	7.1	15.3	18.1
Mobile cellular subscriptions (per 100 people)	7.0	34.2	38.9	67.0
Internet subscribers (per 100 people)	0.5	2.1	6.0	4.5
Personal computers (per 100 people)	1.7	*2.4*	4.6	11.3
Households with a television set (%)	46	63	79	84
Usage				
International voice traffic (minutes/person/month)[b]	1.2	6.7	—	—
Mobile telephone usage (minutes/user/month)	—	—	322	116
Internet users (per 100 people)	1.4	10.5	12.4	26.9
Quality				
Population covered by mobile cellular network (%)	*43*	46	*80*	91
Fixed broadband subscribers (% of total Internet subscrib.)	0.0	17.1	40.4	81.7
International Internet bandwidth (bits/second/person)	2	42	199	1,126
Affordability				
Price basket for residential fixed line (US$/month)	11.6	8.5	7.2	9.5
Price basket for mobile service (US$/month)	—	5.6	9.8	10.4
Price basket for Internet service (US$/month)	—	*12.1*	16.7	25.7
Price of call to United States (US$ for 3 minutes)	2.43	—	2.08	1.21
Trade				
ICT goods exports (% of total goods exports)	0.7	0.1	20.6	11.4
ICT goods imports (% of total goods imports)	7.6	4.9	20.2	15.9
ICT service exports (% of total service exports)	15.9	12.5	15.6	4.7
Applications				
ICT expenditure (% of GDP)	—	5.8	6.5	4.9
E-government Web measure index[c]	—	0.52	0.33	0.44
Secure Internet servers (per 1 million people, Dec. 2008)	*0.6*	3.8	1.8	18.2

Sources: Economic and social context: UIS and World Bank; Sector structure: ITU; Sector efficiency and capacity: ITU and World Bank; Sector performance: Global Insight/WITSA, IMF, ITU, Netcraft, UN Comtrade, UNDESA, UNPAN, Wireless Intelligence and World Bank. Produced by the Global Information and Communication Technologies Department and the Development Economics Data Group. For complete information, see Definitions and Data Sources.

Notes: Use of italics in the column entries indicates years other than those specified. — Not available. GDP = gross domestic product; GNI = gross national income; ICT = information and communication technology; and MDG = Millennium Development Goal.

a. C = competition; M = monopoly; and P = partial competition. **b**. Outgoing and incoming. **c**. Scale of 0–1, where 1 = highest presence. **d**. Millennium Development Goal indicators 8.14, 8.15, and 8.16.

Bosnia and Herzegovina

	Bosnia and Herzegovina 2000	Bosnia and Herzegovina 2007	Lower-middle-income group 2007	Europe & Central Asia Region 2007
Economic and social context				
Population (total, million)	4	4	3,435	446
Urban population (% of total)	43	47	42	64
GNI per capita, World Bank Atlas method (current US$)	1,500	3,790	1,905	6,052
GDP growth, 1995–2000 and 2000–07 (avg. annual %)	25.2	5.3	8.0	6.1
Adult literacy rate (% of ages 15 and older)	97	—	83	98
Gross primary, secondary, tertiary school enrollment (%)	—	—	68	82
Sector structure				
Separate telecommunications regulator	Yes	Yes		
Status of main fixed-line telephone operator	Public	Mixed		
Level of competition[a]				
International long distance service	M	C		
Mobile telephone service	P	C		
Internet service	C	C		
Sector efficiency and capacity				
Telecommunications revenue (% of GDP)	4.3	5.7	*3.1*	2.9
Mobile and fixed-line subscribers per employee	480	657	*624*	*532*
Telecommunications investment (% of revenue)	22.8	13.6	*25.3*	22.0
Sector performance				
Access				
Telephone lines (per 100 people)	21.1	28.2	15.3	25.7
Mobile cellular subscriptions (per 100 people)	2.5	64.9	38.9	95.0
Internet subscribers (per 100 people)	0.4	7.3	6.0	13.6
Personal computers (per 100 people)	*3.9*	*6.4*	*4.6*	*10.6*
Households with a television set (%)	87	96	79	96
Usage				
International voice traffic (minutes/person/month)[b]	7.7	20.1	—	—
Mobile telephone usage (minutes/user/month)	—	—	322	154
Internet users (per 100 people)	1.1	28.0	12.4	21.4
Quality				
Population covered by mobile cellular network (%)	60	99	*80*	92
Fixed broadband subscribers (% of total Internet subscrib.)	0.0	30.9	40.4	32.5
International Internet bandwidth (bits/second/person)	7	530	199	1,114
Affordability				
Price basket for residential fixed line (US$/month)	4.6	*6.3*	*7.2*	*5.8*
Price basket for mobile service (US$/month)	—	*6.6*	*9.8*	*11.8*
Price basket for Internet service (US$/month)	—	*7.6*	*16.7*	*12.0*
Price of call to United States (US$ for 3 minutes)	2.96	*3.62*	*2.08*	*1.63*
Trade				
ICT goods exports (% of total goods exports)	—	0.5	*20.6*	*1.8*
ICT goods imports (% of total goods imports)	—	3.8	*20.2*	*7.0*
ICT service exports (% of total service exports)	—	—	15.6	5.0
Applications				
ICT expenditure (% of GDP)	—	—	6.5	5.0
E-government Web measure index[c]	—	0.29	0.33	0.36
Secure Internet servers (per 1 million people, Dec. 2008)	—	7.4	1.8	23.9

Sources: Economic and social context: UIS and World Bank; Sector structure: ITU; Sector efficiency and capacity: ITU and World Bank; Sector performance: Global Insight/WITSA, IMF, ITU, Netcraft, UN Comtrade, UNDESA, UNPAN, Wireless Intelligence and World Bank. Produced by the Global Information and Communication Technologies Department and the Development Economics Data Group. For complete information, see Definitions and Data Sources.

Notes: Use of italics in the column entries indicates years other than those specified. — Not available. GDP = gross domestic product; GNI = gross national income; ICT = information and communication technology; and MDG = Millennium Development Goal.

[a]. C = competition; M = monopoly; and P = partial competition. [b]. Outgoing and incoming. [c]. Scale of 0–1, where 1 = highest presence. [d]. Millennium Development Goal indicators 8.14, 8.15, and 8.16.

Botswana

	Botswana 2000	Botswana 2007	Upper-middle-income group 2007	Sub-Saharan Africa Region 2007
Economic and social context				
Population (total, million)	2	2	824	800
Urban population (% of total)	53	59	75	36
GNI per capita, World Bank Atlas method (current US$)	3,310	6,120	7,107	951
GDP growth, 1995–2000 and 2000–07 (avg. annual %)	8.6	5.3	4.3	5.1
Adult literacy rate (% of ages 15 and older)	—	83	94	62
Gross primary, secondary, tertiary school enrollment (%)	69	*70*	82	51
Sector structure				
Separate telecommunications regulator	Yes	Yes		
Status of main fixed-line telephone operator	*Public*	*Public*		
Level of competition[a]				
International long distance service	M	P		
Mobile telephone service	C	—		
Internet service	C	C		
Sector efficiency and capacity				
Telecommunications revenue (% of GDP)	3.0	3.0	3.3	4.7
Mobile and fixed-line subscribers per employee	202	1,074	*566*	*499*
Telecommunications investment (% of revenue)	13.9	*139.9*	—	—
Sector performance				
Access				
Telephone lines (per 100 people)	7.9	7.3	22.6	1.6
Mobile cellular subscriptions (per 100 people)	12.9	61.2	84.1	23.0
Internet subscribers (per 100 people)	0.9	—	9.4	1.2
Personal computers (per 100 people)	3.5	4.8	12.4	1.8
Households with a television set (%)	10	9	92	18
Usage				
International voice traffic (minutes/person/month)[b]	3.5	7.8	—	—
Mobile telephone usage (minutes/user/month)	—	—	137	—
Internet users (per 100 people)	2.9	5.3	26.6	4.4
Quality				
Population covered by mobile cellular network (%)	*90*	99	95	56
Fixed broadband subscribers (% of total Internet subscrib.)	0.0	—	47.8	3.1
International Internet bandwidth (bits/second/person)	3	43	1,185	36
Affordability				
Price basket for residential fixed line (US$/month)	4.8	14.4	*10.6*	12.6
Price basket for mobile service (US$/month)	—	8.3	*10.9*	11.6
Price basket for Internet service (US$/month)	—	29.7	*16.4*	43.1
Price of call to United States (US$ for 3 minutes)	*3.64*	*2.88*	*1.55*	*2.43*
Trade				
ICT goods exports (% of total goods exports)	0.7	0.2	*13.5*	*1.1*
ICT goods imports (% of total goods imports)	6.0	5.5	*16.2*	*8.2*
ICT service exports (% of total service exports)	0.1	6.8	*4.6*	*4.2*
Applications				
ICT expenditure (% of GDP)	—	—	5.2	—
E-government Web measure index[c]	—	0.22	0.37	0.16
Secure Internet servers (per 1 million people, Dec. 2008)	—	1.6	26.2	2.9

Sources: Economic and social context: UIS and World Bank; Sector structure: ITU; Sector efficiency and capacity: ITU and World Bank; Sector performance: Global Insight/WITSA, IMF, ITU, Netcraft, UN Comtrade, UNDESA, UNPAN, Wireless Intelligence and World Bank. Produced by the Global Information and Communication Technologies Department and the Development Economics Data Group. For complete information, see Definitions and Data Sources.

Notes: Use of italics in the column entries indicates years other than those specified. — Not available. GDP = gross domestic product; GNI = gross national income; ICT = information and communication technology; and MDG = Millennium Development Goal.

a. C = competition; M = monopoly; and P = partial competition. **b.** Outgoing and incoming. **c.** Scale of 0–1, where 1 = highest presence. **d.** Millennium Development Goal indicators 8.14, 8.15, and 8.16.

World Bank • ICT at a Glance

Brazil

	Brazil 2000	Brazil 2007	Upper-middle-income group 2007	Latin America & the Caribbean Region 2007
Economic and social context				
Population (total, million)	174	192	824	561
Urban population (% of total)	81	85	75	78
GNI per capita, World Bank Atlas method (current US$)	3,870	5,860	7,107	5,801
GDP growth, 1995–2000 and 2000–07 (avg. annual %)	1.7	3.3	4.3	3.6
Adult literacy rate (% of ages 15 and older)	86	90	94	91
Gross primary, secondary, tertiary school enrollment (%)	90	*88*	82	81
Sector structure				
Separate telecommunications regulator	Yes	Yes		
Status of main fixed-line telephone operator	Private	Private		
Level of competition[a]				
International long distance service	P	C		
Mobile telephone service	P	C		
Internet service	C	C		
Sector efficiency and capacity				
Telecommunications revenue (% of GDP)	3.4	4.7	3.3	3.8
Mobile and fixed-line subscribers per employee	516	358	*566*	*530*
Telecommunications investment (% of revenue)	39.8	12.6	—	—
Sector performance				
Access				
Telephone lines (per 100 people)	17.8	20.6	22.6	18.1
Mobile cellular subscriptions (per 100 people)	13.3	63.1	84.1	67.0
Internet subscribers (per 100 people)	1.3	4.1	9.4	4.5
Personal computers (per 100 people)	4.9	*16.1*	*12.4*	*11.3*
Households with a television set (%)	84	*91*	*92*	*84*
Usage				
International voice traffic (minutes/person/month)[b]	0.9	—	—	—
Mobile telephone usage (minutes/user/month)	249	99	137	116
Internet users (per 100 people)	2.9	35.2	26.6	26.9
Quality				
Population covered by mobile cellular network (%)	—	91	95	91
Fixed broadband subscribers (% of total Internet subscrib.)	4.4	86.0	47.8	81.7
International Internet bandwidth (bits/second/person)	5	1,041	1,185	1,126
Affordability				
Price basket for residential fixed line (US$/month)	9.4	*18.3*	*10.6*	*9.5*
Price basket for mobile service (US$/month)	—	*26.2*	*10.9*	*10.4*
Price basket for Internet service (US$/month)	—	*29.0*	*16.4*	*25.7*
Price of call to United States (US$ for 3 minutes)	1.15	*0.71*	*1.55*	*1.21*
Trade				
ICT goods exports (% of total goods exports)	4.6	*3.2*	*13.5*	*11.4*
ICT goods imports (% of total goods imports)	16.4	*14.5*	*16.2*	*15.9*
ICT service exports (% of total service exports)	0.7	1.8	4.6	4.7
Applications				
ICT expenditure (% of GDP)	—	5.8	5.2	4.9
E-government Web measure index[c]	—	0.60	0.37	0.44
Secure Internet servers (per 1 million people, Dec. 2008)	5.8	23.7	26.2	18.2

Sources: Economic and social context: UIS and World Bank; Sector structure: ITU; Sector efficiency and capacity: ITU and World Bank; Sector performance: Global Insight/WITSA, IMF, ITU, Netcraft, UN Comtrade, UNDESA, UNPAN, Wireless Intelligence and World Bank. Produced by the Global Information and Communication Technologies Department and the Development Economics Data Group. For complete information, see Definitions and Data Sources.

Notes: Use of italics in the column entries indicates years other than those specified. — Not available. GDP = gross domestic product; GNI = gross national income; ICT = information and communication technology; and MDG = Millennium Development Goal.

a. C = competition; M = monopoly; and P = partial competition. **b.** Outgoing and incoming. **c.** Scale of 0–1, where 1 = highest presence. **d.** Millennium Development Goal indicators 8.14, 8.15, and 8.16.

World Bank • ICT at a Glance

Bulgaria

	Bulgaria 2000	Bulgaria 2007	Upper-middle-income group 2007	Europe & Central Asia Region 2007
Economic and social context				
Population (total, million)	8	8	824	446
Urban population (% of total)	69	71	75	64
GNI per capita, World Bank Atlas method (current US$)	1,600	4,580	7,107	6,052
GDP growth, 1995–2000 and 2000–07 (avg. annual %)	−0.4	5.7	4.3	6.1
Adult literacy rate (% of ages 15 and older)	*98*	*98*	94	98
Gross primary, secondary, tertiary school enrollment (%)	79	*81*	82	82
Sector structure				
Separate telecommunications regulator	Yes	Yes		
Status of main fixed-line telephone operator	Public	Mixed		
Level of competition[a]				
International long distance service	M	P		
Mobile telephone service	C	P		
Internet service	C	C		
Sector efficiency and capacity				
Telecommunications revenue (% of GDP)	3.1	5.9	3.3	2.9
Mobile and fixed-line subscribers per employee	140	522	*566*	*532*
Telecommunications investment (% of revenue)	14.0	31.4	—	22.0
Sector performance				
Access				
Telephone lines (per 100 people)	35.8	30.0	22.6	25.7
Mobile cellular subscriptions (per 100 people)	9.2	129.2	84.1	95.0
Internet subscribers (per 100 people)	*0.1*	8.5	9.4	13.6
Personal computers (per 100 people)	4.5	8.9	*12.4*	*10.6*
Households with a television set (%)	90	*92*	*92*	*96*
Usage				
International voice traffic (minutes/person/month)[b]	3.3	2.6	—	—
Mobile telephone usage (minutes/user/month)	—	97	137	154
Internet users (per 100 people)	5.3	30.9	26.6	21.4
Quality				
Population covered by mobile cellular network (%)	95	100	95	*92*
Fixed broadband subscribers (% of total Internet subscrib.)	0.0	97.1	47.8	32.5
International Internet bandwidth (bits/second/person)	5	4,909	1,185	1,114
Affordability				
Price basket for residential fixed line (US$/month)	*3.8*	10.7	10.6	5.8
Price basket for mobile service (US$/month)	—	16.2	10.9	11.8
Price basket for Internet service (US$/month)	—	7.4	16.4	12.0
Price of call to United States (US$ for 3 minutes)	2.55	0.57	1.55	1.63
Trade				
ICT goods exports (% of total goods exports)	1.0	*1.8*	13.5	*1.8*
ICT goods imports (% of total goods imports)	5.2	*6.0*	16.2	7.0
ICT service exports (% of total service exports)	2.3	4.4	*4.6*	5.0
Applications				
ICT expenditure (% of GDP)	—	7.7	5.2	5.0
E-government Web measure index[c]	—	0.48	0.37	0.36
Secure Internet servers (per 1 million people, Dec. 2008)	2.3	26.2	26.2	23.9

Sources: Economic and social context: UIS and World Bank; Sector structure: ITU; Sector efficiency and capacity: ITU and World Bank; Sector performance: Global Insight/WITSA, IMF, ITU, Netcraft, UN Comtrade, UNDESA, UNPAN, Wireless Intelligence and World Bank. Produced by the Global Information and Communication Technologies Department and the Development Economics Data Group. For complete information, see Definitions and Data Sources.

Notes: Use of italics in the column entries indicates years other than those specified. — Not available. GDP = gross domestic product; GNI = gross national income; ICT = information and communication technology; and MDG = Millennium Development Goal.

a. C = competition; M = monopoly; and P = partial competition. **b.** Outgoing and incoming. **c.** Scale of 0–1, where 1 = highest presence. **d.** Millennium Development Goal indicators 8.14, 8.15, and 8.16.

World Bank • ICT at a Glance

Burkina Faso

	Burkina Faso 2000	Burkina Faso 2007	Low-income group 2007	Sub-Saharan Africa Region 2007
Economic and social context				
Population (total, million)	12	15	1,296	800
Urban population (% of total)	17	19	32	36
GNI per capita, World Bank Atlas method (current US$)	240	430	574	951
GDP growth, 1995–2000 and 2000–07 (avg. annual %)	6.8	5.8	5.6	5.1
Adult literacy rate (% of ages 15 and older)	—	29	64	62
Gross primary, secondary, tertiary school enrollment (%)	22	29	51	51
Sector structure				
Separate telecommunications regulator	Yes	Yes		
Status of main fixed-line telephone operator	Public	Public		
Level of competition[a]				
International long distance service	M	M		
Mobile telephone service	C	C		
Internet service	C	C		
Sector efficiency and capacity				
Telecommunications revenue (% of GDP)	2.0	4.0	3.3	4.7
Mobile and fixed-line subscribers per employee	62	440	301	499
Telecommunications investment (% of revenue)	38.0	88.6	—	—
Sector performance				
Access				
Telephone lines (per 100 people)	0.4	0.7	4.0	1.6
Mobile cellular subscriptions (per 100 people)	0.2	10.9	21.5	23.0
Internet subscribers (per 100 people)	0.0	0.1	0.8	1.2
Personal computers (per 100 people)	0.1	0.6	1.5	1.8
Households with a television set (%)	8	12	16	18
Usage				
International voice traffic (minutes/person/month)[b]	0.3	0.9	—	—
Mobile telephone usage (minutes/user/month)	—	—	—	—
Internet users (per 100 people)	0.1	0.6	5.2	4.4
Quality				
Population covered by mobile cellular network (%)	22	61	54	56
Fixed broadband subscribers (% of total Internet subscrib.)	0.0	18.6	3.4	3.1
International Internet bandwidth (bits/second/person)	0	15	26	36
Affordability				
Price basket for residential fixed line (US$/month)	11.6	12.0	5.7	12.6
Price basket for mobile service (US$/month)	—	15.5	11.2	11.6
Price basket for Internet service (US$/month)	—	67.8	29.2	43.1
Price of call to United States (US$ for 3 minutes)	3.16	1.14	2.00	2.43
Trade				
ICT goods exports (% of total goods exports)	0.9	0.6	1.4	1.1
ICT goods imports (% of total goods imports)	4.2	3.5	6.7	8.2
ICT service exports (% of total service exports)	7.2	—	—	4.2
Applications				
ICT expenditure (% of GDP)	—	—	—	—
E-government Web measure index[c]	—	0.19	0.11	0.16
Secure Internet servers (per 1 million people, Dec. 2008)	—	0.1	0.5	2.9

Sources: Economic and social context: UIS and World Bank; Sector structure: ITU; Sector efficiency and capacity: ITU and World Bank; Sector performance: Global Insight/WITSA, IMF, ITU, Netcraft, UN Comtrade, UNDESA, UNPAN, Wireless Intelligence and World Bank. Produced by the Global Information and Communication Technologies Department and the Development Economics Data Group. For complete information, see Definitions and Data Sources.

Notes: Use of italics in the column entries indicates years other than those specified. — Not available. GDP = gross domestic product; GNI = gross national income; ICT = information and communication technology; and MDG = Millennium Development Goal.

a. C = competition; M = monopoly; and P = partial competition. b. Outgoing and incoming. c. Scale of 0–1, where 1 = highest presence. d. Millennium Development Goal indicators 8.14, 8.15, and 8.16.

Burundi

	Burundi 2000	Burundi 2007	Low-income group 2007	Sub-Saharan Africa Region 2007
Economic and social context				
Population (total, million)	7	8	1,296	800
Urban population (% of total)	8	10	32	36
GNI per capita, World Bank Atlas method (current US$)	120	110	574	951
GDP growth, 1995–2000 and 2000–07 (avg. annual %)	–0.7	2.7	5.6	5.1
Adult literacy rate (% of ages 15 and older)	59	—	64	62
Gross primary, secondary, tertiary school enrollment (%)	30	38	51	51
Sector structure				
Separate telecommunications regulator	Yes	Yes		
Status of main fixed-line telephone operator	Public	Public		
Level of competition[a]				
International long distance service	M	C		
Mobile telephone service	C	C		
Internet service	C	C		
Sector efficiency and capacity				
Telecommunications revenue (% of GDP)	1.9	—	3.3	4.7
Mobile and fixed-line subscribers per employee	65	234	301	499
Telecommunications investment (% of revenue)	—	—	—	—
Sector performance				
Access				
Telephone lines (per 100 people)	0.3	0.4	4.0	1.6
Mobile cellular subscriptions (per 100 people)	0.2	2.9	21.5	23.0
Internet subscribers (per 100 people)	0.0	—	0.8	1.2
Personal computers (per 100 people)	0.1	0.8	1.5	1.8
Households with a television set (%)	11	15	16	18
Usage				
International voice traffic (minutes/person/month)[b]	0.1	—	—	—
Mobile telephone usage (minutes/user/month)	—	—	—	—
Internet users (per 100 people)	0.1	0.7	5.2	4.4
Quality				
Population covered by mobile cellular network (%)	—	82	54	56
Fixed broadband subscribers (% of total Internet subscrib.)	—	—	3.4	3.1
International Internet bandwidth (bits/second/person)	0	1	26	36
Affordability				
Price basket for residential fixed line (US$/month)	1.9	2.6	5.7	12.6
Price basket for mobile service (US$/month)	—	12.2	11.2	11.6
Price basket for Internet service (US$/month)	—	86.0	29.2	43.1
Price of call to United States (US$ for 3 minutes)	7.35	2.45	2.00	2.43
Trade				
ICT goods exports (% of total goods exports)	—	0.5	1.4	1.1
ICT goods imports (% of total goods imports)	—	2.5	6.7	8.2
ICT service exports (% of total service exports)	—	—	—	4.2
Applications				
ICT expenditure (% of GDP)	—	—	—	—
E-government Web measure index[c]	—	0.01	0.11	0.16
Secure Internet servers (per 1 million people, Dec.2008)	—	0.1	0.5	2.9

Sources: Economic and social context: UIS and World Bank; Sector structure: ITU; Sector efficiency and capacity: ITU and World Bank; Sector performance: Global Insight/WITSA, IMF, ITU, Netcraft, UN Comtrade, UNDESA, UNPAN, Wireless Intelligence and World Bank. Produced by the Global Information and Communication Technologies Department and the Development Economics Data Group. For complete information, see Definitions and Data Sources.

Notes: Use of italics in the column entries indicates years other than those specified. — Not available. GDP = gross domestic product; GNI = gross national income; ICT = information and communication technology; and MDG = Millennium Development Goal.

a. C = competition; M = monopoly; and P = partial competition. b. Outgoing and incoming. c. Scale of 0–1, where 1 = highest presence. d. Millennium Development Goal indicators 8.14, 8.15, and 8.16.

World Bank • ICT at a Glance

Cambodia

	Cambodia 2000	Cambodia 2007	Low-income group 2007	East Asia & Pacific Region 2007
Economic and social context				
Population (total, million)	13	14	1,296	1,912
Urban population (% of total)	17	21	32	43
GNI per capita, World Bank Atlas method (current US$)	300	550	574	2,182
GDP growth, 1995–2000 and 2000–07 (avg. annual %)	7.3	9.8	5.6	9.0
Adult literacy rate (% of ages 15 and older)	*67*	*76*	*64*	*93*
Gross primary, secondary, tertiary school enrollment (%)	51	*60*	*51*	*69*
Sector structure				
Separate telecommunications regulator	—	No		
Status of main fixed-line telephone operator	Public	Public		
Level of competition[a]				
International long distance service	P	P		
Mobile telephone service	P	P		
Internet service	P	P		
Sector efficiency and capacity				
Telecommunications revenue (% of GDP)	0.6	*0.4*	*3.3*	3.0
Mobile and fixed-line subscribers per employee	241	*534*	*301*	546
Telecommunications investment (% of revenue)	—	—	—	—
Sector performance				
Access				
Telephone lines (per 100 people)	0.2	0.3	4.0	23.1
Mobile cellular subscriptions (per 100 people)	1.0	17.9	21.5	43.7
Internet subscribers (per 100 people)	0.0	0.1	*0.8*	9.3
Personal computers (per 100 people)	0.1	0.4	*1.5*	5.6
Households with a television set (%)	*33*	*43*	*16*	*53*
Usage				
International voice traffic (minutes/person/month)[b]	0.3	*0.8*	—	0.8
Mobile telephone usage (minutes/user/month)	—	—	—	333
Internet users (per 100 people)	0.0	0.5	5.2	14.6
Quality				
Population covered by mobile cellular network (%)	80	87	*54*	93
Fixed broadband subscribers (% of total Internet subscrib.)	0.0	60.4	*3.4*	41.8
International Internet bandwidth (bits/second/person)	0	17	26	247
Affordability				
Price basket for residential fixed line (US$/month)	16.7	*9.0*	*5.7*	5.8
Price basket for mobile service (US$/month)	—	5.1	11.2	*5.0*
Price basket for Internet service (US$/month)	—	33.0	29.2	14.4
Price of call to United States (US$ for 3 minutes)	6.00	2.94	2.00	1.16
Trade				
ICT goods exports (% of total goods exports)	0.0	*0.1*	*1.4*	30.9
ICT goods imports (% of total goods imports)	2.4	*2.1*	*6.7*	28.1
ICT service exports (% of total service exports)	—	3.1	—	5.2
Applications				
ICT expenditure (% of GDP)	—	—	—	7.3
E-government Web measure index[c]	—	0.20	0.11	0.18
Secure Internet servers (per 1 million people, Dec. 2008)	*0.2*	0.8	0.5	1.9

Sources: Economic and social context: UIS and World Bank; Sector structure: ITU; Sector efficiency and capacity: ITU and World Bank; Sector performance: Global Insight/WITSA, IMF, ITU, Netcraft, UN Comtrade, UNDESA, UNPAN, Wireless Intelligence and World Bank. Produced by the Global Information and Communication Technologies Department and the Development Economics Data Group. For complete information, see Definitions and Data Sources.

Notes: Use of italics in the column entries indicates years other than those specified. — Not available. GDP = gross domestic product; GNI = gross national income; ICT = information and communication technology; and MDG = Millennium Development Goal.

a. C = competition; M = monopoly; and P = partial competition. **b.** Outgoing and incoming. **c.** Scale of 0–1, where 1 = highest presence. **d.** Millennium Development Goal indicators 8.14, 8.15, and 8.16.

Information and Communications for Development 2009

World Bank • ICT at a Glance

Cameroon

	Cameroon 2000	Cameroon 2007	Lower-middle-income group 2007	Sub-Saharan Africa Region 2007
Economic and social context				
Population (total, million)	16	19	3,435	800
Urban population (% of total)	50	56	42	36
GNI per capita, World Bank Atlas method (current US$)	620	1,050	1,905	951
GDP growth, 1995–2000 and 2000–07 (avg. annual %)	4.8	3.5	8.0	5.1
Adult literacy rate (% of ages 15 and older)	68	—	83	62
Gross primary, secondary, tertiary school enrollment (%)	48	62	68	51
Sector structure				
Separate telecommunications regulator	Yes	Yes		
Status of main fixed-line telephone operator	Public	Public		
Level of competition[a]				
International long distance service	M	M		
Mobile telephone service	P	C		
Internet service	C	C		
Sector efficiency and capacity				
Telecommunications revenue (% of GDP)	1.4	3.1	3.1	4.7
Mobile and fixed-line subscribers per employee	90	1,050	624	499
Telecommunications investment (% of revenue)	62.8	37.4	25.3	—
Sector performance				
Access				
Telephone lines (per 100 people)	0.6	1.0	15.3	1.6
Mobile cellular subscriptions (per 100 people)	0.7	24.5	38.9	23.0
Internet subscribers (per 100 people)	0.0	0.1	6.0	1.2
Personal computers (per 100 people)	0.3	1.1	4.6	1.8
Households with a television set (%)	17	25	79	18
Usage				
International voice traffic (minutes/person/month)[b]	0.5	0.3	—	—
Mobile telephone usage (minutes/user/month)	—	—	322	—
Internet users (per 100 people)	0.3	2.0	12.4	4.4
Quality				
Population covered by mobile cellular network (%)	37	58	80	56
Fixed broadband subscribers (% of total Internet subscrib.)	0.0	1.7	40.4	3.1
International Internet bandwidth (bits/second/person)	0	11	199	36
Affordability				
Price basket for residential fixed line (US$/month)	6.5	14.3	7.2	12.6
Price basket for mobile service (US$/month)	—	14.4	9.8	11.6
Price basket for Internet service (US$/month)	—	48.3	16.7	43.1
Price of call to United States (US$ for 3 minutes)	3.25	—	2.08	2.43
Trade				
ICT goods exports (% of total goods exports)	0.0	0.0	20.6	1.1
ICT goods imports (% of total goods imports)	3.0	3.2	20.2	8.2
ICT service exports (% of total service exports)	1.4	13.0	15.6	4.2
Applications				
ICT expenditure (% of GDP)	—	5.0	6.5	—
E-government Web measure index[c]	—	0.14	0.33	0.16
Secure Internet servers (per 1 million people, Dec. 2008)	—	0.4	1.8	2.9

Sources: Economic and social context: UIS and World Bank; Sector structure: ITU; Sector efficiency and capacity: ITU and World Bank; Sector performance: Global Insight/WITSA, IMF, ITU, Netcraft, UN Comtrade, UNDESA, UNPAN, Wireless Intelligence and World Bank. Produced by the Global Information and Communication Technologies Department and the Development Economics Data Group. For complete information, see Definitions and Data Sources.

Notes: Use of italics in the column entries indicates years other than those specified. — Not available. GDP = gross domestic product; GNI = gross national income; ICT = information and communication technology; and MDG = Millennium Development Goal.

a. C = competition; M = monopoly; and P = partial competition. b. Outgoing and incoming. c. Scale of 0–1, where 1 = highest presence. d. Millennium Development Goal indicators 8.14, 8.15, and 8.16.

World Bank • ICT at a Glance

Canada

	Canada 2000	Canada 2007	High-income group 2007
Economic and social context			
Population (total, million)	31	33	1,056
Urban population (% of total)	80	80	78
GNI per capita, World Bank Atlas method (current US$)	22,130	39,650	37,572
GDP growth, 1995–2000 and 2000–07 (avg. annual %)	4.3	2.7	2.4
Adult literacy rate (% of ages 15 and older)	—	—	99
Gross primary, secondary, tertiary school enrollment (%)	95	*99*	92
Sector structure			
Separate telecommunications regulator	Yes	Yes	
Status of main fixed-line telephone operator	*Private*	*Private*	
Level of competition[a]			
International long distance service	C	C	
Mobile telephone service	P	C	
Internet service	C	C	
Sector efficiency and capacity			
Telecommunications revenue (% of GDP)	2.8	2.7	3.1
Mobile and fixed-line subscribers per employee	331	424	747
Telecommunications investment (% of revenue)	24.0	21.6	14.3
Sector performance			
Access			
Telephone lines (per 100 people)	67.7	55.3	50.0
Mobile cellular subscriptions (per 100 people)	28.4	61.5	100.4
Internet subscribers (per 100 people)	14.1	30.8	25.8
Personal computers (per 100 people)	41.9	*94.3*	67.4
Households with a television set (%)	99	99	98
Usage			
International voice traffic (minutes/person/month)[b]	36.6	—	14.0
Mobile telephone usage (minutes/user/month)	215	351	353
Internet users (per 100 people)	42.2	72.8	65.7
Quality			
Population covered by mobile cellular network (%)	90	98	99
Fixed broadband subscribers (% of total Internet subscrib.)	32.6	89.3	82.6
International Internet bandwidth (bits/second/person)	1,133	16,193	18,242
Affordability			
Price basket for residential fixed line (US$/month)	—	25.0	26.1
Price basket for mobile service (US$/month)	—	9.5	13.0
Price basket for Internet service (US$/month)	—	17.6	22.8
Price of call to United States (US$ for 3 minutes)	—	—	0.81
Trade			
ICT goods exports (% of total goods exports)	8.2	*4.7*	15.2
ICT goods imports (% of total goods imports)	15.0	*10.1*	14.6
ICT service exports (% of total service exports)	9.5	11.1	7.0
Applications			
ICT expenditure (% of GDP)	—	6.4	6.7
E-government Web measure index[c]	—	0.77	0.60
Secure Internet servers (per 1 million people, Dec. 2008)	*162.6*	906.6	662.6

Sources: Economic and social context: UIS and World Bank; Sector structure: ITU; Sector efficiency and capacity: ITU and World Bank; Sector performance: Global Insight/WITSA, IMF, ITU, Netcraft, UN Comtrade, UNDESA, UNPAN, Wireless Intelligence and World Bank. Produced by the Global Information and Communication Technologies Department and the Development Economics Data Group. For complete information, see Definitions and Data Sources.

Notes: Use of italics in the column entries indicates years other than those specified. — Not available. GDP = gross domestic product; GNI = gross national income; ICT = information and communication technology; and MDG = Millennium Development Goal.

a. C = competition; M = monopoly; and P = partial competition. **b.** Outgoing and incoming. **c.** Scale of 0–1, where 1 = highest presence. **d.** Millennium Development Goal indicators 8.14, 8.15, and 8.16.

Information and Communications for Development 2009

World Bank • ICT at a Glance

Central African Republic

	Central African Republic		Low-income group	Sub-Saharan Africa Region
	2000	**2007**	**2007**	**2007**
Economic and social context				
Population (total, million)	4	4	1,296	800
Urban population (% of total)	38	38	32	36
GNI per capita, World Bank Atlas method (current US$)	270	370	574	951
GDP growth, 1995–2000 and 2000–07 (avg. annual %)	3.0	0.0	5.6	5.1
Adult literacy rate (% of ages 15 and older)	49	—	64	62
Gross primary, secondary, tertiary school enrollment (%)	*34*	*30*	*51*	*51*
Sector structure				
Separate telecommunications regulator	Yes	Yes		
Status of main fixed-line telephone operator	Mixed	Mixed		
Level of competition[a]				
International long distance service	M	M		
Mobile telephone service	C	C		
Internet service	—	—		
Sector efficiency and capacity				
Telecommunications revenue (% of GDP)	1.1	*1.1*	3.3	4.7
Mobile and fixed-line subscribers per employee	35	293	301	499
Telecommunications investment (% of revenue)	1.0	—	—	—
Sector performance				
Access				
Telephone lines (per 100 people)	0.2	0.3	4.0	1.6
Mobile cellular subscriptions (per 100 people)	0.1	3.0	21.5	23.0
Internet subscribers (per 100 people)	0.0	0.1	0.8	1.2
Personal computers (per 100 people)	0.2	0.3	1.5	1.8
Households with a television set (%)	3	5	16	18
Usage				
International voice traffic (minutes/person/month)[b]	0.2	—	—	—
Mobile telephone usage (minutes/user/month)	—	—	—	—
Internet users (per 100 people)	0.1	0.3	5.2	4.4
Quality				
Population covered by mobile cellular network (%)	*18*	*19*	54	56
Fixed broadband subscribers (% of total Internet subscrib.)	0.0	0.0	3.4	3.1
International Internet bandwidth (bits/second/person)	*0*	*0*	26	36
Affordability				
Price basket for residential fixed line (US$/month)	—	12.9	5.7	12.6
Price basket for mobile service (US$/month)	—	12.4	11.2	11.6
Price basket for Internet service (US$/month)	—	130.4	29.2	43.1
Price of call to United States (US$ for 3 minutes)	13.31	*1.99*	*2.00*	*2.43*
Trade				
ICT goods exports (% of total goods exports)	0.1	0.1	1.4	1.1
ICT goods imports (% of total goods imports)	2.1	2.7	6.7	8.2
ICT service exports (% of total service exports)	—	—	—	4.2
Applications				
ICT expenditure (% of GDP)	—	—	—	—
E-government Web measure index[c]	—	0.00	0.11	0.16
Secure Internet servers (per 1 million people, Dec. 2008)	—	0.2	0.5	2.9

GNI per Capita, Atlas Method, 2000–07
Current US$

ICT MDG[d] Indicators, 2000–07
Number per 100 people
— Telephone lines
— Mobile cellular subscriptions
— Internet users

Price Basket for Internet Service, 2003–07
US$/month
— Central African Republic
— Sub-Saharan Africa Region

ICT Service Exports, 2000–07
Percentage of total service exports
— Central African Republic (—)
— Sub-Saharan Africa Region

Sources: Economic and social context: UIS and World Bank; Sector structure: ITU; Sector efficiency and capacity: ITU and World Bank; Sector performance: Global Insight/WITSA, IMF, ITU, Netcraft, UN Comtrade, UNDESA, UNPAN, Wireless Intelligence and World Bank. Produced by the Global Information and Communication Technologies Department and the Development Economics Data Group. For complete information, see Definitions and Data Sources.

Notes: Use of italics in the column entries indicates years other than those specified. — Not available. GDP = gross domestic product; GNI = gross national income; ICT = information and communication technology; and MDG = Millennium Development Goal.

a. C = competition; M = monopoly; and P = partial competition. b. Outgoing and incoming. c. Scale of 0–1, where 1 = highest presence. d. Millennium Development Goal indicators 8.14, 8.15, and 8.16.

World Bank • ICT at a Glance

Chad

	Chad 2000	Chad 2007	Low-income group 2007	Sub-Saharan Africa Region 2007
Economic and social context				
Population (total, million)	8	11	1,296	800
Urban population (% of total)	23	26	32	36
GNI per capita, World Bank Atlas method (current US$)	180	540	574	951
GDP growth, 1995–2000 and 2000–07 (avg. annual %)	3.1	12.2	5.6	5.1
Adult literacy rate (% of ages 15 and older)	26	32	64	62
Gross primary, secondary, tertiary school enrollment (%)	32	37	51	51
Sector structure				
Separate telecommunications regulator	Yes	Yes		
Status of main fixed-line telephone operator	Public	Public		
Level of competition[a]				
International long distance service	M	M		
Mobile telephone service	P	—		
Internet service	M	C		
Sector efficiency and capacity				
Telecommunications revenue (% of GDP)	1.4	—	3.3	4.7
Mobile and fixed-line subscribers per employee	31	127	301	499
Telecommunications investment (% of revenue)	—	—	—	—
Sector performance				
Access				
Telephone lines (per 100 people)	0.1	*0.1*	4.0	1.6
Mobile cellular subscriptions (per 100 people)	0.1	8.5	21.5	23.0
Internet subscribers (per 100 people)	0.0	*0.0*	*0.8*	*1.2*
Personal computers (per 100 people)	0.1	*0.2*	*1.5*	*1.8*
Households with a television set (%)	2	4	16	18
Usage				
International voice traffic (minutes/person/month)[b]	0.1	*0.2*	—	—
Mobile telephone usage (minutes/user/month)	—	—	—	—
Internet users (per 100 people)	0.0	*0.6*	5.2	4.4
Quality				
Population covered by mobile cellular network (%)	8	24	54	56
Fixed broadband subscribers (% of total Internet subscrib.)	*0.0*	*0.0*	3.4	3.1
International Internet bandwidth (bits/second/person)	0	1	26	36
Affordability				
Price basket for residential fixed line (US$/month)	16.3	*8.3*	5.7	12.6
Price basket for mobile service (US$/month)	—	16.0	11.2	11.6
Price basket for Internet service (US$/month)	—	105.0	29.2	43.1
Price of call to United States (US$ for 3 minutes)	12.50	—	2.00	2.43
Trade				
ICT goods exports (% of total goods exports)	—	—	*1.4*	*1.1*
ICT goods imports (% of total goods imports)	—	—	*6.7*	*8.2*
ICT service exports (% of total service exports)	—	—	—	4.2
Applications				
ICT expenditure (% of GDP)	—	—	—	—
E-government Web measure index[c]	—	0.01	0.11	0.16
Secure Internet servers (per 1 million people, Dec. 2008)	—	—	0.5	2.9

Sources: Economic and social context: UIS and World Bank; Sector structure: ITU; Sector efficiency and capacity: ITU and World Bank; Sector performance: Global Insight/WITSA, IMF, ITU, Netcraft, UN Comtrade, UNDESA, UNPAN, Wireless Intelligence and World Bank. Produced by the Global Information and Communication Technologies Department and the Development Economics Data Group. For complete information, see Definitions and Data Sources.

Notes: Use of italics in the column entries indicates years other than those specified. — Not available. GDP = gross domestic product; GNI = gross national income; ICT = information and communication technology; and MDG = Millennium Development Goal.

a. C = competition; M = monopoly; and P = partial competition. b. Outgoing and incoming. c. Scale of 0–1, where 1 = highest presence. d. Millennium Development Goal indicators 8.14, 8.15, and 8.16.

Chile

	Chile 2000	Chile 2007	Upper-middle-income group 2007	Latin America & the Caribbean Region 2007
Economic and social context				
Population (total, million)	15	17	824	561
Urban population (% of total)	86	88	75	78
GNI per capita, World Bank Atlas method (current US$)	4,840	8,190	7,107	5,801
GDP growth, 1995–2000 and 2000–07 (avg. annual %)	3.8	4.5	4.3	3.6
Adult literacy rate (% of ages 15 and older)	96	97	94	91
Gross primary, secondary, tertiary school enrollment (%)	78	83	82	81
Sector structure				
Separate telecommunications regulator	—	Yes		
Status of main fixed-line telephone operator	Private	Private		
Level of competition[a]				
International long distance service	C	C		
Mobile telephone service	C	C		
Internet service	—	C		
Sector efficiency and capacity				
Telecommunications revenue (% of GDP)	3.4	—	3.3	3.8
Mobile and fixed-line subscribers per employee	315	1,311	566	530
Telecommunications investment (% of revenue)	44.0	—	—	—
Sector performance				
Access				
Telephone lines (per 100 people)	21.4	20.8	22.6	18.1
Mobile cellular subscriptions (per 100 people)	22.1	84.1	84.1	67.0
Internet subscribers (per 100 people)	3.8	8.2	9.4	4.5
Personal computers (per 100 people)	9.2	14.1	12.4	11.3
Households with a television set (%)	94	97	92	84
Usage				
International voice traffic (minutes/person/month)[b]	2.7	3.4	—	—
Mobile telephone usage (minutes/user/month)	220	147	137	116
Internet users (per 100 people)	16.5	31.1	26.6	26.9
Quality				
Population covered by mobile cellular network (%)	100	100	95	91
Fixed broadband subscribers (% of total Internet subscrib.)	1.3	96.7	47.8	81.7
International Internet bandwidth (bits/second/person)	12	4,086	1,185	1,126
Affordability				
Price basket for residential fixed line (US$/month)	15.7	9.7	10.6	9.5
Price basket for mobile service (US$/month)	—	11.8	10.9	10.4
Price basket for Internet service (US$/month)	—	26.7	16.4	25.7
Price of call to United States (US$ for 3 minutes)	2.45	—	1.55	1.21
Trade				
ICT goods exports (% of total goods exports)	0.2	0.1	13.5	11.4
ICT goods imports (% of total goods imports)	11.0	9.0	16.2	15.9
ICT service exports (% of total service exports)	5.9	2.7	4.6	4.7
Applications				
ICT expenditure (% of GDP)	—	4.2	5.2	4.9
E-government Web measure index[c]	—	0.56	0.37	0.44
Secure Internet servers (per 1 million people, Dec. 2008)	9.0	35.1	26.2	18.2

Sources: Economic and social context: UIS and World Bank; Sector structure: ITU; Sector efficiency and capacity: ITU and World Bank; Sector performance: Global Insight/WITSA, IMF, ITU, Netcraft, UN Comtrade, UNDESA, UNPAN, Wireless Intelligence and World Bank. Produced by the Global Information and Communication Technologies Department and the Development Economics Data Group. For complete information, see Definitions and Data Sources.

Notes: Use of italics in the column entries indicates years other than those specified. — Not available. GDP = gross domestic product; GNI = gross national income; ICT = information and communication technology; and MDG = Millennium Development Goal.

a. C = competition; M = monopoly; and P = partial competition. **b.** Outgoing and incoming. **c.** Scale of 0–1, where 1 = highest presence. **d.** Millennium Development Goal indicators 8.14, 8.15, and 8.16.

China

	China 2000	China 2007	Lower-middle-income group 2007	East Asia & Pacific Region 2007
Economic and social context				
Population (total, million)	1,263	1,318	3,435	1,912
Urban population (% of total)	36	42	42	43
GNI per capita, World Bank Atlas method (current US$)	930	2,370	1,905	2,182
GDP growth, 1995–2000 and 2000–07 (avg. annual %)	8.5	10.3	8.0	9.0
Adult literacy rate (% of ages 15 and older)	91	93	83	93
Gross primary, secondary, tertiary school enrollment (%)	67	69	68	69
Sector structure				
Separate telecommunications regulator	—	No		
Status of main fixed-line telephone operator	Public	Mixed		
Level of competition[a]				
International long distance service	P	P		
Mobile telephone service	P	P		
Internet service	C	C		
Sector efficiency and capacity				
Telecommunications revenue (% of GDP)	3.2	2.9	3.1	3.0
Mobile and fixed-line subscribers per employee	222	1,310	624	546
Telecommunications investment (% of revenue)	69.8	32.0	25.3	—
Sector performance				
Access				
Telephone lines (per 100 people)	11.5	27.7	15.3	23.1
Mobile cellular subscriptions (per 100 people)	6.8	41.5	38.9	43.7
Internet subscribers (per 100 people)	0.7	11.4	6.0	9.3
Personal computers (per 100 people)	1.6	5.7	4.6	5.6
Households with a television set (%)	86	89	79	53
Usage				
International voice traffic (minutes/person/month)[b]	0.3	0.8	—	0.8
Mobile telephone usage (minutes/user/month)	247	393	322	333
Internet users (per 100 people)	1.8	16.1	12.4	14.6
Quality				
Population covered by mobile cellular network (%)	—	97	80	93
Fixed broadband subscribers (% of total Internet subscrib.)	0.3	44.2	40.4	41.8
International Internet bandwidth (bits/second/person)	2	280	199	247
Affordability				
Price basket for residential fixed line (US$/month)	—	2.9	7.2	5.8
Price basket for mobile service (US$/month)	—	3.3	9.8	5.0
Price basket for Internet service (US$/month)	—	5.8	16.7	14.4
Price of call to United States (US$ for 3 minutes)	6.67	2.90	2.08	1.16
Trade				
ICT goods exports (% of total goods exports)	18.9	30.9	20.6	30.9
ICT goods imports (% of total goods imports)	22.5	28.6	20.2	28.1
ICT service exports (% of total service exports)	5.6	4.5	15.6	5.2
Applications				
ICT expenditure (% of GDP)	—	7.9	6.5	7.3
E-government Web measure index[c]	—	0.51	0.33	0.18
Secure Internet servers (per 1 million people, Dec. 2008)	0.1	0.9	1.8	1.9

Sources: Economic and social context: UIS and World Bank; Sector structure: ITU; Sector efficiency and capacity: ITU and World Bank; Sector performance: Global Insight/WITSA, IMF, ITU, Netcraft, UN Comtrade, UNDESA, UNPAN, Wireless Intelligence and World Bank. Produced by the Global Information and Communication Technologies Department and the Development Economics Data Group. For complete information, see Definitions and Data Sources.

Notes: Use of italics in the column entries indicates years other than those specified. — Not available. GDP = gross domestic product; GNI = gross national income; ICT = information and communication technology; and MDG = Millennium Development Goal.

a. C = competition; M = monopoly; and P = partial competition. b. Outgoing and incoming. c. Scale of 0–1, where 1 = highest presence. d. Millennium Development Goal indicators 8.14, 8.15, and 8.16.

World Bank • ICT at a Glance

Colombia

	Colombia 2000	Colombia 2007	Lower-middle-income group 2007	Latin America & the Caribbean Region 2007
Economic and social context				
Population (total, million)	40	44	3,435	561
Urban population (% of total)	72	74	42	78
GNI per capita, World Bank Atlas method (current US$)	2,280	4,100	1,905	5,801
GDP growth, 1995–2000 and 2000–07 (avg. annual %)	0.6	4.9	8.0	3.6
Adult literacy rate (% of ages 15 and older)	—	93	83	91
Gross primary, secondary, tertiary school enrollment (%)	71	75	68	81
Sector structure				
Separate telecommunications regulator	Yes	Yes		
Status of main fixed-line telephone operator	Public	Mixed		
Level of competition[a]				
International long distance service	C	C		
Mobile telephone service	P	C		
Internet service	C	C		
Sector efficiency and capacity				
Telecommunications revenue (% of GDP)	2.8	3.9	*3.1*	3.8
Mobile and fixed-line subscribers per employee	242	—	*624*	*530*
Telecommunications investment (% of revenue)	54.6	—	*25.3*	—
Sector performance				
Access				
Telephone lines (per 100 people)	18.1	18.0	15.3	18.1
Mobile cellular subscriptions (per 100 people)	5.7	77.2	38.9	67.0
Internet subscribers (per 100 people)	0.6	3.1	6.0	4.5
Personal computers (per 100 people)	3.8	8.0	*4.6*	*11.3*
Households with a television set (%)	80	84	79	84
Usage				
International voice traffic (minutes/person/month)[b]	1.9	8.9	—	—
Mobile telephone usage (minutes/user/month)	115	131	322	116
Internet users (per 100 people)	2.2	27.5	12.4	26.9
Quality				
Population covered by mobile cellular network (%)	—	83	80	91
Fixed broadband subscribers (% of total Internet subscrib.)	3.7	87.4	40.4	81.7
International Internet bandwidth (bits/second/person)	16	971	199	1,126
Affordability				
Price basket for residential fixed line (US$/month)	6.6	5.8	7.2	9.5
Price basket for mobile service (US$/month)	—	10.4	9.8	10.4
Price basket for Internet service (US$/month)	—	7.5	16.7	25.7
Price of call to United States (US$ for 3 minutes)	2.00	—	2.08	1.21
Trade				
ICT goods exports (% of total goods exports)	0.2	*0.3*	20.6	11.4
ICT goods imports (% of total goods imports)	10.3	*13.3*	20.2	15.9
ICT service exports (% of total service exports)	9.1	7.9	15.6	4.7
Applications				
ICT expenditure (% of GDP)	—	4.4	6.5	4.9
E-government Web measure index[c]	—	0.56	0.33	0.44
Secure Internet servers (per 1 million people, Dec. 2008)	*1.8*	10.6	1.8	18.2

Sources: Economic and social context: UIS and World Bank; Sector structure: ITU; Sector efficiency and capacity: ITU and World Bank; Sector performance: Global Insight/WITSA, IMF, ITU, Netcraft, UN Comtrade, UNDESA, UNPAN, Wireless Intelligence and World Bank. Produced by the Global Information and Communication Technologies Department and the Development Economics Data Group. For complete information, see Definitions and Data Sources.

Notes: Use of italics in the column entries indicates years other than those specified. — Not available. GDP = gross domestic product; GNI = gross national income; ICT = information and communication technology; and MDG = Millennium Development Goal.

a. C = competition; M = monopoly; and P = partial competition. **b.** Outgoing and incoming. **c.** Scale of 0–1, where 1 = highest presence. **d.** Millennium Development Goal indicators 8.14, 8.15, and 8.16.

Congo, Democratic Republic of

	Congo, Dem. Rep. of 2000	Congo, Dem. Rep. of 2007	Low-income group 2007	Sub-Saharan Africa Region 2007
Economic and social context				
Population (total, million)	51	62	1,296	800
Urban population (% of total)	30	33	32	36
GNI per capita, World Bank Atlas method (current US$)	80	140	574	951
GDP growth, 1995–2000 and 2000–07 (avg. annual %)	−3.8	5.0	5.6	5.1
Adult literacy rate (% of ages 15 and older)	67	—	64	62
Gross primary, secondary, tertiary school enrollment (%)	27	34	51	51
Sector structure				
Separate telecommunications regulator	No	Yes		
Status of main fixed-line telephone operator	Public	Public		
Level of competition[a]				
International long distance service	C	C		
Mobile telephone service	C	C		
Internet service	C	C		
Sector efficiency and capacity				
Telecommunications revenue (% of GDP)	—	7.6	3.3	4.7
Mobile and fixed-line subscribers per employee	—	3,628	301	499
Telecommunications investment (% of revenue)	—	82.5	—	—
Sector performance				
Access				
Telephone lines (per 100 people)	0.0	0.0	4.0	1.6
Mobile cellular subscriptions (per 100 people)	0.0	10.6	21.5	23.0
Internet subscribers (per 100 people)	0.0	0.1	0.8	1.2
Personal computers (per 100 people)	0.0	0.0	1.5	1.8
Households with a television set (%)	1	4	16	18
Usage				
International voice traffic (minutes/person/month)[b]	—	0.3	—	—
Mobile telephone usage (minutes/user/month)	—	—	—	—
Internet users (per 100 people)	0.0	0.4	5.2	4.4
Quality				
Population covered by mobile cellular network (%)	—	50	54	56
Fixed broadband subscribers (% of total Internet subscrib.)	0.0	3.2	3.4	3.1
International Internet bandwidth (bits/second/person)	0	0	26	36
Affordability				
Price basket for residential fixed line (US$/month)	—	—	5.7	12.6
Price basket for mobile service (US$/month)	—	11.2	11.2	11.6
Price basket for Internet service (US$/month)	—	109.5	29.2	43.1
Price of call to United States (US$ for 3 minutes)	—	—	2.00	2.43
Trade				
ICT goods exports (% of total goods exports)	—	—	1.4	1.1
ICT goods imports (% of total goods imports)	—	—	6.7	8.2
ICT service exports (% of total service exports)	—	—	—	4.2
Applications				
ICT expenditure (% of GDP)	—	—	—	—
E-government Web measure index[c]	—	0.09	0.11	0.16
Secure Internet servers (per 1 million people, Dec. 2008)	—	0.1	0.5	2.9

Sources: Economic and social context: UIS and World Bank; Sector structure: ITU; Sector efficiency and capacity: ITU and World Bank; Sector performance: Global Insight/WITSA, IMF, ITU, Netcraft, UN Comtrade, UNDESA, UNPAN, Wireless Intelligence and World Bank. Produced by the Global Information and Communication Technologies Department and the Development Economics Data Group. For complete information, see Definitions and Data Sources.

Notes: Use of italics in the column entries indicates years other than those specified. — Not available. GDP = gross domestic product; GNI = gross national income; ICT = information and communication technology; and MDG = Millennium Development Goal.

a. C = competition; M = monopoly; and P = partial competition. b. Outgoing and incoming. c. Scale of 0–1, where 1 = highest presence. d. Millennium Development Goal indicators 8.14, 8.15, and 8.16.

Congo, Republic of

	Congo, Rep. of 2000	Congo, Rep. of 2007	Lower-middle-income group 2007	Sub-Saharan Africa Region 2007
Economic and social context				
Population (total, million)	3	4	3,435	800
Urban population (% of total)	58	61	42	36
GNI per capita, World Bank Atlas method (current US$)	550	1,540	1,905	951
GDP growth, 1995–2000 and 2000–07 (avg. annual %)	1.9	4.1	8.0	5.1
Adult literacy rate (% of ages 15 and older)	—	—	83	62
Gross primary, secondary, tertiary school enrollment (%)	43	51	68	51
Sector structure				
Separate telecommunications regulator	—	No		
Status of main fixed-line telephone operator	Public	Public		
Level of competition[a]				
International long distance service	C	C		
Mobile telephone service	C	C		
Internet service	—	—		
Sector efficiency and capacity				
Telecommunications revenue (% of GDP)	—	2.9	3.1	4.7
Mobile and fixed-line subscribers per employee	—	—	624	499
Telecommunications investment (% of revenue)	—	—	25.3	—
Sector performance				
Access				
Telephone lines (per 100 people)	0.7	0.4	15.3	1.6
Mobile cellular subscriptions (per 100 people)	2.2	34.2	38.9	23.0
Internet subscribers (per 100 people)	0.0	0.0	6.0	1.2
Personal computers (per 100 people)	0.3	0.5	4.6	1.8
Households with a television set (%)	6	27	79	18
Usage				
International voice traffic (minutes/person/month)[b]	—	—	—	—
Mobile telephone usage (minutes/user/month)	—	—	322	—
Internet users (per 100 people)	0.0	1.9	12.4	4.4
Quality				
Population covered by mobile cellular network (%)	17	53	80	56
Fixed broadband subscribers (% of total Internet subscrib.)	0.0	0.0	40.4	3.1
International Internet bandwidth (bits/second/person)	0	0	199	36
Affordability				
Price basket for residential fixed line (US$/month)	—	—	7.2	12.6
Price basket for mobile service (US$/month)	—	18.8	9.8	11.6
Price basket for Internet service (US$/month)	—	82.7	16.7	43.1
Price of call to United States (US$ for 3 minutes)	—	5.39	2.08	2.43
Trade				
ICT goods exports (% of total goods exports)	—	—	20.6	1.1
ICT goods imports (% of total goods imports)	—	—	20.2	8.2
ICT service exports (% of total service exports)	—	—	15.6	4.2
Applications				
ICT expenditure (% of GDP)	—	—	6.5	—
E-government Web measure index[c]	—	0.07	0.33	0.16
Secure Internet servers (per 1 million people, Dec. 2008)	—	0.3	1.8	2.9

Sources: Economic and social context: UIS and World Bank; Sector structure: ITU; Sector efficiency and capacity: ITU and World Bank; Sector performance: Global Insight/WITSA, IMF, ITU, Netcraft, UN Comtrade, UNDESA, UNPAN, Wireless Intelligence and World Bank. Produced by the Global Information and Communication Technologies Department and the Development Economics Data Group. For complete information, see Definitions and Data Sources.

Notes: Use of italics in the column entries indicates years other than those specified. — Not available. GDP = gross domestic product; GNI = gross national income; ICT = information and communication technology; and MDG = Millennium Development Goal.

a. C = competition; M = monopoly; and P = partial competition. **b.** Outgoing and incoming. **c.** Scale of 0–1, where 1 = highest presence. **d.** Millennium Development Goal indicators 8.14, 8.15, and 8.16.

World Bank • ICT at a Glance

Costa Rica

	Costa Rica 2000	Costa Rica 2007	Upper-middle-income group 2007	Latin America & the Caribbean Region 2007
Economic and social context				
Population (total, million)	4	4	824	561
Urban population (% of total)	59	63	75	78
GNI per capita, World Bank Atlas method (current US$)	3,710	5,520	7,107	5,801
GDP growth, 1995–2000 and 2000–07 (avg. annual %)	5.7	5.4	4.3	3.6
Adult literacy rate (% of ages 15 and older)	95	96	94	91
Gross primary, secondary, tertiary school enrollment (%)	66	*73*	82	81
Sector structure				
Separate telecommunications regulator	Yes	Yes		
Status of main fixed-line telephone operator	*Public*	Public		
Level of competition[a]				
International long distance service	M	M		
Mobile telephone service	M	M		
Internet service	M	P		
Sector efficiency and capacity				
Telecommunications revenue (% of GDP)	1.7	2.2	3.3	3.8
Mobile and fixed-line subscribers per employee	231	470	*566*	*530*
Telecommunications investment (% of revenue)	44.4	53.6	—	—
Sector performance				
Access				
Telephone lines (per 100 people)	22.9	32.2	22.6	18.1
Mobile cellular subscriptions (per 100 people)	5.4	33.8	84.1	67.0
Internet subscribers (per 100 people)	0.9	3.8	9.4	4.5
Personal computers (per 100 people)	15.3	*23.1*	*12.4*	*11.3*
Households with a television set (%)	84	*94*	*92*	*84*
Usage				
International voice traffic (minutes/person/month)[b]	5.2	10.0	—	—
Mobile telephone usage (minutes/user/month)	—	—	137	116
Internet users (per 100 people)	5.8	33.6	26.6	26.9
Quality				
Population covered by mobile cellular network (%)	—	87	95	91
Fixed broadband subscribers (% of total Internet subscrib.)	0.0	74.5	47.8	81.7
International Internet bandwidth (bits/second/person)	91	820	1,185	1,126
Affordability				
Price basket for residential fixed line (US$/month)	7.1	*5.1*	*10.6*	*9.5*
Price basket for mobile service (US$/month)	—	*1.9*	*10.9*	*10.4*
Price basket for Internet service (US$/month)	—	25.7	*16.4*	*25.7*
Price of call to United States (US$ for 3 minutes)	1.93	—	*1.55*	*1.21*
Trade				
ICT goods exports (% of total goods exports)	31.7	29.4	13.5	11.4
ICT goods imports (% of total goods imports)	18.4	25.3	16.2	15.9
ICT service exports (% of total service exports)	7.9	16.4	*4.6*	*4.7*
Applications				
ICT expenditure (% of GDP)	—	3.9	5.2	4.9
E-government Web measure index[c]	—	0.44	0.37	0.44
Secure Internet servers (per 1 million people, Dec. 2008)	13.9	98.8	26.2	18.2

Sources: Economic and social context: UIS and World Bank; Sector structure: ITU; Sector efficiency and capacity: ITU and World Bank; Sector performance: Global Insight/WITSA, IMF, ITU, Netcraft, UN Comtrade, UNDESA, UNPAN, Wireless Intelligence and World Bank. Produced by the Global Information and Communication Technologies Department and the Development Economics Data Group. For complete information, see Definitions and Data Sources.

Notes: Use of italics in the column entries indicates years other than those specified. — Not available. GDP = gross domestic product; GNI = gross national income; ICT = information and communication technology; and MDG = Millennium Development Goal.

a. C = competition; M = monopoly; and P = partial competition. **b.** Outgoing and incoming. **c.** Scale of 0–1, where 1 = highest presence. **d.** Millennium Development Goal indicators 8.14, 8.15, and 8.16.

Information and Communications for Development 2009

Côte d'Ivoire

	Côte d'Ivoire 2000	Côte d'Ivoire 2007	Low-income group 2007	Sub-Saharan Africa Region 2007
Economic and social context				
Population (total, million)	17	19	1,296	800
Urban population (% of total)	44	48	32	36
GNI per capita, World Bank Atlas method (current US$)	630	920	574	951
GDP growth, 1995–2000 and 2000–07 (avg. annual %)	3.4	0.3	5.6	5.1
Adult literacy rate (% of ages 15 and older)	49	—	64	62
Gross primary, secondary, tertiary school enrollment (%)	37	—	51	51
Sector structure				
Separate telecommunications regulator	Yes	Yes		
Status of main fixed-line telephone operator	Mixed	Mixed		
Level of competition[a]				
International long distance service	M	P		
Mobile telephone service	P	P		
Internet service	C	C		
Sector efficiency and capacity				
Telecommunications revenue (% of GDP)	3.5	5.5	3.3	4.7
Mobile and fixed-line subscribers per employee	189	1,442	301	499
Telecommunications investment (% of revenue)	23.7	40.0	—	—
Sector performance				
Access				
Telephone lines (per 100 people)	1.5	1.4	4.0	1.6
Mobile cellular subscriptions (per 100 people)	2.8	36.6	21.5	23.0
Internet subscribers (per 100 people)	0.1	0.1	0.8	1.2
Personal computers (per 100 people)	0.5	1.7	1.5	1.8
Households with a television set (%)	32	35	16	18
Usage				
International voice traffic (minutes/person/month)[b]	0.7	1.4	—	—
Mobile telephone usage (minutes/user/month)	—	—	—	—
Internet users (per 100 people)	0.2	1.6	5.2	4.4
Quality				
Population covered by mobile cellular network (%)	23	59	54	56
Fixed broadband subscribers (% of total Internet subscrib.)	0.0	6.9	3.4	3.1
International Internet bandwidth (bits/second/person)	0	16	26	36
Affordability				
Price basket for residential fixed line (US$/month)	8.5	25.0	5.7	12.6
Price basket for mobile service (US$/month)	—	12.9	11.2	11.6
Price basket for Internet service (US$/month)	—	20.3	29.2	43.1
Price of call to United States (US$ for 3 minutes)	6.07	2.25	2.00	2.43
Trade				
ICT goods exports (% of total goods exports)	0.1	0.4	1.4	1.1
ICT goods imports (% of total goods imports)	3.3	4.2	6.7	8.2
ICT service exports (% of total service exports)	8.2	11.0	—	4.2
Applications				
ICT expenditure (% of GDP)	—	—	—	—
E-government Web measure index[c]	—	0.06	0.11	0.16
Secure Internet servers (per 1 million people, Dec. 2008)	—	0.5	0.5	2.9

Sources: Economic and social context: UIS and World Bank; Sector structure: ITU; Sector efficiency and capacity: ITU and World Bank; Sector performance: Global Insight/WITSA, IMF, ITU, Netcraft, UN Comtrade, UNDESA, UNPAN, Wireless Intelligence and World Bank. Produced by the Global Information and Communication Technologies Department and the Development Economics Data Group. For complete information, see Definitions and Data Sources.

Notes: Use of italics in the column entries indicates years other than those specified. — Not available. GDP = gross domestic product; GNI = gross national income; ICT = information and communication technology; and MDG = Millennium Development Goal.

a. C = competition; M = monopoly; and P = partial competition. b. Outgoing and incoming. c. Scale of 0–1, where 1 = highest presence. d. Millennium Development Goal indicators 8.14, 8.15, and 8.16.

World Bank • ICT at a Glance

Croatia

	Croatia 2000	Croatia 2007	Upper-middle-income group 2007	Europe & Central Asia Region 2007
Economic and social context				
Population (total, million)	4	4	824	446
Urban population (% of total)	56	57	75	64
GNI per capita, World Bank Atlas method (current US$)	4,430	10,460	7,107	6,052
GDP growth, 1995–2000 and 2000–07 (avg. annual %)	3.2	4.8	4.3	6.1
Adult literacy rate (% of ages 15 and older)	*98*	*99*	94	98
Gross primary, secondary, tertiary school enrollment (%)	70	*73*	82	82
Sector structure				
Separate telecommunications regulator	Yes	Yes		
Status of main fixed-line telephone operator	Mixed	Mixed		
Level of competition[a]				
International long distance service	M	C		
Mobile telephone service	C	C		
Internet service	C	C		
Sector efficiency and capacity				
Telecommunications revenue (% of GDP)	*3.9*	*5.3*	3.3	2.9
Mobile and fixed-line subscribers per employee	*178*	*778*	*566*	*532*
Telecommunications investment (% of revenue)	*14.7*	*12.3*	—	22.0
Sector performance				
Access				
Telephone lines (per 100 people)	38.9	41.6	22.6	25.7
Mobile cellular subscriptions (per 100 people)	23.3	113.5	84.1	95.0
Internet subscribers (per 100 people)	4.2	29.9	9.4	13.6
Personal computers (per 100 people)	11.3	18.0	*12.4*	10.6
Households with a television set (%)	85	*94*	92	96
Usage				
International voice traffic (minutes/person/month)[b]	13.8	17.4	—	—
Mobile telephone usage (minutes/user/month)	—	90	137	154
Internet users (per 100 people)	6.8	44.7	26.6	21.4
Quality				
Population covered by mobile cellular network (%)	98	100	95	*92*
Fixed broadband subscribers (% of total Internet subscrib.)	0.0	29.2	47.8	32.5
International Internet bandwidth (bits/second/person)	*41*	3,380	1,185	1,114
Affordability				
Price basket for residential fixed line (US$/month)	*11.9*	13.1	10.6	5.8
Price basket for mobile service (US$/month)	—	14.5	10.9	11.8
Price basket for Internet service (US$/month)	—	16.5	16.4	12.0
Price of call to United States (US$ for 3 minutes)	—	—	1.55	1.63
Trade				
ICT goods exports (% of total goods exports)	3.6	*5.3*	13.5	1.8
ICT goods imports (% of total goods imports)	6.1	*7.5*	16.2	7.0
ICT service exports (% of total service exports)	3.3	4.2	*4.6*	5.0
Applications				
ICT expenditure (% of GDP)	—	—	5.2	5.0
E-government Web measure index[c]	—	0.43	0.37	0.36
Secure Internet servers (per 1 million people, Dec. 2008)	13.7	92.0	26.2	23.9

Sources: Economic and social context: UIS and World Bank; Sector structure: ITU; Sector efficiency and capacity: ITU and World Bank; Sector performance: Global Insight/WITSA, IMF, ITU, Netcraft, UN Comtrade, UNDESA, UNPAN, Wireless Intelligence and World Bank. Produced by the Global Information and Communication Technologies Department and the Development Economics Data Group. For complete information, see Definitions and Data Sources.

Notes: Use of italics in the column entries indicates years other than those specified. — Not available. GDP = gross domestic product; GNI = gross national income; ICT = information and communication technology; and MDG = Millennium Development Goal.

a. C = competition; M = monopoly; and P = partial competition. **b.** Outgoing and incoming. **c.** Scale of 0–1, where 1 = highest presence. **d.** Millennium Development Goal indicators 8.14, 8.15, and 8.16.

Information and Communications for Development 2009

Cuba

	Cuba 2000	Cuba 2007	Upper-middle-income group 2007	Latin America & the Caribbean Region 2007
Economic and social context				
Population (total, million)	11	11	824	561
Urban population (% of total)	76	76	75	78
GNI per capita, World Bank Atlas method (current US$)	—	—	7,107	5,801
GDP growth, 1995–2000 and 2000–07 (avg. annual %)	4.2	*3.4*	4.3	3.6
Adult literacy rate (% of ages 15 and older)	*100*	100	94	91
Gross primary, secondary, tertiary school enrollment (%)	75	*88*	82	81
Sector structure				
Separate telecommunications regulator	—	No		
Status of main fixed-line telephone operator	Mixed	Public		
Level of competition[a]				
International long distance service	M	M		
Mobile telephone service	P	M		
Internet service	C	M		
Sector efficiency and capacity				
Telecommunications revenue (% of GDP)	2.2	*2.6*	3.3	3.8
Mobile and fixed-line subscribers per employee	30	*58*	*566*	*530*
Telecommunications investment (% of revenue)	16.0	7.5	—	—
Sector performance				
Access				
Telephone lines (per 100 people)	4.4	9.3	22.6	18.1
Mobile cellular subscriptions (per 100 people)	0.1	1.8	84.1	67.0
Internet subscribers (per 100 people)	—	0.3	9.4	4.5
Personal computers (per 100 people)	1.2	*3.6*	12.4	11.3
Households with a television set (%)	70	70	92	84
Usage				
International voice traffic (minutes/person/month)[b]	2.4	2.5	—	—
Mobile telephone usage (minutes/user/month)	—	—	137	116
Internet users (per 100 people)	0.5	11.6	26.6	26.9
Quality				
Population covered by mobile cellular network (%)	41	77	95	91
Fixed broadband subscribers (% of total Internet subscrib.)	—	5.8	47.8	81.7
International Internet bandwidth (bits/second/person)	2	19	1,185	1,126
Affordability				
Price basket for residential fixed line (US$/month)	12.4	*13.1*	*10.6*	*9.5*
Price basket for mobile service (US$/month)	—	22.6	*10.9*	*10.4*
Price basket for Internet service (US$/month)	—	32.4	16.4	25.7
Price of call to United States (US$ for 3 minutes)	7.35	7.49	1.55	1.21
Trade				
ICT goods exports (% of total goods exports)	0.2	*1.9*	13.5	11.4
ICT goods imports (% of total goods imports)	5.3	*2.9*	16.2	15.9
ICT service exports (% of total service exports)	—	—	4.6	4.7
Applications				
ICT expenditure (% of GDP)	—	—	5.2	4.9
E-government Web measure index[c]	—	0.21	0.37	0.44
Secure Internet servers (per 1 million people, Dec. 2008)	*0.2*	0.1	26.2	18.2

Sources: Economic and social context: UIS and World Bank; Sector structure: ITU; Sector efficiency and capacity: ITU and World Bank; Sector performance: Global Insight/WITSA, IMF, ITU, Netcraft, UN Comtrade, UNDESA, UNPAN, Wireless Intelligence and World Bank. Produced by the Global Information and Communication Technologies Department and the Development Economics Data Group. For complete information, see Definitions and Data Sources.

Notes: Use of italics in the column entries indicates years other than those specified. — Not available. GDP = gross domestic product; GNI = gross national income; ICT = information and communication technology; and MDG = Millennium Development Goal.

a. C = competition; M = monopoly; and P = partial competition. b. Outgoing and incoming. c. Scale of 0–1, where 1 = highest presence. d. Millennium Development Goal indicators 8.14, 8.15, and 8.16.

World Bank • ICT at a Glance

Czech Republic

	Czech Republic 2000	Czech Republic 2007	High-income group 2007
Economic and social context			
Population (total, million)	10	10	1,056
Urban population (% of total)	74	74	78
GNI per capita, World Bank Atlas method (current US$)	5,800	14,580	37,572
GDP growth, 1995–2000 and 2000–07 (avg. annual %)	1.0	4.6	2.4
Adult literacy rate (% of ages 15 and older)	—	—	99
Gross primary, secondary, tertiary school enrollment (%)	74	83	92
Sector structure			
Separate telecommunications regulator	Yes	Yes	
Status of main fixed-line telephone operator	Mixed	Private	
Level of competition[a]			
International long distance service	M	C	
Mobile telephone service	P	P	
Internet service	C	C	
Sector efficiency and capacity			
Telecommunications revenue (% of GDP)	4.5	3.7	3.1
Mobile and fixed-line subscribers per employee	349	796	747
Telecommunications investment (% of revenue)	47.1	12.3	14.3
Sector performance			
Access			
Telephone lines (per 100 people)	37.7	23.3	50.0
Mobile cellular subscriptions (per 100 people)	42.3	123.1	100.4
Internet subscribers (per 100 people)	4.1	13.6	25.8
Personal computers (per 100 people)	12.2	27.4	67.4
Households with a television set (%)	100	83	98
Usage			
International voice traffic (minutes/person/month)[b]	6.4	6.2	14.0
Mobile telephone usage (minutes/user/month)	145	122	353
Internet users (per 100 people)	9.7	48.3	65.7
Quality			
Population covered by mobile cellular network (%)	99	100	99
Fixed broadband subscribers (% of total Internet subscrib.)	0.6	93.3	82.6
International Internet bandwidth (bits/second/person)	602	7,075	18,242
Affordability			
Price basket for residential fixed line (US$/month)	12.1	25.5	26.1
Price basket for mobile service (US$/month)	—	12.8	13.0
Price basket for Internet service (US$/month)	—	19.9	22.8
Price of call to United States (US$ for 3 minutes)	0.97	1.06	0.81
Trade			
ICT goods exports (% of total goods exports)	7.3	14.2	15.2
ICT goods imports (% of total goods imports)	12.1	15.0	14.6
ICT service exports (% of total service exports)	3.2	8.0	7.0
Applications			
ICT expenditure (% of GDP)	—	7.1	6.7
E-government Web measure index[c]	—	0.65	0.60
Secure Internet servers (per 1 million people, Dec. 2008)	26.7	150.5	662.6

Sources: Economic and social context: UIS and World Bank; Sector structure: ITU; Sector efficiency and capacity: ITU and World Bank; Sector performance: Global Insight/WITSA, IMF, ITU, Netcraft, UN Comtrade, UNDESA, UNPAN, Wireless Intelligence and World Bank. Produced by the Global Information and Communication Technologies Department and the Development Economics Data Group. For complete information, see Definitions and Data Sources.

Notes: Use of italics in the column entries indicates years other than those specified. — Not available. GDP = gross domestic product; GNI = gross national income; ICT = information and communication technology; and MDG = Millennium Development Goal.

a. C = competition; M = monopoly; and P = partial competition. b. Outgoing and incoming. c. Scale of 0–1, where 1 = highest presence. d. Millennium Development Goal indicators 8.14, 8.15, and 8.16.

Denmark

	Denmark 2000	Denmark 2007	High-income group 2007
Economic and social context			
Population (total, million)	5	5	1,056
Urban population (% of total)	85	86	78
GNI per capita, World Bank Atlas method (current US$)	31,850	55,440	37,572
GDP growth, 1995–2000 and 2000–07 (avg. annual %)	2.8	1.8	2.4
Adult literacy rate (% of ages 15 and older)	—	—	99
Gross primary, secondary, tertiary school enrollment (%)	96	*103*	92
Sector structure			
Separate telecommunications regulator	Yes	Yes	
Status of main fixed-line telephone operator	Private	Private	
Level of competition[a]			
International long distance service	C	C	
Mobile telephone service	P	P	
Internet service	C	C	
Sector efficiency and capacity			
Telecommunications revenue (% of GDP)	2.6	2.6	*3.1*
Mobile and fixed-line subscribers per employee	337	512	747
Telecommunications investment (% of revenue)	27.0	15.0	*14.3*
Sector performance			
Access			
Telephone lines (per 100 people)	71.9	51.7	50.0
Mobile cellular subscriptions (per 100 people)	63.0	114.1	100.4
Internet subscribers (per 100 people)	31.6	38.5	25.8
Personal computers (per 100 people)	50.6	54.9	*67.4*
Households with a television set (%)	96	*96*	98
Usage			
International voice traffic (minutes/person/month)[b]	21.2	25.6	14.0
Mobile telephone usage (minutes/user/month)	122	202	353
Internet users (per 100 people)	39.2	80.7	65.7
Quality			
Population covered by mobile cellular network (%)	—	114	*99*
Fixed broadband subscribers (% of total Internet subscrib.)	4.0	93.2	82.6
International Internet bandwidth (bits/second/person)	1,409	34,506	18,242
Affordability			
Price basket for residential fixed line (US$/month)	22.5	*21.6*	26.1
Price basket for mobile service (US$/month)	—	10.9	13.0
Price basket for Internet service (US$/month)	—	9.3	22.8
Price of call to United States (US$ for 3 minutes)	1.30	*0.89*	*0.81*
Trade			
ICT goods exports (% of total goods exports)	8.5	*7.1*	15.2
ICT goods imports (% of total goods imports)	13.3	*11.9*	14.6
ICT service exports (% of total service exports)	—	—	7.0
Applications			
ICT expenditure (% of GDP)	—	5.8	6.7
E-government Web measure index[c]	—	1.00	0.60
Secure Internet servers (per 1 million people, Dec. 2008)	73.9	1,036.5	662.6

Sources: Economic and social context: UIS and World Bank; Sector structure: ITU; Sector efficiency and capacity: ITU and World Bank; Sector performance: Global Insight/WITSA, IMF, ITU, Netcraft, UN Comtrade, UNDESA, UNPAN, Wireless Intelligence and World Bank. Produced by the Global Information and Communication Technologies Department and the Development Economics Data Group. For complete information, see Definitions and Data Sources.

Notes: Use of italics in the column entries indicates years other than those specified. — Not available. GDP = gross domestic product; GNI = gross national income; ICT = information and communication technology; and MDG = Millennium Development Goal.

a. C = competition; M = monopoly; and P = partial competition. b. Outgoing and incoming. c. Scale of 0–1, where 1 = highest presence. d. Millennium Development Goal indicators 8.14, 8.15, and 8.16.

World Bank • ICT at a Glance

Dominican Republic

	Dominican Republic 2000	Dominican Republic 2007	Lower-middle-income group 2007	Latin America & the Caribbean Region 2007
Economic and social context				
Population (total, million)	9	10	3,435	561
Urban population (% of total)	62	68	42	78
GNI per capita, World Bank Atlas method (current US$)	2,050	3,560	1,905	5,801
GDP growth, 1995–2000 and 2000–07 (avg. annual %)	7.8	4.8	8.0	3.6
Adult literacy rate (% of ages 15 and older)	*87*	89	83	91
Gross primary, secondary, tertiary school enrollment (%)	71	74	68	81
Sector structure				
Separate telecommunications regulator	Yes	Yes		
Status of main fixed-line telephone operator	*Private*	*Private*		
Level of competition[a]				
International long distance service	C	C		
Mobile telephone service	C	C		
Internet service	C	C		
Sector efficiency and capacity				
Telecommunications revenue (% of GDP)	—	0.5	3.1	3.8
Mobile and fixed-line subscribers per employee	99	—	624	530
Telecommunications investment (% of revenue)	—	*192.0*	25.3	—
Sector performance				
Access				
Telephone lines (per 100 people)	10.2	9.3	15.3	18.1
Mobile cellular subscriptions (per 100 people)	8.1	56.7	38.9	67.0
Internet subscribers (per 100 people)	0.6	2.7	6.0	4.5
Personal computers (per 100 people)	*1.9*	*3.5*	*4.6*	*11.3*
Households with a television set (%)	74	78	79	84
Usage				
International voice traffic (minutes/person/month)[b]	14.6	—	—	—
Mobile telephone usage (minutes/user/month)	—	—	322	116
Internet users (per 100 people)	3.7	17.2	12.4	26.9
Quality				
Population covered by mobile cellular network (%)	—	90	80	91
Fixed broadband subscribers (% of total Internet subscrib.)	0.0	58.2	40.4	81.7
International Internet bandwidth (bits/second/person)	6	154	199	1,126
Affordability				
Price basket for residential fixed line (US$/month)	18.1	18.2	7.2	9.5
Price basket for mobile service (US$/month)	—	*8.6*	*9.8*	*10.4*
Price basket for Internet service (US$/month)	—	16.7	16.7	25.7
Price of call to United States (US$ for 3 minutes)	—	0.22	2.08	1.21
Trade				
ICT goods exports (% of total goods exports)	0.4	—	20.6	11.4
ICT goods imports (% of total goods imports)	6.6	—	20.2	15.9
ICT service exports (% of total service exports)	3.9	3.7	15.6	4.7
Applications				
ICT expenditure (% of GDP)	—	—	6.5	4.9
E-government Web measure index[c]	—	0.51	0.33	0.44
Secure Internet servers (per 1 million people, Dec. 2008)	0.9	13.2	1.8	18.2

Sources: Economic and social context: UIS and World Bank; Sector structure: ITU; Sector efficiency and capacity: ITU and World Bank; Sector performance: Global Insight/WITSA, IMF, ITU, Netcraft, UN Comtrade, UNDESA, UNPAN, Wireless Intelligence and World Bank. Produced by the Global Information and Communication Technologies Department and the Development Economics Data Group. For complete information, see Definitions and Data Sources.

Notes: Use of italics in the column entries indicates years other than those specified. — Not available. GDP = gross domestic product; GNI = gross national income; ICT = information and communication technology; and MDG = Millennium Development Goal.

a. C = competition; M = monopoly; and P = partial competition. **b.** Outgoing and incoming. **c.** Scale of 0–1, where 1 = highest presence. **d.** Millennium Development Goal indicators 8.14, 8.15, and 8.16.

World Bank • ICT at a Glance

Ecuador

	Ecuador 2000	Ecuador 2007	Lower-middle-income group 2007	Latin America & the Caribbean Region 2007
Economic and social context				
Population (total, million)	12	13	3,435	561
Urban population (% of total)	60	65	42	78
GNI per capita, World Bank Atlas method (current US$)	1,340	3,110	1,905	5,801
GDP growth, 1995–2000 and 2000–07 (avg. annual %)	0.7	5.0	8.0	3.6
Adult literacy rate (% of ages 15 and older)	*91*	84	83	91
Gross primary, secondary, tertiary school enrollment (%)	—	—	68	81
Sector structure				
Separate telecommunications regulator	Yes	Yes		
Status of main fixed-line telephone operator	*Public*	Public		
Level of competition[a]				
International long distance service	M	C		
Mobile telephone service	P	C		
Internet service	P	C		
Sector efficiency and capacity				
Telecommunications revenue (% of GDP)	2.8	4.1	3.1	3.8
Mobile and fixed-line subscribers per employee	244	512	624	530
Telecommunications investment (% of revenue)	9.8	—	25.3	—
Sector performance				
Access				
Telephone lines (per 100 people)	10.0	13.5	15.3	18.1
Mobile cellular subscriptions (per 100 people)	3.9	75.6	38.9	67.0
Internet subscribers (per 100 people)	0.5	*1.6*	6.0	4.5
Personal computers (per 100 people)	2.2	*13.0*	4.6	11.3
Households with a television set (%)	76	87	79	84
Usage				
International voice traffic (minutes/person/month)[b]	3.6	7.5	—	—
Mobile telephone usage (minutes/user/month)	65	69	322	116
Internet users (per 100 people)	1.5	13.2	12.4	26.9
Quality				
Population covered by mobile cellular network (%)	*80*	84	80	91
Fixed broadband subscribers (% of total Internet subscrib.)	3.1	68.4	40.4	81.7
International Internet bandwidth (bits/second/person)	2	324	199	1,126
Affordability				
Price basket for residential fixed line (US$/month)	9.8	7.9	7.2	9.5
Price basket for mobile service (US$/month)	—	18.9	9.8	10.4
Price basket for Internet service (US$/month)	—	37.0	16.7	25.7
Price of call to United States (US$ for 3 minutes)	2.48	—	2.08	1.21
Trade				
ICT goods exports (% of total goods exports)	0.0	0.3	20.6	11.4
ICT goods imports (% of total goods imports)	6.4	7.7	20.2	15.9
ICT service exports (% of total service exports)	7.2	6.2	15.6	4.7
Applications				
ICT expenditure (% of GDP)	—	6.1	6.5	4.9
E-government Web measure index[c]	—	0.44	0.33	0.44
Secure Internet servers (per 1 million people, Dec. 2008)	*0.9*	10.1	1.8	18.2

Sources: Economic and social context: UIS and World Bank; Sector structure: ITU; Sector efficiency and capacity: ITU and World Bank; Sector performance: Global Insight/WITSA, IMF, ITU, Netcraft, UN Comtrade, UNDESA, UNPAN, Wireless Intelligence and World Bank. Produced by the Global Information and Communication Technologies Department and the Development Economics Data Group. For complete information, see Definitions and Data Sources.

Notes: Use of italics in the column entries indicates years other than those specified. — Not available. GDP = gross domestic product; GNI = gross national income; ICT = information and communication technology; and MDG = Millennium Development Goal.

a. C = competition; M = monopoly; and P = partial competition. b. Outgoing and incoming. c. Scale of 0–1, where 1 = highest presence. d. Millennium Development Goal indicators 8.14, 8.15, and 8.16.

World Bank • ICT at a Glance

Egypt, Arab Republic of

	Egypt, Arab Rep. of 2000	Egypt, Arab Rep. of 2007	Lower-middle-income group 2007	Middle East & North Africa Region 2007
Economic and social context				
Population (total, million)	67	75	3,435	313
Urban population (% of total)	43	43	42	57
GNI per capita, World Bank Atlas method (current US$)	1,460	1,580	1,905	2,820
GDP growth, 1995–2000 and 2000–07 (avg. annual %)	5.2	4.3	8.0	4.4
Adult literacy rate (% of ages 15 and older)	—	66	83	73
Gross primary, secondary, tertiary school enrollment (%)	80	77	68	70
Sector structure				
Separate telecommunications regulator	Yes	Yes		
Status of main fixed-line telephone operator	Public	Mixed		
Level of competition[a]				
International long distance service	M	M		
Mobile telephone service	C	C		
Internet service	C	C		
Sector efficiency and capacity				
Telecommunications revenue (% of GDP)	2.8	3.8	3.1	3.1
Mobile and fixed-line subscribers per employee	125	538	624	691
Telecommunications investment (% of revenue)	18.8	24.8	25.3	21.7
Sector performance				
Access				
Telephone lines (per 100 people)	8.2	14.9	15.3	17.0
Mobile cellular subscriptions (per 100 people)	2.0	39.8	38.9	50.7
Internet subscribers (per 100 people)	0.1	3.5	6.0	2.4
Personal computers (per 100 people)	1.2	4.9	4.6	6.3
Households with a television set (%)	86	96	79	94
Usage				
International voice traffic (minutes/person/month)[b]	1.0	3.5	—	2.7
Mobile telephone usage (minutes/user/month)	—	132	322	—
Internet users (per 100 people)	0.7	14.0	12.4	17.1
Quality				
Population covered by mobile cellular network (%)	—	94	80	93
Fixed broadband subscribers (% of total Internet subscrib.)	—	18.0	40.4	—
International Internet bandwidth (bits/second/person)	0	189	199	186
Affordability				
Price basket for residential fixed line (US$/month)	4.3	3.7	7.2	3.9
Price basket for mobile service (US$/month)	—	4.2	9.8	6.5
Price basket for Internet service (US$/month)	—	4.3	16.7	11.6
Price of call to United States (US$ for 3 minutes)	3.33	1.45	2.08	1.45
Trade				
ICT goods exports (% of total goods exports)	—	—	20.6	—
ICT goods imports (% of total goods imports)	—	—	20.2	—
ICT service exports (% of total service exports)	3.4	4.2	15.6	2.6
Applications				
ICT expenditure (% of GDP)	—	5.8	6.5	4.5
E-government Web measure index[c]	—	0.61	0.33	0.22
Secure Internet servers (per 1 million people, Dec. 2008)	0.2	1.1	1.8	1.3

Sources: Economic and social context: UIS and World Bank; Sector structure: ITU; Sector efficiency and capacity: ITU and World Bank; Sector performance: Global Insight/WITSA, IMF, ITU, Netcraft, TeleGeography, UN Comtrade, UNDESA, UNPAN, Wireless Intelligence and World Bank. Produced by the Global Information and Communication Technologies Department and the Development Economics Data Group. For complete information, see Definitions and Data Sources.

Notes: Use of italics in the column entries indicates years other than those specified. — Not available. GDP = gross domestic product; GNI = gross national income; ICT = information and communication technology; and MDG = Millennium Development Goal.

a. C = competition; M = monopoly; and P = partial competition. b. Outgoing and incoming. c. Scale of 0–1, where 1 = highest presence. d. Millennium Development Goal indicators 8.14, 8.15, and 8.16.

El Salvador

	El Salvador 2000	El Salvador 2007	Lower-middle-income group 2007	Latin America & the Caribbean Region 2007
Economic and social context				
Population (total, million)	6	7	3,435	561
Urban population (% of total)	58	60	42	78
GNI per capita, World Bank Atlas method (current US$)	2,030	2,850	1,905	5,801
GDP growth, 1995–2000 and 2000–07 (avg. annual %)	3.3	2.8	8.0	3.6
Adult literacy rate (% of ages 15 and older)	—	82	83	91
Gross primary, secondary, tertiary school enrollment (%)	64	70	68	81
Sector structure				
Separate telecommunications regulator	Yes	Yes		
Status of main fixed-line telephone operator	Mixed	Mixed		
Level of competition[a]				
International long distance service	C	C		
Mobile telephone service	C	C		
Internet service	C	C		
Sector efficiency and capacity				
Telecommunications revenue (% of GDP)	4.3	5.7	3.1	3.8
Mobile and fixed-line subscribers per employee	323	1,657	624	530
Telecommunications investment (% of revenue)	155.8	29.3	25.3	—
Sector performance				
Access				
Telephone lines (per 100 people)	10.1	15.8	15.3	18.1
Mobile cellular subscriptions (per 100 people)	12.0	89.6	38.9	67.0
Internet subscribers (per 100 people)	0.9	1.4	6.0	4.5
Personal computers (per 100 people)	1.9	5.2	4.6	11.3
Households with a television set (%)	82	83	79	84
Usage				
International voice traffic (minutes/person/month)[b]	11.4	42.9	—	—
Mobile telephone usage (minutes/user/month)	—	—	322	116
Internet users (per 100 people)	1.1	11.1	12.4	26.9
Quality				
Population covered by mobile cellular network (%)	85	95	80	91
Fixed broadband subscribers (% of total Internet subscrib.)	0.0	94.3	40.4	81.7
International Internet bandwidth (bits/second/person)	7	18	199	1,126
Affordability				
Price basket for residential fixed line (US$/month)	16.3	2.0	7.2	9.5
Price basket for mobile service (US$/month)	—	8.5	9.8	10.4
Price basket for Internet service (US$/month)	—	22.6	16.7	25.7
Price of call to United States (US$ for 3 minutes)	2.40	2.40	2.08	1.21
Trade				
ICT goods exports (% of total goods exports)	0.6	0.6	20.6	11.4
ICT goods imports (% of total goods imports)	9.3	8.4	20.2	15.9
ICT service exports (% of total service exports)	12.9	9.6	15.6	4.7
Applications				
ICT expenditure (% of GDP)	—	—	6.5	4.9
E-government Web measure index[c]	—	0.58	0.33	0.44
Secure Internet servers (per 1 million people, Dec. 2008)	1.1	9.9	1.8	18.2

Sources: Economic and social context: UIS and World Bank; Sector structure: ITU; Sector efficiency and capacity: ITU and World Bank; Sector performance: Global Insight/WITSA, IMF, ITU, Netcraft, UN Comtrade, UNDESA, UNPAN, Wireless Intelligence and World Bank. Produced by the Global Information and Communication Technologies Department and the Development Economics Data Group. For complete information, see Definitions and Data Sources.

Notes: Use of italics in the column entries indicates years other than those specified. — Not available. GDP = gross domestic product; GNI = gross national income; ICT = information and communication technology; and MDG = Millennium Development Goal.

[a]. C = competition; M = monopoly; and P = partial competition. [b]. Outgoing and incoming. [c]. Scale of 0–1, where 1 = highest presence. [d]. Millennium Development Goal indicators 8.14, 8.15, and 8.16.

World Bank • ICT at a Glance

Eritrea

	Eritrea 2000	Eritrea 2007	Low-income group 2007	Sub-Saharan Africa Region 2007
Economic and social context				
Population (total, million)	4	5	1,296	800
Urban population (% of total)	18	20	32	36
GNI per capita, World Bank Atlas method (current US$)	170	270	574	951
GDP growth, 1995–2000 and 2000–07 (avg. annual %)	1.5	1.4	5.6	5.1
Adult literacy rate (% of ages 15 and older)	*53*	—	64	62
Gross primary, secondary, tertiary school enrollment (%)	33	*35*	*51*	*51*
Sector structure				
Separate telecommunications regulator	Yes	Yes		
Status of main fixed-line telephone operator	*Public*	*Public*		
Level of competition[a]				
International long distance service	*M*	*M*		
Mobile telephone service	*C*	*P*		
Internet service	*C*	*P*		
Sector efficiency and capacity				
Telecommunications revenue (% of GDP)	2.8	2.0	*3.3*	*4.7*
Mobile and fixed-line subscribers per employee	67	105	*301*	*499*
Telecommunications investment (% of revenue)	127.5	47.8	—	—
Sector performance				
Access				
Telephone lines (per 100 people)	0.8	0.8	4.0	1.6
Mobile cellular subscriptions (per 100 people)	0.0	1.7	21.5	23.0
Internet subscribers (per 100 people)	0.0	0.1	*0.8*	*1.2*
Personal computers (per 100 people)	0.2	0.8	1.5	1.8
Households with a television set (%)	11	*18*	16	18
Usage				
International voice traffic (minutes/person/month)[b]	0.5	0.5	—	—
Mobile telephone usage (minutes/user/month)	—	—	—	—
Internet users (per 100 people)	0.1	2.5	5.2	4.4
Quality				
Population covered by mobile cellular network (%)	0	2	*54*	*56*
Fixed broadband subscribers (% of total Internet subscrib.)	0.0	0.0	3.4	3.1
International Internet bandwidth (bits/second/person)	0	2	26	36
Affordability				
Price basket for residential fixed line (US$/month)	5.3	*6.2*	5.7	12.6
Price basket for mobile service (US$/month)	—	*16.8*	11.2	11.6
Price basket for Internet service (US$/month)	—	*28.6*	29.2	43.1
Price of call to United States (US$ for 3 minutes)	5.83	*3.59*	*2.00*	*2.43*
Trade				
ICT goods exports (% of total goods exports)	—	0.2	1.4	1.1
ICT goods imports (% of total goods imports)	—	5.2	6.7	8.2
ICT service exports (% of total service exports)	13.5	—	—	4.2
Applications				
ICT expenditure (% of GDP)	—	—	—	—
E-government Web measure index[c]	—	0.06	0.11	0.16
Secure Internet servers (per 1 million people, Dec. 2008)	—	—	0.5	2.9

Sources: Economic and social context: UIS and World Bank; Sector structure: ITU; Sector efficiency and capacity: ITU and World Bank; Sector performance: Global Insight/WITSA, IMF, ITU, Netcraft, UN Comtrade, UNDESA, UNPAN, Wireless Intelligence and World Bank. Produced by the Global Information and Communication Technologies Department and the Development Economics Data Group. For complete information, see Definitions and Data Sources.

Notes: Use of italics in the column entries indicates years other than those specified. — Not available. GDP = gross domestic product; GNI = gross national income; ICT = information and communication technology; and MDG = Millennium Development Goal.

a. C = competition; M = monopoly; and P = partial competition. **b.** Outgoing and incoming. **c.** Scale of 0–1, where 1 = highest presence. **d.** Millennium Development Goal indicators 8.14, 8.15, and 8.16.

Information and Communications for Development 2009

Estonia

	Estonia 2000	Estonia 2007	High-income group 2007
Economic and social context			
Population (total, million)	1	1	1,056
Urban population (% of total)	69	69	78
GNI per capita, World Bank Atlas method (current US$)	4,190	12,830	37,572
GDP growth, 1995–2000 and 2000–07 (avg. annual %)	5.8	8.1	2.4
Adult literacy rate (% of ages 15 and older)	100	100	99
Gross primary, secondary, tertiary school enrollment (%)	88	*92*	*92*
Sector structure			
Separate telecommunications regulator	Yes	Yes	
Status of main fixed-line telephone operator	Mixed	Mixed	
Level of competition[a]			
International long distance service	M	C	
Mobile telephone service	C	P	
Internet service	C	C	
Sector efficiency and capacity			
Telecommunications revenue (% of GDP)	5.1	4.8	*3.1*
Mobile and fixed-line subscribers per employee	354	707	747
Telecommunications investment (% of revenue)	17.6	12.8	*14.3*
Sector performance			
Access			
Telephone lines (per 100 people)	38.2	36.9	50.0
Mobile cellular subscriptions (per 100 people)	40.7	147.7	100.4
Internet subscribers (per 100 people)	6.0	21.2	25.8
Personal computers (per 100 people)	16.1	52.2	*67.4*
Households with a television set (%)	85	*86*	98
Usage			
International voice traffic (minutes/person/month)[b]	10.7	9.1	*14.0*
Mobile telephone usage (minutes/user/month)	—	—	353
Internet users (per 100 people)	28.6	63.7	65.7
Quality			
Population covered by mobile cellular network (%)	99	100	*99*
Fixed broadband subscribers (% of total Internet subscrib.)	*18.0*	97.7	82.6
International Internet bandwidth (bits/second/person)	137	11,925	18,242
Affordability			
Price basket for residential fixed line (US$/month)	9.4	*15.6*	26.1
Price basket for mobile service (US$/month)	—	*8.6*	13.0
Price basket for Internet service (US$/month)	—	*10.9*	22.8
Price of call to United States (US$ for 3 minutes)	1.62	*0.90*	*0.81*
Trade			
ICT goods exports (% of total goods exports)	26.0	*14.2*	*15.2*
ICT goods imports (% of total goods imports)	20.3	*11.1*	*14.6*
ICT service exports (% of total service exports)	2.8	6.1	7.0
Applications			
ICT expenditure (% of GDP)	—	—	6.7
E-government Web measure index[c]	—	0.71	0.60
Secure Internet servers (per 1 million people, Dec. 2008)	58.6	279.7	662.6

Sources: Economic and social context: UIS and World Bank; Sector structure: ITU; Sector efficiency and capacity: ITU and World Bank; Sector performance: Global Insight/WITSA, IMF, ITU, Netcraft, UN Comtrade, UNDESA, UNPAN, Wireless Intelligence and World Bank. Produced by the Global Information and Communication Technologies Department and the Development Economics Data Group. For complete information, see Definitions and Data Sources.

Notes: Use of italics in the column entries indicates years other than those specified. — Not available. GDP = gross domestic product; GNI = gross national income; ICT = information and communication technology; and MDG = Millennium Development Goal.

a. C = competition; M = monopoly; and P = partial competition. b. Outgoing and incoming. c. Scale of 0–1, where 1 = highest presence. d. Millennium Development Goal indicators 8.14, 8.15, and 8.16.

World Bank • ICT at a Glance

Ethiopia

	Ethiopia 2000	Ethiopia 2007	Low-income group 2007	Sub-Saharan Africa Region 2007
Economic and social context				
Population (total, million)	66	79	1,296	800
Urban population (% of total)	15	17	32	36
GNI per capita, World Bank Atlas method (current US$)	130	220	574	951
GDP growth, 1995–2000 and 2000–07 (avg. annual %)	3.5	7.5	5.6	5.1
Adult literacy rate (% of ages 15 and older)	—	*36*	64	62
Gross primary, secondary, tertiary school enrollment (%)	27	*42*	*51*	*51*
Sector structure				
Separate telecommunications regulator	Yes	Yes		
Status of main fixed-line telephone operator	*Public*	*Public*		
Level of competition[a]				
International long distance service	M	M		
Mobile telephone service	M	M		
Internet service	M	M		
Sector efficiency and capacity				
Telecommunications revenue (% of GDP)	1.1	*2.2*	*3.3*	*4.7*
Mobile and fixed-line subscribers per employee	35	*142*	*301*	*499*
Telecommunications investment (% of revenue)	42.6	*17.8*	—	—
Sector performance				
Access				
Telephone lines (per 100 people)	0.4	1.1	4.0	1.6
Mobile cellular subscriptions (per 100 people)	0.0	1.5	21.5	23.0
Internet subscribers (per 100 people)	0.0	0.0	*0.8*	*1.2*
Personal computers (per 100 people)	0.1	0.7	*1.5*	*1.8*
Households with a television set (%)	2	5	16	18
Usage				
International voice traffic (minutes/person/month)[b]	0.1	*0.3*	—	—
Mobile telephone usage (minutes/user/month)	—	—	—	—
Internet users (per 100 people)	0.0	0.4	5.2	4.4
Quality				
Population covered by mobile cellular network (%)	—	*10*	*54*	*56*
Fixed broadband subscribers (% of total Internet subscrib.)	0.0	1.0	*3.4*	*3.1*
International Internet bandwidth (bits/second/person)	0	3	26	36
Affordability				
Price basket for residential fixed line (US$/month)	3.1	*2.2*	*5.7*	*12.6*
Price basket for mobile service (US$/month)	—	*3.6*	*11.2*	*11.6*
Price basket for Internet service (US$/month)	—	*14.6*	*29.2*	*43.1*
Price of call to United States (US$ for 3 minutes)	7.35	*4.01*	*2.00*	*2.43*
Trade				
ICT goods exports (% of total goods exports)	*0.0*	*0.3*	*1.4*	*1.1*
ICT goods imports (% of total goods imports)	*5.0*	*7.1*	*6.7*	*8.2*
ICT service exports (% of total service exports)	3.6	6.3	—	4.2
Applications				
ICT expenditure (% of GDP)	—	—	—	—
E-government Web measure index[c]	—	0.17	0.11	0.16
Secure Internet servers (per 1 million people, Dec. 2008)	*0.0*	*0.0*	*0.5*	*2.9*

Sources: Economic and social context: UIS and World Bank; Sector structure: ITU; Sector efficiency and capacity: ITU and World Bank; Sector performance: Global Insight/WITSA, IMF, ITU, Netcraft, UN Comtrade, UNDESA, UNPAN, Wireless Intelligence and World Bank. Produced by the Global Information and Communication Technologies Department and the Development Economics Data Group. For complete information, see Definitions and Data Sources.

Notes: Use of italics in the column entries indicates years other than those specified. — Not available. GDP = gross domestic product; GNI = gross national income; ICT = information and communication technology; and MDG = Millennium Development Goal.

a. C = competition; M = monopoly; and P = partial competition. **b**. Outgoing and incoming. **c**. Scale of 0–1, where 1 = highest presence. **d**. Millennium Development Goal indicators 8.14, 8.15, and 8.16.

Finland

	Finland 2000	Finland 2007	High-income group 2007
Economic and social context			
Population (total, million)	5	5	1,056
Urban population (% of total)	61	63	78
GNI per capita, World Bank Atlas method (current US$)	25,480	44,300	37,572
GDP growth, 1995–2000 and 2000–07 (avg. annual %)	4.9	3.0	2.4
Adult literacy rate (% of ages 15 and older)	—	—	99
Gross primary, secondary, tertiary school enrollment (%)	104	101	92
Sector structure			
Separate telecommunications regulator	Yes	Yes	
Status of main fixed-line telephone operator	Mixed	Mixed	
Level of competition[a]			
International long distance service	C	C	
Mobile telephone service	C	C	
Internet service	C	C	
Sector efficiency and capacity			
Telecommunications revenue (% of GDP)	3.3	2.5	3.1
Mobile and fixed-line subscribers per employee	272	584	747
Telecommunications investment (% of revenue)	20.3	—	14.3
Sector performance			
Access			
Telephone lines (per 100 people)	55.0	32.9	50.0
Mobile cellular subscriptions (per 100 people)	72.0	115.0	100.4
Internet subscribers (per 100 people)	11.9	26.8	25.8
Personal computers (per 100 people)	39.6	50.0	67.4
Households with a television set (%)	92	87	98
Usage			
International voice traffic (minutes/person/month)[b]	14.9	—	14.0
Mobile telephone usage (minutes/user/month)	127	283	353
Internet users (per 100 people)	37.2	78.8	65.7
Quality			
Population covered by mobile cellular network (%)	99	99	99
Fixed broadband subscribers (% of total Internet subscrib.)	5.7	57.1	82.6
International Internet bandwidth (bits/second/person)	347	17,221	18,242
Affordability			
Price basket for residential fixed line (US$/month)	19.4	28.7	26.1
Price basket for mobile service (US$/month)	—	11.5	13.0
Price basket for Internet service (US$/month)	—	23.6	22.8
Price of call to United States (US$ for 3 minutes)	1.07	1.80	0.81
Trade			
ICT goods exports (% of total goods exports)	25.4	18.9	15.2
ICT goods imports (% of total goods imports)	18.6	14.4	14.6
ICT service exports (% of total service exports)	5.4	8.4	7.0
Applications			
ICT expenditure (% of GDP)	—	5.2	6.7
E-government Web measure index[c]	—	0.63	0.60
Secure Internet servers (per 1 million people, Dec. 2008)	96.0	684.2	662.6

Sources: Economic and social context: UIS and World Bank; Sector structure: ITU; Sector efficiency and capacity: ITU and World Bank; Sector performance: Global Insight/WITSA, IMF, ITU, Netcraft, UN Comtrade, UNDESA, UNPAN, Wireless Intelligence and World Bank. Produced by the Global Information and Communication Technologies Department and the Development Economics Data Group. For complete information, see Definitions and Data Sources.

Notes: Use of italics in the column entries indicates years other than those specified. — Not available. GDP = gross domestic product; GNI = gross national income; ICT = information and communication technology; and MDG = Millennium Development Goal.

a. C = competition; M = monopoly; and P = partial competition. b. Outgoing and incoming. c. Scale of 0–1, where 1 = highest presence. d. Millennium Development Goal indicators 8.14, 8.15, and 8.16.

France

	France 2000	France 2007	High-income group 2007
Economic and social context			
Population (total, million)	59	62	1,056
Urban population (% of total)	76	77	78
GNI per capita, World Bank Atlas method (current US$)	24,450	38,810	37,572
GDP growth, 1995–2000 and 2000–07 (avg. annual %)	2.9	1.8	2.4
Adult literacy rate (% of ages 15 and older)	—	—	99
Gross primary, secondary, tertiary school enrollment (%)	92	*97*	92
Sector structure			
Separate telecommunications regulator	Yes	Yes	
Status of main fixed-line telephone operator	Mixed	Mixed	
Level of competition[a]			
International long distance service	C	C	
Mobile telephone service	P	C	
Internet service	C	C	
Sector efficiency and capacity			
Telecommunications revenue (% of GDP)	2.1	2.2	3.1
Mobile and fixed-line subscribers per employee	408	695	747
Telecommunications investment (% of revenue)	26.5	14.0	14.3
Sector performance			
Access			
Telephone lines (per 100 people)	57.7	56.4	50.0
Mobile cellular subscriptions (per 100 people)	49.3	89.7	100.4
Internet subscribers (per 100 people)	9.2	27.6	25.8
Personal computers (per 100 people)	30.4	*65.2*	67.4
Households with a television set (%)	94	97	98
Usage			
International voice traffic (minutes/person/month)[b]	14.4	20.3	*14.0*
Mobile telephone usage (minutes/user/month)	140	198	353
Internet users (per 100 people)	14.4	51.2	65.7
Quality			
Population covered by mobile cellular network (%)	99	99	*99*
Fixed broadband subscribers (% of total Internet subscrib.)	3.6	91.2	82.6
International Internet bandwidth (bits/second/person)	1,148	29,466	18,242
Affordability			
Price basket for residential fixed line (US$/month)	18.4	*29.0*	26.1
Price basket for mobile service (US$/month)	—	23.3	13.0
Price basket for Internet service (US$/month)	—	13.7	22.8
Price of call to United States (US$ for 3 minutes)	0.82	*0.84*	0.81
Trade			
ICT goods exports (% of total goods exports)	12.1	*8.0*	15.2
ICT goods imports (% of total goods imports)	13.0	*9.8*	14.6
ICT service exports (% of total service exports)	2.6	4.1	7.0
Applications			
ICT expenditure (% of GDP)	—	5.7	6.7
E-government Web measure index[c]	—	0.83	0.60
Secure Internet servers (per 1 million people, Dec. 2008)	27.7	171.7	662.6

Sources: Economic and social context: UIS and World Bank; Sector structure: ITU; Sector efficiency and capacity: ITU and World Bank; Sector performance: Global Insight/WITSA, IMF, ITU, Netcraft, UN Comtrade, UNDESA, UNPAN, Wireless Intelligence and World Bank. Produced by the Global Information and Communication Technologies Department and the Development Economics Data Group. For complete information, see Definitions and Data Sources.

Notes: Use of italics in the column entries indicates years other than those specified. — Not available. GDP = gross domestic product; GNI = gross national income; ICT = information and communication technology; and MDG = Millennium Development Goal.

a. C = competition; M = monopoly; and P = partial competition. b. Outgoing and incoming. c. Scale of 0–1, where 1 = highest presence. d. Millennium Development Goal indicators 8.14, 8.15, and 8.16.

Information and Communications for Development 2009

Gabon

	Gabon 2000	Gabon 2007	Upper-middle-income group 2007	Sub-Saharan Africa Region 2007
Economic and social context				
Population (total, million)	1	1	824	800
Urban population (% of total)	80	85	75	36
GNI per capita, World Bank Atlas method (current US$)	3,220	7,020	7,107	951
GDP growth, 1995–2000 and 2000–07 (avg. annual %)	0.3	2.0	4.3	5.1
Adult literacy rate (% of ages 15 and older)	—	86	94	62
Gross primary, secondary, tertiary school enrollment (%)	70	—	82	51
Sector structure				
Separate telecommunications regulator	No	Yes		
Status of main fixed-line telephone operator	Public	Public		
Level of competition[a]				
International long distance service	M	C		
Mobile telephone service	P	C		
Internet service	C	C		
Sector efficiency and capacity				
Telecommunications revenue (% of GDP)	2.1	2.0	3.3	4.7
Mobile and fixed-line subscribers per employee	150	244	566	499
Telecommunications investment (% of revenue)	41.8	12.4	—	—
Sector performance				
Access				
Telephone lines (per 100 people)	3.3	2.0	22.6	1.6
Mobile cellular subscriptions (per 100 people)	10.1	87.9	84.1	23.0
Internet subscribers (per 100 people)	0.4	0.8	9.4	1.2
Personal computers (per 100 people)	1.0	3.6	12.4	1.8
Households with a television set (%)	51	58	92	18
Usage				
International voice traffic (minutes/person/month)[b]	3.9	6.2	—	—
Mobile telephone usage (minutes/user/month)	—	—	137	—
Internet users (per 100 people)	1.3	6.2	26.6	4.4
Quality				
Population covered by mobile cellular network (%)	13	79	95	56
Fixed broadband subscribers (% of total Internet subscrib.)	0.0	18.3	47.8	3.1
International Internet bandwidth (bits/second/person)	0	150	1,185	36
Affordability				
Price basket for residential fixed line (US$/month)	23.4	32.4	10.6	12.6
Price basket for mobile service (US$/month)	—	13.7	10.9	11.6
Price basket for Internet service (US$/month)	—	39.2	16.4	43.1
Price of call to United States (US$ for 3 minutes)	14.12	2.77	1.55	2.43
Trade				
ICT goods exports (% of total goods exports)	0.1	0.1	13.5	1.1
ICT goods imports (% of total goods imports)	7.9	6.6	16.2	8.2
ICT service exports (% of total service exports)	0.6	—	4.6	4.2
Applications				
ICT expenditure (% of GDP)	—	—	5.2	—
E-government Web measure index[c]	—	0.08	0.37	0.16
Secure Internet servers (per 1 million people, Dec. 2008)	0.8	4.4	26.2	2.9

Sources: Economic and social context: UIS and World Bank; Sector structure: ITU; Sector efficiency and capacity: ITU and World Bank; Sector performance: Global Insight/WITSA, IMF, ITU, Netcraft, UN Comtrade, UNDESA, UNPAN, Wireless Intelligence and World Bank. Produced by the Global Information and Communication Technologies Department and the Development Economics Data Group. For complete information, see Definitions and Data Sources.

Notes: Use of italics in the column entries indicates years other than those specified. — Not available. GDP = gross domestic product; GNI = gross national income; ICT = information and communication technology; and MDG = Millennium Development Goal.

a. C = competition; M = monopoly; and P = partial competition. **b.** Outgoing and incoming. **c.** Scale of 0–1, where 1 = highest presence. **d.** Millennium Development Goal indicators 8.14, 8.15, and 8.16.

Gambia, The

	Gambia, The 2000	Gambia, The 2007	Low-income group 2007	Sub-Saharan Africa Region 2007
Economic and social context				
Population (total, million)	1	2	1,296	800
Urban population (% of total)	49	56	32	36
GNI per capita, World Bank Atlas method (current US$)	310	320	574	951
GDP growth, 1995–2000 and 2000–07 (avg. annual %)	4.6	4.9	5.6	5.1
Adult literacy rate (% of ages 15 and older)	—	—	64	62
Gross primary, secondary, tertiary school enrollment (%)	45	*50*	*51*	*51*
Sector structure				
Separate telecommunications regulator	No	Yes		
Status of main fixed-line telephone operator	Public	Mixed		
Level of competition[a]				
International long distance service	M	P		
Mobile telephone service	M	C		
Internet service	C	C		
Sector efficiency and capacity				
Telecommunications revenue (% of GDP)	6.4	—	*3.3*	*4.7*
Mobile and fixed-line subscribers per employee	41	481	*301*	*499*
Telecommunications investment (% of revenue)	23.5	—	—	—
Sector performance				
Access				
Telephone lines (per 100 people)	2.4	4.5	4.0	1.6
Mobile cellular subscriptions (per 100 people)	0.4	46.9	21.5	23.0
Internet subscribers (per 100 people)	0.3	0.2	*0.8*	*1.2*
Personal computers (per 100 people)	1.1	3.3	*1.5*	*1.8*
Households with a television set (%)	12	*12*	*16*	*18*
Usage				
International voice traffic (minutes/person/month)[b]	—	—	—	—
Mobile telephone usage (minutes/user/month)	—	—	—	—
Internet users (per 100 people)	0.9	5.9	5.2	4.4
Quality				
Population covered by mobile cellular network (%)	*20*	85	*54*	*56*
Fixed broadband subscribers (% of total Internet subscrib.)	0.0	7.6	*3.4*	*3.1*
International Internet bandwidth (bits/second/person)	0	36	26	36
Affordability				
Price basket for residential fixed line (US$/month)	3.9	4.0	5.7	12.6
Price basket for mobile service (US$/month)	—	6.9	11.2	11.6
Price basket for Internet service (US$/month)	—	17.8	29.2	43.1
Price of call to United States (US$ for 3 minutes)	5.39	1.81	2.00	2.43
Trade				
ICT goods exports (% of total goods exports)	0.5	0.2	*1.4*	*1.1*
ICT goods imports (% of total goods imports)	3.3	4.6	*6.7*	*8.2*
ICT service exports (% of total service exports)	—	—	—	4.2
Applications				
ICT expenditure (% of GDP)	—	—	—	—
E-government Web measure index[c]	—	0.17	0.11	0.16
Secure Internet servers (per 1 million people, Dec. 2008)	—	1.7	0.5	2.9

Sources: Economic and social context: UIS and World Bank; Sector structure: ITU; Sector efficiency and capacity: ITU and World Bank; Sector performance: Global Insight/WITSA, IMF, ITU, Netcraft, UN Comtrade, UNDESA, UNPAN, Wireless Intelligence and World Bank. Produced by the Global Information and Communication Technologies Department and the Development Economics Data Group. For complete information, see Definitions and Data Sources.

Notes: Use of italics in the column entries indicates years other than those specified. — Not available. GDP = gross domestic product; GNI = gross national income; ICT = information and communication technology; and MDG = Millennium Development Goal.

a. C = competition; M = monopoly; and P = partial competition. **b.** Outgoing and incoming. **c.** Scale of 0–1, where 1 = highest presence. **d.** Millennium Development Goal indicators 8.14, 8.15, and 8.16.

Georgia

	Georgia 2000	Georgia 2007	Lower-middle-income group 2007	Europe & Central Asia Region 2007
Economic and social context				
Population (total, million)	5	4	3,435	446
Urban population (% of total)	53	53	42	64
GNI per capita, World Bank Atlas method (current US$)	700	2,120	1,905	6,052
GDP growth, 1995–2000 and 2000–07 (avg. annual %)	5.6	8.3	8.0	6.1
Adult literacy rate (% of ages 15 and older)	—	—	83	98
Gross primary, secondary, tertiary school enrollment (%)	74	76	68	82
Sector structure				
Separate telecommunications regulator	Yes	Yes		
Status of main fixed-line telephone operator	—	Private		
Level of competition[a]				
International long distance service	P	C		
Mobile telephone service	C	C		
Internet service	C	C		
Sector efficiency and capacity				
Telecommunications revenue (% of GDP)	3.5	6.5	3.1	2.9
Mobile and fixed-line subscribers per employee	69	355	624	532
Telecommunications investment (% of revenue)	65.4	30.9	25.3	22.0
Sector performance				
Access				
Telephone lines (per 100 people)	10.8	12.6	15.3	25.7
Mobile cellular subscriptions (per 100 people)	4.1	59.1	38.9	95.0
Internet subscribers (per 100 people)	0.1	6.4	6.0	13.6
Personal computers (per 100 people)	2.4	5.4	4.6	10.6
Households with a television set (%)	81	89	79	96
Usage				
International voice traffic (minutes/person/month)[b]	2.4	4.8	—	—
Mobile telephone usage (minutes/user/month)	—	87	322	154
Internet users (per 100 people)	0.5	8.2	12.4	21.4
Quality				
Population covered by mobile cellular network (%)	79	96	80	92
Fixed broadband subscribers (% of total Internet subscrib.)	25.2	16.6	40.4	32.5
International Internet bandwidth (bits/second/person)	2	745	199	1,114
Affordability				
Price basket for residential fixed line (US$/month)	4.2	9.7	7.2	5.8
Price basket for mobile service (US$/month)	—	44.1	9.8	11.8
Price basket for Internet service (US$/month)	—	9.2	16.7	12.0
Price of call to United States (US$ for 3 minutes)	2.88	—	2.08	1.63
Trade				
ICT goods exports (% of total goods exports)	0.2	0.4	20.6	1.8
ICT goods imports (% of total goods imports)	7.3	7.1	20.2	7.0
ICT service exports (% of total service exports)	—	1.5	15.6	5.0
Applications				
ICT expenditure (% of GDP)	—	—	6.5	5.0
E-government Web measure index[c]	—	0.35	0.33	0.36
Secure Internet servers (per 1 million people, Dec. 2008)	2.1	6.2	1.8	23.9

Sources: Economic and social context: UIS and World Bank; Sector structure: ITU; Sector efficiency and capacity: ITU and World Bank; Sector performance: Global Insight/WITSA, IMF, ITU, Netcraft, UN Comtrade, UNDESA, UNPAN, Wireless Intelligence and World Bank. Produced by the Global Information and Communication Technologies Department and the Development Economics Data Group. For complete information, see Definitions and Data Sources.

Notes: Use of italics in the column entries indicates years other than those specified. — Not available. GDP = gross domestic product; GNI = gross national income; ICT = information and communication technology; and MDG = Millennium Development Goal.

a. C = competition; M = monopoly; and P = partial competition. b. Outgoing and incoming. c. Scale of 0–1, where 1 = highest presence. d. Millennium Development Goal indicators 8.14, 8.15, and 8.16.

Germany

	Germany 2000	Germany 2007	High-income group 2007
Economic and social context			
Population (total, million)	82	82	1,056
Urban population (% of total)	73	74	78
GNI per capita, World Bank Atlas method (current US$)	25,510	38,990	37,572
GDP growth, 1995–2000 and 2000–07 (avg. annual %)	2.0	1.0	2.4
Adult literacy rate (% of ages 15 and older)	—	—	99
Gross primary, secondary, tertiary school enrollment (%)	89	*88*	92
Sector structure			
Separate telecommunications regulator	Yes	Yes	
Status of main fixed-line telephone operator	Mixed	Mixed	
Level of competition[a]			
International long distance service	C	C	
Mobile telephone service	P	C	
Internet service	C	C	
Sector efficiency and capacity			
Telecommunications revenue (% of GDP)	2.7	2.6	3.1
Mobile and fixed-line subscribers per employee	409	703	747
Telecommunications investment (% of revenue)	17.4	*9.8*	14.3
Sector performance			
Access			
Telephone lines (per 100 people)	61.1	65.3	50.0
Mobile cellular subscriptions (per 100 people)	58.6	118.1	100.4
Internet subscribers (per 100 people)	15.8	*24.3*	25.8
Personal computers (per 100 people)	33.6	*65.6*	*67.4*
Households with a television set (%)	97	*94*	98
Usage			
International voice traffic (minutes/person/month)[b]	15.9	—	14.0
Mobile telephone usage (minutes/user/month)	70	140	353
Internet users (per 100 people)	30.2	72.3	65.7
Quality			
Population covered by mobile cellular network (%)	99	100	*99*
Fixed broadband subscribers (% of total Internet subscrib.)	2.0	*54.0*	82.6
International Internet bandwidth (bits/second/person)	848	25,654	18,242
Affordability			
Price basket for residential fixed line (US$/month)	15.1	*26.5*	26.1
Price basket for mobile service (US$/month)	—	21.8	13.0
Price basket for Internet service (US$/month)	—	20.5	22.8
Price of call to United States (US$ for 3 minutes)	0.34	*0.43*	*0.81*
Trade			
ICT goods exports (% of total goods exports)	10.5	*9.6*	15.2
ICT goods imports (% of total goods imports)	13.0	*12.3*	14.6
ICT service exports (% of total service exports)	6.4	7.8	7.0
Applications			
ICT expenditure (% of GDP)	—	6.2	6.7
E-government Web measure index[c]	—	0.58	0.60
Secure Internet servers (per 1 million people, Dec. 2008)	*62.6*	*549.6*	*662.6*

Sources: Economic and social context: UIS and World Bank; Sector structure: ITU; Sector efficiency and capacity: ITU and World Bank; Sector performance: Global Insight/WITSA, IMF, ITU, Netcraft, UN Comtrade, UNDESA, UNPAN, Wireless Intelligence and World Bank. Produced by the Global Information and Communication Technologies Department and the Development Economics Data Group. For complete information, see Definitions and Data Sources.

Notes: Use of italics in the column entries indicates years other than those specified. — Not available. GDP = gross domestic product; GNI = gross national income; ICT = information and communication technology; and MDG = Millennium Development Goal.

a. C = competition; M = monopoly; and P = partial competition. **b.** Outgoing and incoming. **c.** Scale of 0–1, where 1 = highest presence. **d.** Millennium Development Goal indicators 8.14, 8.15, and 8.16.

World Bank • ICT at a Glance

Ghana

	Ghana 2000	Ghana 2007	Low-income group 2007	Sub-Saharan Africa Region 2007
Economic and social context				
Population (total, million)	20	23	1,296	800
Urban population (% of total)	44	49	32	36
GNI per capita, World Bank Atlas method (current US$)	320	590	574	951
GDP growth, 1995–2000 and 2000–07 (avg. annual %)	4.4	5.5	5.6	5.1
Adult literacy rate (% of ages 15 and older)	58	65	64	62
Gross primary, secondary, tertiary school enrollment (%)	46	53	51	51
Sector structure				
Separate telecommunications regulator	Yes	Yes		
Status of main fixed-line telephone operator	Mixed	Public		
Level of competition[a]				
International long distance service	P	P		
Mobile telephone service	C	P		
Internet service	C	C		
Sector efficiency and capacity				
Telecommunications revenue (% of GDP)	1.8	—	3.3	4.7
Mobile and fixed-line subscribers per employee	91	1,261	301	499
Telecommunications investment (% of revenue)	29.5	—	—	—
Sector performance				
Access				
Telephone lines (per 100 people)	1.1	1.6	4.0	1.6
Mobile cellular subscriptions (per 100 people)	0.6	32.4	21.5	23.0
Internet subscribers (per 100 people)	0.1	0.1	0.8	1.2
Personal computers (per 100 people)	0.3	0.6	1.5	1.8
Households with a television set (%)	23	25	16	18
Usage				
International voice traffic (minutes/person/month)[b]	0.9	0.1	—	—
Mobile telephone usage (minutes/user/month)	50	111	—	—
Internet users (per 100 people)	0.1	3.8	5.2	4.4
Quality				
Population covered by mobile cellular network (%)	—	68	54	56
Fixed broadband subscribers (% of total Internet subscrib.)	0.0	69.2	3.4	3.1
International Internet bandwidth (bits/second/person)	0	21	26	36
Affordability				
Price basket for residential fixed line (US$/month)	—	6.3	5.7	12.6
Price basket for mobile service (US$/month)	—	5.7	11.2	11.6
Price basket for Internet service (US$/month)	—	9.4	29.2	43.1
Price of call to United States (US$ for 3 minutes)	1.65	0.39	2.00	2.43
Trade				
ICT goods exports (% of total goods exports)	—	0.0	1.4	1.1
ICT goods imports (% of total goods imports)	—	6.3	6.7	8.2
ICT service exports (% of total service exports)	—	—	—	4.2
Applications				
ICT expenditure (% of GDP)	—	—	—	—
E-government Web measure index[c]	—	0.29	0.11	0.16
Secure Internet servers (per 1 million people, Dec. 2008)	0.0	0.7	0.5	2.9

Sources: Economic and social context: UIS and World Bank; Sector structure: ITU; Sector efficiency and capacity: ITU and World Bank; Sector performance: Global Insight/WITSA, IMF, ITU, Netcraft, UN Comtrade, UNDESA, UNPAN, Wireless Intelligence and World Bank. Produced by the Global Information and Communication Technologies Department and the Development Economics Data Group. For complete information, see Definitions and Data Sources.

Notes: Use of italics in the column entries indicates years other than those specified. — Not available. GDP = gross domestic product; GNI = gross national income; ICT = information and communication technology; and MDG = Millennium Development Goal.

a. C = competition; M = monopoly; and P = partial competition. b. Outgoing and incoming. c. Scale of 0–1, where 1 = highest presence. d. Millennium Development Goal indicators 8.14, 8.15, and 8.16.

Greece

	Greece 2000	Greece 2007	High-income group 2007
Economic and social context			
Population (total, million)	11	11	1,056
Urban population (% of total)	60	61	78
GNI per capita, World Bank Atlas method (current US$)	12,560	25,740	37,572
GDP growth, 1995–2000 and 2000–07 (avg. annual %)	3.5	4.3	2.4
Adult literacy rate (% of ages 15 and older)	*96*	97	99
Gross primary, secondary, tertiary school enrollment (%)	81	*99*	*92*
Sector structure			
Separate telecommunications regulator	Yes	Yes	
Status of main fixed-line telephone operator	Mixed	Mixed	
Level of competition[a]			
International long distance service	M	C	
Mobile telephone service	P	P	
Internet service	C	C	
Sector efficiency and capacity			
Telecommunications revenue (% of GDP)	3.7	3.7	*3.1*
Mobile and fixed-line subscribers per employee	451	802	747
Telecommunications investment (% of revenue)	42.3	15.3	*14.3*
Sector performance			
Access			
Telephone lines (per 100 people)	51.8	53.7	50.0
Mobile cellular subscriptions (per 100 people)	54.3	109.8	100.4
Internet subscribers (per 100 people)	2.5	10.0	25.8
Personal computers (per 100 people)	6.9	*9.4*	*67.4*
Households with a television set (%)	97	*100*	98
Usage			
International voice traffic (minutes/person/month)[b]	12.8	*15.1*	14.0
Mobile telephone usage (minutes/user/month)	89	170	353
Internet users (per 100 people)	9.2	32.9	65.7
Quality			
Population covered by mobile cellular network (%)	99	100	*99*
Fixed broadband subscribers (% of total Internet subscrib.)	0.0	91.3	82.6
International Internet bandwidth (bits/second/person)	51	4,537	18,242
Affordability			
Price basket for residential fixed line (US$/month)	10.2	*21.1*	26.1
Price basket for mobile service (US$/month)	—	23.1	13.0
Price basket for Internet service (US$/month)	—	16.5	22.8
Price of call to United States (US$ for 3 minutes)	0.69	*1.09*	*0.81*
Trade			
ICT goods exports (% of total goods exports)	4.4	*3.3*	15.2
ICT goods imports (% of total goods imports)	8.3	6.0	14.6
ICT service exports (% of total service exports)	1.8	1.6	7.0
Applications			
ICT expenditure (% of GDP)	—	5.4	6.7
E-government Web measure index[c]	—	0.41	0.60
Secure Internet servers (per 1 million people, Dec. 2008)	*10.6*	61.2	662.6

Sources: Economic and social context: UIS and World Bank; Sector structure: ITU; Sector efficiency and capacity: ITU and World Bank; Sector performance: Global Insight/WITSA, IMF, ITU, Netcraft, UN Comtrade, UNDESA, UNPAN, Wireless Intelligence and World Bank. Produced by the Global Information and Communication Technologies Department and the Development Economics Data Group. For complete information, see Definitions and Data Sources.

Notes: Use of italics in the column entries indicates years other than those specified. — Not available. GDP = gross domestic product; GNI = gross national income; ICT = information and communication technology; and MDG = Millennium Development Goal.

a. C = competition; M = monopoly; and P = partial competition. b. Outgoing and incoming. c. Scale of 0–1, where 1 = highest presence. d. Millennium Development Goal indicators 8.14, 8.15, and 8.16.

Guatemala

	Guatemala 2000	Guatemala 2007	Lower-middle-income group 2007	Latin America & the Caribbean Region 2007
Economic and social context				
Population (total, million)	11	13	3,435	561
Urban population (% of total)	45	48	42	78
GNI per capita, World Bank Atlas method (current US$)	1,730	2,450	1,905	5,801
GDP growth, 1995–2000 and 2000–07 (avg. annual %)	4.1	3.6	8.0	3.6
Adult literacy rate (% of ages 15 and older)	*69*	*73*	83	91
Gross primary, secondary, tertiary school enrollment (%)	58	67	68	81
Sector structure				
Separate telecommunications regulator	Yes	Yes		
Status of main fixed-line telephone operator	Private	Private		
Level of competition[a]				
International long distance service	C	C		
Mobile telephone service	C	C		
Internet service	C	C		
Sector efficiency and capacity				
Telecommunications revenue (% of GDP)	2.2	—	3.1	3.8
Mobile and fixed-line subscribers per employee	434	—	624	530
Telecommunications investment (% of revenue)	—	—	25.3	—
Sector performance				
Access				
Telephone lines (per 100 people)	6.0	*10.4*	15.3	18.1
Mobile cellular subscriptions (per 100 people)	7.6	76.0	38.9	67.0
Internet subscribers (per 100 people)	—	—	6.0	4.5
Personal computers (per 100 people)	1.2	2.1	4.6	11.3
Households with a television set (%)	39	50	79	84
Usage				
International voice traffic (minutes/person/month)[b]	—	—	—	—
Mobile telephone usage (minutes/user/month)	—	—	322	116
Internet users (per 100 people)	0.7	10.1	12.4	26.9
Quality				
Population covered by mobile cellular network (%)	54	*76*	*80*	91
Fixed broadband subscribers (% of total Internet subscrib.)	—	—	40.4	81.7
International Internet bandwidth (bits/second/person)	1	187	199	1,126
Affordability				
Price basket for residential fixed line (US$/month)	11.0	9.8	7.2	9.5
Price basket for mobile service (US$/month)	—	6.1	9.8	10.4
Price basket for Internet service (US$/month)	—	53.3	16.7	25.7
Price of call to United States (US$ for 3 minutes)	0.76	1.21	2.08	1.21
Trade				
ICT goods exports (% of total goods exports)	0.1	*0.5*	20.6	11.4
ICT goods imports (% of total goods imports)	9.2	*9.1*	20.2	15.9
ICT service exports (% of total service exports)	0.5	14.8	15.6	4.7
Applications				
ICT expenditure (% of GDP)	—	—	6.5	4.9
E-government Web measure index[c]	—	0.47	0.33	0.44
Secure Internet servers (per 1 million people, Dec. 2008)	*1.0*	8.0	1.8	18.2

Sources: Economic and social context: UIS and World Bank; Sector structure: ITU; Sector efficiency and capacity: ITU and World Bank; Sector performance: Global Insight/WITSA, IMF, ITU, Netcraft, UN Comtrade, UNDESA, UNPAN, Wireless Intelligence and World Bank. Produced by the Global Information and Communication Technologies Department and the Development Economics Data Group. For complete information, see Definitions and Data Sources.

Notes: Use of italics in the column entries indicates years other than those specified. — Not available. GDP = gross domestic product; GNI = gross national income; ICT = information and communication technology; and MDG = Millennium Development Goal.

a. C = competition; M = monopoly; and P = partial competition. b. Outgoing and incoming. c. Scale of 0–1, where 1 = highest presence. d. Millennium Development Goal indicators 8.14, 8.15, and 8.16.

Guinea

	Guinea 2000	Guinea 2007	Low-income group 2007	Sub-Saharan Africa Region 2007
Economic and social context				
Population (total, million)	8	9	1,296	800
Urban population (% of total)	31	34	32	36
GNI per capita, World Bank Atlas method (current US$)	410	400	574	951
GDP growth, 1995–2000 and 2000–07 (avg. annual %)	4.4	2.8	5.6	5.1
Adult literacy rate (% of ages 15 and older)	—	29	64	62
Gross primary, secondary, tertiary school enrollment (%)	31	45	51	51
Sector structure				
Separate telecommunications regulator	Yes	Yes		
Status of main fixed-line telephone operator	Mixed	Mixed		
Level of competition[a]				
International long distance service	M	P		
Mobile telephone service	C	C		
Internet service	C	C		
Sector efficiency and capacity				
Telecommunications revenue (% of GDP)	0.9	—	3.3	4.7
Mobile and fixed-line subscribers per employee	82	—	301	499
Telecommunications investment (% of revenue)	17.8	—	—	—
Sector performance				
Access				
Telephone lines (per 100 people)	0.3	0.5	4.0	1.6
Mobile cellular subscriptions (per 100 people)	0.5	21.3	21.5	23.0
Internet subscribers (per 100 people)	0.0	*0.1*	0.8	1.2
Personal computers (per 100 people)	0.4	*0.5*	1.5	1.8
Households with a television set (%)	9	10	16	18
Usage				
International voice traffic (minutes/person/month)[b]	0.4	—	—	—
Mobile telephone usage (minutes/user/month)	—	—	—	—
Internet users (per 100 people)	0.1	*0.5*	5.2	4.4
Quality				
Population covered by mobile cellular network (%)	—	80	54	56
Fixed broadband subscribers (% of total Internet subscrib.)	0.0	0.0	3.4	3.1
International Internet bandwidth (bits/second/person)	0	0	26	36
Affordability				
Price basket for residential fixed line (US$/month)	10.6	—	5.7	12.6
Price basket for mobile service (US$/month)	—	3.8	11.2	11.6
Price basket for Internet service (US$/month)	—	*17.8*	29.2	43.1
Price of call to United States (US$ for 3 minutes)	5.15	—	2.00	2.43
Trade				
ICT goods exports (% of total goods exports)	0.1	—	*1.4*	*1.1*
ICT goods imports (% of total goods imports)	1.6	—	*6.7*	*8.2*
ICT service exports (% of total service exports)	2.2	—	—	4.2
Applications				
ICT expenditure (% of GDP)	—	—	—	—
E-government Web measure index[c]	—	0.07	0.11	0.16
Secure Internet servers (per 1 million people, Dec. 2008)	—	0.1	0.5	2.9

Sources: Economic and social context: UIS and World Bank; Sector structure: ITU; Sector efficiency and capacity: ITU and World Bank; Sector performance: Global Insight/WITSA, IMF, ITU, Netcraft, UN Comtrade, UNDESA, UNPAN, Wireless Intelligence and World Bank. Produced by the Global Information and Communication Technologies Department and the Development Economics Data Group. For complete information, see Definitions and Data Sources.

Notes: Use of italics in the column entries indicates years other than those specified. — Not available. GDP = gross domestic product; GNI = gross national income; ICT = information and communication technology; and MDG = Millennium Development Goal.

a. C = competition; M = monopoly; and P = partial competition. **b.** Outgoing and incoming. **c.** Scale of 0–1, where 1 = highest presence. **d.** Millennium Development Goal indicators 8.14, 8.15, and 8.16.

Information and Communications for Development 2009

Guinea-Bissau

	Guinea-Bissau 2000	Guinea-Bissau 2007	Low-income group 2007	Sub-Saharan Africa Region 2007
Economic and social context				
Population (total, million)	1	2	1,296	800
Urban population (% of total)	30	30	32	36
GNI per capita, World Bank Atlas method (current US$)	160	200	574	951
GDP growth, 1995–2000 and 2000–07 (avg. annual %)	−2.7	0.4	5.6	5.1
Adult literacy rate (% of ages 15 and older)	—	—	64	62
Gross primary, secondary, tertiary school enrollment (%)	37	—	51	51
Sector structure				
Separate telecommunications regulator	Yes	Yes		
Status of main fixed-line telephone operator	Mixed	Mixed		
Level of competition[a]				
International long distance service	M	M		
Mobile telephone service	C	P		
Internet service	C	C		
Sector efficiency and capacity				
Telecommunications revenue (% of GDP)	—	—	3.3	4.7
Mobile and fixed-line subscribers per employee	46	—	301	499
Telecommunications investment (% of revenue)	—	—	—	—
Sector performance				
Access				
Telephone lines (per 100 people)	0.8	0.3	4.0	1.6
Mobile cellular subscriptions (per 100 people)	0.0	17.5	21.5	23.0
Internet subscribers (per 100 people)	0.0	—	0.8	1.2
Personal computers (per 100 people)	0.2	0.2	1.5	1.8
Households with a television set (%)	20	31	16	18
Usage				
International voice traffic (minutes/person/month)[b]	0.7	—	—	—
Mobile telephone usage (minutes/user/month)	—	—	—	—
Internet users (per 100 people)	0.2	2.2	5.2	4.4
Quality				
Population covered by mobile cellular network (%)	—	65	54	56
Fixed broadband subscribers (% of total Internet subscrib.)	0.0	—	3.4	3.1
International Internet bandwidth (bits/second/person)	0	1	26	36
Affordability				
Price basket for residential fixed line (US$/month)	—	—	5.7	12.6
Price basket for mobile service (US$/month)	—	11.3	11.2	11.6
Price basket for Internet service (US$/month)	—	75.0	29.2	43.1
Price of call to United States (US$ for 3 minutes)	—	—	2.00	2.43
Trade				
ICT goods exports (% of total goods exports)	—	—	1.4	1.1
ICT goods imports (% of total goods imports)	—	—	6.7	8.2
ICT service exports (% of total service exports)	—	—	—	4.2
Applications				
ICT expenditure (% of GDP)	—	—	—	—
E-government Web measure index[c]	—	0.02	0.11	0.16
Secure Internet servers (per 1 million people, Dec. 2008)	—	—	0.5	2.9

Sources: Economic and social context: UIS and World Bank; Sector structure: ITU; Sector efficiency and capacity: ITU and World Bank; Sector performance: Global Insight/WITSA, IMF, ITU, Netcraft, UN Comtrade, UNDESA, UNPAN, Wireless Intelligence and World Bank. Produced by the Global Information and Communication Technologies Department and the Development Economics Data Group. For complete information, see Definitions and Data Sources.

Notes: Use of italics in the column entries indicates years other than those specified. — Not available. GDP = gross domestic product; GNI = gross national income; ICT = information and communication technology; and MDG = Millennium Development Goal.

a. C = competition; M = monopoly; and P = partial competition. b. Outgoing and incoming. c. Scale of 0–1, where 1 = highest presence. d. Millennium Development Goal indicators 8.14, 8.15, and 8.16.

World Bank • ICT at a Glance

Haiti

	Haiti 2000	Haiti 2007	Low-income group 2007	Latin America & the Caribbean Region 2007
Economic and social context				
Population (total, million)	9	10	1,296	561
Urban population (% of total)	36	45	32	78
GNI per capita, World Bank Atlas method (current US$)	470	520	574	5,801
GDP growth, 1995–2000 and 2000–07 (avg. annual %)	2.4	0.2	5.6	3.6
Adult literacy rate (% of ages 15 and older)	—	—	64	91
Gross primary, secondary, tertiary school enrollment (%)	—	—	51	81
Sector structure				
Separate telecommunications regulator	Yes	Yes		
Status of main fixed-line telephone operator	Mixed	Mixed		
Level of competition[a]				
International long distance service	M	P		
Mobile telephone service	P	P		
Internet service	C	C		
Sector efficiency and capacity				
Telecommunications revenue (% of GDP)	—	—	3.3	3.8
Mobile and fixed-line subscribers per employee	32	92	301	530
Telecommunications investment (% of revenue)	—	—	—	—
Sector performance				
Access				
Telephone lines (per 100 people)	0.8	1.1	4.0	18.1
Mobile cellular subscriptions (per 100 people)	0.6	26.0	21.5	67.0
Internet subscribers (per 100 people)	0.1	1.0	0.8	4.5
Personal computers (per 100 people)	0.1	5.2	1.5	11.3
Households with a television set (%)	23	27	16	84
Usage				
International voice traffic (minutes/person/month)[b]	—	—	—	—
Mobile telephone usage (minutes/user/month)	—	—	—	116
Internet users (per 100 people)	0.2	10.4	5.2	26.9
Quality				
Population covered by mobile cellular network (%)	—	32	54	91
Fixed broadband subscribers (% of total Internet subscrib.)	0.0	0.0	3.4	81.7
International Internet bandwidth (bits/second/person)	5	17	26	1,126
Affordability				
Price basket for residential fixed line (US$/month)	—	—	5.7	9.5
Price basket for mobile service (US$/month)	—	4.5	11.2	10.4
Price basket for Internet service (US$/month)	—	70.3	29.2	25.7
Price of call to United States (US$ for 3 minutes)	—	2.15	2.00	1.21
Trade				
ICT goods exports (% of total goods exports)	—	—	1.4	11.4
ICT goods imports (% of total goods imports)	—	—	6.7	15.9
ICT service exports (% of total service exports)	17.4	4.9	—	4.7
Applications				
ICT expenditure (% of GDP)	—	—	—	4.9
E-government Web measure index[c]	—	0.06	0.11	0.44
Secure Internet servers (per 1 million people, Dec. 2008)	0.1	0.8	0.5	18.2

Sources: Economic and social context: UIS and World Bank; Sector structure: ITU; Sector efficiency and capacity: ITU and World Bank; Sector performance: Global Insight/WITSA, IMF, ITU, Netcraft, UN Comtrade, UNDESA, UNPAN, Wireless Intelligence and World Bank. Produced by the Global Information and Communication Technologies Department and the Development Economics Data Group. For complete information, see Definitions and Data Sources.

Notes: Use of italics in the column entries indicates years other than those specified. — Not available. GDP = gross domestic product; GNI = gross national income; ICT = information and communication technology; and MDG = Millennium Development Goal.

a. C = competition; M = monopoly; and P = partial competition. **b.** Outgoing and incoming. **c.** Scale of 0–1, where 1 = highest presence. **d.** Millennium Development Goal indicators 8.14, 8.15, and 8.16.

World Bank • ICT at a Glance

Honduras

	Honduras 2000	Honduras 2007	Lower-middle-income group 2007	Latin America & the Caribbean Region 2007
Economic and social context				
Population (total, million)	6	7	3,435	561
Urban population (% of total)	44	47	42	78
GNI per capita, World Bank Atlas method (current US$)	940	1,590	1,905	5,801
GDP growth, 1995–2000 and 2000–07 (avg. annual %)	2.8	5.3	8.0	3.6
Adult literacy rate (% of ages 15 and older)	*80*	*84*	83	91
Gross primary, secondary, tertiary school enrollment (%)	62	*71*	68	81
Sector structure				
Separate telecommunications regulator	Yes	Yes		
Status of main fixed-line telephone operator	*Public*	*Public*		
Level of competition[a]				
International long distance service	C	M		
Mobile telephone service	P	M		
Internet service	C	—		
Sector efficiency and capacity				
Telecommunications revenue (% of GDP)	4.3	6.6	*3.1*	3.8
Mobile and fixed-line subscribers per employee	107	391	*624*	*530*
Telecommunications investment (% of revenue)	16.2	41.2	*25.3*	—
Sector performance				
Access				
Telephone lines (per 100 people)	4.8	11.6	15.3	18.1
Mobile cellular subscriptions (per 100 people)	2.5	58.9	38.9	67.0
Internet subscribers (per 100 people)	0.3	0.5	6.0	4.5
Personal computers (per 100 people)	1.1	2.0	*4.6*	*11.3*
Households with a television set (%)	44	*61*	*79*	*84*
Usage				
International voice traffic (minutes/person/month)[b]	3.9	2.8	—	—
Mobile telephone usage (minutes/user/month)	—	—	322	116
Internet users (per 100 people)	1.2	6.0	12.4	26.9
Quality				
Population covered by mobile cellular network (%)	*83*	90	*80*	91
Fixed broadband subscribers (% of total Internet subscrib.)	*0.0*	*0.0*	40.4	81.7
International Internet bandwidth (bits/second/person)	2	244	199	1,126
Affordability				
Price basket for residential fixed line (US$/month)	7.0	5.9	7.2	9.5
Price basket for mobile service (US$/month)	—	10.8	9.8	10.4
Price basket for Internet service (US$/month)	—	33.3	16.7	25.7
Price of call to United States (US$ for 3 minutes)	3.97	2.52	2.08	1.21
Trade				
ICT goods exports (% of total goods exports)	0.0	*0.3*	20.6	11.4
ICT goods imports (% of total goods imports)	0.5	*6.9*	20.2	15.9
ICT service exports (% of total service exports)	—	11.5	15.6	4.7
Applications				
ICT expenditure (% of GDP)	—	11.2	6.5	4.9
E-government Web measure index[c]	—	0.37	0.33	0.44
Secure Internet servers (per 1 million people, Dec. 2008)	*0.6*	6.4	1.8	18.2

Sources: Economic and social context: UIS and World Bank; Sector structure: ITU; Sector efficiency and capacity: ITU and World Bank; Sector performance: Global Insight/WITSA, IMF, ITU, Netcraft, UN Comtrade, UNDESA, UNPAN, Wireless Intelligence and World Bank. Produced by the Global Information and Communication Technologies Department and the Development Economics Data Group. For complete information, see Definitions and Data Sources.

Notes: Use of italics in the column entries indicates years other than those specified. — Not available. GDP = gross domestic product; GNI = gross national income; ICT = information and communication technology; and MDG = Millennium Development Goal.

a. C = competition; M = monopoly; and P = partial competition. **b.** Outgoing and incoming. **c.** Scale of 0–1, where 1 = highest presence. **d.** Millennium Development Goal indicators 8.14, 8.15, and 8.16.

World Bank • ICT at a Glance

Hong Kong, China

	Hong Kong, China 2000	Hong Kong, China 2007	High-income group 2007
Economic and social context			
Population (total, million)	7	7	1,056
Urban population (% of total)	100	100	78
GNI per capita, World Bank Atlas method (current US$)	26,570	31,560	37,572
GDP growth, 1995–2000 and 2000–07 (avg. annual %)	1.8	5.2	2.4
Adult literacy rate (% of ages 15 and older)	—	—	99
Gross primary, secondary, tertiary school enrollment (%)	—	76	92
Sector structure			
Separate telecommunications regulator	—	No	
Status of main fixed-line telephone operator	—	—	
Level of competition[a]			
International long distance service	—	—	
Mobile telephone service	—	—	
Internet service	—	—	
Sector efficiency and capacity			
Telecommunications revenue (% of GDP)	4.4	3.5	3.1
Mobile and fixed-line subscribers per employee	243	813	747
Telecommunications investment (% of revenue)	12.0	12.6	14.3
Sector performance			
Access			
Telephone lines (per 100 people)	58.9	59.6	50.0
Mobile cellular subscriptions (per 100 people)	81.7	155.2	100.4
Internet subscribers (per 100 people)	40.0	41.4	25.8
Personal computers (per 100 people)	40.2	68.6	67.4
Households with a television set (%)	98	100	98
Usage			
International voice traffic (minutes/person/month)[b]	62.8	115.6	14.0
Mobile telephone usage (minutes/user/month)	353	491	353
Internet users (per 100 people)	27.8	57.2	65.7
Quality			
Population covered by mobile cellular network (%)	100	100	99
Fixed broadband subscribers (% of total Internet subscrib.)	16.7	66.3	82.6
International Internet bandwidth (bits/second/person)	627	15,892	18,242
Affordability			
Price basket for residential fixed line (US$/month)	12.6	8.4	26.1
Price basket for mobile service (US$/month)	—	2.6	13.0
Price basket for Internet service (US$/month)	—	25.4	22.8
Price of call to United States (US$ for 3 minutes)	2.62	0.77	0.81
Trade			
ICT goods exports (% of total goods exports)	27.3	42.1	15.2
ICT goods imports (% of total goods imports)	30.1	41.8	14.6
ICT service exports (% of total service exports)	1.0	1.6	7.0
Applications			
ICT expenditure (% of GDP)	—	4.7	6.7
E-government Web measure index[c]	—	—	0.60
Secure Internet servers (per 1 million people, Dec. 2008)	80.1	287.5	662.6

Sources: Economic and social context: UIS and World Bank; Sector structure: ITU; Sector efficiency and capacity: ITU and World Bank; Sector performance: Global Insight/WITSA, IMF, ITU, Netcraft, UN Comtrade, UNDESA, UNPAN, Wireless Intelligence and World Bank. Produced by the Global Information and Communication Technologies Department and the Development Economics Data Group. For complete information, see Definitions and Data Sources.

Notes: Use of italics in the column entries indicates years other than those specified. — Not available. GDP = gross domestic product; GNI = gross national income; ICT = information and communication technology; and MDG = Millennium Development Goal.

a. C = competition; M = monopoly; and P = partial competition. **b.** Outgoing and incoming. **c.** Scale of 0–1, where 1 = highest presence. **d.** Millennium Development Goal indicators 8.14, 8.15, and 8.16.

Information and Communications for Development 2009

Hungary

	Hungary 2000	Hungary 2007	High-income group 2007
Economic and social context			
Population (total, million)	10	10	1,056
Urban population (% of total)	65	67	78
GNI per capita, World Bank Atlas method (current US$)	4,660	11,680	37,572
GDP growth, 1995–2000 and 2000–07 (avg. annual %)	4.2	4.0	2.4
Adult literacy rate (% of ages 15 and older)	—	99	99
Gross primary, secondary, tertiary school enrollment (%)	80	*89*	92
Sector structure			
Separate telecommunications regulator	Yes	Yes	
Status of main fixed-line telephone operator	*Private*	*Private*	
Level of competition[a]			
International long distance service	M	C	
Mobile telephone service	P	P	
Internet service	C	C	
Sector efficiency and capacity			
Telecommunications revenue (% of GDP)	6.7	4.2	*3.1*
Mobile and fixed-line subscribers per employee	330	1,009	747
Telecommunications investment (% of revenue)	16.9	8.5	*14.3*
Sector performance			
Access			
Telephone lines (per 100 people)	37.2	32.3	50.0
Mobile cellular subscriptions (per 100 people)	30.1	109.7	100.4
Internet subscribers (per 100 people)	2.2	14.8	25.8
Personal computers (per 100 people)	8.5	25.6	*67.4*
Households with a television set (%)	100	*101*	98
Usage			
International voice traffic (minutes/person/month)[b]	5.5	10.0	*14.0*
Mobile telephone usage (minutes/user/month)	181	164	353
Internet users (per 100 people)	7.0	51.9	65.7
Quality			
Population covered by mobile cellular network (%)	95	99	*99*
Fixed broadband subscribers (% of total Internet subscrib.)	1.5	95.8	82.6
International Internet bandwidth (bits/second/person)	100	4,773	18,242
Affordability			
Price basket for residential fixed line (US$/month)	14.4	*23.6*	26.1
Price basket for mobile service (US$/month)	—	*12.1*	13.0
Price basket for Internet service (US$/month)	—	*10.5*	22.8
Price of call to United States (US$ for 3 minutes)	1.28	1.01	*0.81*
Trade			
ICT goods exports (% of total goods exports)	27.7	26.1	15.2
ICT goods imports (% of total goods imports)	23.7	19.9	14.6
ICT service exports (% of total service exports)	3.2	6.8	7.0
Applications			
ICT expenditure (% of GDP)	—	5.9	6.7
E-government Web measure index[c]	—	0.62	0.60
Secure Internet servers (per 1 million people, Dec. 2008)	12.5	83.5	662.6

Sources: Economic and social context: UIS and World Bank; Sector structure: ITU; Sector efficiency and capacity: ITU and World Bank; Sector performance: Global Insight/WITSA, IMF, ITU, Netcraft, UN Comtrade, UNDESA, UNPAN, Wireless Intelligence and World Bank. Produced by the Global Information and Communication Technologies Department and the Development Economics Data Group. For complete information, see Definitions and Data Sources.

Notes: Use of italics in the column entries indicates years other than those specified. — Not available. GDP = gross domestic product; GNI = gross national income; ICT = information and communication technology; and MDG = Millennium Development Goal.

a. C = competition; M = monopoly; and P = partial competition. b. Outgoing and incoming. c. Scale of 0–1, where 1 = highest presence. d. Millennium Development Goal indicators 8.14, 8.15, and 8.16.

World Bank • ICT at a Glance

India

	India 2000	India 2007	Lower-middle-income group 2007	South Asia Region 2007
Economic and social context				
Population (total, million)	1,016	1,125	3,435	1,522
Urban population (% of total)	28	29	42	29
GNI per capita, World Bank Atlas method (current US$)	450	950	1,905	880
GDP growth, 1995–2000 and 2000–07 (avg. annual %)	5.9	7.8	8.0	7.3
Adult literacy rate (% of ages 15 and older)	*61*	*66*	83	63
Gross primary, secondary, tertiary school enrollment (%)	55	*64*	68	60
Sector structure				
Separate telecommunications regulator	Yes	Yes		
Status of main fixed-line telephone operator	Public	Mixed		
Level of competition[a]				
International long distance service	M	C		
Mobile telephone service	P	C		
Internet service	C	C		
Sector efficiency and capacity				
Telecommunications revenue (% of GDP)	1.5	*2.0*	3.1	2.1
Mobile and fixed-line subscribers per employee	85	—	624	660
Telecommunications investment (% of revenue)	49.3	—	25.3	—
Sector performance				
Access				
Telephone lines (per 100 people)	3.2	3.5	15.3	3.2
Mobile cellular subscriptions (per 100 people)	0.4	20.8	38.9	22.8
Internet subscribers (per 100 people)	0.3	1.2	6.0	1.3
Personal computers (per 100 people)	0.5	3.3	*4.6*	3.3
Households with a television set (%)	30	*53*	79	42
Usage				
International voice traffic (minutes/person/month)[b]	0.2	—	—	—
Mobile telephone usage (minutes/user/month)	*191*	447	322	364
Internet users (per 100 people)	0.5	7.2	12.4	6.6
Quality				
Population covered by mobile cellular network (%)	*21*	*61*	80	61
Fixed broadband subscribers (% of total Internet subscrib.)	0.0	23.2	40.4	18.9
International Internet bandwidth (bits/second/person)	1	32	199	31
Affordability				
Price basket for residential fixed line (US$/month)	6.0	*3.3*	7.2	*4.0*
Price basket for mobile service (US$/month)	—	2.5	9.8	2.4
Price basket for Internet service (US$/month)	—	6.6	16.7	8.0
Price of call to United States (US$ for 3 minutes)	3.36	*1.19*	2.08	2.02
Trade				
ICT goods exports (% of total goods exports)	1.6	1.3	20.6	1.2
ICT goods imports (% of total goods imports)	6.4	8.3	20.2	8.1
ICT service exports (% of total service exports)	31.9	*41.6*	15.6	*39.0*
Applications				
ICT expenditure (% of GDP)	—	5.6	6.5	5.7
E-government Web measure index[c]	—	0.48	0.33	0.37
Secure Internet servers (per 1 million people, Dec. 2008)	*0.1*	1.3	1.8	1.1

Sources: Economic and social context: UIS and World Bank; Sector structure: ITU; Sector efficiency and capacity: ITU and World Bank; Sector performance: Global Insight/WITSA, IMF, ITU, Netcraft, UN Comtrade, UNDESA, UNPAN, Wireless Intelligence and World Bank. Produced by the Global Information and Communication Technologies Department and the Development Economics Data Group. For complete information, see Definitions and Data Sources.

Notes: Use of italics in the column entries indicates years other than those specified. — Not available. GDP = gross domestic product; GNI = gross national income; ICT = information and communication technology; and MDG = Millennium Development Goal.

a. C = competition; M = monopoly; and P = partial competition. b. Outgoing and incoming. c. Scale of 0–1, where 1 = highest presence. d. Millennium Development Goal indicators 8.14, 8.15, and 8.16.

World Bank • ICT at a Glance

Indonesia

	Indonesia 2000	Indonesia 2007	Lower-middle-income group 2007	East Asia & Pacific Region 2007
Economic and social context				
Population (total, million)	206	226	3,435	1,912
Urban population (% of total)	42	50	42	43
GNI per capita, World Bank Atlas method (current US$)	590	1,650	1,905	2,182
GDP growth, 1995–2000 and 2000–07 (avg. annual %)	−0.6	5.1	8.0	9.0
Adult literacy rate (% of ages 15 and older)	—	92	83	93
Gross primary, secondary, tertiary school enrollment (%)	63	68	68	69
Sector structure				
Separate telecommunications regulator	No	Yes		
Status of main fixed-line telephone operator	Mixed	Mixed		
Level of competition[a]				
International long distance service	P	P		
Mobile telephone service	C	C		
Internet service	C	C		
Sector efficiency and capacity				
Telecommunications revenue (% of GDP)	1.4	2.2	3.1	3.0
Mobile and fixed-line subscribers per employee	259	1,095	624	546
Telecommunications investment (% of revenue)	11.4	28.3	25.3	—
Sector performance				
Access				
Telephone lines (per 100 people)	3.2	7.9	15.3	23.1
Mobile cellular subscriptions (per 100 people)	1.8	36.3	38.9	43.7
Internet subscribers (per 100 people)	0.2	1.4	6.0	9.3
Personal computers (per 100 people)	1.0	2.0	4.6	5.6
Households with a television set (%)	54	65	79	53
Usage				
International voice traffic (minutes/person/month)[b]	0.3	0.4	—	0.8
Mobile telephone usage (minutes/user/month)	—	66	322	333
Internet users (per 100 people)	0.9	5.8	12.4	14.6
Quality				
Population covered by mobile cellular network (%)	89	90	80	93
Fixed broadband subscribers (% of total Internet subscrib.)	1.0	8.2	40.4	41.8
International Internet bandwidth (bits/second/person)	1	53	199	247
Affordability				
Price basket for residential fixed line (US$/month)	3.5	5.1	7.2	5.8
Price basket for mobile service (US$/month)	—	7.2	9.8	5.0
Price basket for Internet service (US$/month)	—	21.9	16.7	14.4
Price of call to United States (US$ for 3 minutes)	3.90	2.79	2.08	1.16
Trade				
ICT goods exports (% of total goods exports)	12.6	5.3	20.6	30.9
ICT goods imports (% of total goods imports)	3.0	5.4	20.2	28.1
ICT service exports (% of total service exports)	—	11.9	15.6	5.2
Applications				
ICT expenditure (% of GDP)	—	3.9	6.5	7.3
E-government Web measure index[c]	—	0.33	0.33	0.18
Secure Internet servers (per 1 million people, Dec. 2008)	0.3	1.0	1.8	1.9

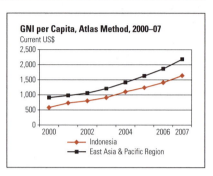

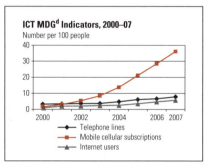

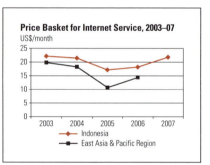

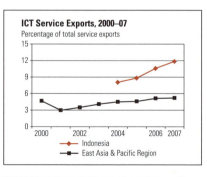

Sources: Economic and social context: UIS and World Bank; Sector structure: ITU; Sector efficiency and capacity: ITU and World Bank; Sector performance: Global Insight/WITSA, IMF, ITU, Netcraft, UN Comtrade, UNDESA, UNPAN, Wireless Intelligence and World Bank. Produced by the Global Information and Communication Technologies Department and the Development Economics Data Group. For complete information, see Definitions and Data Sources.

Notes: Use of italics in the column entries indicates years other than those specified. — Not available. GDP = gross domestic product; GNI = gross national income; ICT = information and communication technology; and MDG = Millennium Development Goal.

a. C = competition; M = monopoly; and P = partial competition. **b.** Outgoing and incoming. **c.** Scale of 0–1, where 1 = highest presence. **d.** Millennium Development Goal indicators 8.14, 8.15, and 8.16.

World Bank • ICT at a Glance

Iran, Islamic Republic of

	Iran, Islamic Rep. of 2000	Iran, Islamic Rep. of 2007	Lower-middle-income group 2007	Middle East & North Africa Region 2007
Economic and social context				
Population (total, million)	64	71	3,435	313
Urban population (% of total)	64	68	42	57
GNI per capita, World Bank Atlas method (current US$)	1,670	3,540	1,905	2,820
GDP growth, 1995–2000 and 2000–07 (avg. annual %)	3.7	5.9	8.0	4.4
Adult literacy rate (% of ages 15 and older)	77	82	83	73
Gross primary, secondary, tertiary school enrollment (%)	69	73	68	70
Sector structure				
Separate telecommunications regulator	—	Yes		
Status of main fixed-line telephone operator	Public	Public		
Level of competition[a]				
International long distance service	M	P		
Mobile telephone service	M	P		
Internet service	M	P		
Sector efficiency and capacity				
Telecommunications revenue (% of GDP)	1.1	1.4	3.1	3.1
Mobile and fixed-line subscribers per employee	221	913	624	691
Telecommunications investment (% of revenue)	6.0	74.5	25.3	21.7
Sector performance				
Access				
Telephone lines (per 100 people)	14.8	33.6	15.3	17.0
Mobile cellular subscriptions (per 100 people)	1.5	41.9	38.9	50.7
Internet subscribers (per 100 people)	0.4	—	6.0	2.4
Personal computers (per 100 people)	6.3	10.6	4.6	6.3
Households with a television set (%)	68	—	79	94
Usage				
International voice traffic (minutes/person/month)[b]	0.6	0.7	—	2.7
Mobile telephone usage (minutes/user/month)	—	—	322	—
Internet users (per 100 people)	1.0	32.4	12.4	17.1
Quality				
Population covered by mobile cellular network (%)	32	95	80	93
Fixed broadband subscribers (% of total Internet subscrib.)	0.1	—	40.4	—
International Internet bandwidth (bits/second/person)	1	153	199	186
Affordability				
Price basket for residential fixed line (US$/month)	10.8	2.1	7.2	3.9
Price basket for mobile service (US$/month)	—	2.7	9.8	6.5
Price basket for Internet service (US$/month)	—	2.0	16.7	11.6
Price of call to United States (US$ for 3 minutes)	7.65	0.55	2.08	1.45
Trade				
ICT goods exports (% of total goods exports)	0.0	0.1	20.6	—
ICT goods imports (% of total goods imports)	5.5	1.9	20.2	—
ICT service exports (% of total service exports)	—	—	15.6	2.6
Applications				
ICT expenditure (% of GDP)	—	3.5	6.5	4.5
E-government Web measure index[c]	—	0.26	0.33	0.22
Secure Internet servers (per 1 million people, Dec. 2008)	0.0	0.3	1.8	1.3

Sources: Economic and social context: UIS and World Bank; Sector structure: ITU; Sector efficiency and capacity: ITU and World Bank; Sector performance: Global Insight/WITSA, IMF, ITU, Netcraft, UN Comtrade, UNDESA, UNPAN, Wireless Intelligence and World Bank. Produced by the Global Information and Communication Technologies Department and the Development Economics Data Group. For complete information, see Definitions and Data Sources.

Notes: Use of italics in the column entries indicates years other than those specified. — Not available. GDP = gross domestic product; GNI = gross national income; ICT = information and communication technology; and MDG = Millennium Development Goal.

a. C = competition; M = monopoly; and P = partial competition. b. Outgoing and incoming. c. Scale of 0–1, where 1 = highest presence. d. Millennium Development Goal indicators 8.14, 8.15, and 8.16.

Iraq

	Iraq 2000	Iraq 2007	Lower-middle-income group 2007	Middle East & North Africa Region 2007
Economic and social context				
Population (total, million)	24	—	3,435	313
Urban population (% of total)	68	67	42	57
GNI per capita, World Bank Atlas method (current US$)	—	—	1,905	2,820
GDP growth, 1995–2000 and 2000–07 (avg. annual %)	17.9	−11.4	8.0	4.4
Adult literacy rate (% of ages 15 and older)	74	—	83	73
Gross primary, secondary, tertiary school enrollment (%)	53	60	68	70
Sector structure				
Separate telecommunications regulator	—	No		
Status of main fixed-line telephone operator	—	Public		
Level of competition[a]				
International long distance service	M	M		
Mobile telephone service	—	—		
Internet service	—	—		
Sector efficiency and capacity				
Telecommunications revenue (% of GDP)	—	—	3.1	3.1
Mobile and fixed-line subscribers per employee	—	941	624	691
Telecommunications investment (% of revenue)	—	—	25.3	21.7
Sector performance				
Access				
Telephone lines (per 100 people)	2.8	—	15.3	17.0
Mobile cellular subscriptions (per 100 people)	0.0	—	38.9	50.7
Internet subscribers (per 100 people)	—	—	6.0	2.4
Personal computers (per 100 people)	—	—	4.6	6.3
Households with a television set (%)	—	—	79	94
Usage				
International voice traffic (minutes/person/month)[b]	—	—	—	2.7
Mobile telephone usage (minutes/user/month)	—	255	322	—
Internet users (per 100 people)	—	—	12.4	17.1
Quality				
Population covered by mobile cellular network (%)	—	72	80	93
Fixed broadband subscribers (% of total Internet subscrib.)	0.0	—	40.4	—
International Internet bandwidth (bits/second/person)	—	—	199	186
Affordability				
Price basket for residential fixed line (US$/month)	—	—	7.2	3.9
Price basket for mobile service (US$/month)	—	2.6	9.8	6.5
Price basket for Internet service (US$/month)	—	—	16.7	11.6
Price of call to United States (US$ for 3 minutes)	—	—	2.08	1.45
Trade				
ICT goods exports (% of total goods exports)	—	—	20.6	—
ICT goods imports (% of total goods imports)	—	—	20.2	—
ICT service exports (% of total service exports)	—	1.9	15.6	2.6
Applications				
ICT expenditure (% of GDP)	—	—	6.5	4.5
E-government Web measure index[c]	—	0.11	0.33	0.22
Secure Internet servers (per 1 million people, Dec. 2008)	—	—	1.8	1.3

Sources: Economic and social context: UIS and World Bank; Sector structure: ITU; Sector efficiency and capacity: ITU and World Bank; Sector performance: Global Insight/WITSA, IMF, ITU, Netcraft, UN Comtrade, UNDESA, UNPAN, Wireless Intelligence and World Bank. Produced by the Global Information and Communication Technologies Department and the Development Economics Data Group. For complete information, see Definitions and Data Sources.

Notes: Use of italics in the column entries indicates years other than those specified. — Not available. GDP = gross domestic product; GNI = gross national income; ICT = information and communication technology; and MDG = Millennium Development Goal.

a. C = competition; M = monopoly; and P = partial competition. **b.** Outgoing and incoming. **c.** Scale of 0–1, where 1 = highest presence. **d.** Millennium Development Goal indicators 8.14, 8.15, and 8.16.

Ireland

	Ireland 2000	Ireland 2007	High-income group 2007
Economic and social context			
Population (total, million)	4	4	1,056
Urban population (% of total)	59	61	78
GNI per capita, World Bank Atlas method (current US$)	23,160	47,610	37,572
GDP growth, 1995–2000 and 2000–07 (avg. annual %)	9.5	5.5	2.4
Adult literacy rate (% of ages 15 and older)	—	—	99
Gross primary, secondary, tertiary school enrollment (%)	91	*100*	92
Sector structure			
Separate telecommunications regulator	No	Yes	
Status of main fixed-line telephone operator	Private	Private	
Level of competition[a]			
International long distance service	C	C	
Mobile telephone service	P	C	
Internet service	C	C	
Sector efficiency and capacity			
Telecommunications revenue (% of GDP)	2.7	2.4	*3.1*
Mobile and fixed-line subscribers per employee	246	*406*	747
Telecommunications investment (% of revenue)	14.3	7.1	14.3
Sector performance			
Access			
Telephone lines (per 100 people)	48.1	48.4	50.0
Mobile cellular subscriptions (per 100 people)	64.7	114.1	100.4
Internet subscribers (per 100 people)	14.5	25.1	25.8
Personal computers (per 100 people)	35.7	*58.2*	67.4
Households with a television set (%)	94	*119*	98
Usage			
International voice traffic (minutes/person/month)[b]	59.1	—	14.0
Mobile telephone usage (minutes/user/month)	—	252	353
Internet users (per 100 people)	17.8	56.1	65.7
Quality			
Population covered by mobile cellular network (%)	98	99	*99*
Fixed broadband subscribers (% of total Internet subscrib.)	0.0	73.6	82.6
International Internet bandwidth (bits/second/person)	241	15,229	18,242
Affordability			
Price basket for residential fixed line (US$/month)	*28.6*	41.5	26.1
Price basket for mobile service (US$/month)	—	*19.3*	13.0
Price basket for Internet service (US$/month)	—	*31.4*	22.8
Price of call to United States (US$ for 3 minutes)	0.80	*0.71*	*0.81*
Trade			
ICT goods exports (% of total goods exports)	34.5	*22.4*	15.2
ICT goods imports (% of total goods imports)	34.0	*24.1*	14.6
ICT service exports (% of total service exports)	45.5	30.1	7.0
Applications			
ICT expenditure (% of GDP)	—	5.9	6.7
E-government Web measure index[c]	—	0.68	0.60
Secure Internet servers (per 1 million people, Dec. 2008)	*90.5*	672.9	662.6

Sources: Economic and social context: UIS and World Bank; Sector structure: ITU; Sector efficiency and capacity: ITU and World Bank; Sector performance: Global Insight/WITSA, IMF, ITU, Netcraft, UN Comtrade, UNDESA, UNPAN, Wireless Intelligence and World Bank. Produced by the Global Information and Communication Technologies Department and the Development Economics Data Group. For complete information, see Definitions and Data Sources.

Notes: Use of italics in the column entries indicates years other than those specified. — Not available. GDP = gross domestic product; GNI = gross national income; ICT = information and communication technology; and MDG = Millennium Development Goal.

a. C = competition; M = monopoly; and P = partial competition. b. Outgoing and incoming. c. Scale of 0–1, where 1 = highest presence. d. Millennium Development Goal indicators 8.14, 8.15, and 8.16.

World Bank • ICT at a Glance

Israel

	Israel 2000	Israel 2007	High-income group 2007
Economic and social context			
Population (total, million)	6	7	1,056
Urban population (% of total)	91	92	78
GNI per capita, World Bank Atlas method (current US$)	17,850	22,170	37,572
GDP growth, 1995–2000 and 2000–07 (avg. annual %)	4.5	3.2	2.4
Adult literacy rate (% of ages 15 and older)	—	—	99
Gross primary, secondary, tertiary school enrollment (%)	88	90	92
Sector structure			
Separate telecommunications regulator	—	No	
Status of main fixed-line telephone operator	Mixed	Mixed	
Level of competition[a]			
International long distance service	C	C	
Mobile telephone service	C	C	
Internet service	C	—	
Sector efficiency and capacity			
Telecommunications revenue (% of GDP)	3.0	4.1	3.1
Mobile and fixed-line subscribers per employee	618	692	747
Telecommunications investment (% of revenue)	16.1	—	14.3
Sector performance			
Access			
Telephone lines (per 100 people)	47.3	42.6	50.0
Mobile cellular subscriptions (per 100 people)	70.0	124.0	100.4
Internet subscribers (per 100 people)	12.8	26.8	25.8
Personal computers (per 100 people)	25.3	24.2	67.4
Households with a television set (%)	92	92	98
Usage			
International voice traffic (minutes/person/month)[b]	22.3	30.3	14.0
Mobile telephone usage (minutes/user/month)	361	351	353
Internet users (per 100 people)	20.2	27.9	65.7
Quality			
Population covered by mobile cellular network (%)	97	100	99
Fixed broadband subscribers (% of total Internet subscrib.)	0.0	75.2	82.6
International Internet bandwidth (bits/second/person)	53	2,003	18,242
Affordability			
Price basket for residential fixed line (US$/month)	11.6	15.1	26.1
Price basket for mobile service (US$/month)	—	10.3	13.0
Price basket for Internet service (US$/month)	—	24.2	22.8
Price of call to United States (US$ for 3 minutes)	—	0.59	0.81
Trade			
ICT goods exports (% of total goods exports)	25.2	10.9	15.2
ICT goods imports (% of total goods imports)	16.4	11.4	14.6
ICT service exports (% of total service exports)	28.7	28.5	7.0
Applications			
ICT expenditure (% of GDP)	—	6.5	6.7
E-government Web measure index[c]	—	0.67	0.60
Secure Internet servers (per 1 million people, Dec. 2008)	46.7	272.7	662.6

Sources: Economic and social context: UIS and World Bank; Sector structure: ITU; Sector efficiency and capacity: ITU and World Bank; Sector performance: Global Insight/WITSA, IMF, ITU, Netcraft, UN Comtrade, UNDESA, UNPAN, Wireless Intelligence and World Bank. Produced by the Global Information and Communication Technologies Department and the Development Economics Data Group. For complete information, see Definitions and Data Sources.

Notes: Use of italics in the column entries indicates years other than those specified. — Not available. GDP = gross domestic product; GNI = gross national income; ICT = information and communication technology; and MDG = Millennium Development Goal.

a. C = competition; M = monopoly; and P = partial competition. b. Outgoing and incoming. c. Scale of 0–1, where 1 = highest presence. d. Millennium Development Goal indicators 8.14, 8.15, and 8.16.

Italy

	Italy 2000	Italy 2007	High-income group 2007
Economic and social context			
Population (total, million)	57	59	1,056
Urban population (% of total)	67	68	78
GNI per capita, World Bank Atlas method (current US$)	20,890	33,490	37,572
GDP growth, 1995–2000 and 2000–07 (avg. annual %)	1.8	1.0	2.4
Adult literacy rate (% of ages 15 and older)	98	99	99
Gross primary, secondary, tertiary school enrollment (%)	81	91	92
Sector structure			
Separate telecommunications regulator	Yes	Yes	
Status of main fixed-line telephone operator	Mixed	Private	
Level of competition[a]			
International long distance service	C	C	
Mobile telephone service	C	C	
Internet service	C	C	
Sector efficiency and capacity			
Telecommunications revenue (% of GDP)	2.2	3.2	3.1
Mobile and fixed-line subscribers per employee	915	1,228	747
Telecommunications investment (% of revenue)	26.7	14.5	14.3
Sector performance			
Access			
Telephone lines (per 100 people)	47.7	45.6	50.0
Mobile cellular subscriptions (per 100 people)	74.2	151.8	100.4
Internet subscribers (per 100 people)	10.2	30.2	25.8
Personal computers (per 100 people)	18.1	36.7	67.4
Households with a television set (%)	98	98	98
Usage			
International voice traffic (minutes/person/month)[b]	13.2	19.7	14.0
Mobile telephone usage (minutes/user/month)	116	116	353
Internet users (per 100 people)	23.2	53.9	65.7
Quality			
Population covered by mobile cellular network (%)	100	100	99
Fixed broadband subscribers (% of total Internet subscrib.)	2.0	38.5	82.6
International Internet bandwidth (bits/second/person)	168	10,302	18,242
Affordability			
Price basket for residential fixed line (US$/month)	18.9	24.9	26.1
Price basket for mobile service (US$/month)	—	14.1	13.0
Price basket for Internet service (US$/month)	—	25.0	22.8
Price of call to United States (US$ for 3 minutes)	0.81	0.79	0.81
Trade			
ICT goods exports (% of total goods exports)	5.3	3.7	15.2
ICT goods imports (% of total goods imports)	9.9	7.0	14.6
ICT service exports (% of total service exports)	3.0	3.5	7.0
Applications			
ICT expenditure (% of GDP)	—	5.8	6.7
E-government Web measure index[c]	—	0.51	0.60
Secure Internet servers (per 1 million people, Dec. 2008)	18.3	92.8	662.6

Sources: Economic and social context: UIS and World Bank; Sector structure: ITU; Sector efficiency and capacity: ITU and World Bank; Sector performance: Global Insight/WITSA, IMF, ITU, Netcraft, UN Comtrade, UNDESA, UNPAN, Wireless Intelligence and World Bank. Produced by the Global Information and Communication Technologies Department and the Development Economics Data Group. For complete information, see Definitions and Data Sources.

Notes: Use of italics in the column entries indicates years other than those specified. — Not available. GDP = gross domestic product; GNI = gross national income; ICT = information and communication technology; and MDG = Millennium Development Goal.

a. C = competition; M = monopoly; and P = partial competition. **b.** Outgoing and incoming. **c.** Scale of 0–1, where 1 = highest presence. **d.** Millennium Development Goal indicators 8.14, 8.15, and 8.16.

World Bank • ICT at a Glance

Jamaica

	Jamaica 2000	Jamaica 2007	Upper-middle-income group 2007	Latin America & the Caribbean Region 2007
Economic and social context				
Population (total, million)	3	3	824	561
Urban population (% of total)	52	53	75	78
GNI per capita, World Bank Atlas method (current US$)	2,930	3,330	7,107	5,801
GDP growth, 1995–2000 and 2000–07 (avg. annual %)	−0.1	1.0	4.3	3.6
Adult literacy rate (% of ages 15 and older)	*80*	*86*	94	91
Gross primary, secondary, tertiary school enrollment (%)	74	*78*	82	81
Sector structure				
Separate telecommunications regulator	Yes	Yes		
Status of main fixed-line telephone operator	*Private*	*Private*		
Level of competition[a]				
International long distance service	M	C		
Mobile telephone service	C	P		
Internet service	C	C		
Sector efficiency and capacity				
Telecommunications revenue (% of GDP)	5.9	*4.9*	3.3	3.8
Mobile and fixed-line subscribers per employee	268	*678*	*566*	*530*
Telecommunications investment (% of revenue)	32.5	*36.7*	—	—
Sector performance				
Access				
Telephone lines (per 100 people)	19.1	13.6	22.6	18.1
Mobile cellular subscriptions (per 100 people)	14.2	100.0	84.1	67.0
Internet subscribers (per 100 people)	—	*3.2*	9.4	4.5
Personal computers (per 100 people)	4.6	*6.8*	12.4	11.3
Households with a television set (%)	71	*70*	92	84
Usage				
International voice traffic (minutes/person/month)[b]	12.9	19.5	—	—
Mobile telephone usage (minutes/user/month)	—	—	137	116
Internet users (per 100 people)	3.1	56.1	26.6	26.9
Quality				
Population covered by mobile cellular network (%)	*80*	*95*	95	91
Fixed broadband subscribers (% of total Internet subscrib.)	—	*92.9*	47.8	81.7
International Internet bandwidth (bits/second/person)	28	*19,151*	1,185	1,126
Affordability				
Price basket for residential fixed line (US$/month)	5.0	*9.1*	10.6	9.5
Price basket for mobile service (US$/month)	—	*7.5*	10.9	10.4
Price basket for Internet service (US$/month)	—	*34.3*	16.4	25.7
Price of call to United States (US$ for 3 minutes)	—	*0.87*	1.55	1.21
Trade				
ICT goods exports (% of total goods exports)	0.3	0.2	13.5	11.4
ICT goods imports (% of total goods imports)	5.7	3.6	16.2	15.9
ICT service exports (% of total service exports)	12.3	6.8	4.6	4.7
Applications				
ICT expenditure (% of GDP)	—	6.6	5.2	4.9
E-government Web measure index[c]	—	0.32	0.37	0.44
Secure Internet servers (per 1 million people, Dec. 2008)	*1.9*	31.6	26.2	18.2

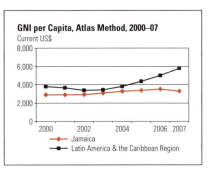

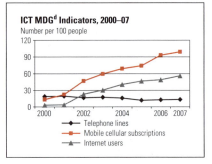

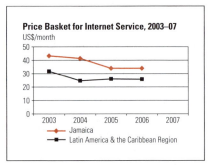

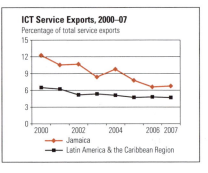

Sources: Economic and social context: UIS and World Bank; Sector structure: ITU; Sector efficiency and capacity: ITU and World Bank; Sector performance: Global Insight/WITSA, IMF, ITU, Netcraft, UN Comtrade, UNDESA, UNPAN, Wireless Intelligence and World Bank. Produced by the Global Information and Communication Technologies Department and the Development Economics Data Group. For complete information, see Definitions and Data Sources.

Notes: Use of italics in the column entries indicates years other than those specified. — Not available. GDP = gross domestic product; GNI = gross national income; ICT = information and communication technology; and MDG = Millennium Development Goal.

a. C = competition; M = monopoly; and P = partial competition. **b.** Outgoing and incoming. **c.** Scale of 0–1, where 1 = highest presence. **d.** Millennium Development Goal indicators 8.14, 8.15, and 8.16.

Japan

	Japan 2000	Japan 2007	High-income group 2007
Economic and social context			
Population (total, million)	127	128	1,056
Urban population (% of total)	65	66	78
GNI per capita, World Bank Atlas method (current US$)	34,620	37,790	37,572
GDP growth, 1995–2000 and 2000–07 (avg. annual %)	0.6	1.7	2.4
Adult literacy rate (% of ages 15 and older)	—	—	99
Gross primary, secondary, tertiary school enrollment (%)	83	*86*	92
Sector structure			
Separate telecommunications regulator	—	No	
Status of main fixed-line telephone operator	Mixed	Private	
Level of competition[a]			
International long distance service	C	C	
Mobile telephone service	C	C	
Internet service	C	C	
Sector efficiency and capacity			
Telecommunications revenue (% of GDP)	2.6	3.1	*3.1*
Mobile and fixed-line subscribers per employee	830	1,334	747
Telecommunications investment (% of revenue)	26.8	13.8	*14.3*
Sector performance			
Access			
Telephone lines (per 100 people)	48.8	40.1	50.0
Mobile cellular subscriptions (per 100 people)	52.6	84.0	100.4
Internet subscribers (per 100 people)	14.3	—	25.8
Personal computers (per 100 people)	31.5	40.7	67.4
Households with a television set (%)	79	99	98
Usage			
International voice traffic (minutes/person/month)[b]	2.5	*3.9*	14.0
Mobile telephone usage (minutes/user/month)	*184*	138	353
Internet users (per 100 people)	30.0	69.0	65.7
Quality			
Population covered by mobile cellular network (%)	99	100	*99*
Fixed broadband subscribers (% of total Internet subscrib.)	4.7	—	82.6
International Internet bandwidth (bits/second/person)	57	3,734	18,242
Affordability			
Price basket for residential fixed line (US$/month)	—	26.1	26.1
Price basket for mobile service (US$/month)	—	29.6	13.0
Price basket for Internet service (US$/month)	—	28.6	22.8
Price of call to United States (US$ for 3 minutes)	1.67	*1.63*	0.81
Trade			
ICT goods exports (% of total goods exports)	25.8	*19.3*	15.2
ICT goods imports (% of total goods imports)	17.6	*13.7*	14.6
ICT service exports (% of total service exports)	3.5	1.2	7.0
Applications			
ICT expenditure (% of GDP)	—	7.2	6.7
E-government Web measure index[c]	—	0.74	0.60
Secure Internet servers (per 1 million people, Dec. 2008)	40.5	471.6	662.6

Sources: Economic and social context: UIS and World Bank; Sector structure: ITU; Sector efficiency and capacity: ITU and World Bank; Sector performance: Global Insight/WITSA, IMF, ITU, Netcraft, UN Comtrade, UNDESA, UNPAN, Wireless Intelligence and World Bank. Produced by the Global Information and Communication Technologies Department and the Development Economics Data Group. For complete information, see Definitions and Data Sources.

Notes: Use of italics in the column entries indicates years other than those specified. — Not available. GDP = gross domestic product; GNI = gross national income; ICT = information and communication technology; and MDG = Millennium Development Goal.

a. C = competition; M = monopoly; and P = partial competition. **b.** Outgoing and incoming. **c.** Scale of 0–1, where 1 = highest presence. **d.** Millennium Development Goal indicators 8.14, 8.15, and 8.16.

Jordan

	Jordan 2000	Jordan 2007	Lower-middle-income group 2007	Middle East & North Africa Region 2007
Economic and social context				
Population (total, million)	5	6	3,435	313
Urban population (% of total)	78	78	42	57
GNI per capita, World Bank Atlas method (current US$)	1,790	2,840	1,905	2,820
GDP growth, 1995–2000 and 2000–07 (avg. annual %)	3.2	6.3	8.0	4.4
Adult literacy rate (% of ages 15 and older)	—	91	83	73
Gross primary, secondary, tertiary school enrollment (%)	76	78	68	70
Sector structure				
Separate telecommunications regulator	Yes	Yes		
Status of main fixed-line telephone operator	Mixed	Private		
Level of competition[a]				
International long distance service	M	C		
Mobile telephone service	P	C		
Internet service	C	C		
Sector efficiency and capacity				
Telecommunications revenue (% of GDP)	6.7	8.3	3.1	3.1
Mobile and fixed-line subscribers per employee	158	1,026	624	691
Telecommunications investment (% of revenue)	38.5	12.4	25.3	21.7
Sector performance				
Access				
Telephone lines (per 100 people)	12.9	10.2	15.3	17.0
Mobile cellular subscriptions (per 100 people)	8.1	83.4	38.9	50.7
Internet subscribers (per 100 people)	0.7	3.9	6.0	2.4
Personal computers (per 100 people)	3.1	6.7	4.6	6.3
Households with a television set (%)	96	96	79	94
Usage				
International voice traffic (minutes/person/month)[b]	6.7	2.7	—	2.7
Mobile telephone usage (minutes/user/month)	—	—	322	—
Internet users (per 100 people)	2.7	19.7	12.4	17.1
Quality				
Population covered by mobile cellular network (%)	99	99	80	93
Fixed broadband subscribers (% of total Internet subscrib.)	0.6	38.2	40.4	—
International Internet bandwidth (bits/second/person)	18	164	199	186
Affordability				
Price basket for residential fixed line (US$/month)	8.8	10.0	7.2	3.9
Price basket for mobile service (US$/month)	—	6.9	9.8	6.5
Price basket for Internet service (US$/month)	—	11.1	16.7	11.6
Price of call to United States (US$ for 3 minutes)	2.86	1.44	2.08	1.45
Trade				
ICT goods exports (% of total goods exports)	3.7	4.8	20.6	—
ICT goods imports (% of total goods imports)	5.9	7.0	20.2	—
ICT service exports (% of total service exports)	—	0.0	15.6	2.6
Applications				
ICT expenditure (% of GDP)	—	9.3	6.5	4.5
E-government Web measure index[c]	—	0.61	0.33	0.22
Secure Internet servers (per 1 million people, Dec. 2008)	0.4	8.8	1.8	1.3

Sources: Economic and social context: UIS and World Bank; Sector structure: ITU; Sector efficiency and capacity: ITU and World Bank; Sector performance: Global Insight/WITSA, IMF, ITU, Netcraft, UN Comtrade, UNDESA, UNPAN, Wireless Intelligence and World Bank. Produced by the Global Information and Communication Technologies Department and the Development Economics Data Group. For complete information, see Definitions and Data Sources.

Notes: Use of italics in the column entries indicates years other than those specified. — Not available. GDP = gross domestic product; GNI = gross national income; ICT = information and communication technology; and MDG = Millennium Development Goal.

[a]. C = competition; M = monopoly; and P = partial competition. [b]. Outgoing and incoming. [c]. Scale of 0–1, where 1 = highest presence. [d]. Millennium Development Goal indicators 8.14, 8.15, and 8.16.

Kazakhstan

	Kazakhstan 2000	Kazakhstan 2007	Upper-middle-income group 2007	Europe & Central Asia Region 2007
Economic and social context				
Population (total, million)	15	15	824	446
Urban population (% of total)	56	58	75	64
GNI per capita, World Bank Atlas method (current US$)	1,270	5,020	7,107	6,052
GDP growth, 1995–2000 and 2000–07 (avg. annual %)	1.9	10.0	4.3	6.1
Adult literacy rate (% of ages 15 and older)	*100*	100	94	98
Gross primary, secondary, tertiary school enrollment (%)	79	94	82	82
Sector structure				
Separate telecommunications regulator	—	No		
Status of main fixed-line telephone operator	Mixed	Mixed		
Level of competition[a]				
International long distance service	C	C		
Mobile telephone service	P	P		
Internet service	—	—		
Sector efficiency and capacity				
Telecommunications revenue (% of GDP)	1.5	2.9	3.3	2.9
Mobile and fixed-line subscribers per employee	66	308	*566*	*532*
Telecommunications investment (% of revenue)	26.7	22.5	—	22.0
Sector performance				
Access				
Telephone lines (per 100 people)	12.3	20.9	22.6	25.7
Mobile cellular subscriptions (per 100 people)	1.3	79.6	84.1	95.0
Internet subscribers (per 100 people)	—	4.4	9.4	13.6
Personal computers (per 100 people)	—	—	*12.4*	*10.6*
Households with a television set (%)	92	—	92	96
Usage				
International voice traffic (minutes/person/month)[b]	1.6	3.9	—	—
Mobile telephone usage (minutes/user/month)	—	100	137	154
Internet users (per 100 people)	0.7	12.3	26.6	21.4
Quality				
Population covered by mobile cellular network (%)	94	81	95	*92*
Fixed broadband subscribers (% of total Internet subscrib.)	—	39.6	47.8	32.5
International Internet bandwidth (bits/second/person)	*1*	129	1,185	1,114
Affordability				
Price basket for residential fixed line (US$/month)	—	4.8	*10.6*	*5.8*
Price basket for mobile service (US$/month)	—	11.4	*10.9*	*11.8*
Price basket for Internet service (US$/month)	—	16.3	*16.4*	*12.0*
Price of call to United States (US$ for 3 minutes)	2.76	—	*1.55*	*1.63*
Trade				
ICT goods exports (% of total goods exports)	0.6	0.1	*13.5*	*1.8*
ICT goods imports (% of total goods imports)	6.9	5.2	*16.2*	*7.0*
ICT service exports (% of total service exports)	3.4	2.5	*4.6*	*5.0*
Applications				
ICT expenditure (% of GDP)	—	—	5.2	5.0
E-government Web measure index[c]	—	0.32	0.37	0.36
Secure Internet servers (per 1 million people, Dec. 2008)	*0.5*	2.0	26.2	23.9

Sources: Economic and social context: UIS and World Bank; Sector structure: ITU; Sector efficiency and capacity: ITU and World Bank; Sector performance: Global Insight/WITSA, IMF, ITU, Netcraft, UN Comtrade, UNDESA, UNPAN, Wireless Intelligence and World Bank. Produced by the Global Information and Communication Technologies Department and the Development Economics Data Group. For complete information, see Definitions and Data Sources.

Notes: Use of italics in the column entries indicates years other than those specified. — Not available. GDP = gross domestic product; GNI = gross national income; ICT = information and communication technology; and MDG = Millennium Development Goal.

a. C = competition; M = monopoly; and P = partial competition. **b.** Outgoing and incoming. **c.** Scale of 0–1, where 1 = highest presence. **d.** Millennium Development Goal indicators 8.14, 8.15, and 8.16.

World Bank • ICT at a Glance

Kenya

	Kenya 2000	Kenya 2007	Low-income group 2007	Sub-Saharan Africa Region 2007
Economic and social context				
Population (total, million)	31	38	1,296	800
Urban population (% of total)	20	21	32	36
GNI per capita, World Bank Atlas method (current US$)	420	640	574	951
GDP growth, 1995–2000 and 2000–07 (avg. annual %)	2.2	4.4	5.6	5.1
Adult literacy rate (% of ages 15 and older)	74	—	64	62
Gross primary, secondary, tertiary school enrollment (%)	53	*61*	*51*	*51*
Sector structure				
Separate telecommunications regulator	Yes	Yes		
Status of main fixed-line telephone operator	Public	Mixed		
Level of competition[a]				
International long distance service	M	C		
Mobile telephone service	P	P		
Internet service	C	C		
Sector efficiency and capacity				
Telecommunications revenue (% of GDP)	2.4	6.1	*3.3*	*4.7*
Mobile and fixed-line subscribers per employee	21	1,782	*301*	*499*
Telecommunications investment (% of revenue)	88.5	38.2	—	—
Sector performance				
Access				
Telephone lines (per 100 people)	0.9	0.7	4.0	1.6
Mobile cellular subscriptions (per 100 people)	0.4	30.2	21.5	23.0
Internet subscribers (per 100 people)	0.1	*0.5*	*0.8*	*1.2*
Personal computers (per 100 people)	0.5	*1.4*	*1.5*	*1.8*
Households with a television set (%)	15	39	*16*	*18*
Usage				
International voice traffic (minutes/person/month)[b]	0.2	0.3	—	—
Mobile telephone usage (minutes/user/month)	—	—	—	—
Internet users (per 100 people)	0.3	8.0	5.2	4.4
Quality				
Population covered by mobile cellular network (%)	—	77	*54*	*56*
Fixed broadband subscribers (% of total Internet subscrib.)	0.0	9.5	*3.4*	*3.1*
International Internet bandwidth (bits/second/person)	0	9	26	36
Affordability				
Price basket for residential fixed line (US$/month)	6.6	19.4	*5.7*	*12.6*
Price basket for mobile service (US$/month)	—	17.8	*11.2*	*11.6*
Price basket for Internet service (US$/month)	—	63.9	*29.2*	*43.1*
Price of call to United States (US$ for 3 minutes)	7.35	*3.00*	*2.00*	*2.43*
Trade				
ICT goods exports (% of total goods exports)	0.1	1.0	*1.4*	*1.1*
ICT goods imports (% of total goods imports)	5.3	5.6	*6.7*	*8.2*
ICT service exports (% of total service exports)	2.2	4.1	—	*4.2*
Applications				
ICT expenditure (% of GDP)	—	8.2	—	—
E-government Web measure index[c]	—	0.30	0.11	0.16
Secure Internet servers (per 1 million people, Dec. 2008)	0.0	1.1	0.5	2.9

Sources: Economic and social context: UIS and World Bank; Sector structure: ITU; Sector efficiency and capacity: ITU and World Bank; Sector performance: Global Insight/WITSA, IMF, ITU, Netcraft, UN Comtrade, UNDESA, UNPAN, Wireless Intelligence and World Bank. Produced by the Global Information and Communication Technologies Department and the Development Economics Data Group. For complete information, see Definitions and Data Sources.

Notes: Use of italics in the column entries indicates years other than those specified. — Not available. GDP = gross domestic product; GNI = gross national income; ICT = information and communication technology; and MDG = Millennium Development Goal.

a. C = competition; M = monopoly; and P = partial competition. b. Outgoing and incoming. c. Scale of 0–1, where 1 = highest presence. d. Millennium Development Goal indicators 8.14, 8.15, and 8.16.

Korea, Republic of

	Korea, Rep. of 2000	Korea, Rep. of 2007	High-income group 2007
Economic and social context			
Population (total, million)	47	48	1,056
Urban population (% of total)	80	81	78
GNI per capita, World Bank Atlas method (current US$)	9,800	19,730	37,572
GDP growth, 1995–2000 and 2000–07 (avg. annual %)	3.5	4.7	2.4
Adult literacy rate (% of ages 15 and older)	—	—	99
Gross primary, secondary, tertiary school enrollment (%)	90	*97*	92
Sector structure			
Separate telecommunications regulator	Yes	Yes	
Status of main fixed-line telephone operator	Mixed	Private	
Level of competition[a]			
International long distance service	C	C	
Mobile telephone service	C	C	
Internet service	C	C	
Sector efficiency and capacity			
Telecommunications revenue (% of GDP)	4.1	5.0	*3.1*
Mobile and fixed-line subscribers per employee	758	637	747
Telecommunications investment (% of revenue)	37.4	*16.2*	*14.3*
Sector performance			
Access			
Telephone lines (per 100 people)	55.0	46.2	50.0
Mobile cellular subscriptions (per 100 people)	57.0	89.8	100.4
Internet subscribers (per 100 people)	10.9	30.4	25.8
Personal computers (per 100 people)	39.6	57.6	*67.4*
Households with a television set (%)	96	*100*	98
Usage			
International voice traffic (minutes/person/month)[b]	3.2	2.4	*14.0*
Mobile telephone usage (minutes/user/month)	149	220	353
Internet users (per 100 people)	40.5	75.9	65.7
Quality			
Population covered by mobile cellular network (%)	99	90	*99*
Fixed broadband subscribers (% of total Internet subscrib.)	75.5	100.0	82.6
International Internet bandwidth (bits/second/person)	48	1,027	18,242
Affordability			
Price basket for residential fixed line (US$/month)	7.1	7.8	*26.1*
Price basket for mobile service (US$/month)	—	18.3	13.0
Price basket for Internet service (US$/month)	—	10.7	22.8
Price of call to United States (US$ for 3 minutes)	1.93	*0.76*	*0.81*
Trade			
ICT goods exports (% of total goods exports)	35.7	*27.2*	15.2
ICT goods imports (% of total goods imports)	24.4	*16.5*	*14.6*
ICT service exports (% of total service exports)	1.3	1.4	7.0
Applications			
ICT expenditure (% of GDP)	—	7.1	6.7
E-government Web measure index[c]	—	0.82	0.60
Secure Internet servers (per 1 million people, Dec. 2008)	7.3	695.7	662.6

Sources: Economic and social context: UIS and World Bank; Sector structure: ITU; Sector efficiency and capacity: ITU and World Bank; Sector performance: Global Insight/WITSA, IMF, ITU, Netcraft, UN Comtrade, UNDESA, UNPAN, Wireless Intelligence and World Bank. Produced by the Global Information and Communication Technologies Department and the Development Economics Data Group. For complete information, see Definitions and Data Sources.

Notes: Use of italics in the column entries indicates years other than those specified. — Not available. GDP = gross domestic product; GNI = gross national income; ICT = information and communication technology; and MDG = Millennium Development Goal.

a. C = competition; M = monopoly; and P = partial competition. **b.** Outgoing and incoming. **c.** Scale of 0–1, where 1 = highest presence. **d.** Millennium Development Goal indicators 8.14, 8.15, and 8.16.

Kuwait

	Kuwait 2000	Kuwait 2007	High-income group 2007
Economic and social context			
Population (total, million)	2	3	1,056
Urban population (% of total)	98	98	78
GNI per capita, World Bank Atlas method (current US$)	16,790	*38,420*	37,572
GDP growth, 1995–2000 and 2000–07 (avg. annual %)	1.8	9.2	2.4
Adult literacy rate (% of ages 15 and older)	—	94	99
Gross primary, secondary, tertiary school enrollment (%)	78	75	92
Sector structure			
Separate telecommunications regulator	—	No	
Status of main fixed-line telephone operator	*Public*	*Public*	
Level of competition[a]			
International long distance service	C	M	
Mobile telephone service	P	M	
Internet service	P	P	
Sector efficiency and capacity			
Telecommunications revenue (% of GDP)	1.7	3.5	3.1
Mobile and fixed-line subscribers per employee	123	372	747
Telecommunications investment (% of revenue)	19.7	27.0	14.3
Sector performance			
Access			
Telephone lines (per 100 people)	21.3	*19.9*	50.0
Mobile cellular subscriptions (per 100 people)	21.7	104.2	100.4
Internet subscribers (per 100 people)	—	11.2	25.8
Personal computers (per 100 people)	11.4	23.7	67.4
Households with a television set (%)	95	95	98
Usage			
International voice traffic (minutes/person/month)[b]	13.5	—	14.0
Mobile telephone usage (minutes/user/month)	—	—	353
Internet users (per 100 people)	6.8	33.8	65.7
Quality			
Population covered by mobile cellular network (%)	100	*100*	99
Fixed broadband subscribers (% of total Internet subscrib.)	—	8.8	82.6
International Internet bandwidth (bits/second/person)	26	871	18,242
Affordability			
Price basket for residential fixed line (US$/month)	9.9	*10.6*	26.1
Price basket for mobile service (US$/month)	—	75.0	13.0
Price basket for Internet service (US$/month)	—	*22.3*	22.8
Price of call to United States (US$ for 3 minutes)	1.94	1.51	0.81
Trade			
ICT goods exports (% of total goods exports)	0.0	—	15.2
ICT goods imports (% of total goods imports)	6.3	—	14.6
ICT service exports (% of total service exports)	—	48.4	7.0
Applications			
ICT expenditure (% of GDP)	—	4.5	6.7
E-government Web measure index[c]	—	0.41	0.60
Secure Internet servers (per 1 million people, Dec. 2008)	1.8	64.9	662.6

Sources: Economic and social context: UIS and World Bank; Sector structure: ITU; Sector efficiency and capacity: ITU and World Bank; Sector performance: Global Insight/WITSA, IMF, ITU, Netcraft, UN Comtrade, UNDESA, UNPAN, Wireless Intelligence and World Bank. Produced by the Global Information and Communication Technologies Department and the Development Economics Data Group. For complete information, see Definitions and Data Sources.

Notes: Use of italics in the column entries indicates years other than those specified. — Not available. GDP = gross domestic product; GNI = gross national income; ICT = information and communication technology; and MDG = Millennium Development Goal.

a. C = competition; M = monopoly; and P = partial competition. **b**. Outgoing and incoming. **c**. Scale of 0–1, where 1 = highest presence. **d**. Millennium Development Goal indicators 8.14, 8.15, and 8.16.

World Bank • ICT at a Glance

Kyrgyz Republic

	Kyrgyz Republic 2000	Kyrgyz Republic 2007	Low-income group 2007	Europe & Central Asia Region 2007
Economic and social context				
Population (total, million)	5	5	1,296	446
Urban population (% of total)	35	36	32	64
GNI per capita, World Bank Atlas method (current US$)	280	610	574	6,052
GDP growth, 1995–2000 and 2000–07 (avg. annual %)	5.4	4.1	5.6	6.1
Adult literacy rate (% of ages 15 and older)	*99*	99	64	98
Gross primary, secondary, tertiary school enrollment (%)	76	*78*	51	82
Sector structure				
Separate telecommunications regulator	Yes	Yes		
Status of main fixed-line telephone operator	*Public*	Mixed		
Level of competition[a]				
International long distance service	M	C		
Mobile telephone service	P	C		
Internet service	P	C		
Sector efficiency and capacity				
Telecommunications revenue (% of GDP)	2.0	*4.8*	*3.3*	2.9
Mobile and fixed-line subscribers per employee	53	311	*301*	532
Telecommunications investment (% of revenue)	8.0	*3.2*	—	22.0
Sector performance				
Access				
Telephone lines (per 100 people)	7.7	9.2	4.0	25.7
Mobile cellular subscriptions (per 100 people)	0.2	41.4	21.5	95.0
Internet subscribers (per 100 people)	0.1	0.4	*0.8*	13.6
Personal computers (per 100 people)	0.5	*1.9*	*1.5*	10.6
Households with a television set (%)	*84*	—	16	96
Usage				
International voice traffic (minutes/person/month)[b]	1.0	2.5	—	—
Mobile telephone usage (minutes/user/month)	—	—	—	154
Internet users (per 100 people)	1.0	14.3	5.2	21.4
Quality				
Population covered by mobile cellular network (%)	—	*24*	54	92
Fixed broadband subscribers (% of total Internet subscrib.)	0.8	14.6	*3.4*	32.5
International Internet bandwidth (bits/second/person)	2	114	26	1,114
Affordability				
Price basket for residential fixed line (US$/month)	5.3	4.7	5.7	5.8
Price basket for mobile service (US$/month)	—	6.4	11.2	11.8
Price basket for Internet service (US$/month)	—	12.0	29.2	12.0
Price of call to United States (US$ for 3 minutes)	9.84	5.40	2.00	1.63
Trade				
ICT goods exports (% of total goods exports)	0.8	0.8	*1.4*	1.8
ICT goods imports (% of total goods imports)	4.6	5.1	*6.7*	7.0
ICT service exports (% of total service exports)	15.7	1.9	—	5.0
Applications				
ICT expenditure (% of GDP)	—	—	—	5.0
E-government Web measure index[c]	—	0.30	0.11	0.36
Secure Internet servers (per 1 million people, Dec. 2008)	*0.4*	0.6	0.5	23.9

GNI per Capita, Atlas Method, 2000–07
Current US$

ICT MDG[d] Indicators, 2000–07
Number per 100 people

Price Basket for Internet Service, 2003–07
US$/month

ICT Service Exports, 2000–07
Percentage of total service exports

Sources: Economic and social context: UIS and World Bank; Sector structure: ITU; Sector efficiency and capacity: ITU and World Bank; Sector performance: Global Insight/WITSA, IMF, ITU, Netcraft, UN Comtrade, UNDESA, UNPAN, Wireless Intelligence and World Bank. Produced by the Global Information and Communication Technologies Department and the Development Economics Data Group. For complete information, see Definitions and Data Sources.

Notes: Use of italics in the column entries indicates years other than those specified. — Not available. GDP = gross domestic product; GNI = gross national income; ICT = information and communication technology; and MDG = Millennium Development Goal.

a. C = competition; M = monopoly; and P = partial competition. **b.** Outgoing and incoming. **c.** Scale of 0–1, where 1 = highest presence. **d.** Millennium Development Goal indicators 8.14, 8.15, and 8.16.

Information and Communications for Development 2009

Lao People's Democratic Republic

	Lao PDR 2000	Lao PDR 2007	Low-income group 2007	East Asia & Pacific Region 2007
Economic and social context				
Population (total, million)	5	6	1,296	1,912
Urban population (% of total)	22	30	32	43
GNI per capita, World Bank Atlas method (current US$)	290	630	574	2,182
GDP growth, 1995–2000 and 2000–07 (avg. annual %)	6.1	6.7	5.6	9.0
Adult literacy rate (% of ages 15 and older)	*69*	*73*	64	93
Gross primary, secondary, tertiary school enrollment (%)	57	*61*	51	69
Sector structure				
Separate telecommunications regulator	—	Yes		
Status of main fixed-line telephone operator	Public	Mixed		
Level of competition[a]				
International long distance service	M	P		
Mobile telephone service	M	P		
Internet service	—	P		
Sector efficiency and capacity				
Telecommunications revenue (% of GDP)	1.5	*1.7*	3.3	3.0
Mobile and fixed-line subscribers per employee	47	748	301	546
Telecommunications investment (% of revenue)	31.3	63.0	—	—
Sector performance				
Access				
Telephone lines (per 100 people)	0.8	1.6	4.0	23.1
Mobile cellular subscriptions (per 100 people)	0.2	25.2	21.5	43.7
Internet subscribers (per 100 people)	0.0	0.1	*0.8*	9.3
Personal computers (per 100 people)	0.3	*1.8*	1.5	5.6
Households with a television set (%)	30	30	16	53
Usage				
International voice traffic (minutes/person/month)[b]	0.4	*0.6*	—	0.8
Mobile telephone usage (minutes/user/month)	—	—	—	333
Internet users (per 100 people)	0.1	1.7	5.2	14.6
Quality				
Population covered by mobile cellular network (%)	—	55	54	93
Fixed broadband subscribers (% of total Internet subscrib.)	0.0	65.7	3.4	41.8
International Internet bandwidth (bits/second/person)	0	32	26	247
Affordability				
Price basket for residential fixed line (US$/month)	—	3.5	5.7	5.8
Price basket for mobile service (US$/month)	—	3.8	11.2	5.0
Price basket for Internet service (US$/month)	—	27.6	29.2	14.4
Price of call to United States (US$ for 3 minutes)	9.20	1.11	2.00	1.16
Trade				
ICT goods exports (% of total goods exports)	—	—	1.4	30.9
ICT goods imports (% of total goods imports)	—	—	6.7	28.1
ICT service exports (% of total service exports)	—	—	—	5.2
Applications				
ICT expenditure (% of GDP)	—	—	—	7.3
E-government Web measure index[c]	—	0.04	0.11	0.18
Secure Internet servers (per 1 million people, Dec. 2008)	—	0.2	0.5	1.9

Sources: Economic and social context: UIS and World Bank; Sector structure: ITU; Sector efficiency and capacity: ITU and World Bank; Sector performance: Global Insight/WITSA, IMF, ITU, Netcraft, UN Comtrade, UNDESA, UNPAN, Wireless Intelligence and World Bank. Produced by the Global Information and Communication Technologies Department and the Development Economics Data Group. For complete information, see Definitions and Data Sources.

Notes: Use of italics in the column entries indicates years other than those specified. — Not available. GDP = gross domestic product; GNI = gross national income; ICT = information and communication technology; and MDG = Millennium Development Goal.

a. C = competition; M = monopoly; and P = partial competition. **b.** Outgoing and incoming. **c.** Scale of 0–1, where 1 = highest presence. **d.** Millennium Development Goal indicators 8.14, 8.15, and 8.16.

World Bank • ICT at a Glance

Latvia

	Latvia 2000	Latvia 2007	Upper-middle-income group 2007	Europe & Central Asia Region 2007
Economic and social context				
Population (total, million)	2	2	824	446
Urban population (% of total)	68	68	75	64
GNI per capita, World Bank Atlas method (current US$)	3,220	9,920	7,107	6,052
GDP growth, 1995–2000 and 2000–07 (avg. annual %)	5.7	9.0	4.3	6.1
Adult literacy rate (% of ages 15 and older)	100	100	94	98
Gross primary, secondary, tertiary school enrollment (%)	84	90	82	82
Sector structure				
Separate telecommunications regulator	No	Yes		
Status of main fixed-line telephone operator	Mixed	Mixed		
Level of competition[a]				
International long distance service	M	C		
Mobile telephone service	C	C		
Internet service	C	C		
Sector efficiency and capacity				
Telecommunications revenue (% of GDP)	4.4	4.0	3.3	2.9
Mobile and fixed-line subscribers per employee	263	697	566	532
Telecommunications investment (% of revenue)	20.9	—	—	22.0
Sector performance				
Access				
Telephone lines (per 100 people)	31.0	28.3	22.6	25.7
Mobile cellular subscriptions (per 100 people)	16.9	97.4	84.1	95.0
Internet subscribers (per 100 people)	1.4	6.4	9.4	13.6
Personal computers (per 100 people)	14.3	32.7	12.4	10.6
Households with a television set (%)	81	80	92	96
Usage				
International voice traffic (minutes/person/month)[b]	5.1	5.6	—	—
Mobile telephone usage (minutes/user/month)	—	—	137	154
Internet users (per 100 people)	6.3	55.0	26.6	21.4
Quality				
Population covered by mobile cellular network (%)	89	99	95	92
Fixed broadband subscribers (% of total Internet subscrib.)	0.8	100.1	47.8	32.5
International Internet bandwidth (bits/second/person)	65	3,537	1,185	1,114
Affordability				
Price basket for residential fixed line (US$/month)	12.5	13.3	10.6	5.8
Price basket for mobile service (US$/month)	—	9.3	10.9	11.8
Price basket for Internet service (US$/month)	—	12.6	16.4	12.0
Price of call to United States (US$ for 3 minutes)	2.05	1.63	1.55	1.63
Trade				
ICT goods exports (% of total goods exports)	1.7	3.4	13.5	1.8
ICT goods imports (% of total goods imports)	7.7	6.9	16.2	7.0
ICT service exports (% of total service exports)	3.5	4.9	4.6	5.0
Applications				
ICT expenditure (% of GDP)	—	—	5.2	5.0
E-government Web measure index[c]	—	0.45	0.37	0.36
Secure Internet servers (per 1 million people, Dec. 2008)	18.2	98.0	26.2	23.9

Sources: Economic and social context: UIS and World Bank; Sector structure: ITU; Sector efficiency and capacity: ITU and World Bank; Sector performance: Global Insight/WITSA, IMF, ITU, Netcraft, UN Comtrade, UNDESA, UNPAN, Wireless Intelligence and World Bank. Produced by the Global Information and Communication Technologies Department and the Development Economics Data Group. For complete information, see Definitions and Data Sources.

Notes: Use of italics in the column entries indicates years other than those specified. — Not available. GDP = gross domestic product; GNI = gross national income; ICT = information and communication technology; and MDG = Millennium Development Goal.

a. C = competition; M = monopoly; and P = partial competition. b. Outgoing and incoming. c. Scale of 0–1, where 1 = highest presence. d. Millennium Development Goal indicators 8.14, 8.15, and 8.16.

World Bank • ICT at a Glance

Lebanon

	Lebanon 2000	Lebanon 2007	Upper-middle-income group 2007	Middle East & North Africa Region 2007
Economic and social context				
Population (total, million)	4	4	824	313
Urban population (% of total)	86	87	75	57
GNI per capita, World Bank Atlas method (current US$)	4,580	5,800	7,107	2,820
GDP growth, 1995–2000 and 2000–07 (avg. annual %)	2.3	3.3	4.3	4.4
Adult literacy rate (% of ages 15 and older)	—	90	94	73
Gross primary, secondary, tertiary school enrollment (%)	78	85	82	70
Sector structure				
Separate telecommunications regulator	—	Yes		
Status of main fixed-line telephone operator	Public	Public		
Level of competition[a]				
International long distance service	M	M		
Mobile telephone service	—	M		
Internet service	C	C		
Sector efficiency and capacity				
Telecommunications revenue (% of GDP)	*3.5*	*8.0*	3.3	3.1
Mobile and fixed-line subscribers per employee	210	—	566	691
Telecommunications investment (% of revenue)	—	—	—	21.7
Sector performance				
Access				
Telephone lines (per 100 people)	15.3	17.0	22.6	17.0
Mobile cellular subscriptions (per 100 people)	19.7	30.8	84.1	50.7
Internet subscribers (per 100 people)	2.2	6.3	9.4	2.4
Personal computers (per 100 people)	4.6	*10.4*	12.4	6.3
Households with a television set (%)	92	95	92	94
Usage				
International voice traffic (minutes/person/month)[b]	*9.0*	*23.3*	—	2.7
Mobile telephone usage (minutes/user/month)	—	—	137	—
Internet users (per 100 people)	8.0	38.3	26.6	17.1
Quality				
Population covered by mobile cellular network (%)	—	*100*	95	93
Fixed broadband subscribers (% of total Internet subscrib.)	0.0	76.9	47.8	—
International Internet bandwidth (bits/second/person)	10	227	1,185	186
Affordability				
Price basket for residential fixed line (US$/month)	21.3	*15.0*	10.6	3.9
Price basket for mobile service (US$/month)	—	*20.1*	10.9	6.5
Price basket for Internet service (US$/month)	—	*10.0*	16.4	11.6
Price of call to United States (US$ for 3 minutes)	4.48	2.19	1.55	1.45
Trade				
ICT goods exports (% of total goods exports)	1.6	*1.2*	13.5	—
ICT goods imports (% of total goods imports)	4.5	*4.0*	16.2	—
ICT service exports (% of total service exports)	1.4	*2.2*	4.6	2.6
Applications				
ICT expenditure (% of GDP)	—	—	5.2	4.5
E-government Web measure index[c]	—	0.39	0.37	0.22
Secure Internet servers (per 1 million people, Dec. 2008)	5.0	13.0	26.2	1.3

Sources: Economic and social context: UIS and World Bank; Sector structure: ITU; Sector efficiency and capacity: ITU and World Bank; Sector performance: Global Insight/WITSA, IMF, ITU, Netcraft, UN Comtrade, UNDESA, UNPAN, Wireless Intelligence and World Bank. Produced by the Global Information and Communication Technologies Department and the Development Economics Data Group. For complete information, see Definitions and Data Sources.

Notes: Use of italics in the column entries indicates years other than those specified. — Not available. GDP = gross domestic product; GNI = gross national income; ICT = information and communication technology; and MDG = Millennium Development Goal.

a. C = competition; M = monopoly; and P = partial competition. b. Outgoing and incoming. c. Scale of 0–1, where 1 = highest presence. d. Millennium Development Goal indicators 8.14, 8.15, and 8.16.

Lesotho

	Lesotho 2000	Lesotho 2007	Lower-middle-income group 2007	Sub-Saharan Africa Region 2007
Economic and social context				
Population (total, million)	2	2	3,435	800
Urban population (% of total)	20	25	42	36
GNI per capita, World Bank Atlas method (current US$)	600	1,030	1,905	951
GDP growth, 1995–2000 and 2000–07 (avg. annual %)	2.3	3.8	8.0	5.1
Adult literacy rate (% of ages 15 and older)	*82*	—	83	62
Gross primary, secondary, tertiary school enrollment (%)	62	*66*	68	51
Sector structure				
Separate telecommunications regulator	Yes	Yes		
Status of main fixed-line telephone operator	Public	Mixed		
Level of competition[a]				
International long distance service	M	C		
Mobile telephone service	M	C		
Internet service	—	C		
Sector efficiency and capacity				
Telecommunications revenue (% of GDP)	1.4	0.6	3.1	4.7
Mobile and fixed-line subscribers per employee	126	*1,111*	624	499
Telecommunications investment (% of revenue)	9.8	20.9	25.3	—
Sector performance				
Access				
Telephone lines (per 100 people)	1.2	2.7	15.3	1.6
Mobile cellular subscriptions (per 100 people)	1.1	22.7	38.9	23.0
Internet subscribers (per 100 people)	*0.1*	*0.1*	6.0	1.2
Personal computers (per 100 people)	*0.1*	*0.3*	4.6	1.8
Households with a television set (%)	12	13	79	18
Usage				
International voice traffic (minutes/person/month)[b]	—	1.5	—	—
Mobile telephone usage (minutes/user/month)	—	—	322	—
Internet users (per 100 people)	0.2	3.5	12.4	4.4
Quality				
Population covered by mobile cellular network (%)	*21*	55	80	56
Fixed broadband subscribers (% of total Internet subscrib.)	0.0	1.8	40.4	3.1
International Internet bandwidth (bits/second/person)	0	2	199	36
Affordability				
Price basket for residential fixed line (US$/month)	9.7	14.5	7.2	12.6
Price basket for mobile service (US$/month)	—	14.0	9.8	11.6
Price basket for Internet service (US$/month)	—	77.5	16.7	43.1
Price of call to United States (US$ for 3 minutes)	2.31	3.28	2.08	2.43
Trade				
ICT goods exports (% of total goods exports)	—	—	20.6	1.1
ICT goods imports (% of total goods imports)	—	—	20.2	8.2
ICT service exports (% of total service exports)	—	—	15.6	4.2
Applications				
ICT expenditure (% of GDP)	—	—	6.5	—
E-government Web measure index[c]	—	0.34	0.33	0.16
Secure Internet servers (per 1 million people, Dec. 2008)	—	0.5	1.8	2.9

Sources: Economic and social context: UIS and World Bank; Sector structure: ITU; Sector efficiency and capacity: ITU and World Bank; Sector performance: Global Insight/WITSA, IMF, ITU, Netcraft, UN Comtrade, UNDESA, UNPAN, Wireless Intelligence and World Bank. Produced by the Global Information and Communication Technologies Department and the Development Economics Data Group. For complete information, see Definitions and Data Sources.

Notes: Use of italics in the column entries indicates years other than those specified. — Not available. GDP = gross domestic product; GNI = gross national income; ICT = information and communication technology; and MDG = Millennium Development Goal.

a. C = competition; M = monopoly; and P = partial competition. **b.** Outgoing and incoming. **c.** Scale of 0–1, where 1 = highest presence. **d.** Millennium Development Goal indicators 8.14, 8.15, and 8.16.

World Bank • ICT at a Glance

Libya

	Libya 2000	Libya 2007	Upper-middle-income group 2007	Middle East & North Africa Region 2007
Economic and social context				
Population (total, million)	5	6	824	313
Urban population (% of total)	76	77	75	57
GNI per capita, World Bank Atlas method (current US$)	—	9,010	7,107	2,820
GDP growth, 1995–2000 and 2000–07 (avg. annual %)	—	3.7	4.3	4.4
Adult literacy rate (% of ages 15 and older)	—	87	94	73
Gross primary, secondary, tertiary school enrollment (%)	95	94	82	70
Sector structure				
Separate telecommunications regulator	—	No		
Status of main fixed-line telephone operator	Public	Public		
Level of competition[a]				
International long distance service	M	M		
Mobile telephone service	M	M		
Internet service	—	—		
Sector efficiency and capacity				
Telecommunications revenue (% of GDP)	—	—	3.3	3.1
Mobile and fixed-line subscribers per employee	46	—	566	691
Telecommunications investment (% of revenue)	—	—	—	21.7
Sector performance				
Access				
Telephone lines (per 100 people)	11.3	14.4	22.6	17.0
Mobile cellular subscriptions (per 100 people)	0.7	73.1	84.1	50.7
Internet subscribers (per 100 people)	—	1.4	9.4	2.4
Personal computers (per 100 people)	2.4	2.2	12.4	6.3
Households with a television set (%)	95	50	92	94
Usage				
International voice traffic (minutes/person/month)[b]	—	5.5	—	2.7
Mobile telephone usage (minutes/user/month)	—	—	137	—
Internet users (per 100 people)	0.2	4.3	26.6	17.1
Quality				
Population covered by mobile cellular network (%)	—	71	95	93
Fixed broadband subscribers (% of total Internet subscrib.)	—	11.7	47.8	—
International Internet bandwidth (bits/second/person)	0	50	1,185	186
Affordability				
Price basket for residential fixed line (US$/month)	—	1.9	10.6	3.9
Price basket for mobile service (US$/month)	—	6.1	10.9	6.5
Price basket for Internet service (US$/month)	—	22.1	16.4	11.6
Price of call to United States (US$ for 3 minutes)	—	—	1.55	1.45
Trade				
ICT goods exports (% of total goods exports)	—	—	13.5	—
ICT goods imports (% of total goods imports)	—	—	16.2	—
ICT service exports (% of total service exports)	2.3	2.5	4.6	2.6
Applications				
ICT expenditure (% of GDP)	—	—	5.2	4.5
E-government Web measure index[c]	—	0.08	0.37	0.22
Secure Internet servers (per 1 million people, Dec. 2008)	—	0.5	26.2	1.3

Sources: Economic and social context: UIS and World Bank; Sector structure: ITU; Sector efficiency and capacity: ITU and World Bank; Sector performance: Global Insight/WITSA, IMF, ITU, Netcraft, UN Comtrade, UNDESA, UNPAN, Wireless Intelligence and World Bank. Produced by the Global Information and Communication Technologies Department and the Development Economics Data Group. For complete information, see Definitions and Data Sources.

Notes: Use of italics in the column entries indicates years other than those specified. — Not available. GDP = gross domestic product; GNI = gross national income; ICT = information and communication technology; and MDG = Millennium Development Goal.

a. C = competition; M = monopoly; and P = partial competition. **b.** Outgoing and incoming. **c.** Scale of 0–1, where 1 = highest presence. **d.** Millennium Development Goal indicators 8.14, 8.15, and 8.16.

Lithuania

	Lithuania 2000	Lithuania 2007	Upper-middle-income group 2007	Europe & Central Asia Region 2007
Economic and social context				
Population (total, million)	3	3	824	446
Urban population (% of total)	67	67	75	64
GNI per capita, World Bank Atlas method (current US$)	3,170	9,770	7,107	6,052
GDP growth, 1995–2000 and 2000–07 (avg. annual %)	4.2	8.0	4.3	6.1
Adult literacy rate (% of ages 15 and older)	*100*	100	94	98
Gross primary, secondary, tertiary school enrollment (%)	88	*91*	82	82
Sector structure				
Separate telecommunications regulator	No	Yes		
Status of main fixed-line telephone operator	Mixed	Private		
Level of competition[a]				
International long distance service	M	C		
Mobile telephone service	C	P		
Internet service	C	C		
Sector efficiency and capacity				
Telecommunications revenue (% of GDP)	2.3	3.1	3.3	2.9
Mobile and fixed-line subscribers per employee	283	—	566	532
Telecommunications investment (% of revenue)	49.8	18.4	—	22.0
Sector performance				
Access				
Telephone lines (per 100 people)	33.9	23.7	22.6	25.7
Mobile cellular subscriptions (per 100 people)	15.0	145.8	84.1	95.0
Internet subscribers (per 100 people)	1.5	15.2	9.4	13.6
Personal computers (per 100 people)	6.9	*18.3*	12.4	10.6
Households with a television set (%)	87	*98*	92	96
Usage				
International voice traffic (minutes/person/month)[b]	3.6	4.5	—	—
Mobile telephone usage (minutes/user/month)	—	103	137	154
Internet users (per 100 people)	6.4	49.2	26.6	21.4
Quality				
Population covered by mobile cellular network (%)	100	100	95	92
Fixed broadband subscribers (% of total Internet subscrib.)	0.0	98.9	47.8	32.5
International Internet bandwidth (bits/second/person)	14	4,656	1,185	1,114
Affordability				
Price basket for residential fixed line (US$/month)	9.5	*17.7*	10.6	5.8
Price basket for mobile service (US$/month)	—	8.9	10.9	11.8
Price basket for Internet service (US$/month)	—	7.3	16.4	12.0
Price of call to United States (US$ for 3 minutes)	3.10	*1.55*	1.55	1.63
Trade				
ICT goods exports (% of total goods exports)	5.2	*4.8*	13.5	1.8
ICT goods imports (% of total goods imports)	5.6	6.4	16.2	7.0
ICT service exports (% of total service exports)	4.8	3.1	4.6	5.0
Applications				
ICT expenditure (% of GDP)	—	—	5.2	5.0
E-government Web measure index[c]	—	0.61	0.37	0.36
Secure Internet servers (per 1 million people, Dec. 2008)	12.4	83.4	26.2	23.9

Sources: Economic and social context: UIS and World Bank; Sector structure: ITU; Sector efficiency and capacity: ITU and World Bank; Sector performance: Global Insight/WITSA, IMF, ITU, Netcraft, UN Comtrade, UNDESA, UNPAN, Wireless Intelligence and World Bank. Produced by the Global Information and Communication Technologies Department and the Development Economics Data Group. For complete information, see Definitions and Data Sources.

Notes: Use of italics in the column entries indicates years other than those specified. — Not available. GDP = gross domestic product; GNI = gross national income; ICT = information and communication technology; and MDG = Millennium Development Goal.

[a]. C = competition; M = monopoly; and P = partial competition. [b]. Outgoing and incoming. [c]. Scale of 0–1, where 1 = highest presence. [d]. Millennium Development Goal indicators 8.14, 8.15, and 8.16.

Macedonia, Former Yugoslav Republic of

	Macedonia, FYR 2000	Macedonia, FYR 2007	Lower-middle-income group 2007	Europe & Central Asia Region 2007
Economic and social context				
Population (total, million)	2	2	3,435	446
Urban population (% of total)	63	66	42	64
GNI per capita, World Bank Atlas method (current US$)	1,840	3,470	1,905	6,052
GDP growth, 1995–2000 and 2000–07 (avg. annual %)	3.0	2.7	8.0	6.1
Adult literacy rate (% of ages 15 and older)	96	97	83	98
Gross primary, secondary, tertiary school enrollment (%)	69	70	68	82
Sector structure				
Separate telecommunications regulator	—	Yes		
Status of main fixed-line telephone operator	Public	Mixed		
Level of competition[a]				
International long distance service	M	M		
Mobile telephone service	M	C		
Internet service	C	C		
Sector efficiency and capacity				
Telecommunications revenue (% of GDP)	5.2	6.8	3.1	2.9
Mobile and fixed-line subscribers per employee	168	1,065	624	532
Telecommunications investment (% of revenue)	26.7	42.5	25.3	22.0
Sector performance				
Access				
Telephone lines (per 100 people)	25.2	22.8	15.3	25.7
Mobile cellular subscriptions (per 100 people)	5.8	95.6	38.9	95.0
Internet subscribers (per 100 people)	1.5	13.4	6.0	13.6
Personal computers (per 100 people)	3.6	36.8	4.6	10.6
Households with a television set (%)	85	98	79	96
Usage				
International voice traffic (minutes/person/month)[b]	9.9	10.4	—	—
Mobile telephone usage (minutes/user/month)	—	85	322	154
Internet users (per 100 people)	2.5	27.3	12.4	21.4
Quality				
Population covered by mobile cellular network (%)	90	100	80	92
Fixed broadband subscribers (% of total Internet subscrib.)	—	36.7	40.4	32.5
International Internet bandwidth (bits/second/person)	25	17	199	1,114
Affordability				
Price basket for residential fixed line (US$/month)	5.3	10.5	7.2	5.8
Price basket for mobile service (US$/month)	—	14.8	9.8	11.8
Price basket for Internet service (US$/month)	—	33.8	16.7	12.0
Price of call to United States (US$ for 3 minutes)	3.95	—	2.08	1.63
Trade				
ICT goods exports (% of total goods exports)	0.3	0.4	20.6	1.8
ICT goods imports (% of total goods imports)	4.4	4.4	20.2	7.0
ICT service exports (% of total service exports)	12.7	14.1	15.6	5.0
Applications				
ICT expenditure (% of GDP)	—	—	6.5	5.0
E-government Web measure index[c]	—	0.36	0.33	0.36
Secure Internet servers (per 1 million people, Dec. 2008)	—	11.8	1.8	23.9

Sources: Economic and social context: UIS and World Bank; Sector structure: ITU; Sector efficiency and capacity: ITU and World Bank; Sector performance: Global Insight/WITSA, IMF, ITU, Netcraft, UN Comtrade, UNDESA, UNPAN, Wireless Intelligence and World Bank. Produced by the Global Information and Communication Technologies Department and the Development Economics Data Group. For complete information, see Definitions and Data Sources.

Notes: Use of italics in the column entries indicates years other than those specified. — Not available. GDP = gross domestic product; GNI = gross national income; ICT = information and communication technology; and MDG = Millennium Development Goal.

a. C = competition; M = monopoly; and P = partial competition. b. Outgoing and incoming. c. Scale of 0–1, where 1 = highest presence. d. Millennium Development Goal indicators 8.14, 8.15, and 8.16.

World Bank • ICT at a Glance

Madagascar

	Madagascar 2000	Madagascar 2007	Low-income group 2007	Sub-Saharan Africa Region 2007
Economic and social context				
Population (total, million)	16	20	1,296	800
Urban population (% of total)	27	29	32	36
GNI per capita, World Bank Atlas method (current US$)	240	320	574	951
GDP growth, 1995–2000 and 2000–07 (avg. annual %)	3.9	3.2	5.6	5.1
Adult literacy rate (% of ages 15 and older)	71	—	64	62
Gross primary, secondary, tertiary school enrollment (%)	43	*60*	51	51
Sector structure				
Separate telecommunications regulator	Yes	Yes		
Status of main fixed-line telephone operator	Mixed	Mixed		
Level of competition[a]				
International long distance service	C	P		
Mobile telephone service	C	P		
Internet service	C	C		
Sector efficiency and capacity				
Telecommunications revenue (% of GDP)	10.3	3.9	3.3	4.7
Mobile and fixed-line subscribers per employee	44	*394*	301	499
Telecommunications investment (% of revenue)	14.0	32.8	—	—
Sector performance				
Access				
Telephone lines (per 100 people)	0.3	0.7	4.0	1.6
Mobile cellular subscriptions (per 100 people)	0.4	11.3	21.5	23.0
Internet subscribers (per 100 people)	0.1	0.1	*0.8*	*1.2*
Personal computers (per 100 people)	0.2	*0.5*	1.5	1.8
Households with a television set (%)	12	*18*	16	18
Usage				
International voice traffic (minutes/person/month)[b]	0.2	0.1	—	—
Mobile telephone usage (minutes/user/month)	—	—	—	—
Internet users (per 100 people)	0.2	*0.6*	5.2	4.4
Quality				
Population covered by mobile cellular network (%)	21	*23*	54	56
Fixed broadband subscribers (% of total Internet subscrib.)	0.0	18.5	3.4	3.1
International Internet bandwidth (bits/second/person)	0	8	26	36
Affordability				
Price basket for residential fixed line (US$/month)	11.3	*10.5*	5.7	12.6
Price basket for mobile service (US$/month)	—	10.2	11.2	11.6
Price basket for Internet service (US$/month)	—	28.9	29.2	43.1
Price of call to United States (US$ for 3 minutes)	8.98	*0.59*	2.00	2.43
Trade				
ICT goods exports (% of total goods exports)	0.9	*0.5*	1.4	1.1
ICT goods imports (% of total goods imports)	3.4	4.7	6.7	8.2
ICT service exports (% of total service exports)	—	0.5	—	4.2
Applications				
ICT expenditure (% of GDP)	—	—	—	—
E-government Web measure index[c]	—	0.24	0.11	0.16
Secure Internet servers (per 1 million people, Dec. 2008)	—	0.2	0.5	2.9

Sources: Economic and social context: UIS and World Bank; Sector structure: ITU; Sector efficiency and capacity: ITU and World Bank; Sector performance: Global Insight/WITSA, IMF, ITU, Netcraft, UN Comtrade, UNDESA, UNPAN, Wireless Intelligence and World Bank. Produced by the Global Information and Communication Technologies Department and the Development Economics Data Group. For complete information, see Definitions and Data Sources.

Notes: Use of italics in the column entries indicates years other than those specified. — Not available. GDP = gross domestic product; GNI = gross national income; ICT = information and communication technology; and MDG = Millennium Development Goal.

a. C = competition; M = monopoly; and P = partial competition. **b.** Outgoing and incoming. **c.** Scale of 0–1, where 1 = highest presence. **d.** Millennium Development Goal indicators 8.14, 8.15, and 8.16.

World Bank • ICT at a Glance

Malawi

	Malawi 2000	Malawi 2007	Low-income group 2007	Sub-Saharan Africa Region 2007
Economic and social context				
Population (total, million)	12	14	1,296	800
Urban population (% of total)	15	18	32	36
GNI per capita, World Bank Atlas method (current US$)	150	250	574	951
GDP growth, 1995–2000 and 2000–07 (avg. annual %)	3.8	3.3	5.6	5.1
Adult literacy rate (% of ages 15 and older)	64	72	64	62
Gross primary, secondary, tertiary school enrollment (%)	69	63	51	51
Sector structure				
Separate telecommunications regulator	Yes	Yes		
Status of main fixed-line telephone operator	Public	Mixed		
Level of competition[a]				
International long distance service	M	M		
Mobile telephone service	M	C		
Internet service	C	C		
Sector efficiency and capacity				
Telecommunications revenue (% of GDP)	1.7	3.3	3.3	4.7
Mobile and fixed-line subscribers per employee	33	—	301	499
Telecommunications investment (% of revenue)	—	—	—	—
Sector performance				
Access				
Telephone lines (per 100 people)	0.4	1.3	4.0	1.6
Mobile cellular subscriptions (per 100 people)	0.4	7.5	21.5	23.0
Internet subscribers (per 100 people)	0.0	0.6	0.8	1.2
Personal computers (per 100 people)	0.1	0.2	1.5	1.8
Households with a television set (%)	1	5	16	18
Usage				
International voice traffic (minutes/person/month)[b]	0.2	—	—	—
Mobile telephone usage (minutes/user/month)	—	—	—	—
Internet users (per 100 people)	0.1	1.0	5.2	4.4
Quality				
Population covered by mobile cellular network (%)	20	93	54	56
Fixed broadband subscribers (% of total Internet subscrib.)	0.0	1.9	3.4	3.1
International Internet bandwidth (bits/second/person)	0	5	26	36
Affordability				
Price basket for residential fixed line (US$/month)	5.8	1.7	5.7	12.6
Price basket for mobile service (US$/month)	—	10.5	11.2	11.6
Price basket for Internet service (US$/month)	—	52.7	29.2	43.1
Price of call to United States (US$ for 3 minutes)	4.32	—	2.00	2.43
Trade				
ICT goods exports (% of total goods exports)	0.2	0.4	1.4	1.1
ICT goods imports (% of total goods imports)	4.6	3.8	6.7	8.2
ICT service exports (% of total service exports)	—	—	—	4.2
Applications				
ICT expenditure (% of GDP)	—	—	—	—
E-government Web measure index[c]	—	0.22	0.11	0.16
Secure Internet servers (per 1 million people, Dec. 2008)	—	0.1	0.5	2.9

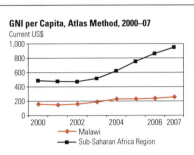

GNI per Capita, Atlas Method, 2000–07
Current US$

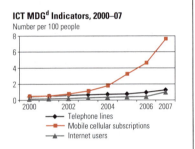

ICT MDG[d] Indicators, 2000–07
Number per 100 people

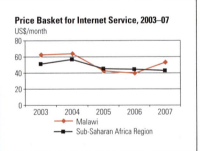

Price Basket for Internet Service, 2003–07
US$/month

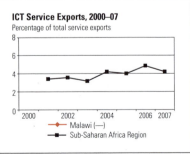

ICT Service Exports, 2000–07
Percentage of total service exports

Sources: Economic and social context: UIS and World Bank; Sector structure: ITU; Sector efficiency and capacity: ITU and World Bank; Sector performance: Global Insight/WITSA, IMF, ITU, Netcraft, UN Comtrade, UNDESA, UNPAN, Wireless Intelligence and World Bank. Produced by the Global Information and Communication Technologies Department and the Development Economics Data Group. For complete information, see Definitions and Data Sources.

Notes: Use of italics in the column entries indicates years other than those specified. — Not available. GDP = gross domestic product; GNI = gross national income; ICT = information and communication technology; and MDG = Millennium Development Goal.

a. C = competition; M = monopoly; and P = partial competition. b. Outgoing and incoming. c. Scale of 0–1, where 1 = highest presence. d. Millennium Development Goal indicators 8.14, 8.15, and 8.16.

Malaysia

	Malaysia 2000	Malaysia 2007	Upper-middle-income group 2007	East Asia & Pacific Region 2007
Economic and social context				
Population (total, million)	23	27	824	1,912
Urban population (% of total)	62	69	75	43
GNI per capita, World Bank Atlas method (current US$)	3,450	6,420	7,107	2,182
GDP growth, 1995–2000 and 2000–07 (avg. annual %)	3.7	5.4	4.3	9.0
Adult literacy rate (% of ages 15 and older)	89	92	94	93
Gross primary, secondary, tertiary school enrollment (%)	70	74	82	69
Sector structure				
Separate telecommunications regulator	Yes	Yes		
Status of main fixed-line telephone operator	Mixed	Mixed		
Level of competition[a]				
International long distance service	P	C		
Mobile telephone service	P	C		
Internet service	C	C		
Sector efficiency and capacity				
Telecommunications revenue (% of GDP)	4.1	4.6	3.3	3.0
Mobile and fixed-line subscribers per employee	394	571	566	546
Telecommunications investment (% of revenue)	27.7	22.5	—	—
Sector performance				
Access				
Telephone lines (per 100 people)	19.9	16.4	22.6	23.1
Mobile cellular subscriptions (per 100 people)	22.0	87.9	84.1	43.7
Internet subscribers (per 100 people)	7.1	18.6	9.4	9.3
Personal computers (per 100 people)	9.5	23.1	12.4	5.6
Households with a television set (%)	82	95	92	53
Usage				
International voice traffic (minutes/person/month)[b]	5.2	—	—	0.8
Mobile telephone usage (minutes/user/month)	204	199	137	333
Internet users (per 100 people)	21.4	55.7	26.6	14.6
Quality				
Population covered by mobile cellular network (%)	95	93	95	93
Fixed broadband subscribers (% of total Internet subscrib.)	0.0	20.5	47.8	41.8
International Internet bandwidth (bits/second/person)	23	998	1,185	247
Affordability				
Price basket for residential fixed line (US$/month)	—	8.5	10.6	5.8
Price basket for mobile service (US$/month)	—	5.5	10.9	5.0
Price basket for Internet service (US$/month)	—	9.3	16.4	14.4
Price of call to United States (US$ for 3 minutes)	2.37	0.71	1.55	1.16
Trade				
ICT goods exports (% of total goods exports)	56.6	41.5	13.5	30.9
ICT goods imports (% of total goods imports)	45.8	36.0	16.2	28.1
ICT service exports (% of total service exports)	1.9	4.9	4.6	5.2
Applications				
ICT expenditure (% of GDP)	—	6.8	5.2	7.3
E-government Web measure index[c]	—	0.68	0.37	0.18
Secure Internet servers (per 1 million people, Dec. 2008)	6.1	27.3	26.2	1.9

Sources: Economic and social context: UIS and World Bank; Sector structure: ITU; Sector efficiency and capacity: ITU and World Bank; Sector performance: Global Insight/WITSA, IMF, ITU, Netcraft, UN Comtrade, UNDESA, UNPAN, Wireless Intelligence and World Bank. Produced by the Global Information and Communication Technologies Department and the Development Economics Data Group. For complete information, see Definitions and Data Sources.

Notes: Use of italics in the column entries indicates years other than those specified. — Not available. GDP = gross domestic product; GNI = gross national income; ICT = information and communication technology; and MDG = Millennium Development Goal.

a. C = competition; M = monopoly; and P = partial competition. b. Outgoing and incoming. c. Scale of 0–1, where 1 = highest presence. d. Millennium Development Goal indicators 8.14, 8.15, and 8.16.

Mali

	Mali 2000	Mali 2007	Low-income group 2007	Sub-Saharan Africa Region 2007
Economic and social context				
Population (total, million)	10	12	1,296	800
Urban population (% of total)	28	32	32	36
GNI per capita, World Bank Atlas method (current US$)	260	500	574	951
GDP growth, 1995–2000 and 2000–07 (avg. annual %)	5.5	5.4	5.6	5.1
Adult literacy rate (% of ages 15 and older)	*19*	*26*	64	62
Gross primary, secondary, tertiary school enrollment (%)	28	*37*	*51*	*51*
Sector structure				
Separate telecommunications regulator	Yes	Yes		
Status of main fixed-line telephone operator	*Public*	*Public*		
Level of competition[a]				
International long distance service	P	P		
Mobile telephone service	P	P		
Internet service	C	C		
Sector efficiency and capacity				
Telecommunications revenue (% of GDP)	2.5	6.0	*3.3*	*4.7*
Mobile and fixed-line subscribers per employee	37	1,490	*301*	*499*
Telecommunications investment (% of revenue)	28.1	29.6	—	—
Sector performance				
Access				
Telephone lines (per 100 people)	0.4	0.6	4.0	1.6
Mobile cellular subscriptions (per 100 people)	0.1	20.5	21.5	23.0
Internet subscribers (per 100 people)	—	0.1	*0.8*	*1.2*
Personal computers (per 100 people)	0.1	0.8	1.5	1.8
Households with a television set (%)	13	*15*	16	18
Usage				
International voice traffic (minutes/person/month)[b]	0.6	0.2	—	—
Mobile telephone usage (minutes/user/month)	—	—	—	—
Internet users (per 100 people)	0.1	0.8	5.2	4.4
Quality				
Population covered by mobile cellular network (%)	15	22	*54*	*56*
Fixed broadband subscribers (% of total Internet subscrib.)	—	45.7	*3.4*	*3.1*
International Internet bandwidth (bits/second/person)	0	17	26	36
Affordability				
Price basket for residential fixed line (US$/month)	—	11.0	*5.7*	*12.6*
Price basket for mobile service (US$/month)	—	14.7	11.2	11.6
Price basket for Internet service (US$/month)	—	43.2	*29.2*	*43.1*
Price of call to United States (US$ for 3 minutes)	12.64	—	*2.00*	*2.43*
Trade				
ICT goods exports (% of total goods exports)	0.2	0.2	*1.4*	*1.1*
ICT goods imports (% of total goods imports)	3.5	4.2	*6.7*	*8.2*
ICT service exports (% of total service exports)	6.2	15.7	—	4.2
Applications				
ICT expenditure (% of GDP)	—	—	—	—
E-government Web measure index[c]	—	0.18	0.11	0.16
Secure Internet servers (per 1 million people, Dec. 2008)	*0.1*	0.6	0.5	2.9

Sources: Economic and social context: UIS and World Bank; Sector structure: ITU; Sector efficiency and capacity: ITU and World Bank; Sector performance: Global Insight/WITSA, IMF, ITU, Netcraft, UN Comtrade, UNDESA, UNPAN, Wireless Intelligence and World Bank. Produced by the Global Information and Communication Technologies Department and the Development Economics Data Group. For complete information, see Definitions and Data Sources.

Notes: Use of italics in the column entries indicates years other than those specified. — Not available. GDP = gross domestic product; GNI = gross national income; ICT = information and communication technology; and MDG = Millennium Development Goal.

a. C = competition; M = monopoly; and P = partial competition. b. Outgoing and incoming. c. Scale of 0–1, where 1 = highest presence. d. Millennium Development Goal indicators 8.14, 8.15, and 8.16.

World Bank • ICT at a Glance

Mauritania

	Mauritania 2000	Mauritania 2007	Low-income group 2007	Sub-Saharan Africa Region 2007
Economic and social context				
Population (total, million)	3	3	1,296	800
Urban population (% of total)	40	41	32	36
GNI per capita, World Bank Atlas method (current US$)	470	840	574	951
GDP growth, 1995–2000 and 2000–07 (avg. annual %)	2.3	5.1	5.6	5.1
Adult literacy rate (% of ages 15 and older)	51	56	64	62
Gross primary, secondary, tertiary school enrollment (%)	43	*46*	*51*	*51*
Sector structure				
Separate telecommunications regulator	Yes	Yes		
Status of main fixed-line telephone operator	Public	Mixed		
Level of competition[a]				
International long distance service	M	C		
Mobile telephone service	M	C		
Internet service	C	C		
Sector efficiency and capacity				
Telecommunications revenue (% of GDP)	2.3	7.5	*3.3*	*4.7*
Mobile and fixed-line subscribers per employee	48	*1,272*	*301*	*499*
Telecommunications investment (% of revenue)	131.7	48.3	—	—
Sector performance				
Access				
Telephone lines (per 100 people)	0.7	1.3	4.0	1.6
Mobile cellular subscriptions (per 100 people)	0.6	41.7	21.5	23.0
Internet subscribers (per 100 people)	0.0	0.2	*0.8*	*1.2*
Personal computers (per 100 people)	1.0	4.6	*1.5*	*1.8*
Households with a television set (%)	19	25	*16*	*18*
Usage				
International voice traffic (minutes/person/month)[b]	1.6	0.4	—	—
Mobile telephone usage (minutes/user/month)	—	—	—	—
Internet users (per 100 people)	0.2	*1.0*	5.2	4.4
Quality				
Population covered by mobile cellular network (%)	—	51	*54*	*56*
Fixed broadband subscribers (% of total Internet subscrib.)	0.0	69.7	*3.4*	*3.1*
International Internet bandwidth (bits/second/person)	0	70	26	36
Affordability				
Price basket for residential fixed line (US$/month)	14.1	11.9	*5.7*	*12.6*
Price basket for mobile service (US$/month)	—	8.9	*11.2*	*11.6*
Price basket for Internet service (US$/month)	—	37.3	*29.2*	*43.1*
Price of call to United States (US$ for 3 minutes)	4.36	—	2.00	2.43
Trade				
ICT goods exports (% of total goods exports)	—	—	*1.4*	*1.1*
ICT goods imports (% of total goods imports)	3.7	2.1	*6.7*	*8.2*
ICT service exports (% of total service exports)	—	—	—	4.2
Applications				
ICT expenditure (% of GDP)	—	—	—	—
E-government Web measure index[c]	—	0.06	0.11	0.16
Secure Internet servers (per 1 million people, Dec. 2008)	*0.4*	1.6	0.5	2.9

GNI per Capita, Atlas Method, 2000–07
Current US$

ICT MDG[d] Indicators, 2000–07
Number per 100 people

Price Basket for Internet Service, 2003–07
US$/month

ICT Service Exports, 2000–07
Percentage of total service exports

Sources: Economic and social context: UIS and World Bank; Sector structure: ITU; Sector efficiency and capacity: ITU and World Bank; Sector performance: Global Insight/WITSA, IMF, ITU, Netcraft, UN Comtrade, UNDESA, UNPAN, Wireless Intelligence and World Bank. Produced by the Global Information and Communication Technologies Department and the Development Economics Data Group. For complete information, see Definitions and Data Sources.

Notes: Use of italics in the column entries indicates years other than those specified. — Not available. GDP = gross domestic product; GNI = gross national income; ICT = information and communication technology; and MDG = Millennium Development Goal.

a. C = competition; M = monopoly; and P = partial competition. **b.** Outgoing and incoming. **c.** Scale of 0–1, where 1 = highest presence. **d.** Millennium Development Goal indicators 8.14, 8.15, and 8.16.

World Bank • ICT at a Glance

Mauritius

	Mauritius 2000	Mauritius 2007	Upper-middle-income group 2007	Sub-Saharan Africa Region 2007
Economic and social context				
Population (total, million)	1	1	824	800
Urban population (% of total)	43	42	75	36
GNI per capita, World Bank Atlas method (current US$)	3,740	5,580	7,107	951
GDP growth, 1995–2000 and 2000–07 (avg. annual %)	5.5	4.0	4.3	5.1
Adult literacy rate (% of ages 15 and older)	84	87	94	62
Gross primary, secondary, tertiary school enrollment (%)	67	75	82	51
Sector structure				
Separate telecommunications regulator	Yes	Yes		
Status of main fixed-line telephone operator	—	—		
Level of competition[a]				
International long distance service	M	C		
Mobile telephone service	P	C		
Internet service	M	C		
Sector efficiency and capacity				
Telecommunications revenue (% of GDP)	3.3	3.6	3.3	4.7
Mobile and fixed-line subscribers per employee	251	492	566	499
Telecommunications investment (% of revenue)	37.6	12.9	—	—
Sector performance				
Access				
Telephone lines (per 100 people)	23.7	28.6	22.6	1.6
Mobile cellular subscriptions (per 100 people)	15.2	73.7	84.1	23.0
Internet subscribers (per 100 people)	2.9	13.2	9.4	1.2
Personal computers (per 100 people)	10.1	17.6	12.4	1.8
Households with a television set (%)	87	96	92	18
Usage				
International voice traffic (minutes/person/month)[b]	6.0	10.4	—	—
Mobile telephone usage (minutes/user/month)	—	—	137	—
Internet users (per 100 people)	7.3	27.0	26.6	4.4
Quality				
Population covered by mobile cellular network (%)	100	99	95	56
Fixed broadband subscribers (% of total Internet subscrib.)	0.0	37.0	47.8	3.1
International Internet bandwidth (bits/second/person)	5	226	1,185	36
Affordability				
Price basket for residential fixed line (US$/month)	5.2	5.3	10.6	12.6
Price basket for mobile service (US$/month)	—	4.1	10.9	11.6
Price basket for Internet service (US$/month)	—	16.4	16.4	43.1
Price of call to United States (US$ for 3 minutes)	4.00	1.59	1.55	2.43
Trade				
ICT goods exports (% of total goods exports)	0.8	4.7	13.5	1.1
ICT goods imports (% of total goods imports)	6.1	6.1	16.2	8.2
ICT service exports (% of total service exports)	1.8	2.8	4.6	4.2
Applications				
ICT expenditure (% of GDP)	—	—	5.2	—
E-government Web measure index[c]	—	0.47	0.37	0.16
Secure Internet servers (per 1 million people, Dec. 2008)	10.0	59.9	26.2	2.9

Sources: Economic and social context: UIS and World Bank; Sector structure: ITU; Sector efficiency and capacity: ITU and World Bank; Sector performance: Global Insight/WITSA, IMF, ITU, Netcraft, UN Comtrade, UNDESA, UNPAN, Wireless Intelligence and World Bank. Produced by the Global Information and Communication Technologies Department and the Development Economics Data Group. For complete information, see Definitions and Data Sources.

Notes: Use of italics in the column entries indicates years other than those specified. — Not available. GDP = gross domestic product; GNI = gross national income; ICT = information and communication technology; and MDG = Millennium Development Goal.

a. C = competition; M = monopoly; and P = partial competition. b. Outgoing and incoming. c. Scale of 0–1, where 1 = highest presence. d. Millennium Development Goal indicators 8.14, 8.15, and 8.16.

Mexico

	Mexico 2000	Mexico 2007	Upper-middle-income group 2007	Latin America & the Caribbean Region 2007
Economic and social context				
Population (total, million)	98	105	824	561
Urban population (% of total)	75	77	75	78
GNI per capita, World Bank Atlas method (current US$)	5,110	9,400	7,107	5,801
GDP growth, 1995–2000 and 2000–07 (avg. annual %)	5.4	2.6	4.3	3.6
Adult literacy rate (% of ages 15 and older)	91	93	94	91
Gross primary, secondary, tertiary school enrollment (%)	71	*76*	*82*	*81*
Sector structure				
Separate telecommunications regulator	Yes	Yes		
Status of main fixed-line telephone operator	*Private*	*Private*		
Level of competition[a]				
International long distance service	C	C		
Mobile telephone service	P	C		
Internet service	C	C		
Sector efficiency and capacity				
Telecommunications revenue (% of GDP)	2.1	2.8	3.3	3.8
Mobile and fixed-line subscribers per employee	*375*	*789*	*566*	*530*
Telecommunications investment (% of revenue)	41.5	11.2	—	—
Sector performance				
Access				
Telephone lines (per 100 people)	12.6	18.8	22.6	18.1
Mobile cellular subscriptions (per 100 people)	14.4	63.2	84.1	67.0
Internet subscribers (per 100 people)	1.2	5.5	9.4	4.5
Personal computers (per 100 people)	5.8	14.4	*12.4*	*11.3*
Households with a television set (%)	90	*98*	*92*	*84*
Usage				
International voice traffic (minutes/person/month)[b]	6.6	15.4	—	—
Mobile telephone usage (minutes/user/month)	83	164	137	116
Internet users (per 100 people)	5.2	22.7	26.6	26.9
Quality				
Population covered by mobile cellular network (%)	86	100	95	91
Fixed broadband subscribers (% of total Internet subscrib.)	1.3	77.9	47.8	81.7
International Internet bandwidth (bits/second/person)	9	178	1,185	1,126
Affordability				
Price basket for residential fixed line (US$/month)	17.5	*16.1*	*10.6*	*9.5*
Price basket for mobile service (US$/month)	—	*13.9*	*10.9*	*10.4*
Price basket for Internet service (US$/month)	—	*20.0*	*16.4*	*25.7*
Price of call to United States (US$ for 3 minutes)	3.01	*0.83*	*1.55*	*1.21*
Trade				
ICT goods exports (% of total goods exports)	23.0	19.6	13.5	11.4
ICT goods imports (% of total goods imports)	20.8	14.9	16.2	15.9
ICT service exports (% of total service exports)	8.8	2.3	*4.6*	*4.7*
Applications				
ICT expenditure (% of GDP)	—	4.0	5.2	4.9
E-government Web measure index[c]	—	0.71	0.37	0.44
Secure Internet servers (per 1 million people, Dec. 2008)	2.6	15.7	26.2	18.2

Sources: Economic and social context: UIS and World Bank; Sector structure: ITU; Sector efficiency and capacity: ITU and World Bank; Sector performance: Global Insight/WITSA, IMF, ITU, Netcraft, UN Comtrade, UNDESA, UNPAN, Wireless Intelligence and World Bank. Produced by the Global Information and Communication Technologies Department and the Development Economics Data Group. For complete information, see Definitions and Data Sources.

Notes: Use of italics in the column entries indicates years other than those specified. — Not available. GDP = gross domestic product; GNI = gross national income; ICT = information and communication technology; and MDG = Millennium Development Goal.

a. C = competition; M = monopoly; and P = partial competition. **b.** Outgoing and incoming. **c.** Scale of 0–1, where 1 = highest presence. **d.** Millennium Development Goal indicators 8.14, 8.15, and 8.16.

World Bank • ICT at a Glance

Moldova

	Moldova 2000	Moldova 2007	Lower-middle-income group 2007	Europe & Central Asia Region 2007
Economic and social context				
Population (total, million)	4	4	3,435	446
Urban population (% of total)	45	42	42	64
GNI per capita, World Bank Atlas method (current US$)	370	1,210	1,905	6,052
GDP growth, 1995–2000 and 2000–07 (avg. annual %)	–2.6	6.5	8.0	6.1
Adult literacy rate (% of ages 15 and older)	—	99	83	98
Gross primary, secondary, tertiary school enrollment (%)	71	70	68	82
Sector structure				
Separate telecommunications regulator	Yes	Yes		
Status of main fixed-line telephone operator	Public	Public		
Level of competition[a]				
International long distance service	M	C		
Mobile telephone service	C	C		
Internet service	C	C		
Sector efficiency and capacity				
Telecommunications revenue (% of GDP)	3.9	10.1	*3.1*	2.9
Mobile and fixed-line subscribers per employee	97	294	*624*	532
Telecommunications investment (% of revenue)	67.9	35.7	25.3	22.0
Sector performance				
Access				
Telephone lines (per 100 people)	14.1	28.4	15.3	25.7
Mobile cellular subscriptions (per 100 people)	3.4	49.5	38.9	95.0
Internet subscribers (per 100 people)	0.3	2.9	6.0	13.6
Personal computers (per 100 people)	1.5	*11.1*	*4.6*	*10.6*
Households with a television set (%)	82	74	79	96
Usage				
International voice traffic (minutes/person/month)[b]	3.3	12.4	—	—
Mobile telephone usage (minutes/user/month)	—	97	322	154
Internet users (per 100 people)	1.3	18.4	12.4	21.4
Quality				
Population covered by mobile cellular network (%)	70	98	*80*	92
Fixed broadband subscribers (% of total Internet subscrib.)	2.3	42.8	40.4	32.5
International Internet bandwidth (bits/second/person)	2	931	199	1,114
Affordability				
Price basket for residential fixed line (US$/month)	2.2	*5.3*	7.2	5.8
Price basket for mobile service (US$/month)	—	*17.1*	9.8	11.8
Price basket for Internet service (US$/month)	—	*22.8*	16.7	12.0
Price of call to United States (US$ for 3 minutes)	4.10	*1.46*	2.08	1.63
Trade				
ICT goods exports (% of total goods exports)	1.6	2.6	20.6	1.8
ICT goods imports (% of total goods imports)	5.5	4.3	20.2	7.0
ICT service exports (% of total service exports)	10.5	15.4	15.6	5.0
Applications				
ICT expenditure (% of GDP)	—	—	6.5	5.0
E-government Web measure index[c]	—	0.31	0.33	0.36
Secure Internet servers (per 1 million people, Dec. 2008)	*0.7*	7.4	1.8	23.9

Sources: Economic and social context: UIS and World Bank; Sector structure: ITU; Sector efficiency and capacity: ITU and World Bank; Sector performance: Global Insight/WITSA, IMF, ITU, Netcraft, UN Comtrade, UNDESA, UNPAN, Wireless Intelligence and World Bank. Produced by the Global Information and Communication Technologies Department and the Development Economics Data Group. For complete information, see Definitions and Data Sources.

Notes: Use of italics in the column entries indicates years other than those specified. — Not available. GDP = gross domestic product; GNI = gross national income; ICT = information and communication technology; and MDG = Millennium Development Goal.

a. C = competition; M = monopoly; and P = partial competition. b. Outgoing and incoming. c. Scale of 0–1, where 1 = highest presence. d. Millennium Development Goal indicators 8.14, 8.15, and 8.16.

World Bank • ICT at a Glance

Mongolia

	Mongolia 2000	Mongolia 2007	Lower-middle-income group 2007	East Asia & Pacific Region 2007
Economic and social context				
Population (total, million)	2	3	3,435	1,912
Urban population (% of total)	57	57	42	43
GNI per capita, World Bank Atlas method (current US$)	410	1,290	1,905	2,182
GDP growth, 1995–2000 and 2000–07 (avg. annual %)	3.0	7.5	8.0	9.0
Adult literacy rate (% of ages 15 and older)	98	97	83	93
Gross primary, secondary, tertiary school enrollment (%)	64	77	68	69
Sector structure				
Separate telecommunications regulator	No	Yes		
Status of main fixed-line telephone operator	Mixed	Mixed		
Level of competition[a]				
International long distance service	M	P		
Mobile telephone service	P	C		
Internet service	C	C		
Sector efficiency and capacity				
Telecommunications revenue (% of GDP)	4.3	3.9	3.1	3.0
Mobile and fixed-line subscribers per employee	60	190	624	546
Telecommunications investment (% of revenue)	4.7	20.1	25.3	—
Sector performance				
Access				
Telephone lines (per 100 people)	4.9	6.1	15.3	23.1
Mobile cellular subscriptions (per 100 people)	6.4	30.0	38.9	43.7
Internet subscribers (per 100 people)	0.3	2.7	6.0	9.3
Personal computers (per 100 people)	1.3	13.9	4.6	5.6
Households with a television set (%)	28	33	79	53
Usage				
International voice traffic (minutes/person/month)[b]	0.8	0.4	—	0.8
Mobile telephone usage (minutes/user/month)	—	—	322	333
Internet users (per 100 people)	1.3	12.3	12.4	14.6
Quality				
Population covered by mobile cellular network (%)	58	41	80	93
Fixed broadband subscribers (% of total Internet subscrib.)	0.5	4.9	40.4	41.8
International Internet bandwidth (bits/second/person)	3	116	199	247
Affordability				
Price basket for residential fixed line (US$/month)	2.6	1.6	7.2	5.8
Price basket for mobile service (US$/month)	—	5.4	9.8	5.0
Price basket for Internet service (US$/month)	—	10.6	16.7	14.4
Price of call to United States (US$ for 3 minutes)	4.92	—	2.08	1.16
Trade				
ICT goods exports (% of total goods exports)	0.1	0.1	20.6	30.9
ICT goods imports (% of total goods imports)	6.8	5.9	20.2	28.1
ICT service exports (% of total service exports)	3.9	3.7	15.6	5.2
Applications				
ICT expenditure (% of GDP)	—	—	6.5	7.3
E-government Web measure index[c]	—	0.42	0.33	0.18
Secure Internet servers (per 1 million people, Dec. 2008)	0.4	8.7	1.8	1.9

Sources: Economic and social context: UIS and World Bank; Sector structure: ITU; Sector efficiency and capacity: ITU and World Bank; Sector performance: Global Insight/WITSA, IMF, ITU, Netcraft, UN Comtrade, UNDESA, UNPAN, Wireless Intelligence and World Bank. Produced by the Global Information and Communication Technologies Department and the Development Economics Data Group. For complete information, see Definitions and Data Sources.

Notes: Use of italics in the column entries indicates years other than those specified. — Not available. GDP = gross domestic product; GNI = gross national income; ICT = information and communication technology; and MDG = Millennium Development Goal.

a. C = competition; M = monopoly; and P = partial competition. **b.** Outgoing and incoming. **c.** Scale of 0–1, where 1 = highest presence. **d.** Millennium Development Goal indicators 8.14, 8.15, and 8.16.

World Bank • ICT at a Glance

Morocco

	Morocco 2000	Morocco 2007	Lower-middle-income group 2007	Middle East & North Africa Region 2007
Economic and social context				
Population (total, million)	28	31	3,435	313
Urban population (% of total)	53	56	42	57
GNI per capita, World Bank Atlas method (current US$)	1,340	2,290	1,905	2,820
GDP growth, 1995–2000 and 2000–07 (avg. annual %)	3.4	5.0	8.0	4.4
Adult literacy rate (% of ages 15 and older)	—	56	83	73
Gross primary, secondary, tertiary school enrollment (%)	51	*58*	68	*70*
Sector structure				
Separate telecommunications regulator	Yes	Yes		
Status of main fixed-line telephone operator	Mixed	Mixed		
Level of competition[a]				
International long distance service	M	C		
Mobile telephone service	C	C		
Internet service	C	C		
Sector efficiency and capacity				
Telecommunications revenue (% of GDP)	3.2	4.8	*3.1*	*3.1*
Mobile and fixed-line subscribers per employee	260	*821*	*624*	*691*
Telecommunications investment (% of revenue)	50.7	17.9	25.3	21.7
Sector performance				
Access				
Telephone lines (per 100 people)	5.0	7.8	15.3	17.0
Mobile cellular subscriptions (per 100 people)	8.2	64.9	38.9	50.7
Internet subscribers (per 100 people)	0.1	1.6	6.0	2.4
Personal computers (per 100 people)	1.2	3.6	4.6	6.3
Households with a television set (%)	77	78	79	94
Usage				
International voice traffic (minutes/person/month)[b]	—	1.9	—	2.7
Mobile telephone usage (minutes/user/month)	—	49	322	—
Internet users (per 100 people)	0.7	21.4	12.4	17.1
Quality				
Population covered by mobile cellular network (%)	95	*98*	*80*	93
Fixed broadband subscribers (% of total Internet subscrib.)	0.0	98.8	40.4	—
International Internet bandwidth (bits/second/person)	4	814	199	186
Affordability				
Price basket for residential fixed line (US$/month)	12.9	22.3	7.2	3.9
Price basket for mobile service (US$/month)	—	19.8	9.8	6.5
Price basket for Internet service (US$/month)	—	15.6	16.7	11.6
Price of call to United States (US$ for 3 minutes)	2.03	*1.69*	2.08	1.45
Trade				
ICT goods exports (% of total goods exports)	*7.4*	5.7	20.6	—
ICT goods imports (% of total goods imports)	*6.9*	6.7	20.2	—
ICT service exports (% of total service exports)	3.7	3.3	15.6	2.6
Applications				
ICT expenditure (% of GDP)	—	8.3	6.5	4.5
E-government Web measure index[c]	—	0.21	0.33	0.22
Secure Internet servers (per 1 million people, Dec. 2008)	0.2	1.4	1.8	1.3

Sources: Economic and social context: UIS and World Bank; Sector structure: ITU; Sector efficiency and capacity: ITU and World Bank; Sector performance: Global Insight/WITSA, IMF, ITU, Netcraft, UN Comtrade, UNDESA, UNPAN, Wireless Intelligence and World Bank. Produced by the Global Information and Communication Technologies Department and the Development Economics Data Group. For complete information, see Definitions and Data Sources.

Notes: Use of italics in the column entries indicates years other than those specified. — Not available. GDP = gross domestic product; GNI = gross national income; ICT = information and communication technology; and MDG = Millennium Development Goal.

a. C = competition; M = monopoly; and P = partial competition. **b.** Outgoing and incoming. **c.** Scale of 0–1, where 1 = highest presence. **d.** Millennium Development Goal indicators 8.14, 8.15, and 8.16.

World Bank • ICT at a Glance

Mozambique

	Mozambique 2000	Mozambique 2007	Low-income group 2007	Sub-Saharan Africa Region 2007
Economic and social context				
Population (total, million)	18	21	1,296	800
Urban population (% of total)	31	36	32	36
GNI per capita, World Bank Atlas method (current US$)	230	330	574	951
GDP growth, 1995–2000 and 2000–07 (avg. annual %)	8.1	8.1	5.6	5.1
Adult literacy rate (% of ages 15 and older)	—	44	64	62
Gross primary, secondary, tertiary school enrollment (%)	37	53	51	51
Sector structure				
Separate telecommunications regulator	Yes	Yes		
Status of main fixed-line telephone operator	Public	Mixed		
Level of competition[a]				
International long distance service	M	M		
Mobile telephone service	C	C		
Internet service	C	C		
Sector efficiency and capacity				
Telecommunications revenue (% of GDP)	2.6	1.2	3.3	4.7
Mobile and fixed-line subscribers per employee	60	980	301	499
Telecommunications investment (% of revenue)	49.6	24.6	—	—
Sector performance				
Access				
Telephone lines (per 100 people)	0.5	0.3	4.0	1.6
Mobile cellular subscriptions (per 100 people)	0.3	15.4	21.5	23.0
Internet subscribers (per 100 people)	0.0	—	0.8	1.2
Personal computers (per 100 people)	0.3	1.4	1.5	1.8
Households with a television set (%)	4	9	16	18
Usage				
International voice traffic (minutes/person/month)[b]	3.0	1.1	—	—
Mobile telephone usage (minutes/user/month)	—	—	—	—
Internet users (per 100 people)	0.1	0.9	5.2	4.4
Quality				
Population covered by mobile cellular network (%)	—	44	54	56
Fixed broadband subscribers (% of total Internet subscrib.)	—	—	3.4	3.1
International Internet bandwidth (bits/second/person)	0	3	26	36
Affordability				
Price basket for residential fixed line (US$/month)	14.6	13.9	5.7	12.6
Price basket for mobile service (US$/month)	—	8.9	11.2	11.6
Price basket for Internet service (US$/month)	—	34.4	29.2	43.1
Price of call to United States (US$ for 3 minutes)	6.21	1.17	2.00	2.43
Trade				
ICT goods exports (% of total goods exports)	0.2	0.1	1.4	1.1
ICT goods imports (% of total goods imports)	5.9	5.1	6.7	8.2
ICT service exports (% of total service exports)	3.2	5.0	—	4.2
Applications				
ICT expenditure (% of GDP)	—	—	—	—
E-government Web measure index[c]	—	0.31	0.11	0.16
Secure Internet servers (per 1 million people, Dec. 2008)	—	0.2	0.5	2.9

GNI per Capita, Atlas Method, 2000–07 (Current US$)

ICT MDG[d] Indicators, 2000–07 (Number per 100 people) — Telephone lines, Mobile cellular subscriptions, Internet users

Price Basket for Internet Service, 2003–07 (US$/month) — Mozambique, Sub-Saharan Africa Region

ICT Service Exports, 2000–07 (Percentage of total service exports) — Mozambique, Sub-Saharan Africa Region

Sources: Economic and social context: UIS and World Bank; Sector structure: ITU; Sector efficiency and capacity: ITU and World Bank; Sector performance: Global Insight/WITSA, IMF, ITU, Netcraft, UN Comtrade, UNDESA, UNPAN, Wireless Intelligence and World Bank. Produced by the Global Information and Communication Technologies Department and the Development Economics Data Group. For complete information, see Definitions and Data Sources.

Notes: Use of italics in the column entries indicates years other than those specified. — Not available. GDP = gross domestic product; GNI = gross national income; ICT = information and communication technology; and MDG = Millennium Development Goal.

a. C = competition; M = monopoly; and P = partial competition. **b.** Outgoing and incoming. **c.** Scale of 0–1, where 1 = highest presence. **d.** Millennium Development Goal indicators 8.14, 8.15, and 8.16.

Myanmar

	Myanmar 2000	Myanmar 2007	Low-income group 2007	East Asia & Pacific Region 2007
Economic and social context				
Population (total, million)	46	49	1,296	1,912
Urban population (% of total)	28	32	32	43
GNI per capita, World Bank Atlas method (current US$)	—	—	574	2,182
GDP growth, 1995–2000 and 2000–07 (avg. annual %)	7.6	*9.2*	5.6	9.0
Adult literacy rate (% of ages 15 and older)	90	—	64	93
Gross primary, secondary, tertiary school enrollment (%)	47	*50*	51	69
Sector structure				
Separate telecommunications regulator	—	No		
Status of main fixed-line telephone operator	Public	Public		
Level of competition[a]				
International long distance service	M	M		
Mobile telephone service	M	M		
Internet service	M	P		
Sector efficiency and capacity				
Telecommunications revenue (% of GDP)	0.2	0.6	*3.3*	3.0
Mobile and fixed-line subscribers per employee	37	81	*301*	546
Telecommunications investment (% of revenue)	25.3	8.6	—	—
Sector performance				
Access				
Telephone lines (per 100 people)	0.6	1.1	4.0	23.1
Mobile cellular subscriptions (per 100 people)	0.0	0.4	21.5	43.7
Internet subscribers (per 100 people)	*0.0*	*0.0*	*0.8*	9.3
Personal computers (per 100 people)	0.2	0.9	1.5	5.6
Households with a television set (%)	3	3	16	53
Usage				
International voice traffic (minutes/person/month)[b]	0.1	0.2	—	0.8
Mobile telephone usage (minutes/user/month)	—	—	—	333
Internet users (per 100 people)	*0.0*	0.1	5.2	14.6
Quality				
Population covered by mobile cellular network (%)	—	10	54	93
Fixed broadband subscribers (% of total Internet subscrib.)	—	10.6	3.4	41.8
International Internet bandwidth (bits/second/person)	0	2	26	247
Affordability				
Price basket for residential fixed line (US$/month)	0.6	1.3	5.7	5.8
Price basket for mobile service (US$/month)	—	141.6	11.2	5.0
Price basket for Internet service (US$/month)	—	46.3	29.2	14.4
Price of call to United States (US$ for 3 minutes)	0.44	0.17	2.00	1.16
Trade				
ICT goods exports (% of total goods exports)	—	—	1.4	30.9
ICT goods imports (% of total goods imports)	—	—	6.7	28.1
ICT service exports (% of total service exports)	—	—	—	5.2
Applications				
ICT expenditure (% of GDP)	—	—	—	7.3
E-government Web measure index[c]	—	0.11	0.11	0.18
Secure Internet servers (per 1 million people, Dec. 2008)	—	0.0	0.5	1.9

Sources: Economic and social context: UIS and World Bank; Sector structure: ITU; Sector efficiency and capacity: ITU and World Bank; Sector performance: Global Insight/WITSA, IMF, ITU, Netcraft, UN Comtrade, UNDESA, UNPAN, Wireless Intelligence and World Bank. Produced by the Global Information and Communication Technologies Department and the Development Economics Data Group. For complete information, see Definitions and Data Sources.

Notes: Use of italics in the column entries indicates years other than those specified. — Not available. GDP = gross domestic product; GNI = gross national income; ICT = information and communication technology; and MDG = Millennium Development Goal.

a. C = competition; M = monopoly; and P = partial competition. **b.** Outgoing and incoming. **c.** Scale of 0–1, where 1 = highest presence. **d.** Millennium Development Goal indicators 8.14, 8.15, and 8.16.

World Bank • ICT at a Glance

Namibia

	Namibia 2000	Namibia 2007	Lower-middle-income group 2007	Sub-Saharan Africa Region 2007
Economic and social context				
Population (total, million)	2	2	3,435	800
Urban population (% of total)	32	36	42	36
GNI per capita, World Bank Atlas method (current US$)	1,880	3,450	1,905	951
GDP growth, 1995–2000 and 2000–07 (avg. annual %)	3.5	4.8	8.0	5.1
Adult literacy rate (% of ages 15 and older)	85	88	83	62
Gross primary, secondary, tertiary school enrollment (%)	68	65	68	51
Sector structure				
Separate telecommunications regulator	Yes	Yes		
Status of main fixed-line telephone operator	Public	Public		
Level of competition[a]				
International long distance service	M	M		
Mobile telephone service	M	P		
Internet service	C	C		
Sector efficiency and capacity				
Telecommunications revenue (% of GDP)	3.6	4.8	3.1	4.7
Mobile and fixed-line subscribers per employee	115	435	624	499
Telecommunications investment (% of revenue)	29.6	7.5	25.3	—
Sector performance				
Access				
Telephone lines (per 100 people)	5.9	6.6	15.3	1.6
Mobile cellular subscriptions (per 100 people)	4.4	38.5	38.9	23.0
Internet subscribers (per 100 people)	0.5	4.3	6.0	1.2
Personal computers (per 100 people)	4.0	24.0	4.6	1.8
Households with a television set (%)	37	41	79	18
Usage				
International voice traffic (minutes/person/month)[b]	5.0	—	—	—
Mobile telephone usage (minutes/user/month)	—	—	322	—
Internet users (per 100 people)	1.6	4.9	12.4	4.4
Quality				
Population covered by mobile cellular network (%)	85	95	80	56
Fixed broadband subscribers (% of total Internet subscrib.)	0.0	0.3	40.4	3.1
International Internet bandwidth (bits/second/person)	2	27	199	36
Affordability				
Price basket for residential fixed line (US$/month)	—	13.0	7.2	12.6
Price basket for mobile service (US$/month)	—	13.1	9.8	11.6
Price basket for Internet service (US$/month)	—	55.8	16.7	43.1
Price of call to United States (US$ for 3 minutes)	4.28	—	2.08	2.43
Trade				
ICT goods exports (% of total goods exports)	1.0	0.5	20.6	1.1
ICT goods imports (% of total goods imports)	8.3	7.3	20.2	8.2
ICT service exports (% of total service exports)	2.7	2.7	15.6	4.2
Applications				
ICT expenditure (% of GDP)	—	—	6.5	—
E-government Web measure index[c]	—	0.17	0.33	0.16
Secure Internet servers (per 1 million people, Dec. 2008)	1.6	8.5	1.8	2.9

Sources: Economic and social context: UIS and World Bank; Sector structure: ITU; Sector efficiency and capacity: ITU and World Bank; Sector performance: Global Insight/WITSA, IMF, ITU, Netcraft, UN Comtrade, UNDESA, UNPAN, Wireless Intelligence and World Bank. Produced by the Global Information and Communication Technologies Department and the Development Economics Data Group. For complete information, see Definitions and Data Sources.

Notes: Use of italics in the column entries indicates years other than those specified. — Not available. GDP = gross domestic product; GNI = gross national income; ICT = information and communication technology; and MDG = Millennium Development Goal.

a. C = competition; M = monopoly; and P = partial competition. **b.** Outgoing and incoming. **c.** Scale of 0–1, where 1 = highest presence. **d.** Millennium Development Goal indicators 8.14, 8.15, and 8.16.

Nepal

	Nepal 2000	Nepal 2007	Low-income group 2007	South Asia Region 2007
Economic and social context				
Population (total, million)	24	28	1,296	1,522
Urban population (% of total)	13	17	32	29
GNI per capita, World Bank Atlas method (current US$)	220	350	574	880
GDP growth, 1995–2000 and 2000–07 (avg. annual %)	4.6	3.4	5.6	7.3
Adult literacy rate (% of ages 15 and older)	*49*	*57*	*64*	*63*
Gross primary, secondary, tertiary school enrollment (%)	56	*58*	*51*	*60*
Sector structure				
Separate telecommunications regulator	Yes	Yes		
Status of main fixed-line telephone operator	Public	Mixed		
Level of competition[a]				
International long distance service	M	P		
Mobile telephone service	M	P		
Internet service	C	C		
Sector efficiency and capacity				
Telecommunications revenue (% of GDP)	1.3	*1.0*	*3.3*	*2.1*
Mobile and fixed-line subscribers per employee	60	565	*301*	660
Telecommunications investment (% of revenue)	26.7	26.7	—	—
Sector performance				
Access				
Telephone lines (per 100 people)	1.1	2.5	4.0	3.2
Mobile cellular subscriptions (per 100 people)	0.0	11.6	21.5	22.8
Internet subscribers (per 100 people)	0.0	0.3	*0.8*	1.3
Personal computers (per 100 people)	0.3	*0.5*	*1.5*	3.3
Households with a television set (%)	3	13	16	42
Usage				
International voice traffic (minutes/person/month)[b]	0.2	*0.5*	—	—
Mobile telephone usage (minutes/user/month)	—	—	—	364
Internet users (per 100 people)	0.2	1.4	5.2	6.6
Quality				
Population covered by mobile cellular network (%)	—	10	54	61
Fixed broadband subscribers (% of total Internet subscrib.)	0.0	12.5	*3.4*	18.9
International Internet bandwidth (bits/second/person)	0	5	26	31
Affordability				
Price basket for residential fixed line (US$/month)	2.6	3.1	*5.7*	*4.0*
Price basket for mobile service (US$/month)	—	2.1	*11.2*	*2.4*
Price basket for Internet service (US$/month)	—	8.0	*29.2*	*8.0*
Price of call to United States (US$ for 3 minutes)	5.28	2.04	2.00	2.02
Trade				
ICT goods exports (% of total goods exports)	0.1	*0.1*	*1.4*	*1.2*
ICT goods imports (% of total goods imports)	3.2	5.4	*6.7*	*8.1*
ICT service exports (% of total service exports)	—	—	—	*39.0*
Applications				
ICT expenditure (% of GDP)	—	—	—	5.7
E-government Web measure index[c]	—	0.29	0.11	0.37
Secure Internet servers (per 1 million people, Dec. 2008)	—	0.9	0.5	1.1

Sources: Economic and social context: UIS and World Bank; Sector structure: ITU; Sector efficiency and capacity: ITU and World Bank; Sector performance: Global Insight/WITSA, IMF, ITU, Netcraft, UN Comtrade, UNDESA, UNPAN, Wireless Intelligence and World Bank. Produced by the Global Information and Communication Technologies Department and the Development Economics Data Group. For complete information, see Definitions and Data Sources.

Notes: Use of italics in the column entries indicates years other than those specified. — Not available. GDP = gross domestic product; GNI = gross national income; ICT = information and communication technology; and MDG = Millennium Development Goal.

a. C = competition; M = monopoly; and P = partial competition. **b.** Outgoing and incoming. **c.** Scale of 0–1, where 1 = highest presence. **d.** Millennium Development Goal indicators 8.14, 8.15, and 8.16.

World Bank • ICT at a Glance

Netherlands

	Netherlands 2000	Netherlands 2007	High-income group 2007
Economic and social context			
Population (total, million)	16	16	1,056
Urban population (% of total)	77	81	78
GNI per capita, World Bank Atlas method (current US$)	26,580	45,650	37,572
GDP growth, 1995–2000 and 2000–07 (avg. annual %)	4.1	1.6	2.4
Adult literacy rate (% of ages 15 and older)	—	—	99
Gross primary, secondary, tertiary school enrollment (%)	99	*98*	92
Sector structure			
Separate telecommunications regulator	Yes	Yes	
Status of main fixed-line telephone operator	Mixed	Private	
Level of competition[a]			
International long distance service	C	C	
Mobile telephone service	C	C	
Internet service	C	C	
Sector efficiency and capacity			
Telecommunications revenue (% of GDP)	3.7	—	3.1
Mobile and fixed-line subscribers per employee	353	—	747
Telecommunications investment (% of revenue)	21.6	—	14.3
Sector performance			
Access			
Telephone lines (per 100 people)	62.1	44.8	50.0
Mobile cellular subscriptions (per 100 people)	67.5	117.7	100.4
Internet subscribers (per 100 people)	37.1	*36.5*	25.8
Personal computers (per 100 people)	39.6	*91.2*	67.4
Households with a television set (%)	96	*99*	98
Usage			
International voice traffic (minutes/person/month)[b]	25.9	—	14.0
Mobile telephone usage (minutes/user/month)	113	*135*	353
Internet users (per 100 people)	44.0	84.2	65.7
Quality			
Population covered by mobile cellular network (%)	100	*100*	99
Fixed broadband subscribers (% of total Internet subscrib.)	4.4	*87.0*	82.6
International Internet bandwidth (bits/second/person)	4,275	78,159	18,242
Affordability			
Price basket for residential fixed line (US$/month)	—	30.3	26.1
Price basket for mobile service (US$/month)	—	12.0	13.0
Price basket for Internet service (US$/month)	—	31.0	22.8
Price of call to United States (US$ for 3 minutes)	0.56	*0.32*	0.81
Trade			
ICT goods exports (% of total goods exports)	22.9	*18.9*	15.2
ICT goods imports (% of total goods imports)	24.1	*19.8*	14.6
ICT service exports (% of total service exports)	5.3	11.0	7.0
Applications			
ICT expenditure (% of GDP)	—	6.6	6.7
E-government Web measure index[c]	—	0.79	0.60
Secure Internet servers (per 1 million people, Dec. 2008)	49.7	1,105.2	662.6

Sources: Economic and social context: UIS and World Bank; Sector structure: ITU; Sector efficiency and capacity: ITU and World Bank; Sector performance: Global Insight/WITSA, IMF, ITU, Netcraft, UN Comtrade, UNDESA, UNPAN, Wireless Intelligence and World Bank. Produced by the Global Information and Communication Technologies Department and the Development Economics Data Group. For complete information, see Definitions and Data Sources.

Notes: Use of italics in the column entries indicates years other than those specified. — Not available. GDP = gross domestic product; GNI = gross national income; ICT = information and communication technology; and MDG = Millennium Development Goal.

a. C = competition; M = monopoly; and P = partial competition. **b.** Outgoing and incoming. **c.** Scale of 0–1, where 1 = highest presence. **d.** Millennium Development Goal indicators 8.14, 8.15, and 8.16.

Information and Communications for Development 2009

New Zealand

	New Zealand 2000	New Zealand 2007	High-income group 2007
Economic and social context			
Population (total, million)	4	4	1,056
Urban population (% of total)	86	86	78
GNI per capita, World Bank Atlas method (current US$)	13,460	27,080	37,572
GDP growth, 1995–2000 and 2000–07 (avg. annual %)	2.5	3.4	2.4
Adult literacy rate (% of ages 15 and older)	—	—	99
Gross primary, secondary, tertiary school enrollment (%)	99	108	92
Sector structure			
Separate telecommunications regulator	—	Yes	
Status of main fixed-line telephone operator	Private	Private	
Level of competition[a]			
International long distance service	C	C	
Mobile telephone service	C	C	
Internet service	C	C	
Sector efficiency and capacity			
Telecommunications revenue (% of GDP)	3.4	3.0	3.1
Mobile and fixed-line subscribers per employee	630	598	747
Telecommunications investment (% of revenue)	16.2	19.9	14.3
Sector performance			
Access			
Telephone lines (per 100 people)	47.5	41.3	50.0
Mobile cellular subscriptions (per 100 people)	40.0	100.5	100.4
Internet subscribers (per 100 people)	13.0	34.3	25.8
Personal computers (per 100 people)	35.8	52.6	67.4
Households with a television set (%)	98	99	98
Usage			
International voice traffic (minutes/person/month)[b]	30.8	25.8	14.0
Mobile telephone usage (minutes/user/month)	87	74	353
Internet users (per 100 people)	47.5	69.2	65.7
Quality			
Population covered by mobile cellular network (%)	97	98	99
Fixed broadband subscribers (% of total Internet subscrib.)	0.9	58.9	82.6
International Internet bandwidth (bits/second/person)	65	4,544	18,242
Affordability			
Price basket for residential fixed line (US$/month)	17.0	28.6	26.1
Price basket for mobile service (US$/month)	—	13.0	13.0
Price basket for Internet service (US$/month)	—	7.3	22.8
Price of call to United States (US$ for 3 minutes)	0.80	1.30	0.81
Trade			
ICT goods exports (% of total goods exports)	2.2	2.3	15.2
ICT goods imports (% of total goods imports)	12.6	9.7	14.6
ICT service exports (% of total service exports)	6.2	5.4	7.0
Applications			
ICT expenditure (% of GDP)	—	5.7	6.7
E-government Web measure index[c]	—	0.64	0.60
Secure Internet servers (per 1 million people, Dec. 2008)	156.9	980.2	662.6

Sources: Economic and social context: UIS and World Bank; Sector structure: ITU; Sector efficiency and capacity: ITU and World Bank; Sector performance: Global Insight/WITSA, IMF, ITU, Netcraft, UN Comtrade, UNDESA, UNPAN, Wireless Intelligence and World Bank. Produced by the Global Information and Communication Technologies Department and the Development Economics Data Group. For complete information, see Definitions and Data Sources.

Notes: Use of italics in the column entries indicates years other than those specified. — Not available. GDP = gross domestic product; GNI = gross national income; ICT = information and communication technology; and MDG = Millennium Development Goal.

a. C = competition; M = monopoly; and P = partial competition. b. Outgoing and incoming. c. Scale of 0–1, where 1 = highest presence. d. Millennium Development Goal indicators 8.14, 8.15, and 8.16.

World Bank • ICT at a Glance

Nicaragua

	Nicaragua 2000	Nicaragua 2007	Lower-middle-income group 2007	Latin America & the Caribbean Region 2007
Economic and social context				
Population (total, million)	5	6	3,435	561
Urban population (% of total)	55	56	42	78
GNI per capita, World Bank Atlas method (current US$)	730	990	1,905	5,801
GDP growth, 1995–2000 and 2000–07 (avg. annual %)	5.0	3.4	8.0	3.6
Adult literacy rate (% of ages 15 and older)	77	78	83	91
Gross primary, secondary, tertiary school enrollment (%)	70	71	68	81
Sector structure				
Separate telecommunications regulator	Yes	Yes		
Status of main fixed-line telephone operator	Public	Private		
Level of competition[a]				
International long distance service	M	C		
Mobile telephone service	C	C		
Internet service	C	C		
Sector efficiency and capacity				
Telecommunications revenue (% of GDP)	2.6	3.7	3.1	3.8
Mobile and fixed-line subscribers per employee	114	334	624	530
Telecommunications investment (% of revenue)	5.9	24.0	25.3	—
Sector performance				
Access				
Telephone lines (per 100 people)	3.2	4.5	15.3	18.1
Mobile cellular subscriptions (per 100 people)	1.8	37.9	38.9	67.0
Internet subscribers (per 100 people)	0.3	0.4	6.0	4.5
Personal computers (per 100 people)	2.3	4.0	4.6	11.3
Households with a television set (%)	59	60	79	84
Usage				
International voice traffic (minutes/person/month)[b]	2.9	5.4	—	—
Mobile telephone usage (minutes/user/month)	—	—	322	116
Internet users (per 100 people)	1.0	2.8	12.4	26.9
Quality				
Population covered by mobile cellular network (%)	—	70	80	91
Fixed broadband subscribers (% of total Internet subscrib.)	4.5	80.6	40.4	81.7
International Internet bandwidth (bits/second/person)	1	144	199	1,126
Affordability				
Price basket for residential fixed line (US$/month)	17.7	9.2	7.2	9.5
Price basket for mobile service (US$/month)	—	15.1	9.8	10.4
Price basket for Internet service (US$/month)	—	26.6	16.7	25.7
Price of call to United States (US$ for 3 minutes)	3.20	3.15	2.08	1.21
Trade				
ICT goods exports (% of total goods exports)	0.1	0.2	20.6	11.4
ICT goods imports (% of total goods imports)	4.1	7.3	20.2	15.9
ICT service exports (% of total service exports)	11.3	8.2	15.6	4.7
Applications				
ICT expenditure (% of GDP)	—	—	6.5	4.9
E-government Web measure index[c]	—	0.29	0.33	0.44
Secure Internet servers (per 1 million people, Dec. 2008)	1.2	6.7	1.8	18.2

Sources: Economic and social context: UIS and World Bank; Sector structure: ITU; Sector efficiency and capacity: ITU and World Bank; Sector performance: Global Insight/WITSA, IMF, ITU, Netcraft, UN Comtrade, UNDESA, UNPAN, Wireless Intelligence and World Bank. Produced by the Global Information and Communication Technologies Department and the Development Economics Data Group. For complete information, see Definitions and Data Sources.

Notes: Use of italics in the column entries indicates years other than those specified. — Not available. GDP = gross domestic product; GNI = gross national income; ICT = information and communication technology; and MDG = Millennium Development Goal.

a. C = competition; M = monopoly; and P = partial competition. **b.** Outgoing and incoming. **c.** Scale of 0–1, where 1 = highest presence. **d.** Millennium Development Goal indicators 8.14, 8.15, and 8.16.

Niger

	Niger 2000	Niger 2007	Low-income group 2007	Sub-Saharan Africa Region 2007
Economic and social context				
Population (total, million)	11	14	1,296	800
Urban population (% of total)	16	16	32	36
GNI per capita, World Bank Atlas method (current US$)	170	280	574	951
GDP growth, 1995–2000 and 2000–07 (avg. annual %)	3.4	3.9	5.6	5.1
Adult literacy rate (% of ages 15 and older)	9	29	64	62
Gross primary, secondary, tertiary school enrollment (%)	15	23	51	51
Sector structure				
Separate telecommunications regulator	—	Yes		
Status of main fixed-line telephone operator	Public	Mixed		
Level of competition[a]				
International long distance service	M	M		
Mobile telephone service	C	C		
Internet service	C	M		
Sector efficiency and capacity				
Telecommunications revenue (% of GDP)	0.9	2.2	3.3	4.7
Mobile and fixed-line subscribers per employee	16	328	301	499
Telecommunications investment (% of revenue)	—	—	—	—
Sector performance				
Access				
Telephone lines (per 100 people)	0.2	0.2	4.0	1.6
Mobile cellular subscriptions (per 100 people)	0.0	6.3	21.5	23.0
Internet subscribers (per 100 people)	0.0	0.0	0.8	1.2
Personal computers (per 100 people)	0.0	0.1	1.5	1.8
Households with a television set (%)	5	7	16	18
Usage				
International voice traffic (minutes/person/month)[b]	0.1	0.2	—	—
Mobile telephone usage (minutes/user/month)	—	—	—	—
Internet users (per 100 people)	0.0	0.3	5.2	4.4
Quality				
Population covered by mobile cellular network (%)	13	45	54	56
Fixed broadband subscribers (% of total Internet subscrib.)	0.0	5.9	3.4	3.1
International Internet bandwidth (bits/second/person)	0	2	26	36
Affordability				
Price basket for residential fixed line (US$/month)	9.4	12.6	5.7	12.6
Price basket for mobile service (US$/month)	—	15.0	11.2	11.6
Price basket for Internet service (US$/month)	—	84.5	29.2	43.1
Price of call to United States (US$ for 3 minutes)	9.03	—	2.00	2.43
Trade				
ICT goods exports (% of total goods exports)	0.2	0.4	1.4	1.1
ICT goods imports (% of total goods imports)	2.0	4.4	6.7	8.2
ICT service exports (% of total service exports)	0.5	32.8	—	4.2
Applications				
ICT expenditure (% of GDP)	—	—	—	—
E-government Web measure index[c]	—	0.07	0.11	0.16
Secure Internet servers (per 1 million people, Dec. 2008)	—	0.3	0.5	2.9

Sources: Economic and social context: UIS and World Bank; Sector structure: ITU; Sector efficiency and capacity: ITU and World Bank; Sector performance: Global Insight/WITSA, IMF, ITU, Netcraft, UN Comtrade, UNDESA, UNPAN, Wireless Intelligence and World Bank. Produced by the Global Information and Communication Technologies Department and the Development Economics Data Group. For complete information, see Definitions and Data Sources.

Notes: Use of italics in the column entries indicates years other than those specified. — Not available. GDP = gross domestic product; GNI = gross national income; ICT = information and communication technology; and MDG = Millennium Development Goal.

a. C = competition; M = monopoly; and P = partial competition. b. Outgoing and incoming. c. Scale of 0–1, where 1 = highest presence. d. Millennium Development Goal indicators 8.14, 8.15, and 8.16.

World Bank • ICT at a Glance

Nigeria

	Nigeria 2000	Nigeria 2007	Low-income group 2007	Sub-Saharan Africa Region 2007
Economic and social context				
Population (total, million)	125	148	1,296	800
Urban population (% of total)	43	48	32	36
GNI per capita, World Bank Atlas method (current US$)	270	920	574	951
GDP growth, 1995–2000 and 2000–07 (avg. annual %)	2.7	6.6	5.6	5.1
Adult literacy rate (% of ages 15 and older)	—	72	64	62
Gross primary, secondary, tertiary school enrollment (%)	49	56	51	51
Sector structure				
Separate telecommunications regulator	Yes	Yes		
Status of main fixed-line telephone operator	Public	Mixed		
Level of competition[a]				
International long distance service	M	P		
Mobile telephone service	C	P		
Internet service	C	P		
Sector efficiency and capacity				
Telecommunications revenue (% of GDP)	0.8	*3.1*	3.3	4.7
Mobile and fixed-line subscribers per employee	50	*256*	*301*	*499*
Telecommunications investment (% of revenue)	37.2	0.4	—	—
Sector performance				
Access				
Telephone lines (per 100 people)	0.4	1.1	4.0	1.6
Mobile cellular subscriptions (per 100 people)	0.0	27.3	21.5	23.0
Internet subscribers (per 100 people)	0.0	*1.4*	*0.8*	*1.2*
Personal computers (per 100 people)	0.6	0.8	1.5	1.8
Households with a television set (%)	26	26	16	18
Usage				
International voice traffic (minutes/person/month)[b]	0.2	—	—	—
Mobile telephone usage (minutes/user/month)	—	—	—	—
Internet users (per 100 people)	0.1	6.8	5.2	4.4
Quality				
Population covered by mobile cellular network (%)	*38*	*60*	*54*	*56*
Fixed broadband subscribers (% of total Internet subscrib.)	0.0	0.0	3.4	3.1
International Internet bandwidth (bits/second/person)	0	5	26	36
Affordability				
Price basket for residential fixed line (US$/month)	—	7.4	5.7	12.6
Price basket for mobile service (US$/month)	—	16.3	11.2	11.6
Price basket for Internet service (US$/month)	—	40.8	29.2	43.1
Price of call to United States (US$ for 3 minutes)	7.15	1.49	2.00	2.43
Trade				
ICT goods exports (% of total goods exports)	0.0	*0.0*	1.4	1.1
ICT goods imports (% of total goods imports)	3.4	*6.9*	6.7	8.2
ICT service exports (% of total service exports)	—	—	—	4.2
Applications				
ICT expenditure (% of GDP)	—	3.4	—	—
E-government Web measure index[c]	—	0.22	0.11	0.16
Secure Internet servers (per 1 million people, Dec. 2008)	0.0	0.8	0.5	2.9

GNI per Capita, Atlas Method, 2000–07
Current US$

ICT MDG[d] Indicators, 2000–07
Number per 100 people
— Telephone lines
— Mobile cellular subscriptions
— Internet users

Price Basket for Internet Service, 2003–07
US$/month
— Nigeria
— Sub-Saharan Africa Region

ICT Service Exports, 2000–07
Percentage of total service exports
— Nigeria (—)
— Sub-Saharan Africa Region

Sources: Economic and social context: UIS and World Bank; Sector structure: ITU; Sector efficiency and capacity: ITU and World Bank; Sector performance: Global Insight/WITSA, IMF, ITU, Netcraft, UN Comtrade, UNDESA, UNPAN, Wireless Intelligence and World Bank. Produced by the Global Information and Communication Technologies Department and the Development Economics Data Group. For complete information, see Definitions and Data Sources.

Notes: Use of italics in the column entries indicates years other than those specified. — Not available. GDP = gross domestic product; GNI = gross national income; ICT = information and communication technology; and MDG = Millennium Development Goal.

a. C = competition; M = monopoly; and P = partial competition. **b.** Outgoing and incoming. **c.** Scale of 0–1, where 1 = highest presence. **d.** Millennium Development Goal indicators 8.14, 8.15, and 8.16.

Norway

	Norway 2000	Norway 2007	High-income group 2007
Economic and social context			
Population (total, million)	4	5	1,056
Urban population (% of total)	76	77	78
GNI per capita, World Bank Atlas method (current US$)	35,870	77,370	37,572
GDP growth, 1995–2000 and 2000–07 (avg. annual %)	3.6	2.4	2.4
Adult literacy rate (% of ages 15 and older)	—	—	99
Gross primary, secondary, tertiary school enrollment (%)	98	99	92
Sector structure			
Separate telecommunications regulator	Yes	Yes	
Status of main fixed-line telephone operator	Public	Mixed	
Level of competition[a]			
International long distance service	C	C	
Mobile telephone service	P	C	
Internet service	C	C	
Sector efficiency and capacity			
Telecommunications revenue (% of GDP)	1.4	1.4	3.1
Mobile and fixed-line subscribers per employee	251	445	747
Telecommunications investment (% of revenue)	89.3	—	14.3
Sector performance			
Access			
Telephone lines (per 100 people)	53.5	42.3	50.0
Mobile cellular subscriptions (per 100 people)	71.8	110.2	100.4
Internet subscribers (per 100 people)	26.2	34.1	25.8
Personal computers (per 100 people)	49.0	62.9	67.4
Households with a television set (%)	97	97	98
Usage			
International voice traffic (minutes/person/month)[b]	17.7	16.1	14.0
Mobile telephone usage (minutes/user/month)	163	232	353
Internet users (per 100 people)	26.7	84.8	65.7
Quality			
Population covered by mobile cellular network (%)	96	—	99
Fixed broadband subscribers (% of total Internet subscrib.)	2.0	89.6	82.6
International Internet bandwidth (bits/second/person)	875	26,904	18,242
Affordability			
Price basket for residential fixed line (US$/month)	25.4	37.9	26.1
Price basket for mobile service (US$/month)	—	15.8	13.0
Price basket for Internet service (US$/month)	—	34.6	22.8
Price of call to United States (US$ for 3 minutes)	0.40	—	0.81
Trade			
ICT goods exports (% of total goods exports)	2.4	1.8	15.2
ICT goods imports (% of total goods imports)	10.6	9.7	14.6
ICT service exports (% of total service exports)	5.4	4.2	7.0
Applications			
ICT expenditure (% of GDP)	—	4.4	6.7
E-government Web measure index[c]	—	0.95	0.60
Secure Internet servers (per 1 million people, Dec. 2008)	81.8	845.0	662.6

Sources: Economic and social context: UIS and World Bank; Sector structure: ITU; Sector efficiency and capacity: ITU and World Bank; Sector performance: Global Insight/WITSA, IMF, ITU, Netcraft, UN Comtrade, UNDESA, UNPAN, Wireless Intelligence and World Bank. Produced by the Global Information and Communication Technologies Department and the Development Economics Data Group. For complete information, see Definitions and Data Sources.

Notes: Use of italics in the column entries indicates years other than those specified. — Not available. GDP = gross domestic product; GNI = gross national income; ICT = information and communication technology; and MDG = Millennium Development Goal.

a. C = competition; M = monopoly; and P = partial competition. **b.** Outgoing and incoming. **c.** Scale of 0–1, where 1 = highest presence. **d.** Millennium Development Goal indicators 8.14, 8.15, and 8.16.

World Bank • ICT at a Glance

Oman

	Oman 2000	Oman 2007	High-income group 2007
Economic and social context			
Population (total, million)	2	3	1,056
Urban population (% of total)	72	72	78
GNI per capita, World Bank Atlas method (current US$)	6,720	12,860	37,572
GDP growth, 1995–2000 and 2000–07 (avg. annual %)	3.2	4.7	2.4
Adult literacy rate (% of ages 15 and older)	—	84	99
Gross primary, secondary, tertiary school enrollment (%)	68	67	92
Sector structure			
Separate telecommunications regulator	No	Yes	
Status of main fixed-line telephone operator	Public	Mixed	
Level of competition[a]			
International long distance service	M	M	
Mobile telephone service	M	P	
Internet service	—	M	
Sector efficiency and capacity			
Telecommunications revenue (% of GDP)	1.9	2.7	3.1
Mobile and fixed-line subscribers per employee	186	858	747
Telecommunications investment (% of revenue)	19.0	63.9	14.3
Sector performance			
Access			
Telephone lines (per 100 people)	9.2	10.3	50.0
Mobile cellular subscriptions (per 100 people)	6.7	96.2	100.4
Internet subscribers (per 100 people)	1.0	2.7	25.8
Personal computers (per 100 people)	3.3	7.1	67.4
Households with a television set (%)	79	79	98
Usage			
International voice traffic (minutes/person/month)[b]	9.2	3.1	14.0
Mobile telephone usage (minutes/user/month)	—	—	353
Internet users (per 100 people)	3.7	13.1	65.7
Quality			
Population covered by mobile cellular network (%)	91	96	99
Fixed broadband subscribers (% of total Internet subscrib.)	0.0	28.9	82.6
International Internet bandwidth (bits/second/person)	16	142	18,242
Affordability			
Price basket for residential fixed line (US$/month)	13.1	8.9	26.1
Price basket for mobile service (US$/month)	—	5.5	13.0
Price basket for Internet service (US$/month)	—	14.5	22.8
Price of call to United States (US$ for 3 minutes)	7.89	1.87	0.81
Trade			
ICT goods exports (% of total goods exports)	0.6	0.8	15.2
ICT goods imports (% of total goods imports)	3.7	3.8	14.6
ICT service exports (% of total service exports)	—	—	7.0
Applications			
ICT expenditure (% of GDP)	—	—	6.7
E-government Web measure index[c]	—	0.48	0.60
Secure Internet servers (per 1 million people, Dec. 2008)	0.8	12.1	662.6

Sources: Economic and social context: UIS and World Bank; Sector structure: ITU; Sector efficiency and capacity: ITU and World Bank; Sector performance: Global Insight/WITSA, IMF, ITU, Netcraft, UN Comtrade, UNDESA, UNPAN, Wireless Intelligence and World Bank. Produced by the Global Information and Communication Technologies Department and the Development Economics Data Group. For complete information, see Definitions and Data Sources.

Notes: Use of italics in the column entries indicates years other than those specified. — Not available. GDP = gross domestic product; GNI = gross national income; ICT = information and communication technology; and MDG = Millennium Development Goal.

a. C = competition; M = monopoly; and P = partial competition. b. Outgoing and incoming. c. Scale of 0–1, where 1 = highest presence. d. Millennium Development Goal indicators 8.14, 8.15, and 8.16.

World Bank • ICT at a Glance

Pakistan

	Pakistan 2000	Pakistan 2007	Low-income group 2007	South Asia Region 2007
Economic and social context				
Population (total, million)	138	162	1,296	1,522
Urban population (% of total)	33	36	32	29
GNI per capita, World Bank Atlas method (current US$)	490	860	574	880
GDP growth, 1995–2000 and 2000–07 (avg. annual %)	3.0	5.6	5.6	7.3
Adult literacy rate (% of ages 15 and older)	*43*	*54*	64	63
Gross primary, secondary, tertiary school enrollment (%)	37	40	51	60
Sector structure				
Separate telecommunications regulator	Yes	Yes		
Status of main fixed-line telephone operator	Mixed	Mixed		
Level of competition[a]				
International long distance service	M	C		
Mobile telephone service	P	C		
Internet service	C	C		
Sector efficiency and capacity				
Telecommunications revenue (% of GDP)	1.8	2.7	*3.3*	*2.1*
Mobile and fixed-line subscribers per employee	50	50	*301*	660
Telecommunications investment (% of revenue)	19.5	1.7	—	—
Sector performance				
Access				
Telephone lines (per 100 people)	2.2	3.0	4.0	3.2
Mobile cellular subscriptions (per 100 people)	0.2	38.7	21.5	22.8
Internet subscribers (per 100 people)	0.1	2.2	*0.8*	1.3
Personal computers (per 100 people)	0.4	—	*1.5*	3.3
Households with a television set (%)	37	47	16	42
Usage				
International voice traffic (minutes/person/month)[b]	0.6	*0.9*	—	—
Mobile telephone usage (minutes/user/month)	—	154	—	364
Internet users (per 100 people)	*1.4*	10.8	5.2	6.6
Quality				
Population covered by mobile cellular network (%)	27	90	*54*	*61*
Fixed broadband subscribers (% of total Internet subscrib.)	0.0	1.3	*3.4*	18.9
International Internet bandwidth (bits/second/person)	0	44	26	31
Affordability				
Price basket for residential fixed line (US$/month)	6.9	*4.1*	*5.7*	*4.0*
Price basket for mobile service (US$/month)	—	2.4	11.2	2.4
Price basket for Internet service (US$/month)	—	9.4	29.2	8.0
Price of call to United States (US$ for 3 minutes)	3.60	*1.03*	2.00	2.02
Trade				
ICT goods exports (% of total goods exports)	—	0.5	*1.4*	1.2
ICT goods imports (% of total goods imports)	—	7.2	*6.7*	8.1
ICT service exports (% of total service exports)	15.4	6.8	—	39.0
Applications				
ICT expenditure (% of GDP)	—	5.6	—	5.7
E-government Web measure index[c]	—	0.42	0.11	0.37
Secure Internet servers (per 1 million people, Dec. 2008)	0.0	0.5	0.5	1.1

Sources: Economic and social context: UIS and World Bank; Sector structure: ITU; Sector efficiency and capacity: ITU and World Bank; Sector performance: Global Insight/WITSA, IMF, ITU, Netcraft, UN Comtrade, UNDESA, UNPAN, Wireless Intelligence and World Bank. Produced by the Global Information and Communication Technologies Department and the Development Economics Data Group. For complete information, see Definitions and Data Sources.

Notes: Use of italics in the column entries indicates years other than those specified. — Not available. GDP = gross domestic product; GNI = gross national income; ICT = information and communication technology; and MDG = Millennium Development Goal.

a. C = competition; M = monopoly; and P = partial competition. b. Outgoing and incoming. c. Scale of 0–1, where 1 = highest presence. d. Millennium Development Goal indicators 8.14, 8.15, and 8.16.

Panama

	Panama 2000	Panama 2007	Upper-middle-income group 2007	Latin America & the Caribbean Region 2007
Economic and social context				
Population (total, million)	3	3	824	561
Urban population (% of total)	66	72	75	78
GNI per capita, World Bank Atlas method (current US$)	3,740	5,500	7,107	5,801
GDP growth, 1995–2000 and 2000–07 (avg. annual %)	5.0	6.0	4.3	3.6
Adult literacy rate (% of ages 15 and older)	92	93	94	91
Gross primary, secondary, tertiary school enrollment (%)	76	*80*	82	81
Sector structure				
Separate telecommunications regulator	Yes	Yes		
Status of main fixed-line telephone operator	*Mixed*	*Mixed*		
Level of competition[a]				
International long distance service	M	C		
Mobile telephone service	P	P		
Internet service	C	C		
Sector efficiency and capacity				
Telecommunications revenue (% of GDP)	3.8	3.5	3.3	3.8
Mobile and fixed-line subscribers per employee	153	229	*566*	*530*
Telecommunications investment (% of revenue)	—	—	—	—
Sector performance				
Access				
Telephone lines (per 100 people)	14.5	14.8	22.6	18.1
Mobile cellular subscriptions (per 100 people)	13.9	90.1	84.1	67.0
Internet subscribers (per 100 people)	1.5	5.0	9.4	4.5
Personal computers (per 100 people)	3.6	4.6	12.4	11.3
Households with a television set (%)	78	87	92	84
Usage				
International voice traffic (minutes/person/month)[b]	4.6	5.5	—	—
Mobile telephone usage (minutes/user/month)	—	—	137	116
Internet users (per 100 people)	6.6	22.3	26.6	26.9
Quality				
Population covered by mobile cellular network (%)	74	81	95	91
Fixed broadband subscribers (% of total Internet subscrib.)	*16.0*	86.9	47.8	81.7
International Internet bandwidth (bits/second/person)	112	15,977	1,185	1,126
Affordability				
Price basket for residential fixed line (US$/month)	—	10.3	10.6	9.5
Price basket for mobile service (US$/month)	—	16.7	10.9	10.4
Price basket for Internet service (US$/month)	—	38.5	16.4	25.7
Price of call to United States (US$ for 3 minutes)	4.36	—	1.55	1.21
Trade				
ICT goods exports (% of total goods exports)	*0.4*	0.0	13.5	11.4
ICT goods imports (% of total goods imports)	8.2	6.7	16.2	15.9
ICT service exports (% of total service exports)	—	4.6	*4.6*	4.7
Applications				
ICT expenditure (% of GDP)	—	5.9	5.2	4.9
E-government Web measure index[c]	—	0.41	0.37	0.44
Secure Internet servers (per 1 million people, Dec. 2008)	9.6	86.9	26.2	18.2

GNI per Capita, Atlas Method, 2000–07
Current US$

ICT MDG[d] Indicators, 2000–07
Number per 100 people
— Telephone lines
— Mobile cellular subscriptions
— Internet users

Price Basket for Internet Service, 2003–07
US$/month
— Panama
— Latin America & the Caribbean Region

ICT Service Exports, 2000–07
Percentage of total service exports
— Panama
— Latin America & the Caribbean Region

Sources: Economic and social context: UIS and World Bank; Sector structure: ITU; Sector efficiency and capacity: ITU and World Bank; Sector performance: Global Insight/WITSA, IMF, ITU, Netcraft, UN Comtrade, UNDESA, UNPAN, Wireless Intelligence and World Bank. Produced by the Global Information and Communication Technologies Department and the Development Economics Data Group. For complete information, see Definitions and Data Sources.

Notes: Use of italics in the column entries indicates years other than those specified. — Not available. GDP = gross domestic product; GNI = gross national income; ICT = information and communication technology; and MDG = Millennium Development Goal.

a. C = competition; M = monopoly; and P = partial competition. **b.** Outgoing and incoming. **c.** Scale of 0–1, where 1 = highest presence. **d.** Millennium Development Goal indicators 8.14, 8.15, and 8.16.

Papua New Guinea

	Papua New Guinea 2000	Papua New Guinea 2007	Low-income group 2007	East Asia & Pacific Region 2007
Economic and social context				
Population (total, million)	5	6	1,296	1,912
Urban population (% of total)	13	13	32	43
GNI per capita, World Bank Atlas method (current US$)	620	850	574	2,182
GDP growth, 1995–2000 and 2000–07 (avg. annual %)	-0.8	2.3	5.6	9.0
Adult literacy rate (% of ages 15 and older)	57	58	64	93
Gross primary, secondary, tertiary school enrollment (%)	41	41	51	69
Sector structure				
Separate telecommunications regulator	Yes	Yes		
Status of main fixed-line telephone operator	Public	Public		
Level of competition[a]				
International long distance service	M	M		
Mobile telephone service	M	M		
Internet service	P	P		
Sector efficiency and capacity				
Telecommunications revenue (% of GDP)	2.3	—	3.3	3.0
Mobile and fixed-line subscribers per employee	41	—	301	546
Telecommunications investment (% of revenue)	82.6	—	—	—
Sector performance				
Access				
Telephone lines (per 100 people)	1.2	0.9	4.0	23.1
Mobile cellular subscriptions (per 100 people)	0.2	4.7	21.5	43.7
Internet subscribers (per 100 people)	0.5	—	0.8	9.3
Personal computers (per 100 people)	5.2	6.4	1.5	5.6
Households with a television set (%)	8	10	16	53
Usage				
International voice traffic (minutes/person/month)[b]	0.7	—	—	0.8
Mobile telephone usage (minutes/user/month)	—	—	—	333
Internet users (per 100 people)	0.8	1.8	5.2	14.6
Quality				
Population covered by mobile cellular network (%)	—	—	54	93
Fixed broadband subscribers (% of total Internet subscrib.)	—	—	3.4	41.8
International Internet bandwidth (bits/second/person)	1	1	26	247
Affordability				
Price basket for residential fixed line (US$/month)	20.1	4.9	5.7	5.8
Price basket for mobile service (US$/month)	—	14.6	11.2	5.0
Price basket for Internet service (US$/month)	—	25.1	29.2	14.4
Price of call to United States (US$ for 3 minutes)	4.32	—	2.00	1.16
Trade				
ICT goods exports (% of total goods exports)	0.0	0.1	1.4	30.9
ICT goods imports (% of total goods imports)	3.6	4.1	6.7	28.1
ICT service exports (% of total service exports)	—	2.2	—	5.2
Applications				
ICT expenditure (% of GDP)	—	—	—	7.3
E-government Web measure index[c]	—	0.09	0.11	0.18
Secure Internet servers (per 1 million people, Dec. 2008)	—	1.1	0.5	1.9

Sources: Economic and social context: UIS and World Bank; Sector structure: ITU; Sector efficiency and capacity: ITU and World Bank; Sector performance: Global Insight/WITSA, IMF, ITU, Netcraft, UN Comtrade, UNDESA, UNPAN, Wireless Intelligence and World Bank. Produced by the Global Information and Communication Technologies Department and the Development Economics Data Group. For complete information, see Definitions and Data Sources.

Notes: Use of italics in the column entries indicates years other than those specified. — Not available. GDP = gross domestic product; GNI = gross national income; ICT = information and communication technology; and MDG = Millennium Development Goal.

a. C = competition; M = monopoly; and P = partial competition. b. Outgoing and incoming. c. Scale of 0–1, where 1 = highest presence. d. Millennium Development Goal indicators 8.14, 8.15, and 8.16.

World Bank • ICT at a Glance

Paraguay

	Paraguay 2000	Paraguay 2007	Lower-middle-income group 2007	Latin America & the Caribbean Region 2007
Economic and social context				
Population (total, million)	5	6	3,435	561
Urban population (% of total)	55	60	42	78
GNI per capita, World Bank Atlas method (current US$)	1,350	1,710	1,905	5,801
GDP growth, 1995–2000 and 2000–07 (avg. annual %)	0.1	3.3	8.0	3.6
Adult literacy rate (% of ages 15 and older)	—	95	83	91
Gross primary, secondary, tertiary school enrollment (%)	70	*69*	68	81
Sector structure				
Separate telecommunications regulator	Yes	Yes		
Status of main fixed-line telephone operator	*Public*	*Public*		
Level of competition[a]				
International long distance service	M	M		
Mobile telephone service	C	C		
Internet service	C	C		
Sector efficiency and capacity				
Telecommunications revenue (% of GDP)	4.6	4.8	3.1	3.8
Mobile and fixed-line subscribers per employee	123	799	624	530
Telecommunications investment (% of revenue)	26.3	—	25.3	—
Sector performance				
Access				
Telephone lines (per 100 people)	5.3	6.4	15.3	18.1
Mobile cellular subscriptions (per 100 people)	15.4	76.7	38.9	67.0
Internet subscribers (per 100 people)	0.5	1.2	6.0	4.5
Personal computers (per 100 people)	1.3	7.8	4.6	11.3
Households with a television set (%)	75	79	79	84
Usage				
International voice traffic (minutes/person/month)[b]	1.6	2.9	—	—
Mobile telephone usage (minutes/user/month)	62	231	322	116
Internet users (per 100 people)	0.7	8.7	12.4	26.9
Quality				
Population covered by mobile cellular network (%)	—	—	80	91
Fixed broadband subscribers (% of total Internet subscrib.)	0.5	69.4	40.4	81.7
International Internet bandwidth (bits/second/person)	2	163	199	1,126
Affordability				
Price basket for residential fixed line (US$/month)	18.0	6.4	7.2	9.5
Price basket for mobile service (US$/month)	—	3.4	9.8	10.4
Price basket for Internet service (US$/month)	—	12.5	16.7	25.7
Price of call to United States (US$ for 3 minutes)	0.97	0.90	2.08	1.21
Trade				
ICT goods exports (% of total goods exports)	0.2	*0.4*	20.6	11.4
ICT goods imports (% of total goods imports)	11.1	*28.6*	20.2	15.9
ICT service exports (% of total service exports)	1.8	2.2	15.6	4.7
Applications				
ICT expenditure (% of GDP)	—	—	6.5	4.9
E-government Web measure index[c]	—	0.44	0.33	0.44
Secure Internet servers (per 1 million people, Dec. 2008)	*0.7*	5.6	1.8	18.2

Sources: Economic and social context: UIS and World Bank; Sector structure: ITU; Sector efficiency and capacity: ITU and World Bank; Sector performance: Global Insight/WITSA, IMF, ITU, Netcraft, UN Comtrade, UNDESA, UNPAN, Wireless Intelligence and World Bank. Produced by the Global Information and Communication Technologies Department and the Development Economics Data Group. For complete information, see Definitions and Data Sources.

Notes: Use of italics in the column entries indicates years other than those specified. — Not available. GDP = gross domestic product; GNI = gross national income; ICT = information and communication technology; and MDG = Millennium Development Goal.

a. C = competition; M = monopoly; and P = partial competition. **b.** Outgoing and incoming. **c.** Scale of 0–1, where 1 = highest presence. **d.** Millennium Development Goal indicators 8.14, 8.15, and 8.16.

World Bank • ICT at a Glance

Peru

	Peru 2000	Peru 2007	Lower-middle-income group 2007	Latin America & the Caribbean Region 2007
Economic and social context				
Population (total, million)	26	28	3,435	561
Urban population (% of total)	71	71	42	78
GNI per capita, World Bank Atlas method (current US$)	2,080	3,410	1,905	5,801
GDP growth, 1995–2000 and 2000–07 (avg. annual %)	2.4	5.4	8.0	3.6
Adult literacy rate (% of ages 15 and older)	—	90	83	91
Gross primary, secondary, tertiary school enrollment (%)	*88*	*86*	*68*	*81*
Sector structure				
Separate telecommunications regulator	Yes	Yes		
Status of main fixed-line telephone operator	Private	Private		
Level of competition[a]				
International long distance service	C	C		
Mobile telephone service	C	C		
Internet service	C	C		
Sector efficiency and capacity				
Telecommunications revenue (% of GDP)	2.7	*2.9*	*3.1*	3.8
Mobile and fixed-line subscribers per employee	473	624	*624*	530
Telecommunications investment (% of revenue)	22.2	*16.6*	25.3	—
Sector performance				
Access				
Telephone lines (per 100 people)	6.7	9.6	15.3	18.1
Mobile cellular subscriptions (per 100 people)	5.0	55.3	38.9	67.0
Internet subscribers (per 100 people)	0.5	*3.7*	6.0	4.5
Personal computers (per 100 people)	4.1	*10.3*	4.6	*11.3*
Households with a television set (%)	67	*73*	*79*	*84*
Usage				
International voice traffic (minutes/person/month)[b]	1.8	*8.3*	—	—
Mobile telephone usage (minutes/user/month)	—	73	322	116
Internet users (per 100 people)	3.1	27.4	12.4	26.9
Quality				
Population covered by mobile cellular network (%)	—	92	*80*	91
Fixed broadband subscribers (% of total Internet subscrib.)	0.8	*47.1*	40.4	81.7
International Internet bandwidth (bits/second/person)	4	2,704	199	1,126
Affordability				
Price basket for residential fixed line (US$/month)	18.9	*18.8*	7.2	9.5
Price basket for mobile service (US$/month)	—	*23.0*	9.8	10.4
Price basket for Internet service (US$/month)	—	*23.2*	16.7	25.7
Price of call to United States (US$ for 3 minutes)	2.08	*1.80*	2.08	1.21
Trade				
ICT goods exports (% of total goods exports)	0.4	0.1	*20.6*	*11.4*
ICT goods imports (% of total goods imports)	10.1	8.0	*20.2*	*15.9*
ICT service exports (% of total service exports)	5.8	2.6	15.6	4.7
Applications				
ICT expenditure (% of GDP)	—	3.9	6.5	4.9
E-government Web measure index[c]	—	0.57	0.33	0.44
Secure Internet servers (per 1 million people, Dec. 2008)	*1.3*	10.2	1.8	18.2

Sources: Economic and social context: UIS and World Bank; Sector structure: ITU; Sector efficiency and capacity: ITU and World Bank; Sector performance: Global Insight/WITSA, IMF, ITU, Netcraft, UN Comtrade, UNDESA, UNPAN, Wireless Intelligence and World Bank. Produced by the Global Information and Communication Technologies Department and the Development Economics Data Group. For complete information, see Definitions and Data Sources.

Notes: Use of italics in the column entries indicates years other than those specified. — Not available. GDP = gross domestic product; GNI = gross national income; ICT = information and communication technology; and MDG = Millennium Development Goal.

a. C = competition; M = monopoly; and P = partial competition. **b.** Outgoing and incoming. **c.** Scale of 0–1, where 1 = highest presence. **d.** Millennium Development Goal indicators 8.14, 8.15, and 8.16.

Philippines

	Philippines 2000	Philippines 2007	Lower-middle-income group 2007	East Asia & Pacific Region 2007
Economic and social context				
Population (total, million)	76	88	3,435	1,912
Urban population (% of total)	59	64	42	43
GNI per capita, World Bank Atlas method (current US$)	1,050	1,620	1,905	2,182
GDP growth, 1995–2000 and 2000–07 (avg. annual %)	3.5	5.1	8.0	9.0
Adult literacy rate (% of ages 15 and older)	93	93	83	93
Gross primary, secondary, tertiary school enrollment (%)	*79*	*81*	*68*	*69*
Sector structure				
Separate telecommunications regulator	Yes	Yes		
Status of main fixed-line telephone operator	Private	Private		
Level of competition[a]				
International long distance service	C	C		
Mobile telephone service	C	C		
Internet service	C	C		
Sector efficiency and capacity				
Telecommunications revenue (% of GDP)	2.9	*4.4*	*3.1*	3.0
Mobile and fixed-line subscribers per employee	482	1,555	*624*	546
Telecommunications investment (% of revenue)	47.4	*24.4*	*25.3*	—
Sector performance				
Access				
Telephone lines (per 100 people)	4.0	4.5	*15.3*	23.1
Mobile cellular subscriptions (per 100 people)	8.5	58.9	*38.9*	43.7
Internet subscribers (per 100 people)	0.5	2.8	*6.0*	9.3
Personal computers (per 100 people)	1.9	*7.3*	*4.6*	5.6
Households with a television set (%)	53	*63*	*79*	53
Usage				
International voice traffic (minutes/person/month)[b]	2.3	2.4	—	0.8
Mobile telephone usage (minutes/user/month)	—	—	*322*	333
Internet users (per 100 people)	2.0	6.0	*12.4*	14.6
Quality				
Population covered by mobile cellular network (%)	70	99	*80*	93
Fixed broadband subscribers (% of total Internet subscrib.)	0.0	19.8	*40.4*	41.8
International Internet bandwidth (bits/second/person)	2	114	199	247
Affordability				
Price basket for residential fixed line (US$/month)	*12.3*	13.3	*7.2*	5.8
Price basket for mobile service (US$/month)	—	5.3	*9.8*	5.0
Price basket for Internet service (US$/month)	—	2.2	*16.7*	14.4
Price of call to United States (US$ for 3 minutes)	2.07	*1.20*	*2.08*	1.16
Trade				
ICT goods exports (% of total goods exports)	69.4	29.1	*20.6*	30.9
ICT goods imports (% of total goods imports)	42.7	20.6	*20.2*	28.1
ICT service exports (% of total service exports)	7.6	7.0	*15.6*	5.2
Applications				
ICT expenditure (% of GDP)	—	5.7	*6.5*	7.3
E-government Web measure index[c]	—	0.51	*0.33*	0.18
Secure Internet servers (per 1 million people, Dec. 2008)	*0.9*	4.6	*1.8*	1.9

Sources: Economic and social context: UIS and World Bank; Sector structure: ITU; Sector efficiency and capacity: ITU and World Bank; Sector performance: Global Insight/WITSA, IMF, ITU, Netcraft, UN Comtrade, UNDESA, UNPAN, Wireless Intelligence and World Bank. Produced by the Global Information and Communication Technologies Department and the Development Economics Data Group. For complete information, see Definitions and Data Sources.

Notes: Use of italics in the column entries indicates years other than those specified. — Not available. GDP = gross domestic product; GNI = gross national income; ICT = information and communication technology; and MDG = Millennium Development Goal.

a. C = competition; M = monopoly; and P = partial competition. b. Outgoing and incoming. c. Scale of 0–1, where 1 = highest presence. d. Millennium Development Goal indicators 8.14, 8.15, and 8.16.

Poland

	Poland 2000	Poland 2007	Upper-middle-income group 2007	Europe & Central Asia Region 2007
Economic and social context				
Population (total, million)	38	38	824	446
Urban population (% of total)	62	61	75	64
GNI per capita, World Bank Atlas method (current US$)	4,570	9,850	7,107	6,052
GDP growth, 1995–2000 and 2000–07 (avg. annual %)	5.4	4.1	4.3	6.1
Adult literacy rate (% of ages 15 and older)	—	99	94	98
Gross primary, secondary, tertiary school enrollment (%)	86	87	82	82
Sector structure				
Separate telecommunications regulator	Yes	Yes		
Status of main fixed-line telephone operator	Mixed	Mixed		
Level of competition[a]				
International long distance service	M	M		
Mobile telephone service	P	C		
Internet service	C	—		
Sector efficiency and capacity				
Telecommunications revenue (% of GDP)	4.1	3.7	3.3	2.9
Mobile and fixed-line subscribers per employee	256	566	566	532
Telecommunications investment (% of revenue)	19.4	19.2	—	22.0
Sector performance				
Access				
Telephone lines (per 100 people)	28.5	27.1	22.6	25.7
Mobile cellular subscriptions (per 100 people)	17.5	108.6	84.1	95.0
Internet subscribers (per 100 people)	2.4	10.5	9.4	13.6
Personal computers (per 100 people)	6.9	16.9	12.4	10.6
Households with a television set (%)	96	89	92	96
Usage				
International voice traffic (minutes/person/month)[b]	4.2	5.1	—	—
Mobile telephone usage (minutes/user/month)	161	101	137	154
Internet users (per 100 people)	7.3	44.0	26.6	21.4
Quality				
Population covered by mobile cellular network (%)	95	99	95	92
Fixed broadband subscribers (% of total Internet subscrib.)	0.0	86.0	47.8	32.5
International Internet bandwidth (bits/second/person)	20	2,748	1,185	1,114
Affordability				
Price basket for residential fixed line (US$/month)	14.0	21.0	10.6	5.8
Price basket for mobile service (US$/month)	—	7.6	10.9	11.8
Price basket for Internet service (US$/month)	—	11.7	16.4	12.0
Price of call to United States (US$ for 3 minutes)	2.92	1.35	1.55	1.63
Trade				
ICT goods exports (% of total goods exports)	4.5	5.6	13.5	1.8
ICT goods imports (% of total goods imports)	10.5	9.6	16.2	7.0
ICT service exports (% of total service exports)	2.8	4.0	4.6	5.0
Applications				
ICT expenditure (% of GDP)	—	6.0	5.2	5.0
E-government Web measure index[c]	—	0.54	0.37	0.36
Secure Internet servers (per 1 million people, Dec. 2008)	8.5	84.7	26.2	23.9

Sources: Economic and social context: UIS and World Bank; Sector structure: ITU; Sector efficiency and capacity: ITU and World Bank; Sector performance: Global Insight/WITSA, IMF, ITU, Netcraft, UN Comtrade, UNDESA, UNPAN, Wireless Intelligence and World Bank. Produced by the Global Information and Communication Technologies Department and the Development Economics Data Group. For complete information, see Definitions and Data Sources.

Notes: Use of italics in the column entries indicates years other than those specified. — Not available. GDP = gross domestic product; GNI = gross national income; ICT = information and communication technology; and MDG = Millennium Development Goal.

a. C = competition; M = monopoly; and P = partial competition. **b.** Outgoing and incoming. **c.** Scale of 0–1, where 1 = highest presence. **d.** Millennium Development Goal indicators 8.14, 8.15, and 8.16.

World Bank • ICT at a Glance

Portugal

	Portugal 2000	Portugal 2007	High-income group 2007
Economic and social context			
Population (total, million)	10	11	1,056
Urban population (% of total)	54	59	78
GNI per capita, World Bank Atlas method (current US$)	11,590	18,950	37,572
GDP growth, 1995–2000 and 2000–07 (avg. annual %)	4.2	0.9	2.4
Adult literacy rate (% of ages 15 and older)	—	95	99
Gross primary, secondary, tertiary school enrollment (%)	92	90	92
Sector structure			
Separate telecommunications regulator	Yes	Yes	
Status of main fixed-line telephone operator	Mixed	Private	
Level of competition[a]			
International long distance service	C	C	
Mobile telephone service	P	C	
Internet service	C	C	
Sector efficiency and capacity			
Telecommunications revenue (% of GDP)	4.5	4.5	3.1
Mobile and fixed-line subscribers per employee	594	1,365	747
Telecommunications investment (% of revenue)	22.7	16.4	14.3
Sector performance			
Access			
Telephone lines (per 100 people)	42.3	39.5	50.0
Mobile cellular subscriptions (per 100 people)	65.2	126.8	100.4
Internet subscribers (per 100 people)	6.3	15.2	25.8
Personal computers (per 100 people)	10.3	17.2	67.4
Households with a television set (%)	100	99	98
Usage			
International voice traffic (minutes/person/month)[b]	11.4	14.8	14.0
Mobile telephone usage (minutes/user/month)	133	120	353
Internet users (per 100 people)	16.4	40.1	65.7
Quality			
Population covered by mobile cellular network (%)	99	99	99
Fixed broadband subscribers (% of total Internet subscrib.)	0.0	94.6	82.6
International Internet bandwidth (bits/second/person)	49	4,790	18,242
Affordability			
Price basket for residential fixed line (US$/month)	17.4	31.8	26.1
Price basket for mobile service (US$/month)	—	23.1	13.0
Price basket for Internet service (US$/month)	—	38.1	22.8
Price of call to United States (US$ for 3 minutes)	0.83	1.04	0.81
Trade			
ICT goods exports (% of total goods exports)	7.8	9.0	15.2
ICT goods imports (% of total goods imports)	9.0	9.3	14.6
ICT service exports (% of total service exports)	2.7	4.8	7.0
Applications			
ICT expenditure (% of GDP)	—	5.7	6.7
E-government Web measure index[c]	—	0.60	0.60
Secure Internet servers (per 1 million people, Dec. 2008)	13.4	115.2	662.6

Sources: Economic and social context: UIS and World Bank; Sector structure: ITU; Sector efficiency and capacity: ITU and World Bank; Sector performance: Global Insight/WITSA, IMF, ITU, Netcraft, UN Comtrade, UNDESA, UNPAN, Wireless Intelligence and World Bank. Produced by the Global Information and Communication Technologies Department and the Development Economics Data Group. For complete information, see Definitions and Data Sources.

Notes: Use of italics in the column entries indicates years other than those specified. — Not available. GDP = gross domestic product; GNI = gross national income; ICT = information and communication technology; and MDG = Millennium Development Goal.

a. C = competition; M = monopoly; and P = partial competition. b. Outgoing and incoming. c. Scale of 0–1, where 1 = highest presence. d. Millennium Development Goal indicators 8.14, 8.15, and 8.16.

World Bank • ICT at a Glance

Puerto Rico

	Puerto Rico 2000	Puerto Rico 2007	High-income group 2007
Economic and social context			
Population (total, million)	4	4	1,056
Urban population (% of total)	95	98	78
GNI per capita, World Bank Atlas method (current US$)	10,560	—	37,572
GDP growth, 1995–2000 and 2000–07 (avg. annual %)	4.4	—	2.4
Adult literacy rate (% of ages 15 and older)	—	—	99
Gross primary, secondary, tertiary school enrollment (%)	—	—	92
Sector structure			
Separate telecommunications regulator	—	No	
Status of main fixed-line telephone operator	—	—	
Level of competition[a]			
International long distance service	—	—	
Mobile telephone service	—	—	
Internet service	—	—	
Sector efficiency and capacity			
Telecommunications revenue (% of GDP)	3.2	—	3.1
Mobile and fixed-line subscribers per employee	387	—	747
Telecommunications investment (% of revenue)	—	—	14.3
Sector performance			
Access			
Telephone lines (per 100 people)	34.0	26.5	50.0
Mobile cellular subscriptions (per 100 people)	24.3	85.7	100.4
Internet subscribers (per 100 people)	6.6	—	25.8
Personal computers (per 100 people)	0.7	0.8	67.4
Households with a television set (%)	87	97	98
Usage			
International voice traffic (minutes/person/month)[b]	—	—	14.0
Mobile telephone usage (minutes/user/month)	—	1,758	353
Internet users (per 100 people)	10.5	25.4	65.7
Quality			
Population covered by mobile cellular network (%)	—	100	99
Fixed broadband subscribers (% of total Internet subscrib.)	8.9	—	82.6
International Internet bandwidth (bits/second/person)	20	511	18,242
Affordability			
Price basket for residential fixed line (US$/month)	—	33.5	26.1
Price basket for mobile service (US$/month)	—	—	13.0
Price basket for Internet service (US$/month)	—	—	22.8
Price of call to United States (US$ for 3 minutes)	—	—	0.81
Trade			
ICT goods exports (% of total goods exports)	—	—	15.2
ICT goods imports (% of total goods imports)	—	—	14.6
ICT service exports (% of total service exports)	—	—	7.0
Applications			
ICT expenditure (% of GDP)	—	—	6.7
E-government Web measure index[c]	—	—	0.60
Secure Internet servers (per 1 million people, Dec. 2008)	16.4	53.6	662.6

Sources: Economic and social context: UIS and World Bank; Sector structure: ITU; Sector efficiency and capacity: ITU and World Bank; Sector performance: Global Insight/WITSA, IMF, ITU, Netcraft, UN Comtrade, UNDESA, UNPAN, Wireless Intelligence and World Bank. Produced by the Global Information and Communication Technologies Department and the Development Economics Data Group. For complete information, see Definitions and Data Sources.

Notes: Use of italics in the column entries indicates years other than those specified. — Not available. GDP = gross domestic product; GNI = gross national income; ICT = information and communication technology; and MDG = Millennium Development Goal.

a. C = competition; M = monopoly; and P = partial competition. **b.** Outgoing and incoming. **c.** Scale of 0–1, where 1 = highest presence. **d.** Millennium Development Goal indicators 8.14, 8.15, and 8.16.

Romania

	Romania 2000	Romania 2007	Upper-middle-income group 2007	Europe & Central Asia Region 2007
Economic and social context				
Population (total, million)	22	22	824	446
Urban population (% of total)	54	54	75	64
GNI per capita, World Bank Atlas method (current US$)	1,690	6,390	7,107	6,052
GDP growth, 1995–2000 and 2000–07 (avg. annual %)	−2.1	6.1	4.3	6.1
Adult literacy rate (% of ages 15 and older)	*97*	*98*	94	98
Gross primary, secondary, tertiary school enrollment (%)	68	*77*	82	82
Sector structure				
Separate telecommunications regulator	No	Yes		
Status of main fixed-line telephone operator	Mixed	Mixed		
Level of competition[a]				
International long distance service	M	C		
Mobile telephone service	C	C		
Internet service	C	C		
Sector efficiency and capacity				
Telecommunications revenue (% of GDP)	2.3	3.5	3.3	2.9
Mobile and fixed-line subscribers per employee	151	617	*566*	*532*
Telecommunications investment (% of revenue)	67.8	21.5	—	22.0
Sector performance				
Access				
Telephone lines (per 100 people)	17.4	19.8	22.6	25.7
Mobile cellular subscriptions (per 100 people)	11.1	106.2	84.1	95.0
Internet subscribers (per 100 people)	*1.7*	10.0	9.4	13.6
Personal computers (per 100 people)	3.2	19.2	*12.4*	*10.6*
Households with a television set (%)	90	*90*	*92*	*96*
Usage				
International voice traffic (minutes/person/month)[b]	2.9	3.4	—	—
Mobile telephone usage (minutes/user/month)	—	289	137	154
Internet users (per 100 people)	3.6	23.9	26.6	21.4
Quality				
Population covered by mobile cellular network (%)	97	98	95	*92*
Fixed broadband subscribers (% of total Internet subscrib.)	*4.3*	90.5	47.8	32.5
International Internet bandwidth (bits/second/person)	4	2,945	1,185	1,114
Affordability				
Price basket for residential fixed line (US$/month)	10.6	*7.2*	10.6	*5.8*
Price basket for mobile service (US$/month)	—	10.5	10.9	*11.8*
Price basket for Internet service (US$/month)	—	*17.0*	16.4	*12.0*
Price of call to United States (US$ for 3 minutes)	2.49	*0.82*	1.55	*1.63*
Trade				
ICT goods exports (% of total goods exports)	5.3	*3.1*	13.5	*1.8*
ICT goods imports (% of total goods imports)	12.0	*7.6*	16.2	*7.0*
ICT service exports (% of total service exports)	10.8	16.3	*4.6*	5.0
Applications				
ICT expenditure (% of GDP)	—	5.3	5.2	5.0
E-government Web measure index[c]	—	0.41	0.37	0.36
Secure Internet servers (per 1 million people, Dec. 2008)	*2.4*	15.5	26.2	23.9

Sources: Economic and social context: UIS and World Bank; Sector structure: ITU; Sector efficiency and capacity: ITU and World Bank; Sector performance: Global Insight/WITSA, IMF, ITU, Netcraft, UN Comtrade, UNDESA, UNPAN, Wireless Intelligence and World Bank. Produced by the Global Information and Communication Technologies Department and the Development Economics Data Group. For complete information, see Definitions and Data Sources.

Notes: Use of italics in the column entries indicates years other than those specified. — Not available. GDP = gross domestic product; GNI = gross national income; ICT = information and communication technology; and MDG = Millennium Development Goal.

a. C = competition; M = monopoly; and P = partial competition. **b.** Outgoing and incoming. **c.** Scale of 0–1, where 1 = highest presence. **d.** Millennium Development Goal indicators 8.14, 8.15, and 8.16.

Russian Federation

	Russian Federation 2000	Russian Federation 2007	Upper-middle-income group 2007	Europe & Central Asia Region 2007
Economic and social context				
Population (total, million)	146	142	824	446
Urban population (% of total)	73	73	75	64
GNI per capita, World Bank Atlas method (current US$)	1,710	7,530	7,107	6,052
GDP growth, 1995–2000 and 2000–07 (avg. annual %)	1.2	6.6	4.3	6.1
Adult literacy rate (% of ages 15 and older)	99	100	94	98
Gross primary, secondary, tertiary school enrollment (%)	—	89	82	82
Sector structure				
Separate telecommunications regulator	—	No		
Status of main fixed-line telephone operator	Mixed	Mixed		
Level of competition[a]				
International long distance service	P	P		
Mobile telephone service	C	C		
Internet service	—	—		
Sector efficiency and capacity				
Telecommunications revenue (% of GDP)	2.0	2.6	3.3	2.9
Mobile and fixed-line subscribers per employee	83	439	566	532
Telecommunications investment (% of revenue)	11.5	—	—	22.0
Sector performance				
Access				
Telephone lines (per 100 people)	21.9	31.1	22.6	25.7
Mobile cellular subscriptions (per 100 people)	2.2	114.9	84.1	95.0
Internet subscribers (per 100 people)	0.3	21.5	9.4	13.6
Personal computers (per 100 people)	6.4	13.3	12.4	10.6
Households with a television set (%)	98	98	92	96
Usage				
International voice traffic (minutes/person/month)[b]	1.1	—	—	—
Mobile telephone usage (minutes/user/month)	137	195	137	154
Internet users (per 100 people)	2.0	21.1	26.6	21.4
Quality				
Population covered by mobile cellular network (%)	—	95	95	92
Fixed broadband subscribers (% of total Internet subscrib.)	0.0	13.1	47.8	32.5
International Internet bandwidth (bits/second/person)	21	573	1,185	1,114
Affordability				
Price basket for residential fixed line (US$/month)	—	9.5	10.6	5.8
Price basket for mobile service (US$/month)	—	5.9	10.9	11.8
Price basket for Internet service (US$/month)	—	13.2	16.4	12.0
Price of call to United States (US$ for 3 minutes)	2.56	2.03	1.55	1.63
Trade				
ICT goods exports (% of total goods exports)	0.8	0.5	13.5	1.8
ICT goods imports (% of total goods imports)	5.6	10.1	16.2	7.0
ICT service exports (% of total service exports)	4.6	6.0	4.6	5.0
Applications				
ICT expenditure (% of GDP)	—	4.1	5.2	5.0
E-government Web measure index[c]	—	0.33	0.37	0.36
Secure Internet servers (per 1 million people, Dec. 2008)	2.0	7.3	26.2	23.9

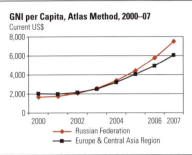

GNI per Capita, Atlas Method, 2000–07
Current US$

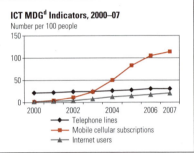

ICT MDG[d] Indicators, 2000–07
Number per 100 people

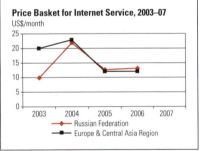

Price Basket for Internet Service, 2003–07
US$/month

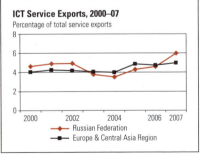

ICT Service Exports, 2000–07
Percentage of total service exports

Sources: Economic and social context: UIS and World Bank; Sector structure: ITU; Sector efficiency and capacity: ITU and World Bank; Sector performance: Global Insight/WITSA, IMF, ITU, Netcraft, UN Comtrade, UNDESA, UNPAN, Wireless Intelligence and World Bank. Produced by the Global Information and Communication Technologies Department and the Development Economics Data Group. For complete information, see Definitions and Data Sources.

Notes: Use of italics in the column entries indicates years other than those specified. — Not available. GDP = gross domestic product; GNI = gross national income; ICT = information and communication technology; and MDG = Millennium Development Goal.

a. C = competition; M = monopoly; and P = partial competition. **b**. Outgoing and incoming. **c**. Scale of 0–1, where 1 = highest presence. **d**. Millennium Development Goal indicators 8.14, 8.15, and 8.16.

Rwanda

	Rwanda 2000	Rwanda 2007	Low-income group 2007	Sub-Saharan Africa Region 2007
Economic and social context				
Population (total, million)	8	10	1,296	800
Urban population (% of total)	14	18	32	36
GNI per capita, World Bank Atlas method (current US$)	240	320	574	951
GDP growth, 1995–2000 and 2000–07 (avg. annual %)	10.1	5.8	5.6	5.1
Adult literacy rate (% of ages 15 and older)	65	—	64	62
Gross primary, secondary, tertiary school enrollment (%)	48	*51*	*51*	*51*
Sector structure				
Separate telecommunications regulator	No	Yes		
Status of main fixed-line telephone operator	*Mixed*	*Mixed*		
Level of competition[a]				
International long distance service	M	P		
Mobile telephone service	M	P		
Internet service	—	C		
Sector efficiency and capacity				
Telecommunications revenue (% of GDP)	1.0	3.2	*3.3*	*4.7*
Mobile and fixed-line subscribers per employee	189	1,040	*301*	*499*
Telecommunications investment (% of revenue)	93.1	15.8	—	—
Sector performance				
Access				
Telephone lines (per 100 people)	0.2	0.2	4.0	1.6
Mobile cellular subscriptions (per 100 people)	0.5	6.5	21.5	23.0
Internet subscribers (per 100 people)	0.0	0.1	*0.8*	*1.2*
Personal computers (per 100 people)	*0.1*	*0.3*	*1.5*	*1.8*
Households with a television set (%)	2	2	*16*	*18*
Usage				
International voice traffic (minutes/person/month)[b]	—	0.9	—	—
Mobile telephone usage (minutes/user/month)	—	—	—	—
Internet users (per 100 people)	0.1	*1.1*	5.2	4.4
Quality				
Population covered by mobile cellular network (%)	50	90	*54*	*56*
Fixed broadband subscribers (% of total Internet subscrib.)	0.0	45.6	*3.4*	*3.1*
International Internet bandwidth (bits/second/person)	*0*	16	26	36
Affordability				
Price basket for residential fixed line (US$/month)	8.4	7.8	*5.7*	*12.6*
Price basket for mobile service (US$/month)	—	11.5	*11.2*	*11.6*
Price basket for Internet service (US$/month)	—	79.7	*29.2*	*43.1*
Price of call to United States (US$ for 3 minutes)	11.23	2.43	*2.00*	*2.43*
Trade				
ICT goods exports (% of total goods exports)	0.0	0.8	*1.4*	*1.1*
ICT goods imports (% of total goods imports)	9.3	8.0	*6.7*	*8.2*
ICT service exports (% of total service exports)	—	1.9	—	4.2
Applications				
ICT expenditure (% of GDP)	—	—	—	—
E-government Web measure index[c]	—	0.27	0.11	0.16
Secure Internet servers (per 1 million people, Dec. 2008)	*0.1*	0.3	0.5	2.9

Sources: Economic and social context: UIS and World Bank; Sector structure: ITU; Sector efficiency and capacity: ITU and World Bank; Sector performance: Global Insight/WITSA, IMF, ITU, Netcraft, UN Comtrade, UNDESA, UNPAN, Wireless Intelligence and World Bank. Produced by the Global Information and Communication Technologies Department and the Development Economics Data Group. For complete information, see Definitions and Data Sources.

Notes: Use of italics in the column entries indicates years other than those specified. — Not available. GDP = gross domestic product; GNI = gross national income; ICT = information and communication technology; and MDG = Millennium Development Goal.

a. C = competition; M = monopoly; and P = partial competition. **b.** Outgoing and incoming. **c.** Scale of 0–1, where 1 = highest presence. **d.** Millennium Development Goal indicators 8.14, 8.15, and 8.16.

Saudi Arabia

	Saudi Arabia 2000	Saudi Arabia 2007	High-income group 2007
Economic and social context			
Population (total, million)	21	24	1,056
Urban population (% of total)	80	83	78
GNI per capita, World Bank Atlas method (current US$)	8,140	15,470	37,572
GDP growth, 1995–2000 and 2000–07 (avg. annual %)	2.3	4.1	2.4
Adult literacy rate (% of ages 15 and older)	79	85	99
Gross primary, secondary, tertiary school enrollment (%)	—	76	92
Sector structure			
Separate telecommunications regulator	No	Yes	
Status of main fixed-line telephone operator	Public	Mixed	
Level of competition[a]			
International long distance service	M	C	
Mobile telephone service	M	C	
Internet service	C	C	
Sector efficiency and capacity			
Telecommunications revenue (% of GDP)	2.4	*3.0*	3.1
Mobile and fixed-line subscribers per employee	189	*933*	747
Telecommunications investment (% of revenue)	33.9	*20.8*	14.3
Sector performance			
Access			
Telephone lines (per 100 people)	14.3	16.5	50.0
Mobile cellular subscriptions (per 100 people)	6.7	117.5	100.4
Internet subscribers (per 100 people)	1.0	*7.6*	25.8
Personal computers (per 100 people)	6.3	14.8	67.4
Households with a television set (%)	93	99	98
Usage			
International voice traffic (minutes/person/month)[b]	7.3	*18.0*	14.0
Mobile telephone usage (minutes/user/month)	—	—	353
Internet users (per 100 people)	2.2	26.4	65.7
Quality			
Population covered by mobile cellular network (%)	*92*	*98*	*99*
Fixed broadband subscribers (% of total Internet subscrib.)	*3.5*	*12.1*	82.6
International Internet bandwidth (bits/second/person)	16	510	18,242
Affordability			
Price basket for residential fixed line (US$/month)	11.7	*9.3*	26.1
Price basket for mobile service (US$/month)	—	*9.7*	13.0
Price basket for Internet service (US$/month)	—	*21.3*	22.8
Price of call to United States (US$ for 3 minutes)	5.20	—	0.81
Trade			
ICT goods exports (% of total goods exports)	0.1	*0.3*	15.2
ICT goods imports (% of total goods imports)	5.1	*7.8*	14.6
ICT service exports (% of total service exports)	—	—	7.0
Applications			
ICT expenditure (% of GDP)	—	4.7	6.7
E-government Web measure index[c]	—	0.46	0.60
Secure Internet servers (per 1 million people, Dec. 2008)	*0.5*	8.3	662.6

Sources: Economic and social context: UIS and World Bank; Sector structure: ITU; Sector efficiency and capacity: ITU and World Bank; Sector performance: Global Insight/WITSA, IMF, ITU, Netcraft, UN Comtrade, UNDESA, UNPAN, Wireless Intelligence and World Bank. Produced by the Global Information and Communication Technologies Department and the Development Economics Data Group. For complete information, see Definitions and Data Sources.

Notes: Use of italics in the column entries indicates years other than those specified. — Not available. GDP = gross domestic product; GNI = gross national income; ICT = information and communication technology; and MDG = Millennium Development Goal.

a. C = competition; M = monopoly; and P = partial competition. b. Outgoing and incoming. c. Scale of 0–1, where 1 = highest presence. d. Millennium Development Goal indicators 8.14, 8.15, and 8.16.

World Bank • ICT at a Glance

Senegal

	Senegal 2000	Senegal 2007	Low-income group 2007	Sub-Saharan Africa Region 2007
Economic and social context				
Population (total, million)	10	12	1,296	800
Urban population (% of total)	41	42	32	36
GNI per capita, World Bank Atlas method (current US$)	490	830	574	951
GDP growth, 1995–2000 and 2000–07 (avg. annual %)	4.4	4.5	5.6	5.1
Adult literacy rate (% of ages 15 and older)	*39*	*42*	64	62
Gross primary, secondary, tertiary school enrollment (%)	33	*40*	51	51
Sector structure				
Separate telecommunications regulator	No	Yes		
Status of main fixed-line telephone operator	Mixed	Mixed		
Level of competition[a]				
International long distance service	M	C		
Mobile telephone service	P	C		
Internet service	C	C		
Sector efficiency and capacity				
Telecommunications revenue (% of GDP)	4.2	9.9	*3.3*	*4.7*
Mobile and fixed-line subscribers per employee	324	1,859	*301*	*499*
Telecommunications investment (% of revenue)	40.4	18.7	—	—
Sector performance				
Access				
Telephone lines (per 100 people)	2.0	2.2	4.0	1.6
Mobile cellular subscriptions (per 100 people)	2.4	29.3	21.5	23.0
Internet subscribers (per 100 people)	0.1	0.3	*0.8*	*1.2*
Personal computers (per 100 people)	1.5	2.1	*1.5*	*1.8*
Households with a television set (%)	26	*41*	16	18
Usage				
International voice traffic (minutes/person/month)[b]	1.4	2.2	—	—
Mobile telephone usage (minutes/user/month)	—	—	—	—
Internet users (per 100 people)	0.4	6.6	5.2	4.4
Quality				
Population covered by mobile cellular network (%)	—	85	*54*	*56*
Fixed broadband subscribers (% of total Internet subscrib.)	0.0	97.5	*3.4*	*3.1*
International Internet bandwidth (bits/second/person)	3	137	26	36
Affordability				
Price basket for residential fixed line (US$/month)	8.7	10.5	5.7	12.6
Price basket for mobile service (US$/month)	—	9.0	11.2	11.6
Price basket for Internet service (US$/month)	—	40.4	29.2	43.1
Price of call to United States (US$ for 3 minutes)	2.23	*1.02*	*2.00*	*2.43*
Trade				
ICT goods exports (% of total goods exports)	0.2	0.5	*1.4*	*1.1*
ICT goods imports (% of total goods imports)	3.2	4.0	*6.7*	*8.2*
ICT service exports (% of total service exports)	13.5	*18.0*	—	4.2
Applications				
ICT expenditure (% of GDP)	—	10.9	—	—
E-government Web measure index[c]	—	0.31	0.11	0.16
Secure Internet servers (per 1 million people, Dec. 2008)	*0.1*	1.0	0.5	2.9

Sources: Economic and social context: UIS and World Bank; Sector structure: ITU; Sector efficiency and capacity: ITU and World Bank; Sector performance: Global Insight/WITSA, IMF, ITU, Netcraft, UN Comtrade, UNDESA, UNPAN, Wireless Intelligence and World Bank. Produced by the Global Information and Communication Technologies Department and the Development Economics Data Group. For complete information, see Definitions and Data Sources.

Notes: Use of italics in the column entries indicates years other than those specified. — Not available. GDP = gross domestic product; GNI = gross national income; ICT = information and communication technology; and MDG = Millennium Development Goal.

a. C = competition; M = monopoly; and P = partial competition. b. Outgoing and incoming. c. Scale of 0–1, where 1 = highest presence. d. Millennium Development Goal indicators 8.14, 8.15, and 8.16.

World Bank • ICT at a Glance

Serbia

	Serbia 2000	Serbia 2007	Upper-middle-income group 2007	Europe & Central Asia Region 2007
Economic and social context				
Population (total, million)	8	7	824	446
Urban population (% of total)	51	52	75	64
GNI per capita, World Bank Atlas method (current US$)	1,470	4,540	7,107	6,052
GDP growth, 1995–2000 and 2000–07 (avg. annual %)	0.0	5.6	4.3	6.1
Adult literacy rate (% of ages 15 and older)	—	—	94	98
Gross primary, secondary, tertiary school enrollment (%)	—	—	82	82
Sector structure				
Separate telecommunications regulator	—	Yes		
Status of main fixed-line telephone operator	—	Mixed		
Level of competition[a]				
International long distance service	—	P		
Mobile telephone service	—	C		
Internet service	—	C		
Sector efficiency and capacity				
Telecommunications revenue (% of GDP)	3.2	5.0	3.3	2.9
Mobile and fixed-line subscribers per employee	247	787	566	532
Telecommunications investment (% of revenue)	28.0	31.2	—	22.0
Sector performance				
Access				
Telephone lines (per 100 people)	32.0	40.6	22.6	25.7
Mobile cellular subscriptions (per 100 people)	17.3	114.5	84.1	95.0
Internet subscribers (per 100 people)	0.4	13.7	9.4	13.6
Personal computers (per 100 people)	3.2	24.4	12.4	10.6
Households with a television set (%)	92	80	92	96
Usage				
International voice traffic (minutes/person/month)[b]	11.3	12.0	—	—
Mobile telephone usage (minutes/user/month)	—	91	137	154
Internet users (per 100 people)	5.3	20.3	26.6	21.4
Quality				
Population covered by mobile cellular network (%)	77	92	95	92
Fixed broadband subscribers (% of total Internet subscrib.)	0.0	32.2	47.8	32.5
International Internet bandwidth (bits/second/person)	1	2,861	1,185	1,114
Affordability				
Price basket for residential fixed line (US$/month)	—	2.6	10.6	5.8
Price basket for mobile service (US$/month)	—	5.8	10.9	11.8
Price basket for Internet service (US$/month)	—	8.9	16.4	12.0
Price of call to United States (US$ for 3 minutes)	—	—	1.55	1.63
Trade				
ICT goods exports (% of total goods exports)	—	—	13.5	1.8
ICT goods imports (% of total goods imports)	—	—	16.2	7.0
ICT service exports (% of total service exports)	—	—	4.6	5.0
Applications				
ICT expenditure (% of GDP)	—	—	5.2	5.0
E-government Web measure index[c]	—	0.35	0.37	0.36
Secure Internet servers (per 1 million people, Dec. 2008)	—	2.4	26.2	23.9

Sources: Economic and social context: UIS and World Bank; Sector structure: ITU; Sector efficiency and capacity: ITU and World Bank; Sector performance: Global Insight/WITSA, IMF, ITU, Netcraft, UN Comtrade, UNDESA, UNPAN, Wireless Intelligence and World Bank. Produced by the Global Information and Communication Technologies Department and the Development Economics Data Group. For complete information, see Definitions and Data Sources.

Notes: Use of italics in the column entries indicates years other than those specified. — Not available. GDP = gross domestic product; GNI = gross national income; ICT = information and communication technology; and MDG = Millennium Development Goal.

a. C = competition; M = monopoly; and P = partial competition. b. Outgoing and incoming. c. Scale of 0–1, where 1 = highest presence. d. Millennium Development Goal indicators 8.14, 8.15, and 8.16.

World Bank • ICT at a Glance

Sierra Leone

	Sierra Leone 2000	Sierra Leone 2007	Low-income group 2007	Sub-Saharan Africa Region 2007
Economic and social context				
Population (total, million)	5	6	1,296	800
Urban population (% of total)	36	37	32	36
GNI per capita, World Bank Atlas method (current US$)	140	260	574	951
GDP growth, 1995–2000 and 2000–07 (avg. annual %)	−5.0	11.2	5.6	5.1
Adult literacy rate (% of ages 15 and older)	—	38	64	62
Gross primary, secondary, tertiary school enrollment (%)	45	74	51	51
Sector structure				
Separate telecommunications regulator	—	Yes		
Status of main fixed-line telephone operator	Public	Public		
Level of competition[a]				
International long distance service	M	P		
Mobile telephone service	C	C		
Internet service	C	P		
Sector efficiency and capacity				
Telecommunications revenue (% of GDP)	—	—	3.3	4.7
Mobile and fixed-line subscribers per employee	31	—	301	499
Telecommunications investment (% of revenue)	—	—	—	—
Sector performance				
Access				
Telephone lines (per 100 people)	0.4	—	4.0	1.6
Mobile cellular subscriptions (per 100 people)	0.3	13.3	21.5	23.0
Internet subscribers (per 100 people)	0.0	—	0.8	1.2
Personal computers (per 100 people)	—	—	1.5	1.8
Households with a television set (%)	4	—	16	18
Usage				
International voice traffic (minutes/person/month)[b]	—	—	—	—
Mobile telephone usage (minutes/user/month)	—	—	—	—
Internet users (per 100 people)	0.1	0.2	5.2	4.4
Quality				
Population covered by mobile cellular network (%)	—	70	54	56
Fixed broadband subscribers (% of total Internet subscrib.)	—	—	3.4	3.1
International Internet bandwidth (bits/second/person)	0	—	26	36
Affordability				
Price basket for residential fixed line (US$/month)	3.0	—	5.7	12.6
Price basket for mobile service (US$/month)	—	19.4	11.2	11.6
Price basket for Internet service (US$/month)	—	10.7	29.2	43.1
Price of call to United States (US$ for 3 minutes)	2.74	—	2.00	2.43
Trade				
ICT goods exports (% of total goods exports)	—	—	1.4	1.1
ICT goods imports (% of total goods imports)	—	—	6.7	8.2
ICT service exports (% of total service exports)	—	0.2	—	4.2
Applications				
ICT expenditure (% of GDP)	—	—	—	—
E-government Web measure index[c]	—	0.06	0.11	0.16
Secure Internet servers (per 1 million people, Dec. 2008)	*0.2*	*0.7*	*0.5*	*2.9*

Sources: Economic and social context: UIS and World Bank; Sector structure: ITU; Sector efficiency and capacity: ITU and World Bank; Sector performance: Global Insight/WITSA, IMF, ITU, Netcraft, UN Comtrade, UNDESA, UNPAN, Wireless Intelligence and World Bank. Produced by the Global Information and Communication Technologies Department and the Development Economics Data Group. For complete information, see Definitions and Data Sources.

Notes: Use of italics in the column entries indicates years other than those specified. — Not available. GDP = gross domestic product; GNI = gross national income; ICT = information and communication technology; and MDG = Millennium Development Goal.

a. C = competition; M = monopoly; and P = partial competition. **b.** Outgoing and incoming. **c.** Scale of 0–1, where 1 = highest presence. **d.** Millennium Development Goal indicators 8.14, 8.15, and 8.16.

Singapore

	Singapore 2000	Singapore 2007	High-income group 2007
Economic and social context			
Population (total, million)	4	5	1,056
Urban population (% of total)	100	100	78
GNI per capita, World Bank Atlas method (current US$)	22,970	32,340	37,572
GDP growth, 1995–2000 and 2000–07 (avg. annual %)	5.7	5.8	2.4
Adult literacy rate (% of ages 15 and older)	93	94	99
Gross primary, secondary, tertiary school enrollment (%)	—	—	92
Sector structure			
Separate telecommunications regulator	Yes	Yes	
Status of main fixed-line telephone operator	Mixed	Mixed	
Level of competition[a]			
International long distance service	C	C	
Mobile telephone service	C	C	
Internet service	C	C	
Sector efficiency and capacity			
Telecommunications revenue (% of GDP)	3.4	2.9	3.1
Mobile and fixed-line subscribers per employee	584	—	747
Telecommunications investment (% of revenue)	14.9	12.6	14.3
Sector performance			
Access			
Telephone lines (per 100 people)	48.3	40.6	50.0
Mobile cellular subscriptions (per 100 people)	68.2	129.1	100.4
Internet subscribers (per 100 people)	21.2	42.2	25.8
Personal computers (per 100 people)	48.2	74.3	67.4
Households with a television set (%)	99	98	98
Usage			
International voice traffic (minutes/person/month)[b]	53.7	127.6	14.0
Mobile telephone usage (minutes/user/month)	411	389	353
Internet users (per 100 people)	32.3	65.7	65.7
Quality			
Population covered by mobile cellular network (%)	100	100	99
Fixed broadband subscribers (% of total Internet subscrib.)	8.1	46.2	82.6
International Internet bandwidth (bits/second/person)	558	22,783	18,242
Affordability			
Price basket for residential fixed line (US$/month)	6.2	7.0	26.1
Price basket for mobile service (US$/month)	—	3.7	13.0
Price basket for Internet service (US$/month)	—	17.7	22.8
Price of call to United States (US$ for 3 minutes)	0.68	0.69	0.81
Trade			
ICT goods exports (% of total goods exports)	56.1	45.6	15.2
ICT goods imports (% of total goods imports)	44.4	38.3	14.6
ICT service exports (% of total service exports)	2.4	3.1	7.0
Applications			
ICT expenditure (% of GDP)	—	6.5	6.7
E-government Web measure index[c]		0.61	0.60
Secure Internet servers (per 1 million people, Dec. 2008)	126.9	390.3	662.6

Sources: Economic and social context: UIS and World Bank; Sector structure: ITU; Sector efficiency and capacity: ITU and World Bank; Sector performance: Global Insight/WITSA, IMF, ITU, Netcraft, UN Comtrade, UNDESA, UNPAN, Wireless Intelligence and World Bank. Produced by the Global Information and Communication Technologies Department and the Development Economics Data Group. For complete information, see Definitions and Data Sources.

Notes: Use of italics in the column entries indicates years other than those specified. — Not available. GDP = gross domestic product; GNI = gross national income; ICT = information and communication technology; and MDG = Millennium Development Goal.

a. C = competition; M = monopoly; and P = partial competition. b. Outgoing and incoming. c. Scale of 0–1, where 1 = highest presence. d. Millennium Development Goal indicators 8.14, 8.15, and 8.16.

Slovak Republic

	Slovak Republic 2000	Slovak Republic 2007	High-income group 2007
Economic and social context			
Population (total, million)	5	5	1,056
Urban population (% of total)	56	56	78
GNI per capita, World Bank Atlas method (current US$)	3,850	11,720	37,572
GDP growth, 1995–2000 and 2000–07 (avg. annual %)	3.8	6.0	2.4
Adult literacy rate (% of ages 15 and older)	—	—	99
Gross primary, secondary, tertiary school enrollment (%)	72	78	92
Sector structure			
Separate telecommunications regulator	Yes	Yes	
Status of main fixed-line telephone operator	Mixed	Mixed	
Level of competition[a]			
International long distance service	M	C	
Mobile telephone service	C	C	
Internet service	C	C	
Sector efficiency and capacity			
Telecommunications revenue (% of GDP)	3.9	3.4	3.1
Mobile and fixed-line subscribers per employee	191	748	747
Telecommunications investment (% of revenue)	22.6	20.9	14.3
Sector performance			
Access			
Telephone lines (per 100 people)	31.5	21.3	50.0
Mobile cellular subscriptions (per 100 people)	23.1	112.4	100.4
Internet subscribers (per 100 people)	1.3	9.9	25.8
Personal computers (per 100 people)	13.7	51.4	67.4
Households with a television set (%)	96	78	98
Usage			
International voice traffic (minutes/person/month)[b]	4.9	8.1	14.0
Mobile telephone usage (minutes/user/month)	148	84	353
Internet users (per 100 people)	9.4	55.9	65.7
Quality			
Population covered by mobile cellular network (%)	98	100	99
Fixed broadband subscribers (% of total Internet subscrib.)	0.0	88.6	82.6
International Internet bandwidth (bits/second/person)	41	5,555	18,242
Affordability			
Price basket for residential fixed line (US$/month)	9.2	17.3	26.1
Price basket for mobile service (US$/month)	—	12.2	13.0
Price basket for Internet service (US$/month)	—	19.8	22.8
Price of call to United States (US$ for 3 minutes)	1.13	1.06	0.81
Trade			
ICT goods exports (% of total goods exports)	3.9	13.2	15.2
ICT goods imports (% of total goods imports)	7.8	10.3	14.6
ICT service exports (% of total service exports)	4.6	6.6	7.0
Applications			
ICT expenditure (% of GDP)	—	6.0	6.7
E-government Web measure index[c]	—	0.47	0.60
Secure Internet servers (per 1 million people, Dec. 2008)	14.7	57.9	662.6

Sources: Economic and social context: UIS and World Bank; Sector structure: ITU; Sector efficiency and capacity: ITU and World Bank; Sector performance: Global Insight/WITSA, IMF, ITU, Netcraft, UN Comtrade, UNDESA, UNPAN, Wireless Intelligence and World Bank. Produced by the Global Information and Communication Technologies Department and the Development Economics Data Group. For complete information, see Definitions and Data Sources.

Notes: Use of italics in the column entries indicates years other than those specified. — Not available. GDP = gross domestic product; GNI = gross national income; ICT = information and communication technology; and MDG = Millennium Development Goal.

a. C = competition; M = monopoly; and P = partial competition. b. Outgoing and incoming. c. Scale of 0–1, where 1 = highest presence. d. Millennium Development Goal indicators 8.14, 8.15, and 8.16.

Slovenia

	Slovenia 2000	Slovenia 2007	High-income group 2007
Economic and social context			
Population (total, million)	2	2	1,056
Urban population (% of total)	51	49	78
GNI per capita, World Bank Atlas method (current US$)	11,090	21,510	37,572
GDP growth, 1995–2000 and 2000–07 (avg. annual %)	4.4	4.3	2.4
Adult literacy rate (% of ages 15 and older)	—	100	99
Gross primary, secondary, tertiary school enrollment (%)	87	94	92
Sector structure			
Separate telecommunications regulator	No	Yes	
Status of main fixed-line telephone operator	Mixed	Mixed	
Level of competition[a]			
International long distance service	M	C	
Mobile telephone service	P	C	
Internet service	C	C	
Sector efficiency and capacity			
Telecommunications revenue (% of GDP)	1.8	3.2	3.1
Mobile and fixed-line subscribers per employee	541	587	747
Telecommunications investment (% of revenue)	97.6	29.5	14.3
Sector performance			
Access			
Telephone lines (per 100 people)	39.5	42.5	50.0
Mobile cellular subscriptions (per 100 people)	61.1	95.6	100.4
Internet subscribers (per 100 people)	7.0	20.7	25.8
Personal computers (per 100 people)	27.6	42.5	67.4
Households with a television set (%)	92	97	98
Usage			
International voice traffic (minutes/person/month)[b]	—	7.6	14.0
Mobile telephone usage (minutes/user/month)	—	134	353
Internet users (per 100 people)	15.1	52.6	65.7
Quality			
Population covered by mobile cellular network (%)	98	100	99
Fixed broadband subscribers (% of total Internet subscrib.)	2.0	82.6	82.6
International Internet bandwidth (bits/second/person)	95	6,720	18,242
Affordability			
Price basket for residential fixed line (US$/month)	10.1	19.5	26.1
Price basket for mobile service (US$/month)	—	10.1	13.0
Price basket for Internet service (US$/month)	—	18.8	22.8
Price of call to United States (US$ for 3 minutes)	0.81	0.65	0.81
Trade			
ICT goods exports (% of total goods exports)	4.6	3.0	15.2
ICT goods imports (% of total goods imports)	6.9	5.3	14.6
ICT service exports (% of total service exports)	4.2	5.2	7.0
Applications			
ICT expenditure (% of GDP)	—	4.7	6.7
E-government Web measure index[c]	—	0.50	0.60
Secure Internet servers (per 1 million people, Dec. 2008)	51.2	170.1	662.6

Sources: Economic and social context: UIS and World Bank; Sector structure: ITU; Sector efficiency and capacity: ITU and World Bank; Sector performance: Global Insight/WITSA, IMF, ITU, Netcraft, UN Comtrade, UNDESA, UNPAN, Wireless Intelligence and World Bank. Produced by the Global Information and Communication Technologies Department and the Development Economics Data Group. For complete information, see Definitions and Data Sources.

Notes: Use of italics in the column entries indicates years other than those specified. — Not available. GDP = gross domestic product; GNI = gross national income; ICT = information and communication technology; and MDG = Millennium Development Goal.

[a]. C = competition; M = monopoly; and P = partial competition. [b]. Outgoing and incoming. [c]. Scale of 0–1, where 1 = highest presence. [d]. Millennium Development Goal indicators 8.14, 8.15, and 8.16.

World Bank • ICT at a Glance

Somalia

	Somalia 2000	Somalia 2007	Low-income group 2007	Sub-Saharan Africa Region 2007
Economic and social context				
Population (total, million)	7	9	1,296	800
Urban population (% of total)	33	36	32	36
GNI per capita, World Bank Atlas method (current US$)	—	—	574	951
GDP growth, 1995–2000 and 2000–07 (avg. annual %)	—	—	5.6	5.1
Adult literacy rate (% of ages 15 and older)	—	—	64	62
Gross primary, secondary, tertiary school enrollment (%)	—	—	51	51
Sector structure				
Separate telecommunications regulator	—	No		
Status of main fixed-line telephone operator	Private	Private		
Level of competition[a]				
International long distance service	C	C		
Mobile telephone service	—	—		
Internet service	—	—		
Sector efficiency and capacity				
Telecommunications revenue (% of GDP)	—	—	3.3	4.7
Mobile and fixed-line subscribers per employee	—	—	301	499
Telecommunications investment (% of revenue)	—	—	—	—
Sector performance				
Access				
Telephone lines (per 100 people)	0.4	1.1	4.0	1.6
Mobile cellular subscriptions (per 100 people)	1.1	6.9	21.5	23.0
Internet subscribers (per 100 people)	0.0	*0.1*	0.8	1.2
Personal computers (per 100 people)	0.1	0.9	1.5	1.8
Households with a television set (%)	8	8	16	18
Usage				
International voice traffic (minutes/person/month)[b]	—	—	—	—
Mobile telephone usage (minutes/user/month)	—	—	—	—
Internet users (per 100 people)	0.2	1.1	5.2	4.4
Quality				
Population covered by mobile cellular network (%)	—	—	54	56
Fixed broadband subscribers (% of total Internet subscrib.)	0.0	0.0	3.4	3.1
International Internet bandwidth (bits/second/person)	0	0	26	36
Affordability				
Price basket for residential fixed line (US$/month)	—	—	5.7	12.6
Price basket for mobile service (US$/month)	—	5.1	11.2	11.6
Price basket for Internet service (US$/month)	—	—	29.2	43.1
Price of call to United States (US$ for 3 minutes)	—	—	2.00	2.43
Trade				
ICT goods exports (% of total goods exports)	—	—	1.4	1.1
ICT goods imports (% of total goods imports)	—	—	6.7	8.2
ICT service exports (% of total service exports)	—	—	—	4.2
Applications				
ICT expenditure (% of GDP)	—	—	—	—
E-government Web measure index[c]	—	0.00	0.11	0.16
Secure Internet servers (per 1 million people, Dec. 2008)	—	0.1	0.5	2.9

Sources: Economic and social context: UIS and World Bank; Sector structure: ITU; Sector efficiency and capacity: ITU and World Bank; Sector performance: Global Insight/WITSA, IMF, ITU, Netcraft, UN Comtrade, UNDESA, UNPAN, Wireless Intelligence and World Bank. Produced by the Global Information and Communication Technologies Department and the Development Economics Data Group. For complete information, see Definitions and Data Sources.

Notes: Use of italics in the column entries indicates years other than those specified. — Not available. GDP = gross domestic product; GNI = gross national income; ICT = information and communication technology; and MDG = Millennium Development Goal.

a. C = competition; M = monopoly; and P = partial competition. b. Outgoing and incoming. c. Scale of 0–1, where 1 = highest presence. d. Millennium Development Goal indicators 8.14, 8.15, and 8.16.

Information and Communications for Development 2009

World Bank • ICT at a Glance

South Africa

	South Africa 2000	South Africa 2007	Upper-middle-income group 2007	Sub-Saharan Africa Region 2007
Economic and social context				
Population (total, million)	44	48	824	800
Urban population (% of total)	57	60	75	36
GNI per capita, World Bank Atlas method (current US$)	3,050	5,720	7,107	951
GDP growth, 1995–2000 and 2000–07 (avg. annual %)	2.5	4.3	4.3	5.1
Adult literacy rate (% of ages 15 and older)	—	88	94	62
Gross primary, secondary, tertiary school enrollment (%)	76	77	82	51
Sector structure				
Separate telecommunications regulator	Yes	Yes		
Status of main fixed-line telephone operator	Mixed	Mixed		
Level of competition[a]				
International long distance service	M	C		
Mobile telephone service	C	P		
Internet service	—	C		
Sector efficiency and capacity				
Telecommunications revenue (% of GDP)	5.1	7.5	3.3	4.7
Mobile and fixed-line subscribers per employee	264	1,145	566	499
Telecommunications investment (% of revenue)	25.5	9.8	—	—
Sector performance				
Access				
Telephone lines (per 100 people)	11.3	9.7	22.6	1.6
Mobile cellular subscriptions (per 100 people)	19.0	88.4	84.1	23.0
Internet subscribers (per 100 people)	1.6	9.1	9.4	1.2
Personal computers (per 100 people)	6.6	8.5	12.4	1.8
Households with a television set (%)	55	59	92	18
Usage				
International voice traffic (minutes/person/month)[b]	2.1	—	—	—
Mobile telephone usage (minutes/user/month)	—	106	137	—
Internet users (per 100 people)	5.5	8.3	26.6	4.4
Quality				
Population covered by mobile cellular network (%)	92	100	95	56
Fixed broadband subscribers (% of total Internet subscrib.)	0.3	3.9	47.8	3.1
International Internet bandwidth (bits/second/person)	8	71	1,185	36
Affordability				
Price basket for residential fixed line (US$/month)	13.3	20.8	10.6	12.6
Price basket for mobile service (US$/month)	—	13.9	10.9	11.6
Price basket for Internet service (US$/month)	—	28.2	16.4	43.1
Price of call to United States (US$ for 3 minutes)	1.98	0.79	1.55	2.43
Trade				
ICT goods exports (% of total goods exports)	2.0	1.8	13.5	1.1
ICT goods imports (% of total goods imports)	13.6	11.3	16.2	8.2
ICT service exports (% of total service exports)	2.9	3.9	4.6	4.2
Applications				
ICT expenditure (% of GDP)	—	9.7	5.2	—
E-government Web measure index[c]	—	0.55	0.37	0.16
Secure Internet servers (per 1 million people, Dec. 2008)	11.6	36.8	26.2	2.9

Sources: Economic and social context: UIS and World Bank; Sector structure: ITU; Sector efficiency and capacity: ITU and World Bank; Sector performance: Global Insight/WITSA, IMF, ITU, Netcraft, UN Comtrade, UNDESA, UNPAN, Wireless Intelligence and World Bank. Produced by the Global Information and Communication Technologies Department and the Development Economics Data Group. For complete information, see Definitions and Data Sources.

Notes: Use of italics in the column entries indicates years other than those specified. — Not available. GDP = gross domestic product; GNI = gross national income; ICT = information and communication technology; and MDG = Millennium Development Goal.

a. C = competition; M = monopoly; and P = partial competition. b. Outgoing and incoming. c. Scale of 0–1, where 1 = highest presence. d. Millennium Development Goal indicators 8.14, 8.15, and 8.16.

World Bank • ICT at a Glance

Spain

	Spain 2000	Spain 2007	High-income group 2007
Economic and social context			
Population (total, million)	40	45	1,056
Urban population (% of total)	76	77	78
GNI per capita, World Bank Atlas method (current US$)	15,420	29,290	37,572
GDP growth, 1995–2000 and 2000–07 (avg. annual %)	4.2	3.4	2.4
Adult literacy rate (% of ages 15 and older)	—	98	99
Gross primary, secondary, tertiary school enrollment (%)	92	98	92
Sector structure			
Separate telecommunications regulator	Yes	Yes	
Status of main fixed-line telephone operator	Private	Private	
Level of competition[a]			
International long distance service	C	C	
Mobile telephone service	C	C	
Internet service	C	C	
Sector efficiency and capacity			
Telecommunications revenue (% of GDP)	3.2	4.2	3.1
Mobile and fixed-line subscribers per employee	638	809	747
Telecommunications investment (% of revenue)	36.7	13.1	14.3
Sector performance			
Access			
Telephone lines (per 100 people)	42.5	45.3	50.0
Mobile cellular subscriptions (per 100 people)	60.3	107.9	100.4
Internet subscribers (per 100 people)	8.0	19.2	25.8
Personal computers (per 100 people)	17.4	39.3	67.4
Households with a television set (%)	97	96	98
Usage			
International voice traffic (minutes/person/month)[b]	11.6	9.7	14.0
Mobile telephone usage (minutes/user/month)	101	152	353
Internet users (per 100 people)	13.6	51.3	65.7
Quality			
Population covered by mobile cellular network (%)	99	99	99
Fixed broadband subscribers (% of total Internet subscrib.)	2.4	93.8	82.6
International Internet bandwidth (bits/second/person)	297	11,008	18,242
Affordability			
Price basket for residential fixed line (US$/month)	14.7	25.8	26.1
Price basket for mobile service (US$/month)	—	23.6	13.0
Price basket for Internet service (US$/month)	—	32.0	22.8
Price of call to United States (US$ for 3 minutes)	1.08	0.60	0.81
Trade			
ICT goods exports (% of total goods exports)	5.4	4.0	15.2
ICT goods imports (% of total goods imports)	9.3	7.9	14.6
ICT service exports (% of total service exports)	5.2	5.4	7.0
Applications			
ICT expenditure (% of GDP)	—	5.5	6.7
E-government Web measure index[c]	—	0.70	0.60
Secure Internet servers (per 1 million people, Dec. 2008)	23.0	170.1	662.6

Sources: Economic and social context: UIS and World Bank; Sector structure: ITU; Sector efficiency and capacity: ITU and World Bank; Sector performance: Global Insight/WITSA, IMF, ITU, Netcraft, UN Comtrade, UNDESA, UNPAN, Wireless Intelligence and World Bank. Produced by the Global Information and Communication Technologies Department and the Development Economics Data Group. For complete information, see Definitions and Data Sources.

Notes: Use of italics in the column entries indicates years other than those specified. — Not available. GDP = gross domestic product; GNI = gross national income; ICT = information and communication technology; and MDG = Millennium Development Goal.

a. C = competition; M = monopoly; and P = partial competition. b. Outgoing and incoming. c. Scale of 0–1, where 1 = highest presence. d. Millennium Development Goal indicators 8.14, 8.15, and 8.16.

Sri Lanka

	Sri Lanka 2000	Sri Lanka 2007	Lower-middle-income group 2007	South Asia Region 2007
Economic and social context				
Population (total, million)	19	20	3,435	1,522
Urban population (% of total)	16	15	42	29
GNI per capita, World Bank Atlas method (current US$)	880	1,540	1,905	880
GDP growth, 1995–2000 and 2000–07 (avg. annual %)	5.1	5.3	8.0	7.3
Adult literacy rate (% of ages 15 and older)	*91*	*91*	83	63
Gross primary, secondary, tertiary school enrollment (%)	*64*	*63*	68	60
Sector structure				
Separate telecommunications regulator	Yes	Yes		
Status of main fixed-line telephone operator	Mixed	Mixed		
Level of competition[a]				
International long distance service	M	M		
Mobile telephone service	P	P		
Internet service	C	P		
Sector efficiency and capacity				
Telecommunications revenue (% of GDP)	1.9	2.5	*3.1*	2.1
Mobile and fixed-line subscribers per employee	101	755	*624*	660
Telecommunications investment (% of revenue)	35.8	*12.2*	25.3	—
Sector performance				
Access				
Telephone lines (per 100 people)	4.1	13.7	15.3	3.2
Mobile cellular subscriptions (per 100 people)	2.3	39.9	38.9	22.8
Internet subscribers (per 100 people)	0.2	1.0	6.0	1.3
Personal computers (per 100 people)	0.7	*3.7*	4.6	3.3
Households with a television set (%)	22	32	79	42
Usage				
International voice traffic (minutes/person/month)[b]	0.8	2.9	—	—
Mobile telephone usage (minutes/user/month)	—	86	322	364
Internet users (per 100 people)	0.6	3.9	12.4	6.6
Quality				
Population covered by mobile cellular network (%)	*58*	90	*80*	*61*
Fixed broadband subscribers (% of total Internet subscrib.)	*0.5*	31.3	40.4	18.9
International Internet bandwidth (bits/second/person)	1	118	199	31
Affordability				
Price basket for residential fixed line (US$/month)	6.9	*8.2*	7.2	4.0
Price basket for mobile service (US$/month)	—	1.2	9.8	2.4
Price basket for Internet service (US$/month)	—	4.4	16.7	8.0
Price of call to United States (US$ for 3 minutes)	3.29	2.11	2.08	2.02
Trade				
ICT goods exports (% of total goods exports)	*2.7*	1.7	20.6	1.2
ICT goods imports (% of total goods imports)	4.2	4.9	20.2	8.1
ICT service exports (% of total service exports)	7.8	10.6	15.6	39.0
Applications				
ICT expenditure (% of GDP)	—	6.0	6.5	5.7
E-government Web measure index[c]	—	0.39	0.33	0.37
Secure Internet servers (per 1 million people, Dec. 2008)	*0.3*	3.2	1.8	1.1

Sources: Economic and social context: UIS and World Bank; Sector structure: ITU; Sector efficiency and capacity: ITU and World Bank; Sector performance: Global Insight/WITSA, IMF, ITU, Netcraft, UN Comtrade, UNDESA, UNPAN, Wireless Intelligence and World Bank. Produced by the Global Information and Communication Technologies Department and the Development Economics Data Group. For complete information, see Definitions and Data Sources.

Notes: Use of italics in the column entries indicates years other than those specified. — Not available. GDP = gross domestic product; GNI = gross national income; ICT = information and communication technology; and MDG = Millennium Development Goal.

a. C = competition; M = monopoly; and P = partial competition. **b.** Outgoing and incoming. **c.** Scale of 0–1, where 1 = highest presence. **d.** Millennium Development Goal indicators 8.14, 8.15, and 8.16.

World Bank • ICT at a Glance

Sudan

	Sudan 2000	Sudan 2007	Lower-middle-income group 2007	Sub-Saharan Africa Region 2007
Economic and social context				
Population (total, million)	33	39	3,435	800
Urban population (% of total)	36	43	42	36
GNI per capita, World Bank Atlas method (current US$)	330	950	1,905	951
GDP growth, 1995–2000 and 2000–07 (avg. annual %)	6.2	7.1	8.0	5.1
Adult literacy rate (% of ages 15 and older)	61	—	83	62
Gross primary, secondary, tertiary school enrollment (%)	31	*37*	68	51
Sector structure				
Separate telecommunications regulator	Yes	Yes		
Status of main fixed-line telephone operator	Mixed	Mixed		
Level of competition[a]				
International long distance service	M	P		
Mobile telephone service	M	P		
Internet service	C	C		
Sector efficiency and capacity				
Telecommunications revenue (% of GDP)	1.1	3.7	3.1	4.7
Mobile and fixed-line subscribers per employee	146	1,557	624	499
Telecommunications investment (% of revenue)	71.2	33.4	25.3	—
Sector performance				
Access				
Telephone lines (per 100 people)	1.2	0.9	15.3	1.6
Mobile cellular subscriptions (per 100 people)	0.1	21.3	38.9	23.0
Internet subscribers (per 100 people)	—	*0.1*	6.0	1.2
Personal computers (per 100 people)	0.3	11.2	4.6	1.8
Households with a television set (%)	17	16	79	18
Usage				
International voice traffic (minutes/person/month)[b]	0.5	0.6	—	—
Mobile telephone usage (minutes/user/month)	—	—	322	—
Internet users (per 100 people)	0.0	9.1	12.4	4.4
Quality				
Population covered by mobile cellular network (%)	20	60	80	56
Fixed broadband subscribers (% of total Internet subscrib.)	—	4.7	40.4	3.1
International Internet bandwidth (bits/second/person)	0	345	199	36
Affordability				
Price basket for residential fixed line (US$/month)	3.8	*6.3*	7.2	12.6
Price basket for mobile service (US$/month)	—	3.8	9.8	11.6
Price basket for Internet service (US$/month)	—	28.9	16.7	43.1
Price of call to United States (US$ for 3 minutes)	42.02	—	2.08	2.43
Trade				
ICT goods exports (% of total goods exports)	0.0	*0.0*	20.6	1.1
ICT goods imports (% of total goods imports)	6.5	*7.5*	20.2	8.2
ICT service exports (% of total service exports)	3.4	5.4	15.6	4.2
Applications				
ICT expenditure (% of GDP)	—	—	6.5	—
E-government Web measure index[c]	—	0.06	0.33	0.16
Secure Internet servers (per 1 million people, Dec. 2008)	—	0.0	1.8	2.9

Sources: Economic and social context: UIS and World Bank; Sector structure: ITU; Sector efficiency and capacity: ITU and World Bank; Sector performance: Global Insight/WITSA, IMF, ITU, Netcraft, UN Comtrade, UNDESA, UNPAN, Wireless Intelligence and World Bank. Produced by the Global Information and Communication Technologies Department and the Development Economics Data Group. For complete information, see Definitions and Data Sources.

Notes: Use of italics in the column entries indicates years other than those specified. — Not available. GDP = gross domestic product; GNI = gross national income; ICT = information and communication technology; and MDG = Millennium Development Goal.

a. C = competition; M = monopoly; and P = partial competition. **b.** Outgoing and incoming. **c.** Scale of 0–1, where 1 = highest presence. **d.** Millennium Development Goal indicators 8.14, 8.15, and 8.16.

World Bank • ICT at a Glance

Swaziland

	Swaziland 2000	Swaziland 2007	Lower-middle-income group 2007	Sub-Saharan Africa Region 2007
Economic and social context				
Population (total, million)	1	1	3,435	800
Urban population (% of total)	23	25	42	36
GNI per capita, World Bank Atlas method (current US$)	1,600	2,560	1,905	951
GDP growth, 1995–2000 and 2000–07 (avg. annual %)	4.3	2.6	8.0	5.1
Adult literacy rate (% of ages 15 and older)	80	—	83	62
Gross primary, secondary, tertiary school enrollment (%)	61	*60*	68	51
Sector structure				
Separate telecommunications regulator	—	No		
Status of main fixed-line telephone operator	Public	Public		
Level of competition[a]				
International long distance service	M	M		
Mobile telephone service	M	M		
Internet service	C	—		
Sector efficiency and capacity				
Telecommunications revenue (% of GDP)	2.6	12.7	3.1	4.7
Mobile and fixed-line subscribers per employee	137	279	624	499
Telecommunications investment (% of revenue)	9.9	13.2	25.3	—
Sector performance				
Access				
Telephone lines (per 100 people)	3.0	*3.9*	15.3	1.6
Mobile cellular subscriptions (per 100 people)	3.2	33.1	38.9	23.0
Internet subscribers (per 100 people)	0.5	*1.8*	6.0	*1.2*
Personal computers (per 100 people)	1.1	3.7	4.6	1.8
Households with a television set (%)	18	18	79	18
Usage				
International voice traffic (minutes/person/month)[b]	3.9	*4.0*	—	—
Mobile telephone usage (minutes/user/month)	—	—	322	—
Internet users (per 100 people)	1.0	3.7	12.4	4.4
Quality				
Population covered by mobile cellular network (%)	70	*90*	80	56
Fixed broadband subscribers (% of total Internet subscrib.)	0.0	0.0	40.4	3.1
International Internet bandwidth (bits/second/person)	1	1	199	36
Affordability				
Price basket for residential fixed line (US$/month)	4.1	5.7	7.2	12.6
Price basket for mobile service (US$/month)	—	13.0	9.8	11.6
Price basket for Internet service (US$/month)	—	39.1	16.7	43.1
Price of call to United States (US$ for 3 minutes)	3.68	2.97	2.08	2.43
Trade				
ICT goods exports (% of total goods exports)	0.4	0.0	20.6	1.1
ICT goods imports (% of total goods imports)	2.8	3.8	20.2	8.2
ICT service exports (% of total service exports)	1.3	1.4	15.6	4.2
Applications				
ICT expenditure (% of GDP)	—	—	6.5	—
E-government Web measure index[c]	—	0.25	0.33	0.16
Secure Internet servers (per 1 million people, Dec. 2008)	*0.9*	*5.2*	*1.8*	*2.9*

Sources: Economic and social context: UIS and World Bank; Sector structure: ITU; Sector efficiency and capacity: ITU and World Bank; Sector performance: Global Insight/WITSA, IMF, ITU, Netcraft, UN Comtrade, UNDESA, UNPAN, Wireless Intelligence and World Bank. Produced by the Global Information and Communication Technologies Department and the Development Economics Data Group. For complete information, see Definitions and Data Sources.

Notes: Use of italics in the column entries indicates years other than those specified. — Not available. GDP = gross domestic product; GNI = gross national income; ICT = information and communication technology; and MDG = Millennium Development Goal.

a. C = competition; M = monopoly; and P = partial competition. **b.** Outgoing and incoming. **c.** Scale of 0–1, where 1 = highest presence. **d.** Millennium Development Goal indicators 8.14, 8.15, and 8.16.

Sweden

	Sweden 2000	Sweden 2007	High-income group 2007
Economic and social context			
Population (total, million)	9	9	1,056
Urban population (% of total)	84	84	78
GNI per capita, World Bank Atlas method (current US$)	29,280	47,870	37,572
GDP growth, 1995–2000 and 2000–07 (avg. annual %)	3.4	3.0	2.4
Adult literacy rate (% of ages 15 and older)	—	—	99
Gross primary, secondary, tertiary school enrollment (%)	113	95	92
Sector structure			
Separate telecommunications regulator	Yes	Yes	
Status of main fixed-line telephone operator	Mixed	Mixed	
Level of competition[a]			
International long distance service	C	C	
Mobile telephone service	P	C	
Internet service	C	C	
Sector efficiency and capacity			
Telecommunications revenue (% of GDP)	3.2	2.7	*3.1*
Mobile and fixed-line subscribers per employee	422	905	747
Telecommunications investment (% of revenue)	31.9	12.7	*14.3*
Sector performance			
Access			
Telephone lines (per 100 people)	64.8	60.2	50.0
Mobile cellular subscriptions (per 100 people)	71.8	113.4	100.4
Internet subscribers (per 100 people)	25.3	44.3	25.8
Personal computers (per 100 people)	50.7	*88.1*	67.4
Households with a television set (%)	99	94	98
Usage			
International voice traffic (minutes/person/month)[b]	—	—	14.0
Mobile telephone usage (minutes/user/month)	112	191	353
Internet users (per 100 people)	45.6	79.7	65.7
Quality			
Population covered by mobile cellular network (%)	*99*	*98*	*99*
Fixed broadband subscribers (% of total Internet subscrib.)	11.1	80.9	82.6
International Internet bandwidth (bits/second/person)	2,098	49,828	18,242
Affordability			
Price basket for residential fixed line (US$/month)	—	26.7	26.1
Price basket for mobile service (US$/month)	—	9.6	13.0
Price basket for Internet service (US$/month)	—	22.8	22.8
Price of call to United States (US$ for 3 minutes)	0.36	*0.41*	*0.81*
Trade			
ICT goods exports (% of total goods exports)	19.1	*11.2*	*15.2*
ICT goods imports (% of total goods imports)	16.3	*12.2*	*14.6*
ICT service exports (% of total service exports)	9.1	13.1	7.0
Applications			
ICT expenditure (% of GDP)	—	6.4	6.7
E-government Web measure index[c]	—	0.98	0.60
Secure Internet servers (per 1 million people, Dec. 2008)	*116.1*	772.0	662.6

Sources: Economic and social context: UIS and World Bank; Sector structure: ITU; Sector efficiency and capacity: ITU and World Bank; Sector performance: Global Insight/WITSA, IMF, ITU, Netcraft, UN Comtrade, UNDESA, UNPAN, Wireless Intelligence and World Bank. Produced by the Global Information and Communication Technologies Department and the Development Economics Data Group. For complete information, see Definitions and Data Sources.

Notes: Use of italics in the column entries indicates years other than those specified. — Not available. GDP = gross domestic product; GNI = gross national income; ICT = information and communication technology; and MDG = Millennium Development Goal.

a. C = competition; M = monopoly; and P = partial competition. **b.** Outgoing and incoming. **c.** Scale of 0–1, where 1 = highest presence. **d.** Millennium Development Goal indicators 8.14, 8.15, and 8.16.

Switzerland

	Switzerland 2000	Switzerland 2007	High-income group 2007
Economic and social context			
Population (total, million)	7	8	1,056
Urban population (% of total)	73	73	78
GNI per capita, World Bank Atlas method (current US$)	40,280	60,820	37,572
GDP growth, 1995–2000 and 2000–07 (avg. annual %)	2.0	1.8	2.4
Adult literacy rate (% of ages 15 and older)	—	—	99
Gross primary, secondary, tertiary school enrollment (%)	84	86	92
Sector structure			
Separate telecommunications regulator	Yes	Yes	
Status of main fixed-line telephone operator	Mixed	Mixed	
Level of competition[a]			
International long distance service	C	C	
Mobile telephone service	C	C	
Internet service	C	C	
Sector efficiency and capacity			
Telecommunications revenue (% of GDP)	3.3	3.2	3.1
Mobile and fixed-line subscribers per employee	409	549	747
Telecommunications investment (% of revenue)	27.2	14.8	14.3
Sector performance			
Access			
Telephone lines (per 100 people)	72.9	65.3	50.0
Mobile cellular subscriptions (per 100 people)	64.6	108.7	100.4
Internet subscribers (per 100 people)	23.2	37.3	25.8
Personal computers (per 100 people)	65.4	91.8	67.4
Households with a television set (%)	93	86	98
Usage			
International voice traffic (minutes/person/month)[b]	55.4	—	14.0
Mobile telephone usage (minutes/user/month)	—	126	353
Internet users (per 100 people)	47.9	76.3	65.7
Quality			
Population covered by mobile cellular network (%)	98	100	99
Fixed broadband subscribers (% of total Internet subscrib.)	3.4	84.6	82.6
International Internet bandwidth (bits/second/person)	2,941	29,417	18,242
Affordability			
Price basket for residential fixed line (US$/month)	22.0	29.5	26.1
Price basket for mobile service (US$/month)	—	31.3	13.0
Price basket for Internet service (US$/month)	—	40.8	22.8
Price of call to United States (US$ for 3 minutes)	0.21	0.32	0.81
Trade			
ICT goods exports (% of total goods exports)	5.8	3.7	15.2
ICT goods imports (% of total goods imports)	11.0	7.4	14.6
ICT service exports (% of total service exports)	—	—	7.0
Applications			
ICT expenditure (% of GDP)	—	8.0	6.7
E-government Web measure index[c]	—	0.56	0.60
Secure Internet servers (per 1 million people, Dec. 2008)	149.2	977.1	662.6

Sources: Economic and social context: UIS and World Bank; Sector structure: ITU; Sector efficiency and capacity: ITU and World Bank; Sector performance: Global Insight/WITSA, IMF, ITU, Netcraft, UN Comtrade, UNDESA, UNPAN, Wireless Intelligence and World Bank. Produced by the Global Information and Communication Technologies Department and the Development Economics Data Group. For complete information, see Definitions and Data Sources.

Notes: Use of italics in the column entries indicates years other than those specified. — Not available. GDP = gross domestic product; GNI = gross national income; ICT = information and communication technology; and MDG = Millennium Development Goal.

a. C = competition; M = monopoly; and P = partial competition. **b.** Outgoing and incoming. **c.** Scale of 0–1, where 1 = highest presence. **d.** Millennium Development Goal indicators 8.14, 8.15, and 8.16.

World Bank • ICT at a Glance

Syrian Arab Republic

	Syrian Arab Republic 2000	Syrian Arab Republic 2007	Lower-middle-income group 2007	Middle East & North Africa Region 2007
Economic and social context				
Population (total, million)	17	20	3,435	313
Urban population (% of total)	52	54	42	57
GNI per capita, World Bank Atlas method (current US$)	960	1,780	1,905	2,820
GDP growth, 1995–2000 and 2000–07 (avg. annual %)	2.2	4.5	8.0	4.4
Adult literacy rate (% of ages 15 and older)	*83*	*83*	83	73
Gross primary, secondary, tertiary school enrollment (%)	*57*	*65*	*68*	*70*
Sector structure				
Separate telecommunications regulator	—	No		
Status of main fixed-line telephone operator	*Public*	*Public*		
Level of competition[a]				
International long distance service	*M*	*M*		
Mobile telephone service	*P*	*P*		
Internet service	—	*P*		
Sector efficiency and capacity				
Telecommunications revenue (% of GDP)	1.9	*3.0*	*3.1*	*3.1*
Mobile and fixed-line subscribers per employee	80	409	*624*	*691*
Telecommunications investment (% of revenue)	58.2	9.1	25.3	21.7
Sector performance				
Access				
Telephone lines (per 100 people)	10.1	17.4	15.3	17.0
Mobile cellular subscriptions (per 100 people)	0.2	31.3	38.9	50.7
Internet subscribers (per 100 people)	0.1	3.5	6.0	2.4
Personal computers (per 100 people)	1.5	9.0	*4.6*	*6.3*
Households with a television set (%)	72	105	79	94
Usage				
International voice traffic (minutes/person/month)[b]	2.0	6.6	—	2.7
Mobile telephone usage (minutes/user/month)	—	—	322	—
Internet users (per 100 people)	0.2	17.4	12.4	17.1
Quality				
Population covered by mobile cellular network (%)	50	96	*80*	93
Fixed broadband subscribers (% of total Internet subscrib.)	0.0	1.0	40.4	—
International Internet bandwidth (bits/second/person)	*1*	53	199	186
Affordability				
Price basket for residential fixed line (US$/month)	3.3	2.4	7.2	3.9
Price basket for mobile service (US$/month)	—	10.0	9.8	6.5
Price basket for Internet service (US$/month)	—	13.7	16.7	11.6
Price of call to United States (US$ for 3 minutes)	4.81	—	2.08	1.45
Trade				
ICT goods exports (% of total goods exports)	*0.0*	*0.1*	20.6	—
ICT goods imports (% of total goods imports)	*1.3*	*2.5*	20.2	—
ICT service exports (% of total service exports)	—	5.8	15.6	2.6
Applications				
ICT expenditure (% of GDP)	—	—	6.5	4.5
E-government Web measure index[c]	—	0.24	0.33	0.22
Secure Internet servers (per 1 million people, Dec. 2008)	0.1	0.1	1.8	1.3

Sources: Economic and social context: UIS and World Bank; Sector structure: ITU; Sector efficiency and capacity: ITU and World Bank; Sector performance: Global Insight/WITSA, IMF, ITU, Netcraft, UN Comtrade, UNDESA, UNPAN, Wireless Intelligence and World Bank. Produced by the Global Information and Communication Technologies Department and the Development Economics Data Group. For complete information, see Definitions and Data Sources.

Notes: Use of italics in the column entries indicates years other than those specified. — Not available. GDP = gross domestic product; GNI = gross national income; ICT = information and communication technology; and MDG = Millennium Development Goal.

a. C = competition; M = monopoly; and P = partial competition. **b.** Outgoing and incoming. **c.** Scale of 0–1, where 1 = highest presence. **d.** Millennium Development Goal indicators 8.14, 8.15, and 8.16.

Tajikistan

	Tajikistan 2000	Tajikistan 2007	Low-income group 2007	Europe & Central Asia Region 2007
Economic and social context				
Population (total, million)	6	7	1,296	446
Urban population (% of total)	27	26	32	64
GNI per capita, World Bank Atlas method (current US$)	160	460	574	6,052
GDP growth, 1995–2000 and 2000–07 (avg. annual %)	1.1	8.8	5.6	6.1
Adult literacy rate (% of ages 15 and older)	99	100	64	98
Gross primary, secondary, tertiary school enrollment (%)	67	71	51	82
Sector structure				
Separate telecommunications regulator	—	No		
Status of main fixed-line telephone operator	Mixed	Mixed		
Level of competition[a]				
International long distance service	M	M		
Mobile telephone service	—	—		
Internet service	—	—		
Sector efficiency and capacity				
Telecommunications revenue (% of GDP)	0.7	2.9	3.3	2.9
Mobile and fixed-line subscribers per employee	45	114	301	532
Telecommunications investment (% of revenue)	1.0	71.4	—	22.0
Sector performance				
Access				
Telephone lines (per 100 people)	3.5	5.0	4.0	25.7
Mobile cellular subscriptions (per 100 people)	0.0	34.9	21.5	95.0
Internet subscribers (per 100 people)	0.0	0.0	0.8	13.6
Personal computers (per 100 people)	—	1.3	1.5	10.6
Households with a television set (%)	80	79	16	96
Usage				
International voice traffic (minutes/person/month)[b]	0.3	0.9	—	—
Mobile telephone usage (minutes/user/month)	—	216	—	154
Internet users (per 100 people)	0.0	7.2	5.2	21.4
Quality				
Population covered by mobile cellular network (%)	0	—	54	92
Fixed broadband subscribers (% of total Internet subscrib.)	0.0	2.2	3.4	32.5
International Internet bandwidth (bits/second/person)	0	0	26	1,114
Affordability				
Price basket for residential fixed line (US$/month)	1.3	0.6	5.7	5.8
Price basket for mobile service (US$/month)	—	23.3	11.2	11.8
Price basket for Internet service (US$/month)	—	14.0	29.2	12.0
Price of call to United States (US$ for 3 minutes)	8.10	7.84	2.00	1.63
Trade				
ICT goods exports (% of total goods exports)	—	—	1.4	1.8
ICT goods imports (% of total goods imports)	—	—	6.7	7.0
ICT service exports (% of total service exports)	9.5	12.6	—	5.0
Applications				
ICT expenditure (% of GDP)	—	—	—	5.0
E-government Web measure index[c]	—	0.04	0.11	0.36
Secure Internet servers (per 1 million people, Dec. 2008)	—	—	0.5	23.9

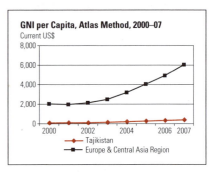

GNI per Capita, Atlas Method, 2000–07
Current US$

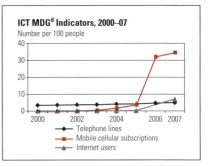

ICT MDG[d] Indicators, 2000–07
Number per 100 people

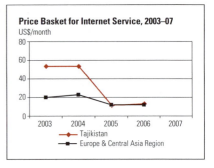

Price Basket for Internet Service, 2003–07
US$/month

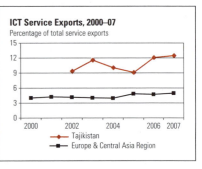

ICT Service Exports, 2000–07
Percentage of total service exports

Sources: Economic and social context: UIS and World Bank; Sector structure: ITU; Sector efficiency and capacity: ITU and World Bank; Sector performance: Global Insight/WITSA, IMF, ITU, Netcraft, UN Comtrade, UNDESA, UNPAN, Wireless Intelligence and World Bank. Produced by the Global Information and Communication Technologies Department and the Development Economics Data Group. For complete information, see Definitions and Data Sources.

Notes: Use of italics in the column entries indicates years other than those specified. — Not available. GDP = gross domestic product; GNI = gross national income; ICT = information and communication technology; and MDG = Millennium Development Goal.

a. C = competition; M = monopoly; and P = partial competition. **b.** Outgoing and incoming. **c.** Scale of 0–1, where 1 = highest presence. **d.** Millennium Development Goal indicators 8.14, 8.15, and 8.16.

World Bank • ICT at a Glance

Tanzania

	Tanzania 2000	Tanzania 2007	Low-income group 2007	Sub-Saharan Africa Region 2007
Economic and social context				
Population (total, million)	34	40	1,296	800
Urban population (% of total)	22	25	32	36
GNI per capita, World Bank Atlas method (current US$)	260	410	574	951
GDP growth, 1995–2000 and 2000–07 (avg. annual %)	3.9	6.7	5.6	5.1
Adult literacy rate (% of ages 15 and older)	*69*	*72*	64	62
Gross primary, secondary, tertiary school enrollment (%)	32	*50*	*51*	*51*
Sector structure				
Separate telecommunications regulator	Yes	Yes		
Status of main fixed-line telephone operator	Mixed	Mixed		
Level of competition[a]				
International long distance service	P	M		
Mobile telephone service	C	C		
Internet service	C	C		
Sector efficiency and capacity				
Telecommunications revenue (% of GDP)	1.9	—	3.3	4.7
Mobile and fixed-line subscribers per employee	78	—	301	499
Telecommunications investment (% of revenue)	12.6	—	—	—
Sector performance				
Access				
Telephone lines (per 100 people)	0.5	0.4	4.0	1.6
Mobile cellular subscriptions (per 100 people)	0.3	20.6	21.5	23.0
Internet subscribers (per 100 people)	0.0	*0.1*	*0.8*	*1.2*
Personal computers (per 100 people)	0.3	0.9	*1.5*	*1.8*
Households with a television set (%)	3	7	*16*	*18*
Usage				
International voice traffic (minutes/person/month)[b]	0.1	0.0	—	—
Mobile telephone usage (minutes/user/month)	—	—	—	—
Internet users (per 100 people)	0.1	1.0	5.2	4.4
Quality				
Population covered by mobile cellular network (%)	*25*	*65*	*54*	*56*
Fixed broadband subscribers (% of total Internet subscrib.)	0.0	0.0	3.4	3.1
International Internet bandwidth (bits/second/person)	0	3	26	36
Affordability				
Price basket for residential fixed line (US$/month)	9.3	11.3	5.7	12.6
Price basket for mobile service (US$/month)	—	9.5	11.2	11.6
Price basket for Internet service (US$/month)	—	19.7	29.2	43.1
Price of call to United States (US$ for 3 minutes)	10.70	3.17	2.00	2.43
Trade				
ICT goods exports (% of total goods exports)	0.4	0.4	*1.4*	*1.1*
ICT goods imports (% of total goods imports)	6.0	6.2	*6.7*	*8.2*
ICT service exports (% of total service exports)	4.3	2.5	—	4.2
Applications				
ICT expenditure (% of GDP)	—	—	—	—
E-government Web measure index[c]	—	0.23	0.11	0.16
Secure Internet servers (per 1 million people, Dec. 2008)	—	0.2	0.5	2.9

Sources: Economic and social context: UIS and World Bank; Sector structure: ITU; Sector efficiency and capacity: ITU and World Bank; Sector performance: Global Insight/WITSA, IMF, ITU, Netcraft, UN Comtrade, UNDESA, UNPAN, Wireless Intelligence and World Bank. Produced by the Global Information and Communication Technologies Department and the Development Economics Data Group. For complete information, see Definitions and Data Sources.

Notes: Use of italics in the column entries indicates years other than those specified. — Not available. GDP = gross domestic product; GNI = gross national income; ICT = information and communication technology; and MDG = Millennium Development Goal.

a. C = competition; M = monopoly; and P = partial competition. b. Outgoing and incoming. c. Scale of 0–1, where 1 = highest presence. d. Millennium Development Goal indicators 8.14, 8.15, and 8.16.

Information and Communications for Development 2009

Thailand

	Thailand 2000	Thailand 2007	Lower-middle-income group 2007	East Asia & Pacific Region 2007
Economic and social context				
Population (total, million)	61	64	3,435	1,912
Urban population (% of total)	31	33	42	43
GNI per capita, World Bank Atlas method (current US$)	2,010	3,400	1,905	2,182
GDP growth, 1995–2000 and 2000–07 (avg. annual %)	–0.7	5.3	8.0	9.0
Adult literacy rate (% of ages 15 and older)	93	94	83	93
Gross primary, secondary, tertiary school enrollment (%)	66	71	68	69
Sector structure				
Separate telecommunications regulator	No	Yes		
Status of main fixed-line telephone operator	Public	Public		
Level of competition[a]				
International long distance service	M	C		
Mobile telephone service	C	C		
Internet service	C	C		
Sector efficiency and capacity				
Telecommunications revenue (% of GDP)	2.6	4.0	3.1	3.0
Mobile and fixed-line subscribers per employee	353	2,808	624	546
Telecommunications investment (% of revenue)	27.0	9.8	25.3	—
Sector performance				
Access				
Telephone lines (per 100 people)	9.2	11.0	15.3	23.1
Mobile cellular subscriptions (per 100 people)	5.0	123.9	38.9	43.7
Internet subscribers (per 100 people)	1.0	—	6.0	9.3
Personal computers (per 100 people)	2.8	7.0	4.6	5.6
Households with a television set (%)	91	92	79	53
Usage				
International voice traffic (minutes/person/month)[b]	0.8	1.2	—	0.8
Mobile telephone usage (minutes/user/month)	—	313	322	333
Internet users (per 100 people)	3.8	21.0	12.4	14.6
Quality				
Population covered by mobile cellular network (%)	—	38	80	93
Fixed broadband subscribers (% of total Internet subscrib.)	0.0	—	40.4	41.8
International Internet bandwidth (bits/second/person)	4	346	199	247
Affordability				
Price basket for residential fixed line (US$/month)	8.4	8.3	7.2	5.8
Price basket for mobile service (US$/month)	—	4.3	9.8	5.0
Price basket for Internet service (US$/month)	—	7.4	16.7	14.4
Price of call to United States (US$ for 3 minutes)	2.19	0.67	2.08	1.16
Trade				
ICT goods exports (% of total goods exports)	29.5	24.2	20.6	30.9
ICT goods imports (% of total goods imports)	25.3	20.0	20.2	28.1
ICT service exports (% of total service exports)	—	—	15.6	5.2
Applications				
ICT expenditure (% of GDP)	—	6.1	6.5	7.3
E-government Web measure index[c]	—	0.51	0.33	0.18
Secure Internet servers (per 1 million people, Dec. 2008)	1.9	9.6	1.8	1.9

Sources: Economic and social context: UIS and World Bank; Sector structure: ITU; Sector efficiency and capacity: ITU and World Bank; Sector performance: Global Insight/WITSA, IMF, ITU, Netcraft, UN Comtrade, UNDESA, UNPAN, Wireless Intelligence and World Bank. Produced by the Global Information and Communication Technologies Department and the Development Economics Data Group. For complete information, see Definitions and Data Sources.

Notes: Use of italics in the column entries indicates years other than those specified. — Not available. GDP = gross domestic product; GNI = gross national income; ICT = information and communication technology; and MDG = Millennium Development Goal.

a. C = competition; M = monopoly; and P = partial competition. b. Outgoing and incoming. c. Scale of 0–1, where 1 = highest presence. d. Millennium Development Goal indicators 8.14, 8.15, and 8.16.

World Bank • ICT at a Glance

Togo

	Togo 2000	Togo 2007	Low-income group 2007	Sub-Saharan Africa Region 2007
Economic and social context				
Population (total, million)	5	7	1,296	800
Urban population (% of total)	37	41	32	36
GNI per capita, World Bank Atlas method (current US$)	270	360	574	951
GDP growth, 1995–2000 and 2000–07 (avg. annual %)	4.2	2.6	5.6	5.1
Adult literacy rate (% of ages 15 and older)	53	—	64	62
Gross primary, secondary, tertiary school enrollment (%)	53	*55*	*51*	*51*
Sector structure				
Separate telecommunications regulator	Yes	Yes		
Status of main fixed-line telephone operator	*Public*	*Public*		
Level of competition[a]				
International long distance service	P	P		
Mobile telephone service	P	P		
Internet service	C	C		
Sector efficiency and capacity				
Telecommunications revenue (% of GDP)	2.9	7.4	*3.3*	*4.7*
Mobile and fixed-line subscribers per employee	86	1,059	*301*	*499*
Telecommunications investment (% of revenue)	45.3	41.1	—	—
Sector performance				
Access				
Telephone lines (per 100 people)	0.8	1.5	4.0	1.6
Mobile cellular subscriptions (per 100 people)	0.9	18.1	21.5	23.0
Internet subscribers (per 100 people)	0.1	*0.2*	*0.8*	*1.2*
Personal computers (per 100 people)	1.9	*3.0*	*1.5*	*1.8*
Households with a television set (%)	11	*14*	*16*	*18*
Usage				
International voice traffic (minutes/person/month)[b]	0.7	0.4	—	—
Mobile telephone usage (minutes/user/month)	—	—	—	—
Internet users (per 100 people)	1.9	*5.0*	5.2	4.4
Quality				
Population covered by mobile cellular network (%)	80	85	*54*	*56*
Fixed broadband subscribers (% of total Internet subscrib.)	0.0	*0.0*	*3.4*	*3.1*
International Internet bandwidth (bits/second/person)	0	4	26	36
Affordability				
Price basket for residential fixed line (US$/month)	10.2	13.9	*5.7*	*12.6*
Price basket for mobile service (US$/month)	—	16.0	11.2	11.6
Price basket for Internet service (US$/month)	—	20.3	*29.2*	*43.1*
Price of call to United States (US$ for 3 minutes)	7.90	*3.98*	*2.00*	*2.43*
Trade				
ICT goods exports (% of total goods exports)	0.0	0.1	*1.4*	*1.1*
ICT goods imports (% of total goods imports)	3.3	4.2	*6.7*	*8.2*
ICT service exports (% of total service exports)	12.5	6.9	—	*4.2*
Applications				
ICT expenditure (% of GDP)	—	—	—	—
E-government Web measure index[c]	—	0.09	0.11	0.16
Secure Internet servers (per 1 million people, Dec. 2008)	—	1.2	0.5	2.9

Sources: Economic and social context: UIS and World Bank; Sector structure: ITU; Sector efficiency and capacity: ITU and World Bank; Sector performance: Global Insight/WITSA, IMF, ITU, Netcraft, UN Comtrade, UNDESA, UNPAN, Wireless Intelligence and World Bank. Produced by the Global Information and Communication Technologies Department and the Development Economics Data Group. For complete information, see Definitions and Data Sources.

Notes: Use of italics in the column entries indicates years other than those specified. — Not available. GDP = gross domestic product; GNI = gross national income; ICT = information and communication technology; and MDG = Millennium Development Goal.

a. C = competition; M = monopoly; and P = partial competition. **b.** Outgoing and incoming. **c.** Scale of 0–1, where 1 = highest presence. **d.** Millennium Development Goal indicators 8.14, 8.15, and 8.16.

Information and Communications for Development 2009

World Bank • ICT at a Glance

Trinidad and Tobago

	Trinidad and Tobago 2000	Trinidad and Tobago 2007	High-income group 2007
Economic and social context			
Population (total, million)	1	1	1,056
Urban population (% of total)	11	13	78
GNI per capita, World Bank Atlas method (current US$)	5,170	14,480	37,572
GDP growth, 1995–2000 and 2000–07 (avg. annual %)	5.0	8.8	2.4
Adult literacy rate (% of ages 15 and older)	—	99	99
Gross primary, secondary, tertiary school enrollment (%)	68	65	92
Sector structure			
Separate telecommunications regulator	No	Yes	
Status of main fixed-line telephone operator	Mixed	Mixed	
Level of competition[a]			
International long distance service	M	C	
Mobile telephone service	C	P	
Internet service	C	C	
Sector efficiency and capacity			
Telecommunications revenue (% of GDP)	3.0	2.6	3.1
Mobile and fixed-line subscribers per employee	157	—	747
Telecommunications investment (% of revenue)	39.2	—	14.3
Sector performance			
Access			
Telephone lines (per 100 people)	24.4	23.1	50.0
Mobile cellular subscriptions (per 100 people)	12.4	113.3	100.4
Internet subscribers (per 100 people)	2.0	6.1	25.8
Personal computers (per 100 people)	6.2	13.2	67.4
Households with a television set (%)	86	88	98
Usage			
International voice traffic (minutes/person/month)[b]	15.1	31.3	14.0
Mobile telephone usage (minutes/user/month)	—	—	353
Internet users (per 100 people)	7.7	16.0	65.7
Quality			
Population covered by mobile cellular network (%)	—	100	99
Fixed broadband subscribers (% of total Internet subscrib.)	0.0	43.4	82.6
International Internet bandwidth (bits/second/person)	46	675	18,242
Affordability			
Price basket for residential fixed line (US$/month)	7.0	16.2	26.1
Price basket for mobile service (US$/month)	—	6.7	13.0
Price basket for Internet service (US$/month)	—	13.4	22.8
Price of call to United States (US$ for 3 minutes)	2.47	2.19	0.81
Trade			
ICT goods exports (% of total goods exports)	0.1	0.2	15.2
ICT goods imports (% of total goods imports)	4.1	5.9	14.6
ICT service exports (% of total service exports)	—	—	7.0
Applications			
ICT expenditure (% of GDP)	—	—	6.7
E-government Web measure index[c]	—	0.44	0.60
Secure Internet servers (per 1 million people, Dec. 2008)	9.2	45.6	662.6

Sources: Economic and social context: UIS and World Bank; Sector structure: ITU; Sector efficiency and capacity: ITU and World Bank; Sector performance: Global Insight/WITSA, IMF, ITU, Netcraft, UN Comtrade, UNDESA, UNPAN, Wireless Intelligence and World Bank. Produced by the Global Information and Communication Technologies Department and the Development Economics Data Group. For complete information, see Definitions and Data Sources.

Notes: Use of italics in the column entries indicates years other than those specified. — Not available. GDP = gross domestic product; GNI = gross national income; ICT = information and communication technology; and MDG = Millennium Development Goal.

a. C = competition; M = monopoly; and P = partial competition. b. Outgoing and incoming. c. Scale of 0–1, where 1 = highest presence. d. Millennium Development Goal indicators 8.14, 8.15, and 8.16.

World Bank • ICT at a Glance

Tunisia

	Tunisia 2000	Tunisia 2007	Lower-middle-income group 2007	Middle East & North Africa Region 2007
Economic and social context				
Population (total, million)	10	10	3,435	313
Urban population (% of total)	63	66	42	57
GNI per capita, World Bank Atlas method (current US$)	2,090	3,210	1,905	2,820
GDP growth, 1995–2000 and 2000–07 (avg. annual %)	5.5	4.8	8.0	4.4
Adult literacy rate (% of ages 15 and older)	—	78	83	73
Gross primary, secondary, tertiary school enrollment (%)	74	76	68	70
Sector structure				
Separate telecommunications regulator	No	Yes		
Status of main fixed-line telephone operator	Public	Mixed		
Level of competition[a]				
International long distance service	M	M		
Mobile telephone service	M	C		
Internet service	C	P		
Sector efficiency and capacity				
Telecommunications revenue (% of GDP)	2.1	*4.3*	3.1	3.1
Mobile and fixed-line subscribers per employee	153	*915*	624	691
Telecommunications investment (% of revenue)	39.8	*23.2*	25.3	21.7
Sector performance				
Access				
Telephone lines (per 100 people)	10.0	12.5	15.3	17.0
Mobile cellular subscriptions (per 100 people)	1.2	76.7	38.9	50.7
Internet subscribers (per 100 people)	0.4	2.5	6.0	2.4
Personal computers (per 100 people)	2.2	7.5	*4.6*	6.3
Households with a television set (%)	87	*93*	79	94
Usage				
International voice traffic (minutes/person/month)[b]	3.7	*6.1*	—	2.7
Mobile telephone usage (minutes/user/month)	—	135	322	—
Internet users (per 100 people)	2.7	16.8	12.4	17.1
Quality				
Population covered by mobile cellular network (%)	*60*	100	80	93
Fixed broadband subscribers (% of total Internet subscrib.)	0.0	45.1	40.4	—
International Internet bandwidth (bits/second/person)	5	303	199	186
Affordability				
Price basket for residential fixed line (US$/month)	4.2	3.0	*7.2*	*3.9*
Price basket for mobile service (US$/month)	—	6.6	*9.8*	*6.5*
Price basket for Internet service (US$/month)	—	11.6	*16.7*	*11.6*
Price of call to United States (US$ for 3 minutes)	2.25	—	*2.08*	*1.45*
Trade				
ICT goods exports (% of total goods exports)	3.4	4.2	20.6	—
ICT goods imports (% of total goods imports)	5.5	5.9	20.2	—
ICT service exports (% of total service exports)	1.2	1.2	15.6	2.6
Applications				
ICT expenditure (% of GDP)	—	6.0	6.5	4.5
E-government Web measure index[c]	—	0.13	0.33	0.22
Secure Internet servers (per 1 million people, Dec. 2008)	*0.4*	10.7	1.8	1.3

Sources: Economic and social context: UIS and World Bank; Sector structure: ITU; Sector efficiency and capacity: ITU and World Bank; Sector performance: Global Insight/WITSA, IMF, ITU, Netcraft, UN Comtrade, UNDESA, UNPAN, Wireless Intelligence and World Bank. Produced by the Global Information and Communication Technologies Department and the Development Economics Data Group. For complete information, see Definitions and Data Sources.

Notes: Use of italics in the column entries indicates years other than those specified. — Not available. GDP = gross domestic product; GNI = gross national income; ICT = information and communication technology; and MDG = Millennium Development Goal.

a. C = competition; M = monopoly; and P = partial competition. **b.** Outgoing and incoming. **c.** Scale of 0–1, where 1 = highest presence. **d.** Millennium Development Goal indicators 8.14, 8.15, and 8.16.

Information and Communications for Development 2009

World Bank • ICT at a Glance

Turkey

	Turkey 2000	Turkey 2007	Upper-middle-income group 2007	Europe & Central Asia Region 2007
Economic and social context				
Population (total, million)	67	74	824	446
Urban population (% of total)	65	68	75	64
GNI per capita, World Bank Atlas method (current US$)	3,930	8,030	7,107	6,052
GDP growth, 1995–2000 and 2000–07 (avg. annual %)	3.5	5.9	4.3	6.1
Adult literacy rate (% of ages 15 and older)	—	89	94	98
Gross primary, secondary, tertiary school enrollment (%)	*68*	*69*	*82*	*82*
Sector structure				
Separate telecommunications regulator	Yes	Yes		
Status of main fixed-line telephone operator	Public	Mixed		
Level of competition[a]				
International long distance service	M	C		
Mobile telephone service	C	P		
Internet service	C	C		
Sector efficiency and capacity				
Telecommunications revenue (% of GDP)	3.0	2.5	3.3	2.9
Mobile and fixed-line subscribers per employee	477	1,782	*566*	*532*
Telecommunications investment (% of revenue)	7.7	11.7	—	22.0
Sector performance				
Access				
Telephone lines (per 100 people)	27.3	24.6	22.6	25.7
Mobile cellular subscriptions (per 100 people)	23.9	83.9	84.1	95.0
Internet subscribers (per 100 people)	2.2	6.3	9.4	13.6
Personal computers (per 100 people)	3.7	*6.0*	12.4	10.6
Households with a television set (%)	96	*112*	92	96
Usage				
International voice traffic (minutes/person/month)[b]	2.4	2.5	—	—
Mobile telephone usage (minutes/user/month)	81	69	137	154
Internet users (per 100 people)	3.7	16.5	26.6	21.4
Quality				
Population covered by mobile cellular network (%)	50	98	95	92
Fixed broadband subscribers (% of total Internet subscrib.)	0.0	97.2	47.8	32.5
International Internet bandwidth (bits/second/person)	9	1,381	1,185	1,114
Affordability				
Price basket for residential fixed line (US$/month)	10.4	13.5	*10.6*	*5.8*
Price basket for mobile service (US$/month)	—	12.7	*10.9*	*11.8*
Price basket for Internet service (US$/month)	—	10.9	*16.4*	*12.0*
Price of call to United States (US$ for 3 minutes)	3.30	2.40	*1.55*	*1.63*
Trade				
ICT goods exports (% of total goods exports)	4.0	*2.0*	*13.5*	*1.8*
ICT goods imports (% of total goods imports)	11.1	*4.0*	*16.2*	*7.0*
ICT service exports (% of total service exports)	—	1.8	*4.6*	*5.0*
Applications				
ICT expenditure (% of GDP)	—	5.5	5.2	5.0
E-government Web measure index[c]	—	0.42	0.37	0.36
Secure Internet servers (per 1 million people, Dec. 2008)	*3.2*	56.5	26.2	23.9

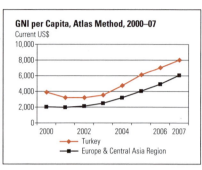

GNI per Capita, Atlas Method, 2000–07
Current US$

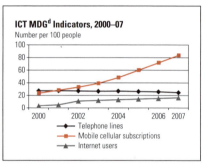

ICT MDG[d] Indicators, 2000–07
Number per 100 people

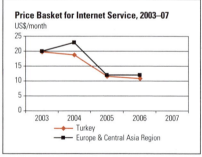

Price Basket for Internet Service, 2003–07
US$/month

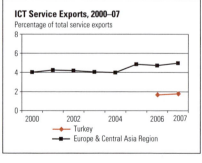

ICT Service Exports, 2000–07
Percentage of total service exports

Sources: Economic and social context: UIS and World Bank; Sector structure: ITU; Sector efficiency and capacity: ITU and World Bank; Sector performance: Global Insight/WITSA, IMF, ITU, Netcraft, UN Comtrade, UNDESA, UNPAN, Wireless Intelligence and World Bank. Produced by the Global Information and Communication Technologies Department and the Development Economics Data Group. For complete information, see Definitions and Data Sources.

Notes: Use of italics in the column entries indicates years other than those specified. — Not available. GDP = gross domestic product; GNI = gross national income; ICT = information and communication technology; and MDG = Millennium Development Goal.

a. C = competition; M = monopoly; and P = partial competition. **b.** Outgoing and incoming. **c.** Scale of 0–1, where 1 = highest presence. **d.** Millennium Development Goal indicators 8.14, 8.15, and 8.16.

World Bank • ICT at a Glance

Turkmenistan

	Turkmenistan 2000	Turkmenistan 2007	Lower-middle-income group 2007	Europe & Central Asia Region 2007
Economic and social context				
Population (total, million)	5	5	3,435	446
Urban population (% of total)	46	48	42	64
GNI per capita, World Bank Atlas method (current US$)	650	—	1,905	6,052
GDP growth, 1995–2000 and 2000–07 (avg. annual %)	4.0	—	8.0	6.1
Adult literacy rate (% of ages 15 and older)	—	100	83	98
Gross primary, secondary, tertiary school enrollment (%)	—	—	68	82
Sector structure				
Separate telecommunications regulator	—	No		
Status of main fixed-line telephone operator	Public	Public		
Level of competition[a]				
International long distance service	M	M		
Mobile telephone service	C	C		
Internet service	—	—		
Sector efficiency and capacity				
Telecommunications revenue (% of GDP)	0.9	0.7	3.1	2.9
Mobile and fixed-line subscribers per employee	50	72	624	532
Telecommunications investment (% of revenue)	9.0	—	25.3	22.0
Sector performance				
Access				
Telephone lines (per 100 people)	8.1	9.2	15.3	25.7
Mobile cellular subscriptions (per 100 people)	0.2	7.0	38.9	95.0
Internet subscribers (per 100 people)	0.0	—	6.0	13.6
Personal computers (per 100 people)	1.5	7.2	4.6	10.6
Households with a television set (%)	0	0	79	96
Usage				
International voice traffic (minutes/person/month)[b]	0.5	—	—	—
Mobile telephone usage (minutes/user/month)	—	282	322	154
Internet users (per 100 people)	0.1	1.4	12.4	21.4
Quality				
Population covered by mobile cellular network (%)	12	14	80	92
Fixed broadband subscribers (% of total Internet subscrib.)	0.0	—	40.4	32.5
International Internet bandwidth (bits/second/person)	0	16	199	1,114
Affordability				
Price basket for residential fixed line (US$/month)	—	1.5	7.2	5.8
Price basket for mobile service (US$/month)	—	17.2	9.8	11.8
Price basket for Internet service (US$/month)	—	69.5	16.7	12.0
Price of call to United States (US$ for 3 minutes)	—	—	2.08	1.63
Trade				
ICT goods exports (% of total goods exports)	0.0	—	20.6	1.8
ICT goods imports (% of total goods imports)	6.3	—	20.2	7.0
ICT service exports (% of total service exports)	—	—	15.6	5.0
Applications				
ICT expenditure (% of GDP)	—	—	6.5	5.0
E-government Web measure index[c]	—	0.05	0.33	0.36
Secure Internet servers (per 1 million people, Dec. 2008)	—	—	1.8	23.9

Sources: Economic and social context: UIS and World Bank; Sector structure: ITU; Sector efficiency and capacity: ITU and World Bank; Sector performance: Global Insight/WITSA, IMF, ITU, Netcraft, UN Comtrade, UNDESA, UNPAN, Wireless Intelligence and World Bank. Produced by the Global Information and Communication Technologies Department and the Development Economics Data Group. For complete information, see Definitions and Data Sources.

Notes: Use of italics in the column entries indicates years other than those specified. — Not available. GDP = gross domestic product; GNI = gross national income; ICT = information and communication technology; and MDG = Millennium Development Goal.

a. C = competition; M = monopoly; and P = partial competition. **b.** Outgoing and incoming. **c.** Scale of 0–1, where 1 = highest presence. **d.** Millennium Development Goal indicators 8.14, 8.15, and 8.16.

Uganda

	Uganda 2000	Uganda 2007	Low-income group 2007	Sub-Saharan Africa Region 2007
Economic and social context				
Population (total, million)	25	31	1,296	800
Urban population (% of total)	12	13	32	36
GNI per capita, World Bank Atlas method (current US$)	260	370	574	951
GDP growth, 1995–2000 and 2000–07 (avg. annual %)	6.4	7.1	5.6	5.1
Adult literacy rate (% of ages 15 and older)	*68*	*74*	64	62
Gross primary, secondary, tertiary school enrollment (%)	66	*63*	51	51
Sector structure				
Separate telecommunications regulator	Yes	Yes		
Status of main fixed-line telephone operator	*Mixed*	*Mixed*		
Level of competition[a]				
International long distance service	C	C		
Mobile telephone service	C	C		
Internet service	C	C		
Sector efficiency and capacity				
Telecommunications revenue (% of GDP)	1.5	3.2	3.3	4.7
Mobile and fixed-line subscribers per employee	79	255	301	499
Telecommunications investment (% of revenue)	26.2	23.4	—	—
Sector performance				
Access				
Telephone lines (per 100 people)	0.2	0.5	4.0	1.6
Mobile cellular subscriptions (per 100 people)	0.5	13.6	21.5	23.0
Internet subscribers (per 100 people)	0.0	0.1	*0.8*	*1.2*
Personal computers (per 100 people)	0.2	1.7	1.5	1.8
Households with a television set (%)	5	10	16	18
Usage				
International voice traffic (minutes/person/month)[b]	*0.1*	0.6	—	—
Mobile telephone usage (minutes/user/month)	—	—	—	—
Internet users (per 100 people)	0.2	2.5	5.2	4.4
Quality				
Population covered by mobile cellular network (%)	16	80	54	56
Fixed broadband subscribers (% of total Internet subscrib.)	0.0	12.0	3.4	3.1
International Internet bandwidth (bits/second/person)	0	11	26	36
Affordability				
Price basket for residential fixed line (US$/month)	14.4	13.2	*5.7*	*12.6*
Price basket for mobile service (US$/month)	—	9.2	11.2	11.6
Price basket for Internet service (US$/month)	—	51.7	29.2	43.1
Price of call to United States (US$ for 3 minutes)	3.63	3.21	2.00	2.43
Trade				
ICT goods exports (% of total goods exports)	1.1	6.9	*1.4*	*1.1*
ICT goods imports (% of total goods imports)	6.1	10.0	*6.7*	8.2
ICT service exports (% of total service exports)	*4.1*	*7.6*	—	4.2
Applications				
ICT expenditure (% of GDP)	—	—	—	—
E-government Web measure index[c]	—	0.27	0.11	0.16
Secure Internet servers (per 1 million people, Dec. 2008)	—	0.2	0.5	2.9

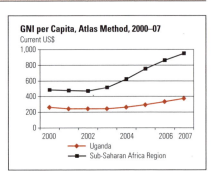

GNI per Capita, Atlas Method, 2000–07
Current US$

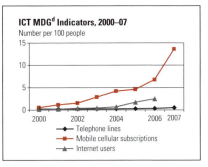

ICT MDG[d] Indicators, 2000–07
Number per 100 people

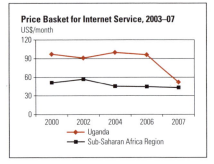

Price Basket for Internet Service, 2003–07
US$/month

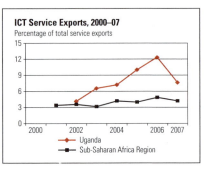

ICT Service Exports, 2000–07
Percentage of total service exports

Sources: Economic and social context: UIS and World Bank; Sector structure: ITU; Sector efficiency and capacity: ITU and World Bank; Sector performance: Global Insight/WITSA, IMF, ITU, Netcraft, UN Comtrade, UNDESA, UNPAN, Wireless Intelligence and World Bank. Produced by the Global Information and Communication Technologies Department and the Development Economics Data Group. For complete information, see Definitions and Data Sources.

Notes: Use of italics in the column entries indicates years other than those specified. — Not available. GDP = gross domestic product; GNI = gross national income; ICT = information and communication technology; and MDG = Millennium Development Goal.

a. C = competition; M = monopoly; and P = partial competition. **b.** Outgoing and incoming. **c.** Scale of 0–1, where 1 = highest presence. **d.** Millennium Development Goal indicators 8.14, 8.15, and 8.16.

World Bank • ICT at a Glance

Ukraine

	Ukraine 2000	Ukraine 2007	Lower-middle-income group 2007	Europe & Central Asia Region 2007
Economic and social context				
Population (total, million)	49	47	3,435	446
Urban population (% of total)	67	68	42	64
GNI per capita, World Bank Atlas method (current US$)	700	2,560	1,905	6,052
GDP growth, 1995–2000 and 2000–07 (avg. annual %)	−1.9	7.6	8.0	6.1
Adult literacy rate (% of ages 15 and older)	*99*	100	83	98
Gross primary, secondary, tertiary school enrollment (%)	84	*86*	68	82
Sector structure				
Separate telecommunications regulator	—	Yes		
Status of main fixed-line telephone operator	Public	Mixed		
Level of competition[a]				
International long distance service	C	C		
Mobile telephone service	C	P		
Internet service	—	C		
Sector efficiency and capacity				
Telecommunications revenue (% of GDP)	3.7	*5.7*	*3.1*	2.9
Mobile and fixed-line subscribers per employee	88	*210*	*624*	*532*
Telecommunications investment (% of revenue)	23.1	*31.4*	*25.3*	22.0
Sector performance				
Access				
Telephone lines (per 100 people)	21.2	27.6	15.3	25.7
Mobile cellular subscriptions (per 100 people)	1.7	118.8	38.9	95.0
Internet subscribers (per 100 people)	0.5	13.8	6.0	13.6
Personal computers (per 100 people)	1.8	4.5	4.6	10.6
Households with a television set (%)	97	97	79	96
Usage				
International voice traffic (minutes/person/month)[b]	1.1	4.7	—	—
Mobile telephone usage (minutes/user/month)	*49*	156	322	154
Internet users (per 100 people)	0.7	21.5	12.4	21.4
Quality				
Population covered by mobile cellular network (%)	*75*	100	*80*	*92*
Fixed broadband subscribers (% of total Internet subscrib.)	0.0	12.5	40.4	32.5
International Internet bandwidth (bits/second/person)	1	206	199	1,114
Affordability				
Price basket for residential fixed line (US$/month)	—	2.6	7.2	5.8
Price basket for mobile service (US$/month)	—	9.4	9.8	11.8
Price basket for Internet service (US$/month)	—	7.6	16.7	12.0
Price of call to United States (US$ for 3 minutes)	—	1.65	2.08	1.63
Trade				
ICT goods exports (% of total goods exports)	*1.3*	1.5	20.6	1.8
ICT goods imports (% of total goods imports)	*3.6*	3.3	20.2	7.0
ICT service exports (% of total service exports)	2.5	3.6	15.6	5.0
Applications				
ICT expenditure (% of GDP)	—	7.1	6.5	5.0
E-government Web measure index[c]	—	0.54	0.33	0.36
Secure Internet servers (per 1 million people, Dec. 2008)	0.9	4.3	1.8	23.9

Sources: Economic and social context: UIS and World Bank; Sector structure: ITU; Sector efficiency and capacity: ITU and World Bank; Sector performance: Global Insight/WITSA, IMF, ITU, Netcraft, UN Comtrade, UNDESA, UNPAN, Wireless Intelligence and World Bank. Produced by the Global Information and Communication Technologies Department and the Development Economics Data Group. For complete information, see Definitions and Data Sources.

Notes: Use of italics in the column entries indicates years other than those specified. — Not available. GDP = gross domestic product; GNI = gross national income; ICT = information and communication technology; and MDG = Millennium Development Goal.

a. C = competition; M = monopoly; and P = partial competition. b. Outgoing and incoming. c. Scale of 0–1, where 1 = highest presence. d. Millennium Development Goal indicators 8.14, 8.15, and 8.16.

Information and Communications for Development 2009

United Arab Emirates

	United Arab Emirates		High-income group
	2000	2007	2007
Economic and social context			
Population (total, million)	3	4	1,056
Urban population (% of total)	78	78	78
GNI per capita, World Bank Atlas method (current US$)	19,270	26,270	37,572
GDP growth, 1995–2000 and 2000–07 (avg. annual %)	5.1	7.7	2.4
Adult literacy rate (% of ages 15 and older)	—	90	99
Gross primary, secondary, tertiary school enrollment (%)	66	60	92
Sector structure			
Separate telecommunications regulator	—	Yes	
Status of main fixed-line telephone operator	Mixed	Mixed	
Level of competition[a]			
International long distance service	M	P	
Mobile telephone service	M	P	
Internet service	M	P	
Sector efficiency and capacity			
Telecommunications revenue (% of GDP)	2.7	2.7	3.1
Mobile and fixed-line subscribers per employee	294	852	747
Telecommunications investment (% of revenue)	28.9	9.8	14.3
Sector performance			
Access			
Telephone lines (per 100 people)	31.4	31.7	50.0
Mobile cellular subscriptions (per 100 people)	44.0	177.1	100.4
Internet subscribers (per 100 people)	6.5	20.7	25.8
Personal computers (per 100 people)	12.3	33.0	67.4
Households with a television set (%)	86	86	98
Usage			
International voice traffic (minutes/person/month)[b]	—	—	14.0
Mobile telephone usage (minutes/user/month)	—	—	353
Internet users (per 100 people)	23.6	51.8	65.7
Quality			
Population covered by mobile cellular network (%)	100	100	99
Fixed broadband subscribers (% of total Internet subscrib.)	0.7	42.0	82.6
International Internet bandwidth (bits/second/person)	5	2,785	18,242
Affordability			
Price basket for residential fixed line (US$/month)	5.0	4.9	26.1
Price basket for mobile service (US$/month)	—	7.7	13.0
Price basket for Internet service (US$/month)	—	13.1	22.8
Price of call to United States (US$ for 3 minutes)	3.51	1.73	0.81
Trade			
ICT goods exports (% of total goods exports)	—	4.3	15.2
ICT goods imports (% of total goods imports)	—	8.6	14.6
ICT service exports (% of total service exports)	—	—	7.0
Applications			
ICT expenditure (% of GDP)	—	5.1	6.7
E-government Web measure index[c]	—	0.72	0.60
Secure Internet servers (per 1 million people, Dec. 2008)	9.1	125.8	662.6

Sources: Economic and social context: UIS and World Bank; Sector structure: ITU; Sector efficiency and capacity: ITU and World Bank; Sector performance: Global Insight/WITSA, IMF, ITU, Netcraft, UN Comtrade, UNDESA, UNPAN, Wireless Intelligence and World Bank. Produced by the Global Information and Communication Technologies Department and the Development Economics Data Group. For complete information, see Definitions and Data Sources.

Notes: Use of italics in the column entries indicates years other than those specified. — Not available. GDP = gross domestic product; GNI = gross national income; ICT = information and communication technology; and MDG = Millennium Development Goal.

a. C = competition; M = monopoly; and P = partial competition. b. Outgoing and incoming. c. Scale of 0–1, where 1 = highest presence. d. Millennium Development Goal indicators 8.14, 8.15, and 8.16.

World Bank • ICT at a Glance

United Kingdom

	United Kingdom 2000	United Kingdom 2007	High-income group 2007
Economic and social context			
Population (total, million)	59	61	1,056
Urban population (% of total)	89	90	78
GNI per capita, World Bank Atlas method (current US$)	25,480	*40,660*	37,572
GDP growth, 1995–2000 and 2000–07 (avg. annual %)	3.2	2.6	2.4
Adult literacy rate (% of ages 15 and older)	—	—	99
Gross primary, secondary, tertiary school enrollment (%)	90	*93*	*92*
Sector structure			
Separate telecommunications regulator	Yes	Yes	
Status of main fixed-line telephone operator	*Private*	*Private*	
Level of competition[a]			
International long distance service	C	C	
Mobile telephone service	C	C	
Internet service	C	C	
Sector efficiency and capacity			
Telecommunications revenue (% of GDP)	3.8	3.7	*3.1*
Mobile and fixed-line subscribers per employee	380	—	747
Telecommunications investment (% of revenue)	30.4	—	*14.3*
Sector performance			
Access			
Telephone lines (per 100 people)	59.8	55.2	50.0
Mobile cellular subscriptions (per 100 people)	73.8	118.0	100.4
Internet subscribers (per 100 people)	14.3	30.0	25.8
Personal computers (per 100 people)	34.3	*80.2*	67.4
Households with a television set (%)	97	*98*	*98*
Usage			
International voice traffic (minutes/person/month)[b]	21.9	—	*14.0*
Mobile telephone usage (minutes/user/month)	150	179	353
Internet users (per 100 people)	26.8	71.7	65.7
Quality			
Population covered by mobile cellular network (%)	99	100	*99*
Fixed broadband subscribers (% of total Internet subscrib.)	0.6	85.4	82.6
International Internet bandwidth (bits/second/person)	1,469	39,650	18,242
Affordability			
Price basket for residential fixed line (US$/month)	25.3	*28.2*	26.1
Price basket for mobile service (US$/month)	—	32.2	13.0
Price basket for Internet service (US$/month)	—	21.4	22.8
Price of call to United States (US$ for 3 minutes)	1.07	*0.77*	*0.81*
Trade			
ICT goods exports (% of total goods exports)	19.8	*20.5*	15.2
ICT goods imports (% of total goods imports)	20.0	*13.6*	*14.6*
ICT service exports (% of total service exports)	5.9	7.8	7.0
Applications			
ICT expenditure (% of GDP)	—	6.7	6.7
E-government Web measure index[c]	—	0.69	0.60
Secure Internet servers (per 1 million people, Dec. 2008)	*109.4*	905.0	662.6

Sources: Economic and social context: UIS and World Bank; Sector structure: ITU; Sector efficiency and capacity: ITU and World Bank; Sector performance: Global Insight/WITSA, IMF, ITU, Netcraft, UN Comtrade, UNDESA, UNPAN, Wireless Intelligence and World Bank. Produced by the Global Information and Communication Technologies Department and the Development Economics Data Group. For complete information, see Definitions and Data Sources.

Notes: Use of italics in the column entries indicates years other than those specified. — Not available. GDP = gross domestic product; GNI = gross national income; ICT = information and communication technology; and MDG = Millennium Development Goal.

a. C = competition; M = monopoly; and P = partial competition. b. Outgoing and incoming. c. Scale of 0–1, where 1 = highest presence. d. Millennium Development Goal indicators 8.14, 8.15, and 8.16.

World Bank • ICT at a Glance

United States

	United States 2000	United States 2007	High-income group 2007
Economic and social context			
Population (total, million)	282	302	1,056
Urban population (% of total)	79	81	78
GNI per capita, World Bank Atlas method (current US$)	34,410	46,040	37,572
GDP growth, 1995–2000 and 2000–07 (avg. annual %)	4.2	2.6	2.4
Adult literacy rate (% of ages 15 and older)	—	—	99
Gross primary, secondary, tertiary school enrollment (%)	92	*93*	*92*
Sector structure			
Separate telecommunications regulator	Yes	Yes	
Status of main fixed-line telephone operator	*Private*	*Private*	
Level of competition[a]			
International long distance service	C	C	
Mobile telephone service	C	C	
Internet service	C	C	
Sector efficiency and capacity			
Telecommunications revenue (% of GDP)	3.0	*3.1*	*3.1*
Mobile and fixed-line subscribers per employee	239	*389*	747
Telecommunications investment (% of revenue)	25.3	*6.6*	*14.3*
Sector performance			
Access			
Telephone lines (per 100 people)	68.2	54.1	50.0
Mobile cellular subscriptions (per 100 people)	38.8	84.7	100.4
Internet subscribers (per 100 people)	*20.0*	—	25.8
Personal computers (per 100 people)	57.1	*80.5*	*67.4*
Households with a television set (%)	96	*95*	*98*
Usage			
International voice traffic (minutes/person/month)[b]	12.7	*23.3*	*14.0*
Mobile telephone usage (minutes/user/month)	387	*748*	353
Internet users (per 100 people)	43.9	73.5	65.7
Quality			
Population covered by mobile cellular network (%)	99	*100*	*99*
Fixed broadband subscribers (% of total Internet subscrib.)	22.4	—	82.6
International Internet bandwidth (bits/second/person)	394	11,277	18,242
Affordability			
Price basket for residential fixed line (US$/month)	21.5	*25.5*	*26.1*
Price basket for mobile service (US$/month)	—	6.7	13.0
Price basket for Internet service (US$/month)	—	20.0	22.8
Price of call to United States (US$ for 3 minutes)	—	—	*0.81*
Trade			
ICT goods exports (% of total goods exports)	23.4	*16.3*	*15.2*
ICT goods imports (% of total goods imports)	18.9	*14.6*	*14.6*
ICT service exports (% of total service exports)	3.3	4.3	7.0
Applications			
ICT expenditure (% of GDP)	—	7.5	6.7
E-government Web measure index[c]	—	0.95	0.60
Secure Internet servers (per 1 million people, Dec. 2008)	274.0	1,173.7	662.6

Sources: Economic and social context: UIS and World Bank; Sector structure: ITU; Sector efficiency and capacity: ITU and World Bank; Sector performance: Global Insight/WITSA, IMF, ITU, Netcraft, UN Comtrade, UNDESA, UNPAN, Wireless Intelligence and World Bank. Produced by the Global Information and Communication Technologies Department and the Development Economics Data Group. For complete information, see Definitions and Data Sources.

Notes: Use of italics in the column entries indicates years other than those specified. — Not available. GDP = gross domestic product; GNI = gross national income; ICT = information and communication technology; and MDG = Millennium Development Goal.

a. C = competition; M = monopoly; and P = partial competition. b. Outgoing and incoming. c. Scale of 0–1, where 1 = highest presence. d. Millennium Development Goal indicators 8.14, 8.15, and 8.16.

World Bank • ICT at a Glance

Uruguay

	Uruguay 2000	Uruguay 2007	Upper-middle-income group 2007	Latin America & the Caribbean Region 2007
Economic and social context				
Population (total, million)	3	3	824	561
Urban population (% of total)	91	92	75	78
GNI per capita, World Bank Atlas method (current US$)	6,220	6,390	7,107	5,801
GDP growth, 1995–2000 and 2000–07 (avg. annual %)	2.2	3.3	4.3	3.6
Adult literacy rate (% of ages 15 and older)	—	98	94	91
Gross primary, secondary, tertiary school enrollment (%)	84	89	82	81
Sector structure				
Separate telecommunications regulator	No	Yes		
Status of main fixed-line telephone operator	Public	Public		
Level of competition[a]				
International long distance service	M	P		
Mobile telephone service	C	C		
Internet service	P	C		
Sector efficiency and capacity				
Telecommunications revenue (% of GDP)	3.9	3.7	3.3	3.8
Mobile and fixed-line subscribers per employee	243	661	*566*	*530*
Telecommunications investment (% of revenue)	13.5	16.4	—	—
Sector performance				
Access				
Telephone lines (per 100 people)	28.1	29.0	22.6	18.1
Mobile cellular subscriptions (per 100 people)	12.4	90.4	84.1	67.0
Internet subscribers (per 100 people)	—	7.3	9.4	4.5
Personal computers (per 100 people)	10.6	13.6	*12.4*	*11.3*
Households with a television set (%)	82	92	92	84
Usage				
International voice traffic (minutes/person/month)[b]	5.1	10.6	—	—
Mobile telephone usage (minutes/user/month)	—	—	137	116
Internet users (per 100 people)	10.6	29.1	26.6	26.9
Quality				
Population covered by mobile cellular network (%)	100	100	95	91
Fixed broadband subscribers (% of total Internet subscrib.)	—	67.8	47.8	81.7
International Internet bandwidth (bits/second/person)	18	903	1,185	1,126
Affordability				
Price basket for residential fixed line (US$/month)	15.6	*10.7*	10.6	9.5
Price basket for mobile service (US$/month)	—	*16.1*	10.9	10.4
Price basket for Internet service (US$/month)	—	23.4	16.4	25.7
Price of call to United States (US$ for 3 minutes)	4.88	*0.52*	1.55	1.21
Trade				
ICT goods exports (% of total goods exports)	0.2	*0.1*	*13.5*	*11.4*
ICT goods imports (% of total goods imports)	7.4	6.5	*16.2*	*15.9*
ICT service exports (% of total service exports)	2.9	8.8	*4.6*	*4.7*
Applications				
ICT expenditure (% of GDP)	—	6.0	5.2	4.9
E-government Web measure index[c]	—	0.51	0.37	0.44
Secure Internet servers (per 1 million people, Dec. 2008)	11.2	42.6	26.2	18.2

Sources: Economic and social context: UIS and World Bank; Sector structure: ITU; Sector efficiency and capacity: ITU and World Bank; Sector performance: Global Insight/WITSA, IMF, ITU, Netcraft, UN Comtrade, UNDESA, UNPAN, Wireless Intelligence and World Bank. Produced by the Global Information and Communication Technologies Department and the Development Economics Data Group. For complete information, see Definitions and Data Sources.

Notes: Use of italics in the column entries indicates years other than those specified. — Not available. GDP = gross domestic product; GNI = gross national income; ICT = information and communication technology; and MDG = Millennium Development Goal.

a. C = competition; M = monopoly; and P = partial competition. **b.** Outgoing and incoming. **c.** Scale of 0–1, where 1 = highest presence. **d.** Millennium Development Goal indicators 8.14, 8.15, and 8.16.

World Bank • ICT at a Glance

Uzbekistan

	Uzbekistan 2000	Uzbekistan 2007	Low-income group 2007	Europe & Central Asia Region 2007
Economic and social context				
Population (total, million)	25	27	1,296	446
Urban population (% of total)	37	37	32	64
GNI per capita, World Bank Atlas method (current US$)	630	730	574	6,052
GDP growth, 1995–2000 and 2000–07 (avg. annual %)	4.1	6.2	5.6	6.1
Adult literacy rate (% of ages 15 and older)	97	—	64	98
Gross primary, secondary, tertiary school enrollment (%)	75	74	51	82
Sector structure				
Separate telecommunications regulator	—	No		
Status of main fixed-line telephone operator	Public	Public		
Level of competition[a]				
International long distance service	P	P		
Mobile telephone service	C	C		
Internet service	—	—		
Sector efficiency and capacity				
Telecommunications revenue (% of GDP)	1.6	2.5	3.3	2.9
Mobile and fixed-line subscribers per employee	66	117	301	532
Telecommunications investment (% of revenue)	25.3	27.9	—	22.0
Sector performance				
Access				
Telephone lines (per 100 people)	6.7	6.8	4.0	25.7
Mobile cellular subscriptions (per 100 people)	0.2	21.9	21.5	95.0
Internet subscribers (per 100 people)	0.0	0.1	0.8	13.6
Personal computers (per 100 people)	—	3.1	1.5	10.6
Households with a television set (%)	93	99	16	96
Usage				
International voice traffic (minutes/person/month)[b]	0.5	1.0	—	—
Mobile telephone usage (minutes/user/month)	—	411	—	154
Internet users (per 100 people)	0.5	4.5	5.2	21.4
Quality				
Population covered by mobile cellular network (%)	75	75	54	92
Fixed broadband subscribers (% of total Internet subscrib.)	0.0	7.4	3.4	32.5
International Internet bandwidth (bits/second/person)	0	9	26	1,114
Affordability				
Price basket for residential fixed line (US$/month)	2.6	0.9	5.7	5.8
Price basket for mobile service (US$/month)	—	1.8	11.2	11.8
Price basket for Internet service (US$/month)	—	5.2	29.2	12.0
Price of call to United States (US$ for 3 minutes)	13.95	—	2.00	1.63
Trade				
ICT goods exports (% of total goods exports)	—	—	1.4	1.8
ICT goods imports (% of total goods imports)	—	—	6.7	7.0
ICT service exports (% of total service exports)	—	—	—	5.0
Applications				
ICT expenditure (% of GDP)	—	—	—	5.0
E-government Web measure index[c]	—	0.27	0.11	0.36
Secure Internet servers (per 1 million people, Dec. 2008)	—	0.2	0.5	23.9

Sources: Economic and social context: UIS and World Bank; Sector structure: ITU; Sector efficiency and capacity: ITU and World Bank; Sector performance: Global Insight/WITSA, IMF, ITU, Netcraft, UN Comtrade, UNDESA, UNPAN, Wireless Intelligence and World Bank. Produced by the Global Information and Communication Technologies Department and the Development Economics Data Group. For complete information, see Definitions and Data Sources.

Notes: Use of italics in the column entries indicates years other than those specified. — Not available. GDP = gross domestic product; GNI = gross national income; ICT = information and communication technology; and MDG = Millennium Development Goal.

a. C = competition; M = monopoly; and P = partial competition. b. Outgoing and incoming. c. Scale of 0–1, where 1 = highest presence. d. Millennium Development Goal indicators 8.14, 8.15, and 8.16.

Venezuela, República Bolivariana de

	Venezuela, R. B. de 2000	Venezuela, R. B. de 2007	Upper-middle-income group 2007	Latin America & the Caribbean Region 2007
Economic and social context				
Population (total, million)	24	27	824	561
Urban population (% of total)	90	93	75	78
GNI per capita, World Bank Atlas method (current US$)	4,100	7,550	7,107	5,801
GDP growth, 1995–2000 and 2000–07 (avg. annual %)	0.6	4.6	4.3	3.6
Adult literacy rate (% of ages 15 and older)	93	95	94	91
Gross primary, secondary, tertiary school enrollment (%)	67	76	82	81
Sector structure				
Separate telecommunications regulator	Yes	Yes		
Status of main fixed-line telephone operator	Mixed	Mixed		
Level of competition[a]				
International long distance service	M	C		
Mobile telephone service	C	C		
Internet service	C	C		
Sector efficiency and capacity				
Telecommunications revenue (% of GDP)	3.3	3.8	3.3	3.8
Mobile and fixed-line subscribers per employee	386	677	566	530
Telecommunications investment (% of revenue)	26.3	18.5	—	—
Sector performance				
Access				
Telephone lines (per 100 people)	10.4	18.5	22.6	18.1
Mobile cellular subscriptions (per 100 people)	22.4	86.7	84.1	67.0
Internet subscribers (per 100 people)	1.1	3.7	9.4	4.5
Personal computers (per 100 people)	4.5	9.3	12.4	11.3
Households with a television set (%)	82	90	92	84
Usage				
International voice traffic (minutes/person/month)[b]	1.9	—	—	—
Mobile telephone usage (minutes/user/month)	—	—	137	116
Internet users (per 100 people)	3.4	20.8	26.6	26.9
Quality				
Population covered by mobile cellular network (%)	—	90	95	91
Fixed broadband subscribers (% of total Internet subscrib.)	1.6	85.3	47.8	81.7
International Internet bandwidth (bits/second/person)	6	628	1,185	1,126
Affordability				
Price basket for residential fixed line (US$/month)	—	6.7	10.6	9.5
Price basket for mobile service (US$/month)	—	1.2	10.9	10.4
Price basket for Internet service (US$/month)	—	23.0	16.4	25.7
Price of call to United States (US$ for 3 minutes)	0.78	0.84	1.55	1.21
Trade				
ICT goods exports (% of total goods exports)	0.1	0.0	13.5	11.4
ICT goods imports (% of total goods imports)	9.4	12.1	16.2	15.9
ICT service exports (% of total service exports)	9.6	11.1	4.6	4.7
Applications				
ICT expenditure (% of GDP)	—	3.9	5.2	4.9
E-government Web measure index[c]	—	0.47	0.37	0.44
Secure Internet servers (per 1 million people, Dec. 2008)	3.7	6.8	26.2	18.2

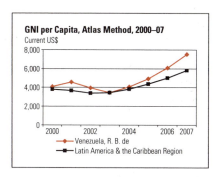

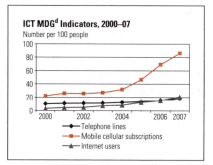

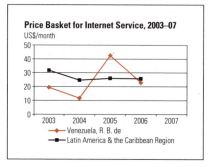

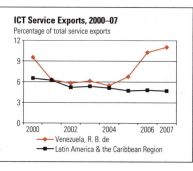

Sources: Economic and social context: UIS and World Bank; Sector structure: ITU; Sector efficiency and capacity: ITU and World Bank; Sector performance: Global Insight/WITSA, IMF, ITU, Netcraft, UN Comtrade, UNDESA, UNPAN, Wireless Intelligence and World Bank. Produced by the Global Information and Communication Technologies Department and the Development Economics Data Group. For complete information, see Definitions and Data Sources.

Notes: Use of italics in the column entries indicates years other than those specified. — Not available. GDP = gross domestic product; GNI = gross national income; ICT = information and communication technology; and MDG = Millennium Development Goal.

a. C = competition; M = monopoly; and P = partial competition. b. Outgoing and incoming. c. Scale of 0–1, where 1 = highest presence. d. Millennium Development Goal indicators 8.14, 8.15, and 8.16.

Vietnam

	Vietnam 2000	Vietnam 2007	Low-income group 2007	East Asia & Pacific Region 2007
Economic and social context				
Population (total, million)	78	85	1,296	1,912
Urban population (% of total)	24	27	32	43
GNI per capita, World Bank Atlas method (current US$)	390	770	574	2,182
GDP growth, 1995–2000 and 2000–07 (avg. annual %)	6.7	7.8	5.6	9.0
Adult literacy rate (% of ages 15 and older)	*90*	—	64	93
Gross primary, secondary, tertiary school enrollment (%)	64	*64*	51	*69*
Sector structure				
Separate telecommunications regulator	—	No		
Status of main fixed-line telephone operator	Public	Mixed		
Level of competition[a]				
International long distance service	M	C		
Mobile telephone service	M	C		
Internet service	P	C		
Sector efficiency and capacity				
Telecommunications revenue (% of GDP)	3.5	*4.7*	*3.3*	3.0
Mobile and fixed-line subscribers per employee	43	79	*301*	*546*
Telecommunications investment (% of revenue)	50.1	—	—	—
Sector performance				
Access				
Telephone lines (per 100 people)	3.3	33.5	4.0	23.1
Mobile cellular subscriptions (per 100 people)	1.0	27.9	21.5	43.7
Internet subscribers (per 100 people)	0.1	6.2	*0.8*	9.3
Personal computers (per 100 people)	0.8	*9.6*	*1.5*	*5.6*
Households with a television set (%)	78	89	16	53
Usage				
International voice traffic (minutes/person/month)[b]	0.6	—	—	0.8
Mobile telephone usage (minutes/user/month)	—	71	—	333
Internet users (per 100 people)	0.3	21.0	5.2	14.6
Quality				
Population covered by mobile cellular network (%)	—	70	*54*	93
Fixed broadband subscribers (% of total Internet subscrib.)	0.0	24.7	3.4	41.8
International Internet bandwidth (bits/second/person)	0	148	26	247
Affordability				
Price basket for residential fixed line (US$/month)	5.4	2.7	5.7	5.8
Price basket for mobile service (US$/month)	—	6.3	11.2	5.0
Price basket for Internet service (US$/month)	—	10.4	29.2	14.4
Price of call to United States (US$ for 3 minutes)	9.29	1.95	2.00	1.16
Trade				
ICT goods exports (% of total goods exports)	4.7	5.1	*1.4*	30.9
ICT goods imports (% of total goods imports)	6.7	7.6	*6.7*	28.1
ICT service exports (% of total service exports)	—	—	—	5.2
Applications				
ICT expenditure (% of GDP)	—	6.1	—	7.3
E-government Web measure index[c]	—	0.44	0.11	0.18
Secure Internet servers (per 1 million people, Dec. 2008)	0.1	1.1	0.5	1.9

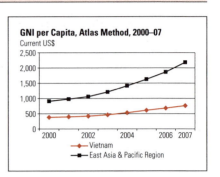

GNI per Capita, Atlas Method, 2000–07
Current US$

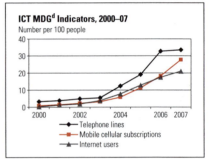

ICT MDG[d] Indicators, 2000–07
Number per 100 people

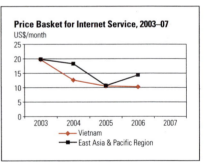

Price Basket for Internet Service, 2003–07
US$/month

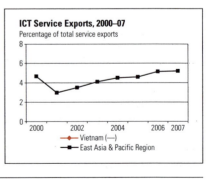

ICT Service Exports, 2000–07
Percentage of total service exports

Sources: Economic and social context: UIS and World Bank; Sector structure: ITU; Sector efficiency and capacity: ITU and World Bank; Sector performance: Global Insight/WITSA, IMF, ITU, Netcraft, UN Comtrade, UNDESA, UNPAN, Wireless Intelligence and World Bank. Produced by the Global Information and Communication Technologies Department and the Development Economics Data Group. For complete information, see Definitions and Data Sources.

Notes: Use of italics in the column entries indicates years other than those specified. — Not available. GDP = gross domestic product; GNI = gross national income; ICT = information and communication technology; and MDG = Millennium Development Goal.

a. C = competition; M = monopoly; and P = partial competition. **b.** Outgoing and incoming. **c.** Scale of 0–1, where 1 = highest presence. **d.** Millennium Development Goal indicators 8.14, 8.15, and 8.16.

World Bank • ICT at a Glance

West Bank and Gaza

	West Bank and Gaza 2000	West Bank and Gaza 2007	Lower-middle-income group 2007	Middle East & North Africa Region 2007
Economic and social context				
Population (total, million)	3	4	3,435	313
Urban population (% of total)	72	72	42	57
GNI per capita, World Bank Atlas method (current US$)	1,600	*1,290*	1,905	2,820
GDP growth, 1995–2000 and 2000–07 (avg. annual %)	7.4	–0.9	8.0	4.4
Adult literacy rate (% of ages 15 and older)	—	94	83	73
Gross primary, secondary, tertiary school enrollment (%)	77	82	68	70
Sector structure				
Separate telecommunications regulator	—	No		
Status of main fixed-line telephone operator	—	—		
Level of competition[a]				
International long distance service	—	—		
Mobile telephone service	—	—		
Internet service	—	—		
Sector efficiency and capacity				
Telecommunications revenue (% of GDP)	0.4	0.8	*3.1*	*3.1*
Mobile and fixed-line subscribers per employee	263	880	*624*	*691*
Telecommunications investment (% of revenue)	46.2	18.4	25.3	21.7
Sector performance				
Access				
Telephone lines (per 100 people)	9.3	9.4	15.3	17.0
Mobile cellular subscriptions (per 100 people)	6.0	27.7	38.9	50.7
Internet subscribers (per 100 people)	0.2	2.8	6.0	2.4
Personal computers (per 100 people)	*3.5*	*5.6*	*4.6*	*6.3*
Households with a television set (%)	85	93	79	94
Usage				
International voice traffic (minutes/person/month)[b]	2.4	5.7	—	2.7
Mobile telephone usage (minutes/user/month)	—	—	322	—
Internet users (per 100 people)	1.2	9.6	12.4	17.1
Quality				
Population covered by mobile cellular network (%)	95	95	*80*	93
Fixed broadband subscribers (% of total Internet subscrib.)	*0.0*	54.5	40.4	—
International Internet bandwidth (bits/second/person)	7	324	199	186
Affordability				
Price basket for residential fixed line (US$/month)	10.3	1.0	7.2	3.9
Price basket for mobile service (US$/month)	—	9.6	9.8	6.5
Price basket for Internet service (US$/month)	—	15.7	16.7	11.6
Price of call to United States (US$ for 3 minutes)	1.11	1.17	2.08	1.45
Trade				
ICT goods exports (% of total goods exports)	—	—	20.6	—
ICT goods imports (% of total goods imports)	—	—	20.2	—
ICT service exports (% of total service exports)	—	—	15.6	2.6
Applications				
ICT expenditure (% of GDP)	—	—	6.5	4.5
E-government Web measure index[c]	—	—	0.33	0.22
Secure Internet servers (per 1 million people, Dec. 2008)	—	1.3	1.8	1.3

Sources: Economic and social context: UIS and World Bank; Sector structure: ITU; Sector efficiency and capacity: ITU and World Bank; Sector performance: Global Insight/WITSA, IMF, ITU, Netcraft, UN Comtrade, UNDESA, UNPAN, Wireless Intelligence and World Bank. Produced by the Global Information and Communication Technologies Department and the Development Economics Data Group. For complete information, see Definitions and Data Sources.

Notes: Use of italics in the column entries indicates years other than those specified. — Not available. GDP = gross domestic product; GNI = gross national income; ICT = information and communication technology; and MDG = Millennium Development Goal.

a. C = competition; M = monopoly; and P = partial competition. **b.** Outgoing and incoming. **c.** Scale of 0–1, where 1 = highest presence. **d.** Millennium Development Goal indicators 8.14, 8.15, and 8.16.

Yemen, Republic of

	Yemen, Rep. of 2000	Yemen, Rep. of 2007	Low-income-group 2007	Middle East & North Africa Region 2007
Economic and social context				
Population (total, million)	18	22	1,296	313
Urban population (% of total)	26	30	32	57
GNI per capita, World Bank Atlas method (current US$)	400	870	574	2,820
GDP growth, 1995–2000 and 2000–07 (avg. annual %)	5.6	4.0	5.6	4.4
Adult literacy rate (% of ages 15 and older)	—	59	64	73
Gross primary, secondary, tertiary school enrollment (%)	50	55	51	70
Sector structure				
Separate telecommunications regulator	—	No		
Status of main fixed-line telephone operator	—	Public		
Level of competition[a]				
International long distance service	M	M		
Mobile telephone service	C	C		
Internet service	M	C		
Sector efficiency and capacity				
Telecommunications revenue (% of GDP)	1.0	1.2	3.3	3.1
Mobile and fixed-line subscribers per employee	72	—	301	691
Telecommunications investment (% of revenue)	51.1	—	—	21.7
Sector performance				
Access				
Telephone lines (per 100 people)	1.9	4.5	4.0	17.0
Mobile cellular subscriptions (per 100 people)	0.2	13.7	21.5	50.7
Internet subscribers (per 100 people)	0.0	0.7	0.8	2.4
Personal computers (per 100 people)	0.2	2.8	1.5	6.3
Households with a television set (%)	43	43	16	94
Usage				
International voice traffic (minutes/person/month)[b]	0.7	—	—	2.7
Mobile telephone usage (minutes/user/month)	—	—	—	—
Internet users (per 100 people)	0.1	1.4	5.2	17.1
Quality				
Population covered by mobile cellular network (%)	—	68	54	93
Fixed broadband subscribers (% of total Internet subscrib.)	0.0	—	3.4	—
International Internet bandwidth (bits/second/person)	0	28	26	186
Affordability				
Price basket for residential fixed line (US$/month)	3.7	2.5	5.7	3.9
Price basket for mobile service (US$/month)	—	4.2	11.2	6.5
Price basket for Internet service (US$/month)	—	11.0	29.2	11.6
Price of call to United States (US$ for 3 minutes)	4.45	2.39	2.00	1.45
Trade				
ICT goods exports (% of total goods exports)	—	0.1	1.4	—
ICT goods imports (% of total goods imports)	—	3.7	6.7	—
ICT service exports (% of total service exports)	28.1	18.9	—	2.6
Applications				
ICT expenditure (% of GDP)	—	—	—	4.5
E-government Web measure index[c]	—	0.07	0.11	0.22
Secure Internet servers (per 1 million people, Dec. 2008)	—	0.2	0.5	1.3

Sources: Economic and social context: UIS and World Bank; Sector structure: ITU; Sector efficiency and capacity: ITU and World Bank; Sector performance: Global Insight/WITSA, IMF, ITU, Netcraft, UN Comtrade, UNDESA, UNPAN, Wireless Intelligence and World Bank. Produced by the Global Information and Communication Technologies Department and the Development Economics Data Group. For complete information, see Definitions and Data Sources.

Notes: Use of italics in the column entries indicates years other than those specified. — Not available. GDP = gross domestic product; GNI = gross national income; ICT = information and communication technology; and MDG = Millennium Development Goal.

[a]. C = competition; M = monopoly; and P = partial competition. [b]. Outgoing and incoming. [c]. Scale of 0–1, where 1 = highest presence. [d]. Millennium Development Goal indicators 8.14, 8.15, and 8.16.

World Bank • ICT at a Glance

Zambia

	Zambia 2000	Zambia 2007	Low-income group 2007	Sub-Saharan Africa Region 2007
Economic and social context				
Population (total, million)	10	12	1,296	800
Urban population (% of total)	35	35	32	36
GNI per capita, World Bank Atlas method (current US$)	300	770	574	951
GDP growth, 1995–2000 and 2000–07 (avg. annual %)	2.2	5.1	5.6	5.1
Adult literacy rate (% of ages 15 and older)	68	71	64	62
Gross primary, secondary, tertiary school enrollment (%)	43	60	51	51
Sector structure				
Separate telecommunications regulator	Yes	Yes		
Status of main fixed-line telephone operator	Public	Public		
Level of competition[a]				
International long distance service	M	M		
Mobile telephone service	C	C		
Internet service	C	C		
Sector efficiency and capacity				
Telecommunications revenue (% of GDP)	2.0	2.5	3.3	4.7
Mobile and fixed-line subscribers per employee	59	175	301	499
Telecommunications investment (% of revenue)	12.3	29.3	—	—
Sector performance				
Access				
Telephone lines (per 100 people)	0.8	0.8	4.0	1.6
Mobile cellular subscriptions (per 100 people)	0.9	22.1	21.5	23.0
Internet subscribers (per 100 people)	0.1	0.1	0.8	1.2
Personal computers (per 100 people)	0.7	1.1	1.5	1.8
Households with a television set (%)	23	—	16	18
Usage				
International voice traffic (minutes/person/month)[b]	0.3	0.6	—	—
Mobile telephone usage (minutes/user/month)	—	—	—	—
Internet users (per 100 people)	0.2	4.2	5.2	4.4
Quality				
Population covered by mobile cellular network (%)	51	50	54	56
Fixed broadband subscribers (% of total Internet subscrib.)	0.3	25.8	3.4	3.1
International Internet bandwidth (bits/second/person)	0	3	26	36
Affordability				
Price basket for residential fixed line (US$/month)	4.6	8.9	5.7	12.6
Price basket for mobile service (US$/month)	—	14.6	11.2	11.6
Price basket for Internet service (US$/month)	—	78.6	29.2	43.1
Price of call to United States (US$ for 3 minutes)	2.57	1.41	2.00	2.43
Trade				
ICT goods exports (% of total goods exports)	0.0	0.1	1.4	1.1
ICT goods imports (% of total goods imports)	6.7	5.1	6.7	8.2
ICT service exports (% of total service exports)	—	8.5	—	4.2
Applications				
ICT expenditure (% of GDP)	—	—	—	—
E-government Web measure index[c]	—	0.00	0.11	0.16
Secure Internet servers (per 1 million people, Dec. 2008)	—	0.3	0.5	2.9

Sources: Economic and social context: UIS and World Bank; Sector structure: ITU; Sector efficiency and capacity: ITU and World Bank; Sector performance: Global Insight/WITSA, IMF, ITU, Netcraft, UN Comtrade, UNDESA, UNPAN, Wireless Intelligence and World Bank. Produced by the Global Information and Communication Technologies Department and the Development Economics Data Group. For complete information, see Definitions and Data Sources.

Notes: Use of italics in the column entries indicates years other than those specified. — Not available. GDP = gross domestic product; GNI = gross national income; ICT = information and communication technology; and MDG = Millennium Development Goal.

a. C = competition; M = monopoly; and P = partial competition. **b.** Outgoing and incoming. **c.** Scale of 0–1, where 1 = highest presence. **d.** Millennium Development Goal indicators 8.14, 8.15, and 8.16.

Zimbabwe

	Zimbabwe 2000	Zimbabwe 2007	Low-income group 2007	Sub-Saharan Africa Region 2007
Economic and social context				
Population (total, million)	13	13	1,296	800
Urban population (% of total)	34	37	32	36
GNI per capita, World Bank Atlas method (current US$)	450	*340*	574	951
GDP growth, 1995–2000 and 2000–07 (avg. annual %)	0.7	−5.7	5.6	5.1
Adult literacy rate (% of ages 15 and older)	—	91	64	62
Gross primary, secondary, tertiary school enrollment (%)	57	*52*	*51*	*51*
Sector structure				
Separate telecommunications regulator	No	Yes		
Status of main fixed-line telephone operator	Public	Private		
Level of competition[a]				
International long distance service	M	P		
Mobile telephone service	C	C		
Internet service	C	C		
Sector efficiency and capacity				
Telecommunications revenue (% of GDP)	2.2	—	*3.3*	*4.7*
Mobile and fixed-line subscribers per employee	112	*381*	*301*	*499*
Telecommunications investment (% of revenue)	90.9	—	—	—
Sector performance				
Access				
Telephone lines (per 100 people)	2.0	2.6	4.0	1.6
Mobile cellular subscriptions (per 100 people)	2.1	9.1	21.5	23.0
Internet subscribers (per 100 people)	0.2	0.7	*0.8*	*1.2*
Personal computers (per 100 people)	1.5	6.5	*1.5*	*1.8*
Households with a television set (%)	18	32	*16*	*18*
Usage				
International voice traffic (minutes/person/month)[b]	0.9	*1.7*	—	—
Mobile telephone usage (minutes/user/month)	—	—	—	—
Internet users (per 100 people)	0.4	10.1	5.2	4.4
Quality				
Population covered by mobile cellular network (%)	—	75	*54*	*56*
Fixed broadband subscribers (% of total Internet subscrib.)	2.2	15.3	*3.4*	*3.1*
International Internet bandwidth (bits/second/person)	1	4	26	36
Affordability				
Price basket for residential fixed line (US$/month)	6.3	*4.3*	*5.7*	*12.6*
Price basket for mobile service (US$/month)	—	*3.4*	*11.2*	*11.6*
Price basket for Internet service (US$/month)	—	*24.6*	*29.2*	*43.1*
Price of call to United States (US$ for 3 minutes)	4.36	—	*2.00*	*2.43*
Trade				
ICT goods exports (% of total goods exports)	0.2	0.3	*1.4*	*1.1*
ICT goods imports (% of total goods imports)	4.0	2.0	*6.7*	*8.2*
ICT service exports (% of total service exports)	—	—	—	*4.2*
Applications				
ICT expenditure (% of GDP)	—	3.5	—	—
E-government Web measure index[c]	—	0.09	0.11	0.16
Secure Internet servers (per 1 million people, Dec. 2008)	*0.1*	0.5	0.5	2.9

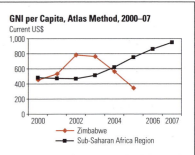

GNI per Capita, Atlas Method, 2000–07
Current US$

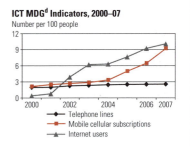

ICT MDG[d] Indicators, 2000–07
Number per 100 people

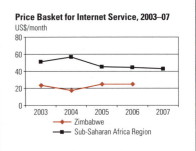

Price Basket for Internet Service, 2003–07
US$/month

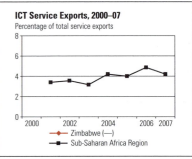

ICT Service Exports, 2000–07
Percentage of total service exports

Sources: Economic and social context: UIS and World Bank; Sector structure: ITU; Sector efficiency and capacity: ITU and World Bank; Sector performance: Global Insight/WITSA, IMF, ITU, Netcraft, UN Comtrade, UNDESA, UNPAN, Wireless Intelligence and World Bank. Produced by the Global Information and Communication Technologies Department and the Development Economics Data Group. For complete information, see Definitions and Data Sources.

Notes: Use of italics in the column entries indicates years other than those specified. — Not available. GDP = gross domestic product; GNI = gross national income; ICT = information and communication technology; and MDG = Millennium Development Goal.

a. C = competition; M = monopoly; and P = partial competition. b. Outgoing and incoming. c. Scale of 0–1, where 1 = highest presence. d. Millennium Development Goal indicators 8.14, 8.15, and 8.16.

World Bank • ICT at a Glance

Key ICT Indicators for Other Economies, 2007

	Population (total, thousand)	GNI per capita, World Bank Atlas method (current US$)	Telephone lines (per 100 people)	Mobile cellular subscriptions (per 100 people)	Internet users (per 100 people)	Personal computers (per 100 people)
American Samoa	65	—a	16.8	3.6	—	—
Andorra	82	—b	45.5	83.3	71.8	—
Antigua and Barbuda	85	11,650	44.7	132.5	70.7	20.8
Aruba	101	—b	38.3	144.5	23.8	9.9
Bahamas, The	331	17,160	40.1	112.9	36.2	12.4
Bahrain	753	17,390	26.3	148.2	33.2	18.3
Barbados	294	—b	46.2	80.9	54.8	13.7
Belize	304	3,760	11.2	38.9	10.5	14.8
Bermuda	64	—b	90.4	94.2	75.0	22.5
Bhutan	657	1,770	4.5	22.7	6.1	2.6
Brunei Darussalam	389	26,740	21.0	78.9	41.7	8.8
Cape Verde	530	2,430	13.8	27.9	7.0	13.0
Cayman Islands	54	—b	—	84.0	41.1	—
Channel Islands	149	68,640	—	—	—	—
Comoros	628	680	3.1	6.4	3.4	0.9
Cyprus	855	24,940	44.9	115.6	38.0	38.3
Djibouti	833	1,090	1.3	5.5	1.3	2.4
Dominica	73	4,030	29.3	58.5	36.6	18.2
Equatorial Guinea	508	12,860	2.1	43.3	1.6	1.9
Faeroe Islands	48	—b	46.4	107.8	77.5	—
Fiji	834	3,750	14.6	63.5	9.6	6.1
French Polynesia	263	—b	20.7	66.6	28.6	11.0
Greenland	57	—b	40.8	117.0	91.6	—
Grenada	106	3,920	26.2	43.7	21.8	15.3
Guam	173	—b	40.2	59.1	38.6	—
Guyana	739	1,250	14.9	38.0	25.7	3.9
Iceland	311	57,750	60.0	105.4	65.0	52.7
Isle of Man	77	45,810	—	—	—	—
Kiribati	95	1,120	4.3	0.7	2.1	1.1
Korea, Democratic People's Republic of	23,783	—c	5.0	0.0	0.0	—
Liberia	3,714	140	0.1	15.2	0.5	—
Liechtenstein	35	—b	55.5	90.8	65.2	—
Luxembourg	480	72,430	51.7	125.9	75.8	67.3
Macao, China	480	—b	37.1	165.4	49.6	40.2
Maldives	305	3,190	10.8	102.7	10.8	20.0
Malta	409	16,680	56.2	90.8	44.7	—
Marshall Islands	58	3,240	8.3	1.2	3.9	9.1
Mayotte	186	—a	—	28.0	—	—
Micronesia, Federated States of	111	2,280	7.8	24.7	13.5	5.5
Monaco	33	—b	104.9	56.0	61.2	—
Montenegro	599	5,270	58.8	107.1	46.7	—
Netherlands Antilles	191	—b	—	108.6	—	—
New Caledonia	242	—b	24.8	72.8	33.5	17.1
Northern Mariana Islands	84	—b	—	26.2	—	—
Palau	20	8,270	37.0	53.0	27.3	—
Qatar	836	—b	28.4	151.2	42.0	19.1
Samoa	181	2,700	10.9	47.4	4.4	2.3
San Marino	31	46,770	68.9	57.2	51.0	80.0

(Table continues on the following page.)

Key ICT Indicators *continued*

	Population (total, thousand)	GNI per capita, World Bank Atlas method (current US$)	Telephone lines (per 100 people)	Mobile cellular subscriptions (per 100 people)	Internet users (per 100 people)	Personal computers (per 100 people)
São Tomé and Principe	158	870	4.8	19.0	14.6	*3.9*
Seychelles	85	8,960	26.7	90.9	37.6	21.2
Solomon Islands	495	750	*1.6*	2.2	*1.7*	*4.7*
St. Kitts and Nevis	49	9,990	*53.2*	21.3	30.7	*23.4*
St. Lucia	168	5,520	—	64.1	65.5	*16.0*
St. Vincent and the Grenadines	120	4,210	19.1	91.8	47.4	*13.8*
Suriname	458	4,730	*17.9*	*70.3*	9.6	*4.4*
Timor-Leste	1,061	1,510	0.2	7.4	*0.1*	—
Tonga	102	2,480	20.6	45.5	8.2	*5.9*
Vanuatu	226	1,840	3.9	11.5	7.5	*1.4*
Virgin Islands (U.S.)	108	—[b]	*66.0*	*73.9*	27.7	*2.8*

Sources: ITU and World Bank. Produced by the Global Information and Communication Technologies Department and the Development Economics Data Group. For complete information, see Definitions and Data Sources.

Notes: Use of italics in the column entries indicates years other than those specified. — Not available. GNI = gross national income; ICT = information and communication technology.

a. Estimated to be upper middle income ($3,706–$11,455).
b. Estimated to be high income ($11,456 or more).
c. Estimated to be low income ($935 or less).

Contributors

Deepak Bhatia is the practice leader for sectoral and e-government applications in the World Bank Group's Global Information and Communication Technologies (GICT) Department, and has more than 25 years of experience managing and implementing information technology services. He has held various senior positions in the World Bank, most recently as manager of the E-government Practice Unit in the Information Solutions Group (ISG). Mr. Bhatia has published and coauthored documents in the area of customs applications, using information and communication technology (ICT) to enable beneficiary verification, and measuring the impact of ICT investment in the public sector, and has been a frequent presenter on e-government topics at international conferences. He holds a bachelor's degree in electronics and telecommunication, a master's in computer applications, and a master's in business administration.

Subhash C. Bhatnagar has been a professor of information systems at the Indian Institute of Management, Ahmedabad (IIMA) since 1975. During his years at IIMA, he has held the CMC Chair Professorship in Information Technology, served as dean, and was a member of the board of governors. From 2000 to 2006, he worked for the World Bank in Washington, D.C., mainstreaming the organization's e-government activities. At present, he divides his time between teaching and research at IIMA and advising various e-government institutions. Professor Bhatnagar's research is focused on ICT for development, e-government, and e-commerce. He has published 80 research papers and seven books, is the principal author of three reports on the impact of e-government projects in India, serves on the editorial boards of seven international journals, and was chief editor of the *Journal of Information Technology for Development*. He is also a recipient of the Silver Core Award from the International Federation for Information Processing (IFIP) and a Fellowship Award from Computer Society of India.

David A. Cieslikowski is a senior economist with the Development Economics Data Group of the World Bank. He has been the team coordinator of the Bank's flagship statistical publication, *World Development Indicators*, and is the team's specialist on data related to the private sector, governance, transport, ICT, and science and technology. Along with Data Group colleagues, he developed the original ICT at-a-glance table, and he collaborates with Bank staff on ICT-related statistical issues and the Partnership on Measuring ICT for Development. He has a master's degree in international relations from the University of Southern California.

Valerie D'Costa is program manager of the World Bank's Information for Development Program (*info*Dev), and had a distinguished career in private legal practice and in government service prior to this position. She leads *info*Dev's efforts to advance the role that ICT can play in fighting poverty and empowering people. Prior to her current role, Ms. D'Costa served as director of the International

Division at the InfoComm Development Authority of Singapore, the government agency charged with the development, promotion, and regulation of Singapore's ICT sector. There, she was responsible for the formulation of the Singapore government's policies on international ICT issues and bilateral relations with other countries on these issues. She holds a bachelor of laws degree from the National University of Singapore and a master of laws from University College, University of London.

Philippe Dongier is sector manager of the World Bank Group's GICT Department. He has management responsibility for World Bank policy and operational engagements in the telecommunications and ICT sectors globally, working with more than 80 countries. Prior to assuming this role, Mr. Dongier managed World Bank support to Afghanistan's reconstruction and led an initiative on strengthening the Bank's organizational effectiveness. He also played a range of leadership roles in the infrastructure and sustainable development sectors. Before joining the Bank, Mr. Dongier worked for five years with McKinsey & Company in Canada, the United States, Asia, and Europe, advising companies and governments on issues of strategy and organizational reform, including in the telecommunications and information technology (IT) industries. Earlier, he was based in Nepal for six years, where he worked for the Canadian Center for International Cooperation (CECI) in support of community infrastructure and microfinance programs. He has a master's degree in business administration from INSEAD and a bachelor's in economics from McGill University.

Naomi J. Halewood is an operations analyst with the Policy Division of the GICT Department of the World Bank Group. Her current work focuses on ICT development projects in the East Asia and Pacific Region, and she is part of the Monitoring and Evaluation Team, for which she performs data analysis and research. Prior to joining GICT, she worked in the Development Economics Data Group of the World Bank and for the Center for International Political Economy in New York. She has a master's degree in business administration and a master's in international development from American University.

Nagy K. Hanna is an author, educator, and independent consultant on e-strategies, with more than 33 years of development experience across all geographic regions. He is also a senior scholar at the Academy for Leadership of the University of Maryland. He led the World Bank's practice in applying ICT for development and innovation as the Bank's first senior adviser on e-strategies; established the Bank's global community of practice on e-development; pioneered e-Sri Lanka, the first World Bank lending operation in support of comprehensive ICT-enabled development; and authored books on this pace-setting experience. Mr. Hanna also initiated Bank assistance in the early 1990s that helped unleash India's software revolution. He lectures and publishes extensively on e-development, strategic planning, executive education and leadership for innovation, and the knowledge economy. He holds a doctorate from the Wharton School at the University of Pennsylvania and has completed an executive development program at Harvard University.

Mohsen Khalil is joint director of the World Bank and International Finance Corporation (IFC) in charge of the GICT Department. Prior to this appointment, he was director of the IFC's Central Asia, Middle East, and North Africa Department. He also served IFC as chief investment officer in the Telecommunications, Transport, and Utilities Department. Before joining the World Bank and while a professor of business at the American University of Beirut, Mr. Khalil served as chief adviser to the Lebanese minister of Post and Telecommunications, board director of Lebanon's Autonomous Fund for Housing, and adviser to governments and major corporations in the Middle East. He also worked with McKinsey & Company, NASA (National Aeronautics and Space Administration) Goddard Space Flight Center, and MITRE Corporation. Mr. Khalil holds a doctorate in electrical engineering from the University of Southern California, a master's degree from the Massachusetts Institute of Technology (MIT) Sloan School of Management, a master's in electrical engineering from the University of Wisconsin Madison, and a bachelor's in physics from the American University of Beirut.

Kaoru Kimura is an operations analyst with the Policy Division of the GICT Department of the World Bank Group. Her main responsibility is to provide operational support and analytical work for countries in sub-Saharan Africa and China. In addition, she has been actively involved in monitoring and evaluation activities in the ICT sector and has worked on the ICT at-a-glance tables and the *Little Data Book on Information and Communication Technology* series. Prior to joining GICT, Ms. Kimura was at

Nippon Telegraph and Telecommunication (NTT) in Japan, where she worked on marketing, business consulting, and designing communication systems for multinational corporate clients. She has a master's degree in international development studies from the National Graduate Institute for Policy Studies (GRIPS) in Japan.

Siou Chew Kuek is a consultant with the Policy Division of the GICT Department of the World Bank Group. He focuses on policy, operations, and analytical work for the Middle East and North Africa and the South Asia Regions and the e-government and IT/ITES (IT-enabled services) practices. Additionally, he contributes to the global analytical and knowledge agenda for the department. Prior to joining the World Bank, Mr. Kuek spent about 10 years in the private sector, where he worked in a business development, consulting, and implementation capacity on IT projects for multinational corporate clients in Asia. He holds a bachelor's degree in management studies from the University of London and a master's in international development from American University.

Raymond Muhula is a consultant with the Development Economics Data Group of the World Bank. As a member of the *World Development Indicators* team, he works on data related to the private sector, governance, education, transport, ICT, and science and technology. He holds a doctorate in political science from Howard University and a master's in public administration from the University of Texas at Tyler.

William Prince is an information officer in the Development Economics Data Group of the World Bank, where he is responsible for production and content management of electronic data products, including the online and CD-ROM versions of *World Development Indicators* and *Global Development Finance*. He also provides overall data quality management and specialized data support for the Data Group's clients. Prior to joining the Bank, he was a statistical consultant for British Telecom. He has a master's in business administration in decision analysis from Arizona State University.

Christine Zhen-Wei Qiang is the coordinator of the global analytical work program in the GICT Department of the World Bank Group, where she manages the Information and Communication for Development flagship reports. Her main responsibilities include overseeing the World Bank's analytical work on ICT policies, economics, and impact analysis, as well as leading ICT operations and policy dialogue in countries in Asia. She has published over 20 journal articles, book chapters, and reports on ICT for development, economic growth, and productivity. She holds a doctorate in economics and a master's degree in computer science and engineering from Johns Hopkins University.

Siddhartha Raja is a telecommunications policy analyst with the GICT Department of the World Bank Group. His main responsibilities include ICT sector strategy development, telecommunications policy formulation and analysis, and regulatory capacity building. He has published journal articles, book chapters, and conference papers on the political economy of telecommunications, the sociology of technology, and the politics of development. Mr. Raja has a master's degree in infrastructure policy studies from Stanford University, has studied media law and policy at the University of Oxford, and is currently completing his doctorate in telecommunications policy from the University of Illinois.

Carlo M. Rossotto is the regional coordinator for the Middle East and North Africa Region in the GICT Policy Division of the World Bank Group. His main responsibilities include overseeing the Bank's ICT program in the Middle East and North Africa and leading ICT operations and policy dialogue with countries in this region. Before joining the Bank, he consulted extensively with telecommunications, technology, and media companies in Europe on corporate strategy, regulatory affairs, and ICT economics. Mr. Rossotto has authored several publications on ICT, particularly on the subjects of telecommunication competition and regulatory reform. He holds a master's of economics in financial and commercial regulation from the London School of Economics and Political Science and a master's degree in economics and business administration from Bocconi University.

Rajendra Singh is a senior regulatory specialist in the Policy Division of GICT Department in the World Bank Group, where he advises countries in the South Asia, Middle East and North Africa, and East Asia and Pacific Regions on telecommunications policy and regulatory issues. He also leads the analytical work on convergence in the Bank. Before joining the Bank in December 2006, Mr. Singh worked as secretary of the Telecom Regulatory Authority of India (TRAI). He holds a bachelor's degree and a master's degree in technology from the Indian Institute of Technology (IIT), Roorkee and Delhi, respectively, and a master's in business administration from the University of Delhi. He has published around 50 papers in various international journals and conferences.

Peter L. Smith was practice leader for telecommunications policy and regulation in the World Bank Group's GICT Department until 2007. At the Bank, he worked extensively on projects involving telecommunications sector reform, market structure, regulation, and privatization—mainly in Asia but also in Africa and Latin America. He also helped to lead work on *info*Dev's ICT Regulation Toolkit. Prior to joining the Bank, he was senior economist with the Canadian Radio-television and Telecommunications Commission. Mr. Smith holds a master's degree in economics from Carleton University, Ottawa, Canada.

Randeep Sudan is a lead ICT policy specialist with the GICT Department of the World Bank Group, where he is working on ICT projects in Africa, Europe and Central Asia, East Asia, Latin America, and South Asia. Prior to joining the Bank, he held senior government positions in India, as a member of the Indian Administrative Service. He was special secretary to the chief minister and ex officio secretary of information technology in the state of Andhra Pradesh. He has also served as chief executive of APFIRST, an organization focused on promoting investments in the ICT sector, and of AP Technology Services, a company specializing in the use of ICT in government. Mr. Sudan holds a master's degree in social policy and planning from the London School of Economics and a master's in economics from the Jawaharlal Nehru University. He has also been a visiting faculty member in the Department of Informatics at the University of Oslo.

Eric Swanson is program manager in the Development Economics Data Group of the World Bank, where he leads a team responsible for producing the *World Development Indicators* and other statistical publications and databases. He works with the United Nations and its specialized agencies to coordinate the compilation and dissemination of statistical information for the monitoring of the Millennium Development Goals. He holds a doctorate in economics from the State University of New York at Buffalo.

Jiro Tominaga, a former senior information officer in the e-Government Practice Unit in the Information Solutions Group of the World Bank, has been responsible for a wide range of operational and analytical work related to the use of ICT in developing countries. His work focused primarily on ICT applications in service delivery and government administration in the South Asia and Europe and Central Asia Regions. He is currently special assistant in the Independent Evaluation Group of the World Bank. He holds a master's degree in economics from the London School of Economics.

Björn Wellenius is an independent consultant on telecommunications policy, regulation, and economics in developing countries. He advises the World Bank, law and consulting firms in Europe and North America, and governments and regulatory authorities. Until 1999, he was the World Bank's telecommunications adviser. His publications include five books on telecommunications and economic development, as well as book chapters, best practice notes, and technical papers on universal service, rural infrastructure financing, spectrum management, and regulatory capacity building. Before joining the Bank, he was professor of telecommunications at Universidad de Chile and currently is an adjunct professor at Michigan State University. Mr. Wellenius has a doctorate in physical sciences (telecommunications) from the University of Essex and a degree in electronics and telecommunications engineering from Universidad de Chile.

Mark D. J. Williams is a senior economist in the GICT Department of the World Bank Group, where he specializes in the economics and regulation of telecommunications networks. He has worked in over 30 countries, advising governments, regulators, and private companies. Recently, he has advised on market liberalization in the Middle East and on the development of core telecommunications infrastructure in Africa. He regularly publishes articles in journals and contributes to books on telecommunications, including authoring a chapter in *Telecommunications Law and Regulation*, a standard legal textbook on the subject. Prior to joining the Bank, Mr. Williams was an economist at Frontier Economics in London specializing in the economics of the telecommunications and postal industries. He holds a bachelor's degree in politics and economics from Oxford University and a master's in quantitative economics from the University of Warwick.

ECO-AUDIT
Environmental Benefits Statement

The World Bank is committed to preserving endangered forests and natural resources. The Office of the Publisher has chosen to print *Information and Communications for Development 2009* on recycled paper with 30 percent postconsumer fiber in accordance with the recommended standards for paper usage set by the Green Press Initiative, a nonprofit program supporting publishers in using fiber that is not sourced from endangered forests. For more information, visit www.greenpressinitiative.org.

Saved:
- 44 trees
- 30 million Btu of total energy
- 3,831 lb. of net greenhouse gases
- 15,901 gal. of waste water
- 2,042 lb. of solid waste